Methods in Molecular Genetics

Volume 4

Molecular Virology Techniques

Part A

Methods in Molecular Genetics

Edited by

Kenneth W. Adolph

Department of Biochemistry
University of Minnesota Medical School
Minneapolis, Minnesota

Volume 4

Molecular Virology Techniques
Part A

ACADEMIC PRESS

San Diego   New York   Boston   London   Sydney   Tokyo   Toronto

Front cover photograph: HIV DNA in CD4+ lymphocytes in a section of a lymph node. CD4+ lymphocytes have been labeled immunocytochemically and can be identified in the photograph as cells with brown-stained cytoplasm. The section was subsequently carried through *in situ* amplification and hybridization procedures with HIV-specific primers and probes labeled with ^{35}S. When the developed radioautograph is illuminated with epipolarized light, the silver grains appear green (specular reflectance). The cells with HIV DNA in their nuclei that bound the HIV probe will have many grains and therefore appear green. This exceptionally sensitive double-label PCR *in situ* technique revealed HIV DNA in surprisingly large numbers of CD4+ lymphocytes relatively early in the course of HIV infection. Photo courtesy of Professor Ashley T. Haase, Department of Microbiology, University of Minnesota Medical School, Minneapolis, MN.

Academic Press, Inc.
A Division of Harcourt Brace & Company
525 B Street, Suite 1900, San Diego, California 92101

United Kingdom Edition published by
Academic Press Limited
24–28 Oval Road, London NW1 7DX

International Standard Serial Number: 1067-2389

International Standard Book Number: 0-12-044306-6

PRINTED IN THE UNITED STATES OF AMERICA
94 95 96 97 98 99   EB   9 8 7 6 5 4 3 2 1

Table of Contents

Section II Retroviruses/RNA Viruses

Section III DNA Viruses

Contributors to Volume 4

Article numbers are in parentheses following the names of contributors. Affiliations listed are current.

MICHAEL B. AGY (12), Regional Primate Research Center, University of Washington, Seattle, Washington 98195

ANNA ALDOVINI (1), Whitehead Institute for Biomedical Research, Cambridge, Massachusetts 02142

GLEN N. BARBER (12), Department of Microbiology, University of Washington School of Medicine, Seattle, Washington 98195

RALF BARTENSCHLAGER (25), F. Hoffmann-LaRoche Ltd., Pharmaceutical Research-New Technologies, 4002 Basel, Switzerland

DAVID H. L. BISHOP (18), Institute of Virology and Environmental Microbiology, Oxford, OX1 3SR, United Kingdom

JANET S. BUTEL (19), Division of Molecular Virology, Baylor College of Medicine, Houston, Texas 77030

J. PATRICK CARD (23), Department of Behavioral Neuroscience, University of Pittsburgh, Pittsburgh, Pennsylvania 15260

DAYUE CHEN (14), Division of Molecular Virology, Baylor College of Medicine, Houston, Texas 77030

JIN CHEN (11), Department of Microbiology and Immunology, Vanderbilt University School of Medicine, Nashville, Tennessee 37232

RONALD COLLMAN (6), Departments of Medicine and Microbiology, Pulmonary and Critical Care Division, University of Pennsylvania Medical Center, Philadelphia, Pennsylvania 19104

DAVID G. COOK (8), Departments of Pathology and Laboratory Medicine, University of Pennsylvania Medical Center, Philadelphia, Pennsylvania 19104

CHRISTINE DEBOUCK (6), Molecular Genetics Department, SmithKline Beecham Pharmaceuticals, King of Prussia, Pennsylvania 19406

JAMES DEGREGORI (11), Department of Microbiology and Immunology, Vanderbilt University School of Medicine, Nashville, Tennessee 37232

ULRIKE DELLING (4), Department of Biochemistry, and McGill Cancer Center, McGill University, Montreal, Quebec, Canada H3G 1Y6

JAMES P. DOOHAN (15), Department of Biological Sciences, and Graduate Program of Biochemistry and Molecular Biology, University of California at Santa Barbara, Santa Barbara, California 93106

R. DUGGAL (16), Institute of Developmental and Molecular Biology, Texas A&M University, College Station, Texas 77843

LYNN W. ENQUIST (23), Department of Molecular Biology, Princeton University, Princeton, New Jersey 08544

JACQUES FANTINI (8), INSERM Biologie Cellulaire des Systemes Organises, Faculté de Médecine Nord, 13916 Marseille, France

HELMUT FICKENSCHER (22), Institut für Klinische und Molekulare Virologie, D-91054 Erlangen, Germany

BERNHARD FLECKENSTEIN (22), Institut für Klinische und Molekulare Virologie, D-91054 Erlangen, Germany

FERNANDO GARCÍA-ARENAL (17), Departamento de Patologia Vegetal, E.T.S.I. Agrónomos, Ciudad Universitaria, 28040 Madrid, Spain

ANNE GATIGNOL (2), Laboratory of Molecular Microbiology, National Institute of Allergy and Infectious Diseases, National Institutes of Health, Bethesda, Maryland 20892

FRANCISCO GONZALEZ-SCARANO (6, 8, 9), Departments of Neurology and Microbiology, University of Pennsylvania Medical Center, Philadelphia, Pennsylvania 19104

JAAP GOUDSMIT (10), Department of Virology, Academic Medical Center, University of Amsterdam, 1105 AZ Amsterdam, The Netherlands, and Aaron Diamond AIDS Research Center, New York, New York 10016

SUSAN GREGORY (7), Departments of Medicine and Microbiology, Pulmonary and Critical Care Division, University of Pennsylvania School of Medicine, Philadelphia, Pennsylvania 19106

T. C. HALL (16), Institute of Developmental and Molecular Biology, Texas A&M University, College Station, Texas 77843

JANET M. HAROUSE (9), Graduate Group in Molecular Biology, University of Pennsylvania Medical Center, Philadelphia, Pennsylvania 19104

GEOFF HICKS (11), Department of Microbiology and Immunology, Vanderbilt University School of Medicine, Nashville, Tennessee 37232

WILLIAM JAMES (7), Sir William Dunn School of Pathology, Oxford University, Oxford OX1 3RE, United Kingdom

lands, access to the half-timbered cottages where the farm workers and their families lived. Already the labourers' wives were returning from the village, long loaves tucked under their arms. Some rode ancient and solid bicycles. Others trudged along in wellingtons, a bulging canvas bag in each hand. Léonie knew them all by sight. She greeted each one.

The road wound away from the Martin farm into the outskirts of the village. Léonie plodded between small houses that leaned together. Through their open doors she caught flashes of patterned lino floor, a corner of tablecloth, a chair leg. Voices inside called out, responded. She went on, past the blacksmith's, past the corrugated-iron public lavatory on the corner, plastered with faded posters that peeled and flapped, across the road, through the stepped gap between two old houses whose lath and plaster upper storeys almost touched overhead, and into the little square.

She walked very slowly across it. She told herself she would not be shy, she would not blush when addressed as *la petite Anglaise*, she would not mind having her fluency admiringly remarked upon, she would not care that everyone in the shop would turn round and stare at her, the foreigner. Today would be different.

The two bakeries stood side by side. Almost identical, if you had not been brought up by Victorine to know that one was good and the other bad: both had wide shop windows displaying shelves of apple tarts, turnovers, puffs; striped awnings above; tiled steps.

Léonie stared at the two shops and came to a decision. She would go into the bad woman's shop. The collaborator's shop. Just to see what it was like. Surely nobody at home would notice if the bread, just once, tasted a bit different?

She pushed open the glass door, muttered good morning, and took her place in the queue. She couldn't believe that she wasn't in the baker's next door. Same list of icecream flavours hung on the wall and bowl of aniseed lollipops on the counter, same gilt baskets of *croissants* and racks like umbrella stands packed with tall loaves. The woman in the grey overall serving customers looked just like anyone else. She snatched up a square of tissue paper, deftly swung and twisted it round a fat *brioche*.

Léonie bought two *baguettes*. The shop-woman smiled at her and asked after the bereaved family. Léonie felt her cheeks go red. She gabbled something polite and slunk out of the door, just remembering in time to bid everybody goodbye.

She slung her long loaves over her shoul[der?] its warm heel in her joined palms. She tur[ned] stone steps leading out of the square.

Baptiste, four other boys behind him, flowed from Léonie's armpits down to her

Eengleesh peeg, Baptiste yelled at her.

The boys whistled. Léonie wheeled, baker's, and fled along the boulevard edg the church and the walled cemetery, to th

The boys jeered as she ran away. She let vanished. And she still had the bread, clu off her like a jacket. Her heartbeat slack

The drizzle had stopped. Sun shone on the muddy ruts of the overgrown lane. G she didn't care. Delicious, the coolness Brambles hooked her shorts, unloaded Ferns of brilliant green slapped her legs. bread under one arm, and swished it ove of cattle flies that buzzed there.

The lane was deserted, quiet. She re hardly anyone used it any more. It was a the road. She would be late, even if she move on. Her pace slowed even more as and sour with grief. And it would be her late. She'd get a scolding from Victorin most likely.

The strap of one of her sandals l stopped, shook her foot experimentall same moment Baptiste and his gang the corner.

Boys behind, the woods in front. scattering loaves. Freed of her burde the ditch opposite the Martins' orcha the undergrowth. The woods receive her, dense green water.

They hadn't followed her in. Jee faded. Léonie sat up, rubbing her el and smarting. Her foot, when she e a couple of thorns stuck in it and extra trouble for losing the sandal.

back
to re-
She
centre
white
be fill
remen
She
not kr
Som
head j
and go
woma
Late
when s
Under
betraye
into th
fading
memor
Some
explori
she ha
pleasur
across
Then
forgotte
Deeper
it spoke
then cl
forced t
the pull
fine spr
she had
always
by her
Time
a jerk,
Victor
to have

look for it? Suppose the boys were hiding, waiting for her
merge?

got up and limped forward. She was in a little clearing, in the
of which was a tall heap of large stones. Behind it was a small
cliff, the bubble of water. She paused. Memory of a basket to
d with blackberries. Victorine's tale of . . . she couldn't quite
ber what.

stood and shivered. It had begun to rain again. She hovered,
owing what to do, afraid to start crying here all by herself.
thing like a rough finger stroked the back of her neck. Her
rked up. Then she saw it. Saw the fine rainy air become solid
den and red, form itself into the shape of a living and breathing

when she blurted out to Victorine what she had seen,
ne tried to describe it, she struggled with inadequate words.
Victorine's mocking questioning she understood that she had
d her vision by mentioning it. Its red and gold brilliance sank
darkness of her imagination like a firework blazing then
the black night sky. All she had left to clutch at was the
of how she had felt.

thing outside her, mysterious and huge, put out a kindly
g hand and touched her. Something was restored to her which
lost and believed she would never find again. The deepest
she had ever known possessed her. It started in her toes and
r shoulders and squirmed through her, aching, sweet.
she remembered it. A language she once knew but had
about, forgotten ever hearing, forgotten she could speak.
han English or French; not foreign; her own. She had heard
long ago. She heard it now, at first far off, thin gold,
e, warm. The secret language, the underground stream that
rough her like a river, that rose and danced inside her like
g jet of a fountain, that wetted her face and hands like
, that joined her back to what she had lost, to something
nce intimately known, that she could hardly believe would
there as it was now, which waited for her and called her
me.

illed, and suspended itself. In the cool drizzle. Then, with
e world went on again.
ne and the others were interested solely in what she claimed
en. Léonie tried to tell them. It came out all confused.

THE BREAD-BASKET

Thérèse no longer shared the first job of the day with her
cousin. Léonie went alone, now, to fetch the bread from the
baker's in the village square. On her return she laid the cloth, collected
the knives and plates, sounded the gong for breakfast. Thérèse was
let off these tasks because she was in mourning. She slumped in bed,
miserable and full of headache.

Poor child, Madeleine said: no wonder.

Thérèse had not once cried in public for her mother. Louis wept
openly and could not be comforted. Not with caresses, not with
sweets.

Leave him be, for the moment, Madeleine advised: leave him alone
and he'll be all right.

Two days since the funeral. Léonie was delighted to get out of the
house. She shut the kitchen door behind her, turned to study the strip
of seaweed nailed there, to check the weather. It suggested wet. It was
right. As she emerged into the yard fine rain dampened her face. The
air smelled of salt and the sky was grey.

She sauntered around the side of the house, hands in the pockets
of her shorts, trod over the white gravel to the gates. She heard a
thin crowing of cocks. Her bare feet slipped in her sandals, which
were already wet, and her cotton shirt felt chilly. Too much bother
to fetch a raincoat. She went on out to the main road.

The ditch on her left swirled with rainwater. The bank above it
was a tangled slope of late-flowering mallow and campion, bright
purple-pink in the long grass. On her right the green meadow was
full of cows. Every so often there was a gap in the bank of beech
trees that guarded the farm from wind and storms, a muddy track
between red brick gateposts affording entry for tractors into the farm

The red lady. The golden woman in red. She swam up slowly. She developed, like a photograph. She composed herself, a red and gold figure on a red ground.

She referred to the apparition as *that person*. It seemed to her polite to do so. Also there was no other way to express her sense of something having arrived from somewhere else, something normally invisible to the eye choosing to put on a human form. It was Victorine, listening with one ear in the kitchen with the others while she sawed a *baguette* into chunks, who laughed and said *Madame la sainte Vierge* I suppose.

Usually when she got home with the bread Léonie dumped it on the kitchen table. Then she opened the bread drawer of the dresser, took out the bread knife and the oval basket of plaited straw with its red check cotton lining. She would cut up yesterday's leftover bread, staling now but to the taste of Madeleine, and carry that and the new *baguette* along the corridor to the dining-room. Then she would open the doors of the *buffet* and take out the tablecloth.

This morning she was in another of her daydreams, it looked like to the exasperated Victorine. There she was standing in front of the *buffet* with her eyes half-closed and no sign of the table being laid. Well after half-past eight and that poor man upstairs needing his breakfast.

Victorine was confused by the tale Baptiste had told her just now when he and his mother turned up at the back door with Léonie's sandal and an armful of bread. Was it true? she demanded of the silly girl: that she'd run away from the boys when they were only trying to be friendly?

She intended to restore discipline. She took Léonie into the kitchen, where Rose and Baptiste waited. How small and cowed he looked next to his mother. That cheered Léonie up.

Victorine put Léonie on to a chair, the bread on to a board, the knife to the crust.

So. What happened to make you so late back? Tell me the truth, mind.

Thérèse appeared in the doorway. She fingered the edge of her dressing-gown. She pouted.

Oh. I just came to see if the coffee was ready for me to take up to Papa.

Rose gave Baptiste's earlobe a hard pinch. His cheeks were very red. Both girls quickly turned their faces away, just in case he would want to take revenge on them later for having witnessed his humiliation.

Rose said: was he bullying you? Because if so I'll beat the hide off him, I've told him so, he's not too big to be beaten and if Monsieur Martin won't do it I will.

Léonie said very fast: what happened was I saw this person all in red.

Victorine laughed.

A person in red. Some excuse. Why not simply admit it, that you were dawdling again. Tell us her name then.

Léonie insisted, watching the blade fall quickly through the bread, that she did not know the person's name.

Well then, retorted her amused audience: you had better find it out hadn't you?

It must have been someone from the village, Rose said: who else would be walking in the woods at that hour? The Parisians only come here for walks at weekends.

Léonie leaned her head on her hand, drew patterns with one finger in the litter of crumbs around the breadboard. Victorine took the kettle off the flame, poured a stream of boiling water into the top of the mottled blue enamel coffee-pot. A cloud of steam arose, the perfume of coffee.

Was she beautiful? Thérèse asked.

Yes, Léonie said.

As beautiful, Victorine asked: as Marie Guérin? Who had ridden as village queen on a flower-hung cart only the previous week in the procession to celebrate the feast of the Assumption.

Far more beautiful, replied Léonie. Unsuccessfully hiding her scorn of Victorine's own style of beauty: frizzy blonde hair and small blue eyes, plucked eyebrows and plump white calves, big feet crammed into white winkle-pickers.

So what did she look like then? demanded Victorine. Glancing with irritation at the cocky girl who wasn't even properly French but a *bourgeoise* English snob with no clue about what was what.

She, that person, had a wide mouth, with plump lips, like cushions. Dark eyes under feathery black brows. A lot of black hair that curled down her back. She was very young. Not fat but not thin. And everything about her, her long nightdress and the mist she was held in, was so golden-red that even the dark gold skin of her lovely face had a reddish tinge to it. As though she were made of fire. And she had yellow stars around her head.

She was coloured? Victorine roared with sorrowful mirth: oh what a story, well that certainly cuts out the Blessed Virgin.

She scooped up coffee-pot, bread-basket, jam-pots, Camembert, on to the tin tray, and made for the door. She shot her diagnosis over her shoulder. Indistinct rumble of words as the door swung shut.

Just an excuse. Or your imagination. Overtired. Staying awake too late chatting to poor Thérèse.

Rose said: suppose it *was* the Virgin? Isn't that exactly where the old shrine used to be, in those woods?

Thérèse laughed.

How could it possibly be Our Lady! She only wears blue. She's been making it up.

Baptiste flicked a look at Léonie. Comrades. They were not going to give each other away. Then he went on gazing at Thérèse, voluptuous in flowered chintz and blue ribbons.

Léonie picked up a fistful of knives and trailed after Victorine. She promised herself that she would never mention the lady again.

THE QUIMPER DISH

*T*he room was too warm. It smelled of Thérèse's stale breath and sweat. Sour, sweetish. Léonie lowered herself on to the chair beside the bed, her magazine on her lap. She tried not to look at the glass on the pink marble top of the wash-stand, empty, a sticky red stain at the bottom. A fly circled its smeared rim.

Thérèse's face glistened. Her hair flopped on the pillow. This morning she was just a limp nightdress that needed washing. Last night she'd tried to toss the covers off, then cried weakly, tears leaking down the sides of her nose on to the sheet. Léonie hadn't been able to hush her.

No one could help Thérèse. She turned her head away and refused to recognize the women who hung over her, imploring. Thérèse, please eat. You must eat. Thérèse wielded great power in her illness. Léonie knew it. The doctor, flustered, did not. A chill, he suggested, and then: influenza. A fever. Which took Thérèse by the throat and shook her, rattled her speechless, dumped her in bed. A nervous collapse. Yes, that was true, Léonie thought: but what they all refused to see was the willpower, the rage. How she lay there, the focus of their anxious attention, and rejected them. They called it being ill. How long could that game go on?

Léonie could not say this to anyone. A criticism which would hurt. She sat still and said nothing. Thérèse was scared to be left alone. She whispered that the Devil hid behind the curtains, a red devil with a feathery red tail. Her fretfulness eased if someone stayed near her, in view. She got her way. Léonie hunched on the bedside chair and lapped up comics, while Thérèse dozed.

Léonie still could not let herself cry for her aunt. She noted that there was now an absence, but got no farther. She twisted the comic into a tube on her lap and clenched it like a truncheon. Thérèse opened her eyes.

Calmly Léonie unrolled her weapon and smoothed it over her knees.

She said: I'll have to go downstairs in a minute. I've got to lay the table for lunch.

Thérèse took this in. Her eyes still sluggish but indicating: go on.

There's no one to help me at the moment, Léonie added: so I have to get all the china out of the *bonnetière* all by myself. The Quimper dish too of course.

She stood up.

See you, then.

Five minutes later Léonie walked carefully down the stone-flagged passage into the kitchen. She carried the Quimper dish on her upturned hands. They sank under its cool weight. She set it down on the kitchen table, stroked it with one finger. A big dish, roughly oblong in shape, with rounded shoulders. Its thickness and heaviness were emphasized by the bold strokes of its painted decoration, dark orange, dark pink, and navy blue. In its centre a squat Breton countrywoman in white bonnet and striped blue gown planted her saboted feet on a clump of vivid grass. The glazed surface of the dish was a network of fine cracks. Its two handles, flattened, curled like pigs' ears, were striped in yellow and blue.

Antoinette had loved this dish, had used it every day for serving the fruit, piled on vine leaves, that ended the *déjeuner*. Thérèse and Léonie loved it too, quarrelling over whose turn it was to load it with grapes and plums, arranged in blue and green pyramids, and carry it in.

Now that Thérèse was ill in bed, Léonie could handle the Quimper dish every day. Now that Antoinette was dead, there was no one to repeat to her to be careful, not to drop it, to try at least to walk like a lady, not to plonk it down like that.

Léonie went to the orchard to pick some vine leaves. The vine clawed its way up the wall at the end. It dragged a train of pointed green and scarlet leaves. Ornamental, Victorine always said: planted for no good reason. The grapes it produced were small and hard, inedible. How could it be otherwise, in this northern climate with its rain and storms? It just proved how Monsieur Martin, with his jumped-up-gentleman's fancies, was not a real farmer at all. Not as she, Victorine, the daughter of proper peasants, understood that word.

Perhaps, Léonie thought: Louis simply liked the colours of the vine's leaves, which turned, as autumn deepened, to a fine flare of pure red. She stood, indolent, in the wild grass of the orchard. It was almost lunch-time, but the sun on her face stroked her into lingering, into staying still. A few yellow leaves lay at her feet. The trees had begun to loosen themselves of leaves as they had of golden pears, rough-skinned and scarred, a little while before. Léonie saw a dropped pear rotting in the grass, bruised amber, its creamy flesh exposed, its grainy core.

Without thinking about it she was climbing over the wall. She hopped across the lane and the ditch, and entered the woods. Wandered along the path that led to the little clearing.

She startled fully awake, alert. That touch again, its velvet insistence. As though some enormous beast nuzzled her then picked her up in its mouth by the scruff of her neck. She dangled in free air, then was put down. She closed her eyes against the play of gold and green light. Reopened them.

The lady, yes, she certainly was a woman, stood barefoot on top of the white outcrop of rock above the spring. She held an overhanging branch with one hand, as though to steady herself, and put up the other in a gesture of greeting. She smiled. Summoned Léonie with that look of interest and tenderness. Drew her, as surely as if she had her on a string, unresisting, across the grass.

Thérèse tossed in bed. For some reason she found herself obsessed with the idea that she should get up to lay the table. To help Léonie. She was helping no one lying here. In fact causing them extra trouble, extra work. She had offered up her illness for her mother's soul. And her willingness to suffer more if need be. Was that enough? But could she do more when she felt so weak?

Victorine shoved her out of her worries, opening the bedroom door without knocking and clattering in.

Oh. Isn't Léonie here then? Where's she got to I'd like to know? I need her to go and fetch the fruit from the *grenier* and then there's the table to lay and I don't know what.

Thérèse sat up in bed.

I'll go and see what she's up to, she told Victorine: don't worry, I'll find her.

Victorine opened her mouth. Thérèse jumped out of bed and hurled on some clothes. She felt capable and important. She hurried out of the room, before the astonished Victorine had completed her first sentence of protest.

Thérèse was afraid of the *grenier* where the fruit was kept. It was a high little barn on wooden legs, half-timbered and thatched, reached by a rickety wooden ladder. It was on the far side of the yard, opposite the kitchen door. Between the wooden legs Louis had built chicken coops. Thérèse hated their smell. The *grenier* above them was so dark that, stumbling across the floor, she was always afraid that it had rotted away into gaping holes. Through these she would fall into the muddy embrace, dirty with droppings, of the chicken coop below. Hens had horrible faces close to. They wanted to peck you all over, out of sheer spite. If she had Léonie with her it wouldn't be so bad. So she made for the orchard first.

She sped down the little path leading between the long vegetable beds of the kitchen-garden. How good to run, her legs wobbly at first but then miraculously able to move easily and fast. The sandy path flowed towards and under her. Tennis shoes went *toc-toc-toc*. She galloped past the rabbit hutches, the ducks swaying towards the pond. She leaned against the wooden door of the orchard, which Léonie had left slightly ajar, and put a hand to the stitch in her side. Then she pushed it wider open and went through it, calling out her cousin's name.

Léonie was not in the orchard. Thérèse looked at the red-tinged green brilliance of the spreadeagled vine. Then she understood. She ran to the wall and began to climb.

When she got to the little clearing she halted, seeing Léonie there on her knees, looking up at the outcrop of rock. Then she too knelt down.

THE DUSTPAN

*T*he two girls walked back down the salmon-coloured path towards the house, each with a bouquet of vine leaves. On one side of them was a sprawl of turquoise cabbages, frilly and tight-waisted, ready to bolt. On the other, the big green umbrellas of the courgette plants, the orange swell of pumpkins. They did not speak. Léonie ran up the stairs to the *grenier* and reappeared holding a basket of plums. They crossed the yard and reached the kitchen door. Thérèse put out her hand and opened it.

The Quimper dish lay in pieces upon the floor. Violence measured the distance of one fragment from another. Painted jigsaw bits. The Breton lady had been dismembered. Her head lay near a table-leg. Her flower-clasping hands rested at the foot of the stove.

Madeleine and Victorine awaited them.

The voices all jumped out at once.

I didn't do it. It's a miracle. Whatever are you doing out of bed? Heavens above. Just look at you. You're supposed to be ill.

Thérèse looked at the pale Léonie.

I did it, she said in a clear voice: I heard someone calling me so I got up and came downstairs. The voice was calling from outside. I ran through the kitchen. The dish was on the edge of the table. I knocked it off as I ran past. It was an accident. I'm sorry.

Heroic Thérèse, standing up straight to make her confession with shining eyes, await due punishment.

Who called you? Madeleine asked: Victorine? Léonie?

Thérèse blushed pink. She trembled. But she stood up as eager as Joan of Arc before the judges at her infamous trial.

No, it was neither of them. Something forced me to go to the orchard. So I went. The voice called me to go there. And then. It was a miracle! Aunt, look, I'm cured. It was Our Lady.

Madeleine hesitated. Her hand twitched on the rolled edge of the black silk scarf about her neck, clearly longed to rise and slap someone.

What's the point of all this fuss? she said at last: what's done is done. I'm glad you're feeling better, Thérèse. It's much better you should be up, not brooding in bed. So come along then. You can join us for lunch.

Aunt, Thérèse whispered: it was, the lady in the woods I mean, when I got there, I saw her. It was the Virgin Mary. Just like –

Victorine had become busy with dustpan and brush. Madeleine interrupted Thérèse with a sharp cry.

No, *imbécile*, pick them up piece by piece. Just in case it's worth mending. No, with your hands, not with the dustpan.

Victorine's face went bright red. Her voice was muffled when she crawled under the oilcloth-covered table to search for Quimper bits.

So wasn't she black, then? Not the same one Miss Léonie saw?

Madeleine stiffened and frowned. But Thérèse recited happily.

She had on a long blue dress. Her hair, which was long and fair, was almost entirely covered by her white veil. Her hands were clasped, and she carried a crystal rosary over one arm. Her feet were bare, and there was a golden rose resting on the toes of each one.

Léonie could see her mother making a big effort not to lose her temper.

Yes, Thérèse, just like the statue at the side of your bed. You made it up, didn't you? Your imagination. You and Léonie, what a pair. You've been ill, very feverish, so we'll say no more about it.

Thérèse flung her arms wide and broke into sobs.

I want my mummy and she's dead!

She clutched Madeleine around the waist. Her aunt melted, patted her.

Poor child, poor child.

Léonie lifted her sleeve and wiped her nose on it. Just to test whether anyone was watching her. They weren't. She bent down and picked out a fragment of Quimper from the dustpan. The joined painted hands that held flowers. Then she slid out into the hall, fist clenched over the treasure in her pocket. She seized the black leather strap dangling near the hall mirror and beat the gong for lunch. Then she ran upstairs, to put the piece of Quimper in a safe hiding-place.

She tackled Thérèse in the bath that night. Her cousin lay under a quilt of white bubbles thick as fur. She had drawn it up to her chin. Her toes poked out, and her knees. Léonie perched on the cold wet edge of the white bath and leaned forwards.

Why did you say it was Our Lady we saw? How did you know?

Thérèse smiled at the taps.

She told me. I asked her her name just in case it was the Devil in disguise and she told me who she was.

She turned her head and studied Léonie. Gave a wondering smile which made Léonie want to hit her.

Didn't you hear her? I did, loud and clear.

Léonie prodded the foam with one finger. She longed to say: you're lying, you made the whole thing up so you wouldn't get scolded for breaking the dish.

She said: she didn't say a word to me. It wasn't about talking. Not that sort anyway.

She's going to come back, Thérèse said: she told me so. I'll know when. I'll just get the feeling that I've got to go to the woods. You can come with me if you like.

She glanced at Léonie from under her eyelashes.

Would you go now please? I'm going to get out. Oh, just hand me my towel would you? Thanks.

THE ORANGES

Counting could be done by means of magpies. When you were in England. The rhyme went one for sorrow, two for joy. That made you anxious to see two. Léonie's problem was that she could not count. Not these simple numbers. When did one magpie become part of two? If you spotted a smart black-and-white magpie in a field, that was certainly one, but if, when you'd walked into the next field, you saw another, was that two or another one? How much time, spent in climbing over stiles and so on, had to elapse before you could proclaim one, and then one, rather than two? How many fields, hedges and stiles were needed, what thickness thereof, to separate one from getting mixed into two?

Two was an odd word anyway. It did not express twoness. It was as short, round and compact as one. Léonie's formula was: one magpie in the same field as another magpie, both in view at once, makes two magpies. She preferred saying one-and-one to two. She knew what she meant. Two was blurry and made her anxious. She did not have this problem with fields because they had names and did not need counting. Similarly with the cows on the farm. If magpies had names all would be well. Her joy perched on trees above her head, laughed at her, flew away.

She sat on the kitchen doorstep, knees wide apart, frowning. She was juggling with two oranges. They whirled through the air in front of her, an oval streak of orange. One. Then she made them slow down, and two oranges again spun between her hands.

One red lady. One blue.

One? Or two?

THE GREEN SCARF

*L*éonie and Thérèse gave the scarf jointly to Victorine on her birthday. A square of thin green wool scattered with small paisley shapes in yellow and red. Victorine protested that she did not need a new scarf and that green was not her colour. She said she'd rather have had some lilac bathsalts, the sort that crumpled to silk in the water. A good soak in fragrant steam. To ease her aching legs. In the end she yielded, declared that she would wear their gift, tucked inside her raincoat. Her old scarf, purple rayon printed with black flowers, she would demote to second-best.

Now the green scarf would become part of Victorine. Like her blue overall for cooking, her little gold earrings, her grey tweed overcoat for winter, her beige mac that doubled as her coat in summer. She took care of her things, she made them last. She folded her jumpers around sheets of tissue paper after ironing so that they wouldn't crease on the shelf. She hung old cotton sheets around her Sunday dress, her pleated terylene skirt. She slapped, brushed, shook, mended, darned. She kept things in good repair. She made sure that they went on being there. She saved them. She hadn't been able to save Antoinette.

As it's my birthday, Victorine announced: I'm going to visit Rose. I'll have supper there.

Her words shot out in a clatter. Her tone was odd. Almost excited.

I'll be back late, she said to Madeleine: I'll lock up when I come in.

She kissed them good night, quick smacks on the cheeks, one, two, they offered her.

Thérèse and Léonie were laying the table. Antoinette's place had to be laid opposite her husband's. He insisted. Her silver fork with the monogram engraved on the handle, her silver napkin ring. A form of remembrance. Madeleine called it ghoulish, but obeyed her brother-in-law's wishes. Now, hands on hips, she was checking the girls' work.

Don't forget Louis's pills, he has to take them with his food. And a bottle of wine, open one just in case he feels like a glass.

Victorine had on her beige raincoat, the green scarf, a woollen shawl, and her stout boots. As she left the dining-room her hand went to the bulge in her pocket. A lot of clothes just for crossing the farmyard on a warm autumn night. Not yet dark. They saw her plainly from the kitchen window when they ran to watch. Blue dusk began to blur shapes together, but the beige mackintosh was visible. Skipping past the poultry shed, excited, furtive. From their bedroom window they saw Victorine knock on Rose's door. The beige mackintosh was embraced by a blue one. Arm in arm they made off towards the kitchen-garden. Stealthy and eager as two crooks.

Supper was dismal. Like all the other meals nowadays. Louis sat listlessly in his place opposite his dead wife's high-backed chair and empty plate. He poked at his soup. The two girls were accustomed to not talking at meals. They clattered their spoons dutifully around their plates. Madeleine chattered to her brother-in-law. The weather, news of the farm, the village. He grunted. From time to time her voice trailed off and she became her grief, heavy as a stone. She carried it without complaining and without any comfort. She stared at her glass. Then she revived into cheering Louis.

Eat, she coaxed him: it'll do you good.

The two girls were swift as hares ushering the plates in and out. Louis's untasted food was scraped into the dogs' bowl, the baked leeks shut into the fridge, the cloth whisked off, shaken, folded away in the *buffet*. Thérèse flew at the washing-up while Léonie clattered coffee and *tisane* on to the silver tray and took it through into the white *salon*. Madeleine had on her spectacles and was reading the newspaper while Louis dozed on the sofa. She sighed as the two girls stood over her to kiss her good night.

Off to bed already? Sleep well then.

She no longer came in to kiss them once they were in bed. They were too old for that now. Teeth-cleaning, proper washing, saying of prayers, they were trusted to do those on their own. So their empty beds would not be noticed.

They copied Victorine and put on thick jumpers and raincoats and boots, just in case it was cold and wet in the woods. Léonie said that was where they were going. She just knew. Thérèse had the pocket torch she used for reading under the bedclothes. Léonie had a candle, candlestick, box of matches. She had a big bulge in her pocket, exactly like Victorine's.

They took their usual night-time route out of the house, via the backstairs and the dark kitchen. They glided out of the back door and leapt into a pit of shadow.

Tiptoe through the mud in the chilly dark. Crickets, owls, their own hasty breathing. No lights on in Rose's cottage. It thrust out at them, a black angle sharp as an elbow, as they went by. Stars overhead were frostily clear, Orion in front of them straddling the path.

In the furthest orchard they paused in front of the wall where the vine grew, and listened. Feet came along smartly over the sand and dirt of the path, were cushioned briefly by the grass of the bank into inaudibility, then snapped over twigs and beech-husks. Loud whispers, some giggles. From far inside the woods a thin music, a voice singing. Plaintive contralto. Victorine?

They pulled themselves to the top of the wall, sat there on the rough stone, shivering. Love made Léonie speechless. She thought: just suppose I saw her again. She did not dare say more. Love made her sharp-eyed too. She peered into the darkness of the lane. Thérèse beside her was calm and still.

People were coming from the village in groups of two and three, a stealthy procession. They knew their way all right. Steady and determined. Bulky shapes hatted and scarved against the cool night. Murmuring to each other, helping each other across the ditch then vanishing into the woods. In the close thickness of trees lights glimmered and swayed. Lanterns, Léonie thought: candles. She slipped from the wall down into the lane, pulled Thérèse after her. They were received by the group of villagers they joined with no comment other than a whispered *bonsoir*. Léonie recognized the blacksmith and his wife, their two teenage children. So they were all to be anonymous. Very well then. She crept forwards after them.

The little clearing in the heart of the woods looked different at night. Larger, its edges melted to blackness. Stars snapped overhead, through the branches. Thérèse and Léonie stood crushed up against adults they did not know. So many people here. Strangers, surely, from other farms, neighbouring villages. Damp wool and gaberdine in their nostrils, wide dark shoulders of men's coats looming above them, their noses pressed against foreign backs. Léonie loved it, this intimacy in the crowded dark. She was a spy, noting the spurt and lick of match-flame as someone lit a cigarette, the scent of rough tobacco, the light in the cupped hands, then the red tip of ash. Love made her reckless. She towed Thérèse through the pack of skirts and trouserlegs. They knocked against linked bodies, which parted to let them pass. Excuse me, excuse me. Now they were stumbling into the front row of people, on to rough grass. Now they were recognized, pushed forwards, blinking, to the glare of flames.

A bonfire blazed inside a small circle of stones. Tended by a woman with a long stick. She kept it tidy, the fire, pushing fallen embers into its heart, knocking its flaming hands together, greedy, bright. It was Rose. She nodded to them, went on with her stirring and poking.

Candles in jam-jars were set around the heap of boulders. Tongues of the holy spirit of the woods, that trembled but did not go out. Women knelt there, in the mud, among stones, candles clasped between their praying hands, their faces dramatically lit from underneath, their eyes turned towards the rock, the spring. A sigh shivered through them as Léonie and Thérèse joined them and knelt down.

Victorine was there too. She plucked the green scarf from the throat of her raincoat, spread it on top of the heap of boulders. She dressed her makeshift altar with a necklace of twisted corn, two jam-jars stuffed with bouquets of oats and barley.

Thérèse agitated her knees and rattled the rosary she'd produced from her pocket.

She whispered: she shouldn't be doing that.

Victorine gave a brisk nod, came to kneel down next to them. After a moment or two of uncertain waiting, someone started the rosary. Léonie was bored. She thought: she won't come, too many people.

Beside her Thérèse gasped. Became rigid. Her face tipped up, radiant. She opened her arms and smiled. The glow of the bonfire outlined her in gold.

That was the sign. Everyone, even the men, sank to their knees. All heads turned towards Thérèse, while the *Ave Maria* shot from their lips. Léonie closed her eyes and felt nothing.

Thérèse's ecstasy lasted through three decades of the rosary. Then, in the middle of a *Gloria*, she brought her outstretched hands back together, gave a funny little bow, and got stiffly to her feet. She seemed to know exactly what to do. She was poised, cheerful even. She lifted her hand, waved towards the crowd to get them standing up, and began the *Salve Regina*. Rose and Victorine stood one on each side of the heap of stones. Rose conducted, Victorine led the singing. They all knew the hymn. They sang it every week in church. *Mater misericordia*. A tune of much sweetness because so familiar.

But the tune changed halfway through. Now it leapt up and down, it gambolled. A man standing under the trees a little way off had produced an accordion and begun to play. The Latin words died, replaced by ones in *patois*.

People stretched, shook themselves, smiled. They looked at each other with a sudden lightheartedness that had gravity mixed into it. They bowed to each other like Thérèse had to the Virgin. Then they clasped hands and arms, began dancing. Léonie stood on a boulder to get a good view.

It was old-fashioned dancing. Like at the recent fête on the fifteenth of August, when rockets dashed upwards from the dark field outside the village and the Virgin ascended to heaven as a shower of fiery red sparks. Men and women dancing in formal couples, plenty of women together too. The sort of dancing where you hold your body straight and move from the knees. Much twinkling of booted feet, much strict twirling, heads held high, hands splayed in the small of backs, firm grips and spins. The accordion kept them at it, energetic, sweet. Here was Rose, she drew Léonie in.

Dancing was a consolation for not seeing the lady. At first they bumped each other, Rose's soft bosom a cushion for Léonie's chin. This made Léonie giggle; she rather liked it. But Rose was serious, hissing in her ear to behave properly, just concentrate. Between them they invented a sort of polka, in which their feet did not get too mixed up. They took turns at leading, each of them having begun by trying to be the man. Now they held each other in a practised and relaxed way, their heavy coats flying out behind them as they swept back and forth, back and forth. They wove in and out of other couples. The moon slid from the clouds and burned silver. A boy dangled a paper lantern over the dancers' heads, a globe of pleated paper with a lit candle inside, bobbing at the end of a pole, a second moon. It lurched in a sudden eddy of air, the dancers pushed on.

That dance stopped, and another began. Panting, delighted, Léonie leant against a tree, watching Rose trot, turn and swivel in the embrace of – was it? – yes, the postman. The secret party conferred adult status on her. She felt tall and languid, heated as though she had drunk wine. When a dark shape swam up at her and blotted the dancers from her view, she smiled, began to say yes of course I'll dance.

It was Baptiste. He clutched her against him too hard and too close. His breath warmed her ear, the side of her neck. He definitely did not believe in taking turns to be the man. He steered her in between the other villagers with fierce concentration. She thought oh well I'll let you, and her feet stopped insisting and darted in and out of his. She let herself press back against him. It was all right, because it was dark and they didn't have to talk. Two thicknesses of coat between them, she couldn't feel much. Rose and Victorine went past with a smile and a nod, feet severely in tune. Where was Thérèse? She had vanished. Léonie lifted her head from her partner's serge shoulder and stared about.

Baptiste apparently could read her mind.

She's gone home to bed, he growled in her ear: my mother says I'm to see you back.

Léonie was disappointed. And she had a sense of being got rid of. Though all about her the dancers were stopping, bowing and smiling, the atmosphere had heated up. The accordion had ceased. Bodies poised alertly. They expected. They knew what would happen next. And first, they waited for the outsiders to go, and for the sleepy children to slip off, melt away through the trees. The candles on the ground burned on.

Come on.

Baptiste took her hand and tugged her after him. Rose's face loomed suddenly, ecstatic and blank. She waved. She put her hand on Victorine's shoulder. Léonie couldn't watch them any more, she was being pulled on to the narrow track between the trees.

She knew the way back. Of course she did. Baptiste was making sure she didn't return to the clearing. No spying on adult secrets. At first she was blind in the dark, away from the candlelight, the soft glow of the bobbing lantern. Baptiste felt her stumble and slowed down so abruptly that she ran into him, grabbed at him to keep her balance.

Everyone had vanished. They were alone in the woods. Léonie's knees shook. Her stomach jumped up and down. Baptiste gripped her by the lapels, steered her off the path, shouldered her up against a tree. He stood over her and fumbled at her coat, opening it. His hands felt her rapidly. His mouth surrounded hers. Grazing it. She turned her head aside, met his cheek. He kissed her again. One hand on each side of her waist, he pinioned her, slid his tongue into her mouth. Warm and soft, he was in her. Léonie had one clear thought: he is pretending I am Thérèse. So she felt free to yield, enjoy herself: it's not me he's kissing, she explained to the nun in her head. His nose bobbed against hers. He worked his lips up and down, up and down, over hers. It seemed very businesslike. Soft, rubbery touches. Then his hands left her waist and burrowed under her jumper.

Baptiste stroked her tenderly, attentively. She thought of the drowning kittens and wanted to laugh. Both of them were breathing hard, such a hurry and scramble of lips, hands, such a heat inside and between them, he stuffed one small breast into his mouth and then the other, giving her the same churn of astonished guilty pleasure as when she lay on her bed in the afternoons and read the forbidden books she smuggled home from the library. Those books where so little, finally, was spelt out that Léonie's imagination took hold, she pressed down, down, on the pillows so carefully arranged underneath her, till the wave of warm sweetness toppled, fell, receded, and she lay back curled up eyes closed almost in pain. Those books Thérèse read too but pretended she didn't because then she'd have to go to confession about them and spell out exactly what she'd thought.

Now Léonie was thinking about Thérèse, how unfair it was boys wanted her when anyone could see, however well-developed her figure was, she wasn't interested in all that. In what Baptiste was doing now to her, Léonie whom he thought of as Thérèse, lifting her skirt, sliding his hand between her thighs, sliding his hand further up to touch her wet knickers. Léonie sighed. The nun in her head shouted out. That was as far as the books said you should go. Quite a lot farther. Stop *now*. She shook her head and sighed.

I want to go home. I'm going home now.

She walked in front of him. They climbed over the cold wall, into the orchard. Baptiste jerked his head at her, then disappeared into the darkness in the direction of Rose's cottage. Léonie picked her way up the kitchen-garden towards the house. Was that a light burning in one of the windows of the second floor? It flickered, then went out. Thérèse's room. Thérèse pretending to be asleep.

She stood outside the back door for a moment. Familiar smell, in the darkness, of salt-laden air, manure, compost. The strip of seaweed nailed by the door was slimy to her touch. A dog barked in one of the cottages. A car swished by on the main road. Everything was exactly as it always was. Léonie had just been kissed for the first time by a boy. More than just kissed, really. The sort of passionate encounter she had read about in novels. She had thought of that first male kiss as opening a door inside her, through which she would step towards adulthood, real sex. She expected to feel utterly changed. The possessor, at last, of secret knowledge. She lifted her feet, one by one, pressed them down on to the boot-scraper that stood by the back door. Lumps of mud fell off. She longed for a thick ham sandwich and her bed.

The kitchen door flew open. Madeleine looked out.

Where have you *been*?

Léonie was tempted to confess everything. To watch her mother go first pale with shock, then stiff with rage. To hear her stutter, then screech. You let a farm boy maul you about?! Probably there wouldn't be words sufficient to express the outrage. Léonie decided to reserve the pleasure of baiting Madeleine for later. She sounded as casual as possible.

I went for a walk in the woods. It was all right, there were lots of people there. Victorine. Rose and Baptiste.

Madeleine turned away.

I've been worried to death. I'm very disappointed in you.

THE WATER-BOTTLE

*L*éonie had agreed to help Victorine strip the tomato plants. Hauled from the corner where she was writing up her diary, she complained. Victorine insisted. Too much work and not enough people to do it. Where was Thérèse? That child always vanished just when you needed her.

On the way to the tomato patch they made a detour to Louis's workshop to collect some baskets. Big osier ones, flat, that you could crook over your arm or set on the ground. Her uncle sat at his workbench looking vague, as though he'd forgotten why he was there. His brown fingers played with one of the pieces of the broken Quimper dish. His eyes were watery, tears collected at the corners. Victorine pushed past him to collect the baskets.

She said: you should throw those bits away. That dish is unmendable. You're just wasting your time. Anyway there's a piece lost.

They made their stooping way up and down the rows of tomato plants. Two osier baskets each: one for red tomatoes and one for green. To save having to sort them later. With the green ones, every year, Louis made what he called frog jam. This year, Victorine snorted: she and Léonie would have to do it.

Why not chutney? Léonie enquired: like we do in England?

But she was too warm and lazy for a proper argument. The heat of the sun in this sheltered place made her not want to think about anything. She moved her hands among the leaves of the tomato plants, liking their roughness and hairiness, their harsh earthy smell, a bit sour, like that of geraniums. She liked breaking off the plump red fruits, each one crowned with a stalk like a star. Some were dusty and cracked, oozing juice they were so ripe. The green ones held on tighter, didn't plop into her palms but had to be tugged.

Here she is, said Victorine: here's *mademoiselle*.

Thérèse made her way through the neat kitchen-garden. She wore a striped blue and brown dress, dangled a net veil in one hand, clutched a bunch of red and purple gladioli in the other.

I'm going to put some flowers on Maman's grave, she said to Léonie: would you like to come?

Victorine measured the last two rows of unpicked tomatoes with her sharp blue eyes. She waved towards the baskets at her feet.

We've nearly finished. Go on, you might as well. Take these in for me first.

Léonie decided there was no time to put on a dress. But she stood in front of the large square mirror in the hall and lifted the gilt-backed brush she found there. She dabbed at her hair, smoothing it. She pulled her blouse down, found a cardigan.

The cemetery was a square plot of ground, enclosed by a high wall with ornamental turrets at the corners. The dead lay inside this fortified enclosure in rows as neat as those in the Martins' kitchen-garden. They rotted quietly, like the dropped fruit you found hidden under the leaves of the tomato plants. The older graves had crosses and stone angels, low wrought-iron fences. The modern ones were like fallen doors. Thick slabs of shiny granite. On top might be a porcelain book open at a bit of holy text, plus a photograph, or a porcelain cluster of pink and red roses with sharp frilled edges. Plastic pots of bright flowers were less ugly than the tombs, made the cemetery messy, even cheerful.

Here and there women tended the graves, just as they tended their houses. Swept and polished plots had their weeds removed, their flowers replaced, their carpets of blue and green marble chips pulled straight. One or two ancient graves were neglected, rusty iron crosses dangling broken beadwork bouquets. But most looked like dolls' houses, Léonie thought, where the women played at rearranging the clean furniture. Why did it matter so much where you were buried? That people knew, and came to visit you?

Thérèse knew that it mattered, but she did not say why. She was busy at the tap over in the far corner of the cemetery, filling the empty bottle she had brought with her from the house. Léonie stood by her aunt's grave, scuffing her sandal in the dust of the path and reciting the Hail Mary under her breath because she felt she should. Antoinette lay in the earth, crushed by a smooth black slab. Incised letters, filled in with gold, spelt out her name and dates. Like the cover of a book. Which was supposed to open and flap back at the end of the world and let her out, resurrected. Léonie looked around. A garden of books, each with its title and date. Characters staggering out from between the pages on the Day of Judgment, brushing earth from their lips. People in books did not die.

She chanted aloud: now and at the hour of our death amen.

Thérèse returned with her water-bottle, knelt down to arrange her flowers. Gladioli blazing out of their thick green sheaths, sprigs of speckled green and yellow laurel. She crossed herself, closed her eyes. Her expression became dreamy, ecstatic. Léonie looked away and coughed.

They walked out through the wrought-iron gates. Léonie dragged these to behind her.

She said: I'm going to go to the woods again tonight. Will you come too? I want to see more of what's going on.

Thérèse upended the empty bottle she carried, shaking it. A couple of drops flew out.

I don't know, she said: I don't think I want to. I didn't like it when it turned into a dance. I couldn't find you so I went home.

Oh, Léonie said: I was dancing with Baptiste.

Were you! Thérèse said.

She marched ahead, very fast, her back straight, the bottle grasped by the neck, like a club. Léonie caught her up, tapped her shoulder.

Eh, eh, calm down will you.

She tweaked Thérèse's earlobe: slow down!

Thérèse shook her off: let go of me!

Léonie wheeled. Instead of following Thérèse down the road that led to the centre of the village, out again, and so, eventually, to the Martin farm, she went in the other direction. The long way round. The path through the fields, that ran past the back of the farm, and the woods. She refused to look round, even though she was sure that the angry Thérèse, determined to miss nothing, was behind her. She counted the telegraph poles, cows behind the electric fence, her own heartbeats. This was one of Thérèse's record silences if you didn't include the ones during her illness.

Thérèse drew level with her.

The thing is, Léonie, it's rather vulgar, isn't it? There's Our Lady appeared to us and you let yourself be pawed by some boy from the farm. It's not really on you know.

You weren't really asleep when I came in last night, were you? Léonie said: you were just pretending. And you told Maman that I was out in the woods.

She peeped sideways at Thérèse's breasts, hips, calves. Never mind, Madeleine always said to the flinching Thérèse: wide hips are best for having babies. And to Léonie: never mind, you'll see when you grow up, it's fashionable to be thin.

They reached the back of the Martin property, the orchard wall.

Anyhow it's you he fancies, Léonie announced.

Thérèse was pink as a boiled prawn.

I don't care! she shouted.

She jumped across the ditch that marked the entrance to the woods. She turned back towards Léonie, with a recomposed face.

We've come to say our prayers to Our Lady, haven't we? To see whether she'll appear again? So let's be like we are in church, all right?

The wood was full of noises. Crashes in the undergrowth. The whine of psalms as squeezed out of the village choirboys. The tinkling of a gold bell. Loud Latin words in a clerical voice. Léonie sniffed at a current of incense that drifted past, cold and pungent. It smelled all wrong in these woods.

Rose barred their way. She seemed to have jumped up from nowhere. Out of a bush. She raised her finger at them.

It's *Monsieur le Curé*. Don't let him see you. It'll only mean trouble.

She spoke in a whisper. Thérèse whispered back.

But what's he doing? That's Our Lady's shrine. Is he saying Mass?

No, Rose said: he's been doing an exorcism. He says the apparitions are the work of the Devil. He's been casting the Devil out.

She put her hand on Thérèse's arm.

Don't go and look. You'll only get upset.

The girls ran past her, into the clearing. Word of the priest's intentions must have spread very fast, for many of the villagers were there, shop workers and farm labourers all mixed up, just like last night. In church they kept to their own groups, but here they jostled shoulder to shoulder, watchful and sulky. Last night's jam-jars, with their stubs of candle, had been knocked over. The posies of corn had been trampled in the mud. Victorine's green scarf, torn almost in two, flapped from a nearby branch.

The wobbly singing of the little choir stopped. One of the altar-boys looked about uncertainly, then went on swinging his censer. Another struggled to relight his taper, extinguished by the wind. A third held a bucket of holy water towards the priest. *Monsieur le Curé* controlled his wind-whipped surplice, stole and skirts with one angry hand. With the other he dipped what looked like a hearthbrush into the bucket then flung its load of holy-water drops all over the pile of stones. He glared, and crossed himself. Loudly he intoned some short Latin prayer, then crossed himself again and turned. He came across the clearing followed by his retinue.

The people drew back to let him pass. They bent their heads and turned away, muttering in *patois*. Only Thérèse stood fast, shoulders back and chin up. She barred his way. Léonie beside her wanted to run away from this red-faced furious man. Thérèse dropped to her knees and tried to kiss his ring. He jerked his hand away. You only kissed bishops' rings, even Léonie knew that, Thérèse must have forgotten.

The priest shouted, so that everyone could hear.

I've given my orders. By tomorrow night that pile of stones will have been pulled down and this pagan nonsense completely done away with.

He stared at Thérèse.

As for you, you'd better come to see me this afternoon. Tell your father to come with you, I want a word with him too.

He was off in a swirl of lace and black skirts. Léonie felt very relieved. She didn't have to go too and be grilled. The priest didn't know about the golden-red lady. She was safe. She ran across the clearing and tugged down Victorine's scarf from the tree.

THE ROSARY

Monsieur le Curé kept himself shut away from his parishioners in a large stone house. A long thin garden of lawn and ornamental cypresses, enclosed by shrubs and then by iron railings, separated it from the boulevard between the cemetery and the church.

People were not encouraged to call at his house. None of the Martins, as far as Thérèse knew, had ever been inside it. The priest was left in peace outside his sorties forth to say Mass and hear confessions and have lunch with what Madeleine, waving her cigarette, called *the better families*. He hadn't been to the Martins' for lunch since Antoinette died. A relief to Léonie, not to have to observe him sitting knees crossed in the best armchair daintily sipping his *apéritif*, one hand going up to smooth his shining black hair. In between courses he would wash his hands in the air and speak with a pulpit authority laced with a show of charm. It was hard to decide whether she loathed him more inside church or out.

You can't imagine him going into one of the labourers' cottages can you, Thérèse said, pulling on her nylons.

Nor having a shit, said Léonie: nor lying in bed. What kind of pyjamas does he wear I wonder? Purple and gold bri-nylon, with a dog-collar. While you're there you'll be able to find out. And he won't dare be too rude to you because you're not a peasant. Not like Saint Bernadette.

Thérèse looked noble for a minute, like one of her favourite virgin martyrs about to be broken on the wheel. Then she relapsed into gloom. Léonie picked up the brush and got going on her cousin's hair.

Just don't let him bully you that's all.

There was a truce between them at the moment. The priest had done it with orders for the pile of stones to be pulled down. Léonie smiled at Thérèse in the mirror and zipped her up at the back. She fastened the clasp of the gold crucifix Thérèse always wore now on a thin gold chain around her neck. Antoinette had left it to her, plus some gold bangles and a string of pearls. To Léonie she left an ivory bracelet and a silver brooch.

Thérèse, poured into a tube of blue linen, sighed. She got up, picked up her white gloves, her shoulder-bag of blue quilted plastic with a gilt chain. Shoulders back at the angle recommended by *L'Echo de la mode*, chin up, stomach sucked in. On her high heels she teetered downstairs to join her father.

Louis had put on his grey suit, and carried his beret. Thérèse sprang at him, kissed him on both cheeks. They went out to the car. Léonie watched them from her window upstairs, her hand wound in the muslin curtain.

Monsieur le Curé's house, pinkish-grey stone faced with white, was dark inside. The housekeeper showed them into an antechamber bleak as a dentist's waiting-room. The mottled tiles of the floor were slimy purple like uncooked liver. Curtains of closely woven cotton lace were stretched across the windows, fastened so tightly they kept out both air and sun. A single aspidistra gleamed in one corner in a lime-green ceramic pot. The walls were papered dim brown. Christ writhed on a knobbly wooden crucifix next to a photograph of the Pope. A pleated white paper fan perched in the empty grate below.

The room smelled of winter and of washing. Thérèse imagined the housekeeper wiping the aspidistra with a soapy sponge, rinsing down the ivory Christ, loincloth and all, holding him in her arms like a baby with a wet nappy. *Monsieur le Curé* sat in his bath, a huge holy-water stoup, still holding his biretta, while the housekeeper washed his bristly black hair. These were thoughts sent by the Devil. She shut her eyes and prayed furiously for deliverance.

She opened them to find that the priest had come in and was blandly regarding her and her father. She did the bob they were taught at school, the deep version. The two men sat down in red velvet chairs on opposite sides of the cold fireplace, glared at by Christ and the Pope. Thérèse stood between them, hands clasped. Her shoes began to pinch her toes, like a warning to run away, and she was suddenly desperate to go to the lavatory. More distractions from the Devil.

Monsieur le Curé moved his long fingers, which he clearly admired, over the skirts of his black soutane. He sounded, when he spoke, as deliberately bored as possible.

Well then, my dear friend, so what's all this gossip in the village about visions of Our Blessed Lady? I must say I'd have thought I could rely on you to calm people down.

Louis spoke, letting go of his beret to spread his hands. The black armband strained on one sleeve. He was humble, he shrugged, he mentioned that Thérèse was a truthful child.

The priest listened, head on one side. His confession pose. The heron profile that she saw through the grille. She hadn't been to confession since the vision. Couldn't. His fingers were clasped now, supporting his chin. The most delicate of yawns. His voice, when finally he deigned to open his mouth, was smooth. Thérèse jumped. The cool grey eyes swerved, rested on her.

Farradiddle. Yes my girl that's what I said. I've never heard of such nonsense.

Thérèse dared to interrupt his strictures when he paused for breath.

But Father I did see her. And she told me her name when I asked, she said Mary Mother of God.

The priest looked at her with contempt.

What do you suppose you are, some sort of little saint?

He went on, more to her father now than to her, in his light drawling voice she had to listen to in church every Sunday of her life telling them all what was what because he had the God-given power to do so and they had to listen and obey. That was what being a Catholic meant. Thérèse tried to remember that the priest represented Christ and must be right. The blood rushed to her cheeks, swelled and thickened them. Her chin wobbled, she couldn't control it, however hard she tried not to cry as his words reached her. Young girls of a certain age. Impressionable, heated imaginations, hysteria. Romanticize. Idealistic. Thérèse stared at her black patent shoes, those twin vices crushing her toes, telling her to take her feet away. She tried to concentrate on that particular pain. She gritted her teeth. She wiped her wet palms on the back of her dress. Between her thighs, stuck together, moisture slipped.

The priest glanced at her and threw up his hands.

Oh no, no tears, I beg of you. You see? You're just an ordinary little girl. A true visionary wouldn't crumple at the first sign of opposition, would take all this in her stride.

He turned back to Louis, who sat calm and polite in his chair, who didn't turn to look at Thérèse. The thread that bound them had snapped with one expert tug. The priest's tone grew stern.

What convinces me that this child is deluded is the conflicting reports of the so-called apparition. Some say she is robed in blue, others in yellow and red. Red, I ask you! Some say she has long fair hair, others that she's as dusky a beauty as you'd find among the *pieds-noirs*!

Thérèse interrupted him.

No, that's not true, Father. It's Our Lady, dressed in blue. She's got long fair hair but she's got a veil over it.

She lowered her eyes to avoid the priest's glance. He laughed and went on, addressing Louis again.

All this is extremely bad for your daughter's reputation my dear friend. She's making herself a laughing-stock. Dangerous pagan nonsense. Idiotic outdated folklore. In these days of religious renewal. Low sort of gossip. Pernicious tales put about by the likes of Rose Taillé and her layabout son.

Thérèse's tears splashed on to the tiles at her feet. The priest looked pleased at her collapse. He tapped her on the shoulder. She felt his fingers spring away again, relieved. He picked up a rosary of cheap light beads that lay coiled in an alabaster ashtray on a side table near the door.

Here, child. Take this with you, it's been blessed by the Holy Father and it will do you good to use it. Say a decade of the rosary next time you're tempted to imagine you're seeing visions and the Devil won't be able to harm you. On your knees for half an hour every night with your rosary, that's the right thing for young girls chasing holiness. One last thing: I forbid you to go to the woods again.

They were dismissed. Bladder clenched, Thérèse limped down the drive behind her silent father.

Next day, Sunday, she sat between him and Léonie in church. She still had not been to confession. She decided not to go to communion. Suppose, suppose, she were in a state of mortal sin? Or that the priest refused her?

The stone interior of the church held darkness as a curved shell holds water. The statues of the saints were ranged along the walls in clumps of two and three, like groups of chatty friends. God hid in the domed box on the altar. Don't abandon me, Thérèse silently begged: don't.

The priest ascended the steps to the pulpit. Below him, the altar-boys settled on their bench, hands in their sleeves, as scornful as he who drilled them week after week. Wrapped in lace and linen, they stared at the women and girls forbidden to enter the sanctuary.

The theme of the sermon was reverence and obedience. Waywardness of certain elements of the youth in the parish. Authority of our Holy Mother the Church vested in me. Regular attendance at Mass and the sacraments, especially confession, as ordained by Holy Church. The sheep guided by the shepherd. Undesirable elements of individualism and mysticism, undesirable attempts at originality, to be weeded out. Adolescent frailty and need of guidance. Saint Paul and Saint Augustine on women.

Thérèse sweated with shame. She stared at her lap where her hands, gloved in tight white net, gripped her missal. She pressed her knees together. If she pressed very hard then her mouth would not open to scream. Torrents of lava would not tumble out to force fire down his throat, torch his tongue. She was red and liquid and dangerous. She would damage that priestly flesh, oh yes, scorch it, she would tear his head very slowly from his neck and laugh as the blood gushed. She would shut him up, trample him down, stop up his mouth for ever with hot red mud.

She couldn't hear him any longer. White peace descended and surrounded her like a tent of cool white gauze. God was in the tent. God was the tent. He surrounded her and wrapped her up in the folds of his silence, his mystery. His great heartbeat near hers. She lay on his heart and did nothing. She let go of her hot grief, which dissolved and became the dew on his garments. He was the coolness at the centre of the fire and his look purged and freed her. She was with him, unafraid. She was nothing and she was love and she was a voice singing in the desert, full of trust, full of gladness at her deliverance.

Beside her Léonie shifted from one buttock to the other and coughed loudly. The priest glared at her then swept on with his discourse. Thérèse didn't hear him. She didn't feel Léonie's elbow jab hers. She was rapt in a frame of fiery clouds. She no longer spoke to anyone except God. She chose silence, obscurity, poverty. She chose him who was everything, her hollow in the rock, her desert refuge. She rose to meet her future self. She flew towards it.

Thérèse's faint in the middle of the priest's sermon spoilt its effect. Though he continued his tirade, people were distracted. They craned their necks to watch the white-faced girl, limp as a corpse, carried out in her father's arms.

It was a sign, but of what kind? Was it the Virgin rescuing her own or the Devil showing he was beaten? Had Thérèse had a fit? Was she indeed making it all up? The little knots of people that formed as usual outside the church after Mass was over and the bells were tumbling out their soft dismissal song had plenty to discuss. Léonie stood amongst them, next to Madeleine and Rose. Victorine was the one who brought the news. The men destroying the heap of stones in the woods, on the orders of the priest, had found that it covered a shallow grave. Inside this was a mess of human skulls and bones.

THE CAKE TIN

Madeleine was seated at the little writing-desk in the back corridor on the first floor. From time to time she sighed. She was checking a column of figures, Léonie could see. She hovered, peering over her mother's shoulder to see how she was getting on. Madeleine started adding up again, nodding her head as her eyes strayed down the page, moving her lips. This time she was satisfied. She ticked the sum and underlined the total. It was a bill sent by the village undertaker. *Pompes Funèbres* in queer modern lettering, thick and black. She sighed again.

Léonie galloped in with her question. *The* question.

But whose bones *were* they?

Madeleine tucked a curl behind her ear. She lit a cigarette, sucked expertly. Her words drifted like smoke.

Oh. From the war.

But why bury them there? Léonie persisted: why not in the cemetery?

Men from the village had fought in two world wars. The names of those who had died in battle were inscribed on a stone roll of honour on the war memorial by the cemetery. Every village had one. Blémont's was a heartily built stone woman in clinging robes carrying a wreath and a scroll. Her hair tumbled loose, her breasts were pointed. She wasn't Our Lady, Madeleine had explained to the children years ago: she was *La France*, also *La République*. It was on Our Lady's feast-day, though, that the village remembered its dead. On the fifteenth of August, after High Mass. The village band, all discordant trumpetings and squeaks, led the congregation in procession to the war memorial. The national anthem was played. Everyone bowed their heads. Later on that day there would be the fair, fireworks, and dancing, but now they were still, quiet. Baptiste played in the band. He wore a brown uniform and carried a cornet in shining yellow brass. Our Lady ascended to heaven on the fifteenth of August and looked after *La République*. When you were in France you were certain Our Lady was French. You felt that she passionately cared about the French soldiers and made sure they won the war. You felt, like Victorine, that the French were the best.

Madeleine said: they couldn't be buried in the cemetery if people had forgotten all about them and didn't know where they were, could they?

But why had people forgotten? Léonie asked.

Timidly she tacked on another question that bothered her.

Why didn't any women die in the war?

Madeleine put down her pen and knocked her cigarette ash into a blue Limoges dish like a saucer.

Look. You and your questions can't you see I'm busy?

Léonie waited. She could see there was more to come.

Madeleine said in a rush: one more thing. You're not to go to the woods any more. I absolutely forbid it. You'll only get yourself into trouble. You don't understand what these people are like. I don't want you mixing with children like Baptiste. You know I've always believed that people from different backgrounds shouldn't mix, it's not fair on the children. You're much too young to be thinking about boys, when I was your age I went around in a big friendly group, plenty of time for all that later on.

Léonie wandered downstairs. She imagined the children of mixed marriages: striped black and white, like badgers. In the kitchen she found Victorine tightly absorbed in skimming the thick skin off the saucepan of milk she had just boiled. She cocked an eye at Léonie, grunted.

I'm going to make *gâteau à la peau de lait* for supper. You can help if you want.

It was a cake you never got in England, because there the milk did not have to be boiled and so you could never collect a bowlful of creamy skin from successive goes. Victorine opened the fridge door and took out a fluted blue bowl clotted with white. She measured the amount with one quick look.

Yes. Just enough.

The cake was everyone's favourite. A sort of sponge, low and crusty and golden, which they ate with apricot jam. Léonie tied on an apron and hoisted herself on to a kitchen chair.

She asked: but whose bones *were* they?

Victorine steadily beat the cream in the bowl. Léonie sifted silky flour on to the wide brass pan of the scales.

Well, Victorine said at last: possibly Jews. And possibly someone who was trying to help them. Escape from the Germans I mean. They got caught and so they were shot.

Léonie plucked out the brass weights from their deep plush-lined nests in the wooden box. They fitted into these exactly. She grasped each one by its golden nipple and set it on the scales. She liked the way they were graded in the box by size, tiny to large. Like children herded into line for a family snap. She used a combination of small weights, to have the fun of adding them up.

She remarked: but in the war the Germans were fighting the French. Jews aren't French, they're Jews.

Victorine tipped flour into cream and beat again.

A lot of the Jews were French citizens. The point was the Germans hated them and wanted to get rid of them. Surely you know that. Don't they teach you anything in English schools?

Léonie said: the war's too modern for us to do it in history. We're doing the Tudors and Stuarts.

She watched Victorine grease the cake tin. Rapid strokes of a pastry brush dipped into golden oil.

She said: but why did the Germans hate the Jews?

Victorine balanced the cake tin on the palm of her outstretched hand and frowned at it.

A lot of people did in those days. Things were different then.

Even French people? Léonie asked.

Some French people. Yes.

Did you? Léonie persisted.

Victorine slapped the cake tin down on to the table and poured cake mixture into it, helping the thick flow along with a rubber spatula. Her voice was a shrug.

Of course not. As long as they weren't making trouble and thinking they were better than us.

She pushed the empty bowl across the table towards Léonie.

Here. Lick it if you like. But shut up for a bit will you, I've got to get on.

Léonie hesitated. She was a bit too old for licking bowls, wasn't she? She tapped her forefinger on the bowl's edge.

But why weren't they buried in the cemetery after they were shot?

Victorine clattered the cake into the oven.

The Germans took them away in secret, at night. Then in the early morning, just before dawn, they took them into the woods and shot them. They buried the bodies themselves. Nobody knew where. The grave was never found.

Léonie asked: but how do you know?

Victorine shouted: I'm busy, be quiet will you?

THE PACKAGE

The postbox was made of white wood. It hung on the outside of the small wrought-iron gate set in the brick wall at the main entrance. People and cars always came in through the big double gates. The small one was wreathed in weeds and brambles no one bothered to clear away. Even though it meant that going to fetch the letters gave you scratches and nettle-stings. Madeleine kept repeating that she'd get down to hacking back the brambles one of these days but on the other hand she didn't have the time. The girls did not volunteer for the job. Thérèse because she was indifferent to the acid jab of nettles, Léonie because she liked to see that little corner of the grounds turned untidy and wild, the gateposts swarming with weeds and grass in their crevices, clumps of Michaelmas daisies.

The postman usually bicycled up around mid-morning, a time when everyone in the house was busy or pretending to be. They rarely noticed him arrive. By the time Madeleine remembered to go and check the box it was nearly lunch-time and she was making *hors d'œuvres* with one hand and pouring *apéritifs* with the other. In the days just after Antoinette's funeral, however, the postman brought bundles of letters to the kitchen door. Too many to cram down the slit of the little wooden box. He brought village news, too, titbits that Victorine indicated she had already gleaned in her trips to the shops. He would float these across the kitchen to her while implacably she checked the meat roasting in the oven for lunch. Oh yes? she'd say, polite and distant: oh really? And she'd nod meaningfully in the direction of the hovering Léonie: not in front of the children please. All Léonie managed to discover from the postman's hints was that the priest had had the bones taken away for a quick burial in the cemetery with as few people present as possible, and that people were still visiting the site of the apparitions, waiting to see whether Thérèse would go back.

Léonie had her first period on the day that Madeleine finished replying to all the letters of condolence sent to Louis. He wasn't expected by her or Victorine to reply to them, Léonie saw. He wrote business letters and that was it. Madeleine stuck on a final stamp then took Léonie upstairs. She kitted her out in a belt and a thick wad of gauze. This felt soft, rather comfortable, a bulky caress between the legs. Léonie held herself straight so that no one should know her secret. She felt different but didn't know how to express it. Not walking wounded. Not more grown-up. It was like putting on a costume for a play, or running in the three-legged race. Madeleine gave her a quick kiss, then tossed her the postbox keys.

All right darling? Go and check for letters will you? I must see if Louis is all right. It'll do you good to get a breath of air. You look a bit pale.

Léonie collected Thérèse and they went across the white gravel in the sunshine. Léonie strutted stiff-legged like a cowboy. She felt wetness leave her and sink into the towel. Thérèse didn't seem to have noticed. Léonie wondered if she were extremely pale or just moderately so. It was the kind of thing Antoinette would have commented on. Now Madeleine had taken over doing that. Her stomach clenched itself and ached. Between her legs the wetness gushed again.

Thérèse insisted it was her right, as the daughter of the house, to fit the little black iron key into the lock and turn it. Léonie lifted up the wooden flap and peered into what always seemed to her like a bird-house in which they might find golden eggs. Two letters for Louis. Cheap envelopes of rough paper, one blue, one greeny-grey. The colour of thrushes' eggs. No trace of magical peacocks. No phoenixes rising reborn from the raging red fire. A Catholic newsletter addressed to Antoinette, *L'Echo de la mode* for Madeleine, and some sort of bill in a white envelope with a pearly paper window.

Léonie had hardly written to her schoolfriends in England that summer. She did not know how to describe her life in France, had never done so. It required a language that her English friends did not share. Froggy, they put on their postcards to her, rocked with laughter at the thought of her eating horse steaks and snails. Now that she was going to stay on in France for a while and not go back to her old school, she'd already let go of those friends. They didn't matter. They faded out, a blur of English sounds, hearty, absurd, foreign. Still she wished someone had written to her. She wished she had a letter. She scowled at her sandals and scuffed them on the gravel. With Thérèse at her side she began to walk back towards the house.

A shout made them look up. The postman was pedalling back towards them down the road. He drew up on the other side of the tall white gates and fished in his grey sack.

Sorry, he called: I forgot this one.

He held a package in his hands. He raised it, threw it over the small postbox gate. It dropped into Thérèse's hands. He waved, and bicycled off again.

The parcel was substantial, tightly wrapped in brown paper, crisscrossed by waxed brown twine with many knots. One white label stuck on the front and one on the back.

It's for me, Thérèse exclaimed: it's from the nuns in Sœur Dosithée's old convent.

She opened it upstairs, in her bedroom, where no one except Léonie would see. She wanted a secret. Also to examine one. She cut through the tight bonds of twine, the shiny brown wrappings, with Léonie's penknife. A crust of brown corrugated paper required the knife again. Like slicing into pastry, Léonie thought: to see what kind of pie you'd got. Four and twenty blackbirds. Or magpies. Four and twenty black and white nuns singing the praises of God baked in a pie.

Thérèse lifted out bunches of letters held together by rubber bands. All of them, she saw as she flicked through them, addressed to Sœur Dosithée, and all of them signed *your loving sister Antoinette*. Blue, white, cream, grey envelopes, that had been neatly slit open. A calm nun's hand wielding a paper-knife. Self-control: waiting until the hour of recreation, when letters from home were allowed to be read. Not so would Léonie treat a letter. If someone sent me one, she thought: I'd tear it open straight away, I couldn't wait.

The gong for lunch sounded two floors down.

But why did they send them to you? Léonie said: and not to Madeleine. It must be a mistake.

It says Mademoiselle Martin, Thérèse said: and that's me. It's quite right I should have them. They must have realized how I'd cherish them. What use would your mother have for them? She'd probably throw them away. She only likes modern things.

She collected up the letters and stuffed them back, higgledy-piggledy, into their brown paper and cardboard nest.

I won't read them just yet. First I'll find them a good hiding-place.

They ran downstairs together. On the first landing Thérèse uttered a shriek when Léonie overtook her.

Do something. Quick. Don't let them see. You've got a huge red stain all down the back of your shorts.

Thérèse found Léonie a fresh towel. The pad like a hammock, Léonie thought, held in its net of gauze. She lowered herself on to it, snug.

You should wear a skirt, Thérèse explained: so that no one can see. Otherwise you might have a bulge showing.

She showed Léonie how to roll up the used towel in a paper bag, smuggle it down the backstairs into the kitchen, bury it in the red heart of the range. They walked into the dining-room together only five minutes late. Clean white half-moons of nails held out for inspection, hands reddened from hot water and soap, hair brushed. Proper *jeunes filles*. Which meant having secrets.

THE IRONING-BOARD

O n Mondays the wash was hung out to dry on the clothes-lines at the end of the kitchen-garden nearest to the house. By mid-afternoon the first batch was almost dry, still slightly damp, just right for ironing, and could be fetched in by Victorine and whichever child she could catch to help her. It took two people quite a time to take the big sheets down one by one, flap and fold them, holding the corners tight. Then you had to unpeg all the other linen and clean clothes and cart it all upstairs to the *lingerie* in round two-handled baskets.

Rose had gone on coming in to help Victorine. She came when it suited her, when she could spare the time from her other work, but she always popped in on Monday afternoons to make a start on the ironing.

Is Rose coming today? Léonie asked: she missed last Monday.

She and Victorine held on to a great square sheet. As they tugged it straight the wind bellied it out, taut as a sail.

She was at the funeral, Victorine said: she would go, poor thing. She'll be along soon. She told me she'd come today.

I'll help you, shall I? Léonie said to Rose: I'd like to give you a hand.

Rose was wearing a dull black cardigan over a black dress. Although she was a widow, she didn't usually wear black. She wore blue checks, or brown. For gardening she wore a scarlet jumper. Today her nose and cheeks showed red under the powder, and her one good eye was red-rimmed, as though she'd been crying. Her green glass eye didn't care. It shone as usual. She joked with Victorine in the kitchen as quick as ever, then drank the last of her cup of coffee and got up.

All right then, she said to Léonie: if you want to.

They lugged the baskets of laundry upstairs together to the *lingerie*. It was at the far end of the attics. A small room. A box for linen. In the centre was the ironing-board made out of an old table covered with a blanket and a sheet. These were marked with brown triangles. Traces of an overheated iron allowed to pause face-down. Shirts and sheets they hung from the ceiling, draping them on lines and hangers. Damp, fragrant veils that made a cool tent around the central space of heat where Rose began to ply the heavy iron. The hot smell of ironing mixed with that of garden-dried linen. Added in was Rose's own smell, warm animal washed with lily-of-the-valley soap. Fresh air and sweat and scorched cloth: that was the smell of their conversation.

Rose, as she listened to Léonie's questions, took off her black cardigan, rolled up her black sleeves. Her face, as she worked and talked, reddened, till she was as rosy as her name.

The Germans were highly organized, she said, pushing the blue-and-white enamelled iron back and forth: they hunted up all the Jews they could find. Regular round-ups. The really big one in Paris, we called it *la grande rafle*.

She finished the collar, began on a sleeve. The cuff first. Dancing the iron over its crisp edge.

They kept the Jews in a sports stadium outside Paris. Packed in with hardly any food or water. Of course lots of them died. Then they were sent to the camp at Drancy, and from there they were put on trains and sent to Auschwitz to be gassed.

To send, Léonie thought: letters, packets. How did you send people in their thousands from one side of Europe to another? She concentrated on this problem so that she did not have to imagine the people themselves. Sent off to become parcels of ash.

Freight trains, Rose said: people jammed in standing up, into trucks. Without food or water. For a journey that took three days. Afterwards, we found out. Those who wanted to know that is.

The sleeve was too dry. She banged the iron down on to its tin rest, took up a small bowl of water, sprinkled the sleeve. The spots of damp hissed under her iron.

Anyway, some Jews did manage to escape the round-ups. One family from Rouen, a couple with their young daughter, turned up here in the village. Henri and I sheltered them for a while, but of course it was very risky with the Germans billeted everywhere except the smallest cottages.

She finished the second sleeve, began on the front. She frowned in concentration but at the same time she suddenly looked very tired, as though she should stop and sit down. The arm moving the iron dragged itself forward and back.

Somebody from the village must have betrayed us, because the Germans came in the night and took the Jewish family away. They took Henri as well. In the morning, just before it was light, they took them into the woods and shot them. They hid the grave so that we couldn't give the bodies a proper burial. They must have guessed we wouldn't look under the shrine. We'd all kept away from it ever since the priest had had it pulled down the month before. We never thought they'd dare put the bodies there.

Rose's voice was so dry that Léonie kept quiet. She watched the nose of the iron poke between the pearl buttons on Louis's shirt, thick cotton of silvery blue with a faint red stripe. Every time afterwards when she saw her uncle wear that shirt she remembered the words that Rose spoke.

None of us knew for certain who the informer was. Some people swore it must have been someone in the Martin family. Others said it must have been that girl who worked here, the one who was the German officer's mistress, she knew everything that was going on around here. Well, we all did. No one ever found out who it was. And then once the war was over people wanted to forget. But I can't.

Rose whipped the sleeves of the shirt behind its back, folded them like double-jointed elbows, bent the shirt in two, and gave it a final press on top.

I was expecting a baby at the time. I went into labour early. The baby was born dead.

THE ONYX ASHTRAY

*T*hey all assumed that Louis's stroke was the result of his grief. His wife, his strong prop, pulled away, and he leaned on air. Victorine said he had run down, like a watch. He had broken.

The girls were allowed to visit him in the clinic. They tiptoed along corridors shining and antiseptic, peeped at him where he lay, inert, in his white bed. He could not speak to them, though his mouth tried to. His lips worked, strained, around his lolling tongue. Something in his face fought, clawed, was smothered. Tears slid down his cheeks. Thérèse cried too, and had to go out. The nurse in charge, as coldly white as the bed and the room, said they were too noisy. They disturbed the other patients and should not come again. All the way home in the car Thérèse wept for her father. Léonie edged away from her on the warm plastic seat. Her thighs itched, encased in scratchy bronze nylons. She closed her eyes, tried to summon the image that consoled. But the golden-red lady did not come. Léonie could not remember her face. It dissolved into a black blindfold, a black gag.

The priest came to condole with Madeleine. Thérèse had to offer him the *apéritif*, Léonie the fluted glass dish of polished Japanese crackers. When Madeleine offered him a cigarette, he hesitated, then took one. Thérèse glanced away in disdain. No wonder enclosed nuns had to pray night and day for priests when they were so worldly.

Madeleine pushed a green onyx ashtray within his reach.

You'll have to forgive me, Father, but I can't repress my curiosity.

Her way of leaning forwards, eyes sparkling, fingers almost touching the priest's black sleeve, made Thérèse squirm. Really, women should not assume such familiar manners with priests. It suggested they did not sufficiently respect the vow of celibacy. It made them temptresses. It wasn't fair on the priests.

Madeleine said: what's going to happen with the shrine in the woods? Do tell us. We're all so longing to know.

Well, Thérèse supposed: Madeleine was a widow. She lacked male company. She'd no hope of remarrying if she stayed here, there was no one suitable in the village. Women of that age, she'd read somewhere, often made a push for one final fling. Madeleine was too old, though, to be behaving so girlishly. And her skirt was too short. When she crossed her legs you could almost see her stocking-tops. Luckily the priest had very good manners. He appeared not to notice his hostess's flirtatious twinkle. There was even a certain warmth when he glanced at Thérèse, as though he understood how embarrassed she was by her aunt's lack of dignity. She'd been living in England such a long time, of course, it was a bit too free and easy over there.

He said: the Bishop is coming down tomorrow to see me. I feel that some sort of ceremony of blessing of the place may be in order, now that the remains have been given a proper burial, and I have asked for his opinion. In all things I desire to be ruled by Our Holy Mother the Church.

He glanced severely at the two girls.

Won't you allow the shrine to be rebuilt? Madeleine asked: wouldn't that be a good thing for everyone?

He hesitated.

Some sort of stone memorial, at the place where they fell, to heroes who gave their lives for France, yes, the Bishop thinks that may be in order.

He crushed out his cigarette.

And how is my dear friend Monsieur Martin?

Good news, Madeleine said: he's coming home tomorrow. It was only a slight stroke, after all. But he can't talk yet.

He can't talk? the priest murmured: how very distressing.

He took his leave. Madeleine grimaced after she'd shut the door on him. Two small whiskies and her words were unguarded.

It looks better, doesn't it, if he agrees with what the Bishop wants. When of course you can see he wants the whole thing hushed up.

Why? asked Léonie.

Madeleine was instantly vague.

Oh. I don't know. I must go and see what Victorine's up to.

THE FISH-KETTLE

The ambulance was expected at midday. Thérèse scooted through breakfast to give herself time for everything she had to do. Get Louis's room ready. Prepare lunch. Put on her best frock. Hastily she swallowed her coffee and bread, wiped the crumbs from her mouth and got up.

Hearing noises in the little white *salon* as she passed its door, she poked her head in to see what was going on. A bed had been put in one corner. Madeleine was whipping a stiff white sheet on to it, mitring the corners, folding back the top edge with its embroidered monogram.

Why are you moving him down here? Thérèse cried: I was just going to do his room upstairs.

He'll feel more in touch with things, Madeleine said: and it'll be easier to move him in and out. If he wants to sit outdoors sometimes or eat with us.

Madeleine's hands seized the pillow and held it to her in a quick embrace. She squeezed it, shook it, plumped it up. Tenderly she laid it on the bed. Her hands went to the sheets again. They stroked, smoothed. The sheets lay back, flat, unresistant.

I was going to do that, Thérèse said: I thought of that.

Madeleine was wearing a dress Thérèse hadn't seen before. Yellow piqué with little cap sleeves. Wedge-heeled white sandals completed her outfit. She'd painted her toenails red.

Nothing more to do in here, Madeleine said: that's it, finished.

She indicated the vase of daisies on the table by the bed, the pile of books, the radio. She put her hands on her jaunty yellow hips, patted the crisp flare of her skirts. She smiled at Thérèse.

If you want to help, perhaps you'd give Victorine a hand with washing up the breakfast things? I want to get on with preparing lunch. I want to have everything ready in good time.

I want to make lunch, Thérèse said: let me do it.

The white bed looked very comfortable and cool. Madeleine gave it one more glance, then shooed Thérèse out in front of her.

Are you sure you'll be able to manage? she asked: I thought you loathed cooking. It's fish. D'you know how to do it?

Of course, Thérèse said.

Well, Madeleine fussed: if you're sure.

Thérèse triumphed. She hustled Victorine and Madeleine from the kitchen, then tied on an apron and set to. For some reason she felt like crying. She held the feeling in tight, she squashed it down under the heavy pasteboard covers of the recipe book. Her tears died like a flattened insect, quick smear on the table-top.

She decided to make a herb sauce to go with the cold poached mackerel, rather than the mayonnaise Madeleine had planned. Tastier, more original. Her father might fancy that. She'd coax him to try a bit. Egg yolks mixed with Dijon mustard, thickened with both melted butter and olive oil, flavoured with tarragon.

She fetched a handful of the aniseed-smelling leaves from the kitchen-garden, stripped them, limp and thin, from their stems. She began to mince them into a fragrant green hill on the board in front of her. She tried to pray for her father at the same time, but Madeleine danced across her invocation, Madeleine provocative in yellow piqué and high-heeled sandals disrupted her holy words.

A correct *liaison* between the egg yolks and the butter is obtained, Madeleine cooed: incorrect. A liaison.

Thérèse shook her head, to clear it. Madeleine vanished. From the curdled phrases sprang a message.

Fetch your mother's letters and read them.

That hoard of words was still in the biscuit tin at the back of the *buffet*. Unless stupid interfering Madeleine had found it and removed it.

Louis was the King, and Thérèse was his little queen. He'd always called her that. Till he got ill, and lost his speech. When he gets well again, Thérèse thought: Madeleine won't let him remember, she doesn't know the right words to say.

Little flower. Little queen. Madeleine picked up the epithets and laughed at them. She dropped them on the chopping-board and crushed them under the blade of her knife. She sniffed them, wrinkled up her nose in disgust. Thérèse shouted out the Holy Name of Jesus and wiped Madeleine from her mind, wiped her hands on her apron. She ran to the dining-room and opened the doors of the *buffet*. She took out the tin box of letters and carried it back with her into the kitchen.

Thérèse knew that she had been sent away from her parents at the age of two months, to be fed by Rose, that she had lived with Rose for sixteen months. It wasn't a secret. Antoinette had once or twice talked of it, and Thérèse herself had one or two dim memories of that time. A shallow stone sink in the corner where Rose unbuttoned the front of her dress to wash herself. The brown of her sunburned arms that ended above her elbows, abruptly as gloves, the startling white of her shoulders and breast.

There was a secret somewhere, though, that Thérèse had to get hold of and understand. Secrets were what lay underneath. Like when a woman slowly unbuttoned her yellow clothes, let them fall down around her waist. While Louis watched. It was disgusting. Antoinette was pure. Surely she would never have behaved like that.

Thérèse was fishing out letters at random, glancing at them while she worked. Antoinette's cool voice sounded faintly in her imagination. As though she were buried under layers of white leaves.

Do you mind that I continue to call you Marie-Jo? You're still my sister, after all! I feel so close to you. But now that *this* has happened I'll never be able to join you in the convent. My dream of the religious life is shattered. I'm so ashamed.

The letters sounded as though they were weeping, Thérèse thought.

Those filthy Germans. Destroying everything. Taking everything they've a mind to. Overfamiliarity. Illbred.

Too much for Thérèse to take in at one go. She dropped the flimsy pages on to the table-top, concentrated on whisking two egg yolks in a china bowl. Her poor father might still like his lunch on time, even though he could no longer hold a fork and had to be helped to eat. She had watched Madeleine feed him in hospital, patiently push soup between his lips. A man turned into a baby, who bobbed his head and wept. But a man still, who could betray one woman with another.

Words from the letters banged about in Thérèse's head, just above her eyebrows she pictured them, as she dripped melted butter from the little saucepan on to the yellow puddle of egg yolks. Cellar. Hiding it. He found me. Dark. Held. Couldn't escape.

She put down the saucepan, prodded the pile of letters, tugged one out from farther down.

Léonie's turned out the difficult one, the neat script ran on: I tell Madeleine she should be stricter with her but of course I mustn't criticize her too much. She's got English ways that's all. Spoils that child. But I'll never forget how good she was to me all through the war, took so much off my hands when I just couldn't cope. I'm worried about Léonie, who's been sleepwalking again. Up and down she goes, talking gibberish. You can imagine what it reminds me of, it's terribly upsetting. But Madeleine won't let me wake her up, she insists on waiting until she goes back to bed of her own accord. I think we should tie her in, but Madeleine won't hear of it. But it solved the problem of Thérèse's tantrums at night, it cured them in no time.

I had a very happy childhood! Thérèse called out to the wooden spoon clotted with Dijon mustard. She banged it on the chopping-board. Mustard splashed about. One big drop hit a letter, sank in, an oily stain on the crinkled paper. She dotted her forefinger into a thick yellow pool on the table-top, licked it, winced at its fiery and concentrated taste. Her mind was as hot as the mustard, words wanting to spill out, dirty the front of a yellow piqué dress.

What did he do is Léonie really my sister what did he do?

The questions swam about the kitchen. Not calm and cool like poached mackerel. This was a school of monster fish, hungrily alive. Baring their white teeth and smiling.

Thérèse mopped up the mustard, swept the chopped tarragon into the mayonnaise, mixed it well together. She could not get the image of a yellow piqué dress crumpled on a white bed out of her mind.

The love of human beings, Thérèse knew from the lives of the saints, was unreliable and let you down. Only God, as she'd found out herself, was an inexhaustible source of love. He never failed you. Even when he hid in the darkness he was just teasing. You could be sure he was there. He never went away. He simply waited for sinners to return to him. And sometimes, as had happened to her last Sunday in church, he came in person and snatched them up in his everlasting arms.

Go to the woods.

Thérèse started. The voice came from inside or outside? She didn't care. What mattered was obeying it. Her conviction that the Virgin was calling her was so strong that she felt able to defy the priest.

She would go back to the shrine. And perhaps the Virgin would be there.

She snatched at the strings of her apron, fingers fretting at the lumpy knot in the small of her back. She hung the apron over the back of a chair, shoved the sauce under a covering plate, turned off the gas under the fish-kettle in which herb-scented water had begun to simmer. I'll only be gone for half an hour or so, she reassured herself: plenty of time to finish it all when I get back. She opened the back door and ran across the yard.

THE DUST

*L*éonie did not want to see the ambulance from the clinic
arrive. She kept clear of everybody in the house in case
they reminded her of the ill person her uncle had become. At first
she hovered in the farmyard, then she wandered over to the stables
and sat on the outside steps up to Louis's workroom. From here she
could see into the fields. Cows grazed with a placid slapping of jaws.
Leaves parted from the beech trees, whirled in eddies of wind. She
was seated in blue sky, the clouds at her back, hearing the church
bells roll out the quarter-hour. Thérèse, rushing across the back yard,
didn't notice her. She vanished at full tilt towards the kitchen-garden
and the orchards. Then Baptiste appeared.

He strolled, hands in pockets, whistling. Was obviously pretending he didn't know she was there. Léonie sat up straight,
tucked her feet to one side, put her hands round her knees. Her
stomach had gone all loose. Baptiste was standing on the bottom
step of the wooden staircase, affecting surprise at the sight of
her.

Oh. Hello. Where's Thérèse?

Léonie managed to reply: hello.

Then she added: I don't know. I haven't seen her.

Both stared at the ground. At least he'd stopped calling her an
English pig. Baptiste coughed and she looked up.

He said: I've got something to show you. Come and see. Then you
can tell Thérèse.

What is it?

I'm not telling you. Come and see.

Léonie stood up. She straightened the skirt that Thérèse had lent
her, dusted it with both hands, then descended the stairs.

It's in your house, Baptiste said: upstairs. My mother told me about it. Let's go and see.

Léonie stood still.

My uncle.

They won't take any notice of us, Baptiste said: the ambulance is round the front, I saw it, and they're carrying him into the downstairs bit. You've got a backstairs haven't you?

She'd done this so often with Thérèse. Slipping in without the grown-ups noticing. Not necessarily *en route* to mischief. Simply the pleasure of not being seen and bothered with. Up the backstairs in bare feet so that no one would hear and interrupt them. Baptiste was trying to take the lead, but that wouldn't do. He didn't know the house, didn't know which doors and floorboards creaked. She motioned him with her head and he swung in behind. This was a betrayal of Thérèse all right. Letting Baptiste in where the two of them had reigned together for so long. She wanted to do it. To hurt Thérèse back for robbing her of the lady in the woods. She would show her. So she beckoned to Baptiste and he crept up the stairs after her, his shoes in his hand.

She paused in the doorway leading to the grey corridor that ran across the back of the house.

Where to now then?

Baptiste looked about. Clearly doing calculations in his head. Screwing up his eyes as he attempted to remember what Rose had said. He pointed.

In there. Maman said it was at the back, the other way from the bathroom end of the corridor.

It can't be, Léonie said: not in there. What is it anyway?

She leaned against the door of her old room. The one she had slept in for years before moving upstairs this summer to share with Thérèse. There was nothing in this room she didn't know. No secrets. It was too bare and plain for that.

It was just around the time you were born, Baptiste said: I don't remember it either. I was too young.

They stood close together on the wooden floor. Carpet and bits of furniture were gone. Sunlight fell across the windowsill, on to their feet. There was a closed smell. Stuffiness and dust.

My parents hid the Jews in their apple loft for two months, Baptiste said: then when the Germans found them they locked them up in here. On the night before they took them out to be shot. They did it on purpose, Maman says. To make people think someone in the Martin family was the informer. They chose this room because it's nearest the backstairs. They weren't going to have Jews using the front ones.

Somewhere a massive pendulum swung to and fro. It counted the minutes before the dawn. There was no escape from it. So heavy it would crush you as it pushed from side to side. It was the blood in Léonie's chest. Her heart pumped so strongly she felt she'd burst. There was a heartbeat in her neck, in her head, on her tongue.

The Jews, the Jews, she said: didn't they have names then?

Some foreign name, Baptiste said: Maman told it me but I can't remember. She said she always used their false name anyway, never their real one. They weren't from round here. They were refugees from Rouen.

He struck a heroic attitude. Like one of the soldiers on the war memorial rallying *La République* with fixed bayonets.

Those pigs of Germans shot my father. *Vive la France!*

Léonie pointed her toe and wrote an imaginary signature in the dust. Then she looked at Baptiste. Her blood was slowing down, but her head still felt strange. Full of something thick. Dizzy with the memory of bad dreams.

Did they keep your father in here too? she asked: before he was shot?

Baptiste nodded.

You know what, he said in a rush: at the funeral the Monday before last, they buried my father and the Jews all together in the same grave in the cemetery, they couldn't tell whose bones were whose.

He sounded ashamed.

Monsieur le Curé said the sooner we buried them properly the better, he's not going to tell anyone. So no one can make a fuss. Jews might if they found out. So we're just going to have a plain headstone, with my father's names and dates. *Monsieur le Curé* said he'd pay for it.

Léonie frowned. Something was wrong with this rattled-off speech. Too much of it, perhaps. A pile of leftover words. Scraps of words, old bones of words. Like the sawed bloodied pieces of shin and gristle in the butcher's, shoved into a sacking bag and taken home to feed the dogs. That's what a grave was: a dump for torn flesh, broken bones. The Jews were back in the ground again. Mixed up more than ever before. She wanted to laugh. She felt sick. She leaned on the handle of the door. Coldness of brass, that was solid, refreshing. Grown-ups' secrets. She was sick of them.

She shouldn't have to be bothered. She was only a girl, she was too young.

I'll show you how to juggle if you like, she told Baptiste: come downstairs and we'll get some potatoes off Victorine and I'll show you how to juggle.

Magic tricks. To make things vanish you threw them in the air then cooked and ate them. You could do it with bones too. Léonie left the Jews behind her in the room. She closed the door on them. They could not escape, but she could. She was a mongrel, only half-French, but she wasn't Jewish. She had a larder with baskets of potatoes, she would not starve, she would not burn.

Victorine was in the kitchen, ripping the silvery-blue skin off mackerel. The air was warm with the scent of hot fish stock, *bouquet garni*. The biscuit tin supported the open recipe book.

Where's Thérèse? Victorine cried out as soon as she saw them: that wretched child, I knew it would be hopeless letting her help, it's too much, I'll never be done in time.

She swivelled her eyes to the potatoes Léonie was fetching from their earthy resting-place on the larder floor.

Good idea. You get on with them. Baptiste, run and ask your mother to come and give me a hand, would you?

Léonie felt comfortable again. Back with what she knew. She watched Victorine lay the cooling fillets one by one on a flat dish. The bones lifted off easily, a spiky white all-in-one. Victorine tossed them into the plastic waste bowl that stood by the side of the sink. She ladled a little of the *court bouillon* over the fat bits of mackerel. Peppercorns and parsley, a good fish and white wine stink. She threw her ladle down. It clanged against the biscuit tin.

What on earth is that box of biscuits doing out at this time of day? Victorine shouted: go on, put it away will you? Hurry up. Then come back and we'll make a start on the potatoes.

It isn't biscuits, Léonie said: it's letters. Thérèse's. I'd better not move it or she'll think it's lost.

Victorine's voice softened.

She's probably crept in to be with her father. His first day back after all. It's what you'd expect.

THE BLUE SKIRT

*T*hérèse came to slowly. Sunlight dazzling through the trees forced her eyelids open. She brought her outstretched arms down, clasped her hands together. The light rested on her cheeks as heavily as tears. The people kneeling around her sighed all together, a breath of wonder that rippled through them. Wind over the wheat, she thought: over the leaves on the vine. The bread and the wine. The harvest's gifts. Those were the words she had to remember. The bread and the wine.

Behind her she heard the Bishop's voice.

That little girl's no fake. I don't know when I've seen anything so heavenly as that gesture of welcome she made, that smile that she mirrored.

Thérèse came to kneel at his feet. She hitched up her new blue skirt to avoid dirtying it. Her bare knees met muddy grass, stones. The Bishop's shoes were black and shiny as hearses. The Bishop's hand, with its jewelled gold ring, was extended for her to kiss. Beside him stood the village priest, his face carefully bland.

What an exquisite honour, remarked the priest: that our little ceremony of blessing should be graced with the presence of Our Lady. Really so completely unexpected!

He glared at Thérèse. She bowed her head to the Bishop's purple and gold. The blue and silver glow of the vision had pushed him out of her consciousness, but now she saw him she had to admit how very handsome and grand he was in his mitre and robes.

He said: well child, what has Our Lady got to say to me?

His eyes were wrinkled up. Perhaps against the sun. But they looked wet. He wanted something, a share in what she'd seen. Everyone did. That was why the small crowd of villagers stood near, to catch a word, an expression on her face, to touch her hand or sleeve. Not as many as had first come, of course. These country people did not love bishops, the paraphernalia of church hierarchy. Still, enough of them were present to make sure the Virgin's message would be spread all round the village by supper-time.

The Bishop was taller than Thérèse, yet, even kneeling, she had the sensation of looking down on him. With tenderness. Poor man, shut out, as they all were except her, from that glimpse of heaven's dazzling sweetness. At the same time he was a great lord. With real power in the world. If he commanded then people jumped to do his bidding.

She raised her eyes shyly. She clasped her hands.

She said: Our Blessed Lady asked me to ask you to have a small chapel built here in her honour. On this spot where the shrine was before. She wants a statue set up inside the chapel. She wants to be known as Our Lady of Blémont-la-Fontaine, and her feast-day is to be in early September on the day we have the harvest festival.

A demanding lot, these women, murmured the *curé*: they don't ask much!

The Bishop smiled.

After all, it may be for the best. In all ways. A sign of the renewal of faith in these dark materialistic times.

He put his hand under Thérèse's chin and made her rise. He stared into her face. She blushed faintly and gazed back. He patted her cheek. Raised his hand to bless her. As she made the sign of the cross he spoke in loud and solemn tones.

Tell the lady next time you see her that she can have her chapel. I am convinced it will do nothing but good.

He turned to the *curé*.

My dear Father, you were mentioning lunch?

Thérèse clapped her hand to her mouth.

I must go. I'm cooking fish for lunch for my father. He's come home today from hospital.

The Bishop took off his mitre and handed it to an adjacent altar-boy. He ran his hand through his flattened hair.

Fish? What kind of fish?

Fish soup with *rouille*, Thérèse said: followed by poached mackerel with *sauce à l'estragon*.

I shall come and lunch with your family, the Bishop declared: nothing formal, eh? A simple meal *en famille*. I feel I must meet, as part of my pastoral duties, the noble father of such a modest little visionary.

He turned to the priest at his side.

My dear Father. You will accompany me of course. Your faithful housekeeper can have the day off. I shall enjoy her admirable turnip stew another time.

He waved his beringed hand at Thérèse.

Run along child. We'll be with you at one o'clock sharp.

She ran.

THE SLOTTED SPOON

R ose and Victorine reacted to the word *Bishop* in a practical
way. They sped off to the dining-room to organize putting
an extra leaf in the table. Thérèse strolled from the back door over to
the stove where the fish-kettle still stood, awash with cooking juice.
She dipped a finger in, licked it.

Isn't lunch ready yet? she enquired.

Léonie peered at her. She was pink-cheeked, smiling. But she
wouldn't catch Léonie's eye. She hummed a tune and pretended
to care about tasting a fragment of fish she'd pinched up out of the
herby broth. Léonie waited a second. Then she pulled the biscuit tin
towards her across the table, yanked off the lid. She lifted a sheet of
writing and studied it. Thérèse hurriedly wiped her hand against her
mouth, lunged for the letter.

Leave it alone! They're mine!

She scrambled the piece of paper back into its sugar-and-vanilla-
scented tin box. She held the lid as though it were a tambourine.

I don't think you should read them. You haven't, have you? You'd
be upset.

Léonie was alert. She reflected. She blurted out her offer.

I'll tell you my secret if you'll tell me yours.

Thérèse asked: what secret? I'm not interested in Baptiste if that's
what you mean.

She jerked her chin at the box of letters.

Go on then. Read them if you want to. Read the ones on top first.
Only don't say I didn't warn you.

When shall I tell you my secret then? Léonie asked.

Thérèse shrugged.

Oh. I don't care. It'll keep I expect.

She picked up a slotted spoon from the table and beat gently on the tin lid.

What have you done with the sauce I made? It was here when I went out, I left it here. In a bowl. Under a plate.

In the fridge of course, Léonie said: if you mean that tarragon stuff.

She was exasperated. But she felt she had to obey Thérèse. Choosing a letter would have to be done on the same principle as choosing a biscuit. Good manners forbade you to riffle through the pile, lift the layers until you found one you liked the look of. You took what your hostess offered. So Léonie plucked out the letter Thérèse indicated, the one that she had replaced on top. December 1941. Antoinette to Sœur Dosithée. Her eyes wandered around the loops and curlicues of the writing. She began to read.

Thérèse laughed. She was dancing the slotted spoon in and out of the fish-kettle, a clatter of aluminium.

You call this stuff fish soup? I'd better do something about improving it. Can't give this to the Bishop. What can Madeleine be thinking of?

Léonie lifted her head.

We made the soup while you were out. It's in the larder, so the cats don't get at it. I was just going to wash up the kettle when you came in.

Too good to waste, Thérèse said.

She poured the leftover liquid into a small bowl. She drank it thirstily, holding back the bits of herb and onion with a teaspoon. She watched Léonie finish reading the letter.

THE WASHING-UP BOWL

Madeleine and Victorine had dressed Louis in his best blue jersey for lunch. The stroke had flattened and aged him. His body had become plump and loose. His hands dangled over the sides of the wheelchair as they took him to the *salon* to meet the Bishop. He was like a middle-aged child, puzzled and sweet, his blue eyes full of tears. His hair, which had not been cut for weeks, fell in silky waves around his face.

The priest and the Bishop stood in front of the fireplace, looking down at him. Madeleine and Victorine stood behind his wheelchair, like nurses, while the two girls huddled on the windowseat. Rose came in with a tray of glasses and bottles. The two clerics brightened up, and accepted a whisky each.

At lunch Louis was wedged, lolling, into his chair. He peered over a rampart of cushions. Madeleine fastened his napkin around his neck. She pegged it on to his blue jersey with the tiny plastic clothes-pins she used for hanging up her dripping stockings in the bathroom to dry overnight. She sat next to him so that she could cut up his food and discreetly help his wavering hand shove it into his mouth. From time to time he dribbled and then wept tears of rage and humiliation. The guests looked the other way and talked more loudly.

Thérèse sat with lowered eyes. Looking like a sort of holy pig, Léonie thought. Too rapt in awe of the Bishop to bother with her food. She dipped her spoon into her plate of fish soup as though it were gutter water she was forcing herself to swallow, and she wouldn't take any of the little circles of toast spread with delicious fiery *rouille*. The Bishop nodded his approval of this asceticism before accepting the offer of a second helping from Madeleine. The lid lifted off the tureen. The fragrance of fish, tomato and garlic gushed out.

Léonie followed his example and had some more too. She eyed the last piece of toast and *rouille*, decided she didn't dare look as though she wanted it. Today she was on best behaviour. She'd put on a dress for lunch. The Bishop reached out a casual hand and took it. Madeleine smiled her satisfaction.

The two girls' job was to clear the table between courses, fetch and carry dishes. They ferried plates to and fro from the kitchen, where Rose and Victorine supervised what was to come next. They'd taken no notice of Thérèse reporting that the Bishop expected only a simple meal. They'd got out the best china and crystal glasses, the damask napkins, the ebony-handled knives. The best cider for the Bishop, and the best wine. Certainly the Bishop did not seem impatient of the efforts expended on his behalf. He praised the poached mackerel and its delicate herb-scented sauce. He accepted a third glass of Muscadet. He raised it to the blushing Thérèse.

My compliments, child, on your cooking.

She looked at him imploringly.

God isn't calling me to be a cook, Your Excellency, he's calling me to be a contemplative. Please, Your Excellency, I beg of you, give me your permission to enter the convent as soon as I'm sixteen.

The Bishop roared with laughter.

Absurd child. That's far too young. Wait until you're eighteen and then we'll see.

Louis turned his head towards his daughter, his tongue protruding as he tried to speak, failed, wept. She swivelled her face from his. She smiled at the Bishop.

Really Thérèse, Madeleine scolded her: to upset him like that on his first day at home! How dare you talk such nonsense!

The Bishop wiped a speck of sauce from his chin.

Not nonsense, my dear *Madame*. A religious vocation is the highest call anyone may receive. And for a young girl who has been privileged to see, with the eyes of her soul, the Mother of Our Lord, it is perhaps the only vocation that she should contemplate. That those eyes which have witnessed the Divine Motherhood should henceforward be for ever lowered in contemplation in an enclosed cloister seems to me entirely fitting. Heaven forbid that this child should ever be tainted by the grossness of the world!

Madeleine mopped Louis's eyes with the corner of his napkin. She murmured to him and put her hand over his.

A good religious boarding-school, the *curé* announced: that might be the answer for now. For certainly, once the chapel is built and consecrated, there will be pilgrims coming from all over Normandy. The shrine may become famous throughout France. Who knows, perhaps we'll even see a miracle or two.

He beamed at Thérèse.

To me, dear child, you confided your secret. To me you told the story of the apparitions. As your parish priest, it is my duty to shield you from any overenthusiasm on the part of visitors hoping to catch a glimpse of a visionary. We must remember how much poor Saint Bernadette suffered. A boarding-school, that is undoubtedly the best course for now.

Léonie, Madeleine said: go and fetch the salad will you?

The Bishop peered at her.

Léonie. Yes. I heard tales, let me see, about a second child having visions, of a somewhat different sort?

Completely discredited, the *curé* said: a most unfortunate mistake. She's half-English I'm afraid. There was a mix-up of languages. A problem of translation. She was trying to report what Thérèse told her and she got it wrong.

Léonie waited for her mother to laugh and assert that her daughter spoke perfect French. But Madeleine didn't seem to have heard. She was stroking Louis's cheek and whispering to him reassuringly. Her pet, to be slipped titbits and fondled when no one was looking. What kind of pet? Léonie got up after a moment and went out to get the salad.

After the *dessert* and the fruit came the coffee. Leaving the men and Madeleine sipping Calvados and smoking, the two girls brushed the crumbs off the glossy white cloth and took themselves to the kitchen. Victorine and Rose had gone, leaving tottering piles of dirty plates, cutlery soaking in jugs, saucepans and pots lined up at the side of the sink.

Léonie and Thérèse did the washing-up together. Léonie, with her sleeves rolled up and Victorine's big blue overall wrapped round her, held the smeared plates under the hot tap, dunked them into the soapy water in the yellowing plastic bowl in the enamel sink, stacked them in the wooden rack on the tin draining-board. Thérèse, armed with a thin red and white linen cloth, dried, polished, sorted.

At first they did not speak to each other. Léonie tried, but each time she opened her mouth an invisible hand blocked it and shut her up. What it was that wanted to come out she did not know. An animal's yelp of pain or a shriek of amusement. She clutched the soft sponge in her palm, feathery with water and soap. She let the cooling suds tickle her wrists, worked her white-locked mop in and out of glasses.

Thérèse picked with her fingernail at a crust of grease in the bottom of a saucepan. Still she kept silence. Then she could not restrain herself, pounced with a sharp cry on a smear under the handle of the colander. Léonie grabbed both pots back, rebaptized them in a squirt of lemony green goo, hot water.

Holy holy holy, she sang: Lord God of Bishops. Thérèse you can't mean it, you're not really going to go and be a nun?

Thérèse swabbed the lid of the saucepan. Thin and battered, made of tin, with a wooden button to lift it by.

I am. You and Madeleine will go on living here. You can look after Papa. You don't need me.

Léonie upended her bowl and let the water swish round the sink. It left a tidemark of greasy bits as it sank away down the plughole.

We're never going back to England? How d'you know?

Thérèse folded her tea-towel and hung it on the silver bar in front of the range.

Madeleine's going to marry Papa, isn't she? It's obvious. It's far too soon but still. Don't be stupid. Use your eyes. It's really sick and disgusting I think, she should be ashamed of herself.

Thérèse's eyes glittered. They said: grow up. Don't make one single sound of grief. Pain squatted Thérèse, Léonie could see. A stretch, an ache. But her voice was cool and light.

I'd like to go to the missions abroad, Thérèse mused: that would be the most exciting. Real adventures. The African bush. Danger. But my vocation might be to enter the convent here in Normandy. It's hard to tell. I'll have to pray for guidance.

Did saints ever bat their eyelids and look sleepily self-satisfied as cats? Thérèse, lowering her lashes like a lacy brown veil and trying not to smile too obviously, did not look modest. It was the same look she'd directed at the men all through lunch and they'd loved it. Léonie thought men were stupid to be so easily taken in. Look flutteringly at them, pout with all your maidenly charm, above all don't say a word, and they were yours. She vowed that never would she resort to such cheap tricks. She would die rather than roll her eyes and wriggle and blush.

God speaks to me, Thérèse said: I can't explain it, you wouldn't understand.

Why not, Léonie cried out: why wouldn't I?

You're the sort of girl, Thérèse pronounced: who'll probably not have much of a career. You'll get married very young I expect. You're a bit of a tart really aren't you. Like Madeleine.

And you're not I suppose, Léonie shouted: just a holy tart, making eyes at the Bishop to feel important. God I hate you Thérèse Martin.

Thérèse picked up the biscuit box from the table. Her voice was calm but her hand shook.

Unchastity is a mortal sin. It means you go to hell. My poor mother. It's even worse than what happened to Saint Maria Goretti. At least *she* died defending her honour. It was the only thing she could do. It's worse when you don't die. Everybody whispers about you. It's disgusting.

Thérèse's eyes glared in her pale face. Sweat from the washing-up misted her forehead and nose.

I don't know what you're talking about, Léonie lied: I don't know what's the matter with you.

Thérèse clasped the biscuit tin in the crook of her arm.

It was a terrible sin, she said: which means we've got to pray for forgiveness. To make reparation. You ask Madeleine. She'll have to admit it, it's all true.

She won't tell me anything, Léonie said: she never does. Did she tell you?

She didn't need to, Thérèse said: it's all there, in the letters. Of course we were never meant to find out.

She was trying not to cry.

They've betrayed me. I don't want you as my sister. I want Papa.

She didn't care what Léonie felt. She didn't care who was watching. She wrenched the lid off the biscuit tin and tipped her mother's letters into the range, stuffing them well down with the poker until there was nothing left of them but black ash.

Explain to me, Léonie cried: tell me what you mean.

THE VASE

*A*fter the lunch guests had gone the house settled down into quiet. Louis slept in his bed in the little white *salon*. Madeleine sat and sewed next door. Within earshot in case he woke and needed her. Thérèse had stormed off towards the village.

Léonie didn't know where to put herself and the knowledge dumped on her by Thérèse. This secret could not be shared with the grown-ups. They had kept it all these years. They would be distraught if they discovered that now she and Thérèse *knew*. Better to keep quiet, protect them. The words stuck in her throat, like a large crumb, or a grape-pip she wanted to spit out and couldn't.

She wandered into the dining-room. So empty and undisturbed you couldn't imagine the bustle of the lunch party only an hour before. Chairs decorously ranged in their places, the white cloth gone. A faint scent of tobacco and fish lingered, even though one of the windows had been opened. The thin white curtain, pulled back, fluttered in the cool air. The only sound was the soft tick of the tall clock in the corner, the cooing of pigeons and doves outside.

Mother. Father. All you had to go on was what they told you. You had to believe that Madeleine told the truth. Now you suspected she'd been lying to you for thirteen years.

She felt clumsy, too big. She put out a hand to touch one of the ribbed blue and yellow vases on the marble mantelpiece. Easy to break it. Thin, hollow porcelain. Fragile. She wanted the vase to lift up and off, potter through the air in a slow-motion arc, sprinkle itself in tinkling bits on the glazed tiles surrounding the empty fireplace.

She shouted inside herself to Thérèse: I don't want you to be my sister either so there, see if *I* care.

She jammed her fists into her pockets. Her surprised fingers encountered something hard. A sharp earthenware edge. She brought it out to look at it. The fragment of the Quimper dish she had picked up from the dustpan on the kitchen floor that day when she and Thérèse had seen, when she saw, when the lady had shown herself for the second time. She'd hidden her stolen relic at the back of the bedroom cupboard in the pocket of this dress that she hardly ever wore. Then she'd forgotten all about it.

The painted hands of the Quimper lady, joined about her tiny posy of flowers, were severed at the wrist. What had her face looked like? Léonie could not remember. She curled her fingers around the shard in her palm, slipped it back into the thin pouch of cotton hidden inside the skirts of her dress. Then she went over to the *buffet* and stood in front of it. She traced with her forefinger the silky whorls of the carved garland of oak leaves that swung across both doors.

The only thing to do was to go for a walk. She ambled towards the orchards, the boundary wall.

The small green space in the centre of the woods looked like a chapel already. People had piled the fallen stones into the shape of an altar, and had put a white lace-edged cloth on top. On this were arranged thick candles in red glass jars, two vases of dahlias, a small plaster statue of Our Lady. Strips of old carpet had been laid down in rows, like pews. Scattered about was exactly the sort of clutter you found in the parish church: holy pictures of saints, missals, rosaries, little bottles of holy water.

Léonie curled up on one of the bits of carpet. She laid her head on her arm. She squinted. Yes. It was Baptiste, walking across the clearing and lowering himself to sit next to her. Everybody else in the village was presumably either still eating lunch or else dozing. There were just the two of them. The hairs on his forearm brushed hers, he sat so close.

He said: I thought you were asleep.

Léonie shook her head. Doing this dislodged the words stuck in her throat. They flew about inside her like magpies in an orchard, then settled in a new pattern. When she opened her mouth they darted out, glossy and black and white.

She said: Thérèse has found out that she and I are really sisters, not cousins. Twins. My mother's really my aunt. She adopted me. She's not my mother at all.

Baptiste looked stupefied. He was lipreading as well as listening. Needing all his wits to follow her. Failing.

What? What did you say?

I'm not half-English at all, Léonie said: Maurice wasn't my father. They just told me he was. They lied to me.

Baptiste tugged out a thread from the edge of the carpet. He rolled it into a ball, put it into his mouth, chewed.

Thérèse is crazy. She's made it up. It's just one of her mad stories. Like seeing the Virgin Mary.

Léonie frowned. Baptiste spat out his bit of thread and laughed.

So you're not *half*-French. You're *French*. Now that suits you much better. Little French girl.

He rolled his rrrs to make her laugh too. Léonie sprawled lower on the muddy carpet and looked up at the beech trees spread against the sky.

She said: Thérèse says it was in the war. After people hid their cider and wine from the Germans in our cellar. They covered the bottles with a great heap of sand.

I know that, Baptiste protested: everyone knows that. Try telling me a story I don't know!

The sky and the trees pressed down on to Léonie's face. She recited. Words that were not hers. They tripped out neat and pat. Arranged in rows like magpies on a fence.

Thérèse says it was all there in the letters she got back from the convent. One of the Germans stationed in the house got suspicious of Antoinette hanging about near the cellar door and made her give him the key and go down there with him so he could see what she was hiding. Then later on she found she was going to have a baby.

It's a story, Baptiste declared: she got from some stupid women's magazine or other. You're telling me she gave herself to the German to stop him finding the wine? She seduced him?

Léonie's arms and legs dissolved with relief now the words were out, had flown off squawking.

Thérèse says the letters didn't put it very clearly. You could hardly say that to a nun could you? But that's what they *meant*.

Half-French and half-*German*? Baptiste enquired.

He examined her, as though her skin, her eyelashes, her hair, could tell the whole truth and nothing but the truth. He wouldn't like her so much now. It was all Thérèse's fault.

Antoinette married Louis, Léonie explained: so that everyone would think the baby was his. She'd been thinking about becoming a nun but she couldn't be with a baby. Then when she had twins she gave one to her married sister. That was me.

It didn't matter. Nothing mattered. She kept her head down, let her mouth brush the fibres of the carpet, caked with mud and dust. When Baptiste's hand touched her arm she jumped.

You're not half-German, he said: I don't believe it. It's a pack of lies.

Baptiste lay down beside her. His mouth brushed her ear.

What do *you* think happened?

The end of Thérèse's story was that Antoinette had been found out. Proof: her stomach had swelled. Shame brought upon her family. Outrage. Disgrace.

Léonie turned her head so that she could look at Baptiste. She was pleased at how low her voice was, how steady.

Oh, I think I'm wholly French, she said: I've been working it out. I don't think it was a German soldier at all. I'm sure Thérèse made that up. I think it was Louis all along.

For Louis, Antoinette had kicked off her buttoned high-heeled shoes by the wine racks and lain down on gritty sand. Was it better to be raped by a Nazi than seduced by a Frenchman? Would you be forgiven quicker? The priest would say neither. But anyway Antoinette had had to get married. She'd been found out.

The pain burrowed through Léonie, tore at her with sharp claws. She'd been cast out, first by one woman and then by another. Tossed between them like a broken toy fit only for the dustbin. Really it was all Thérèse's fault. She'd insisted on telling Léonie. She'd taken everything from her. Then she'd said they were sisters. Like a slap. Léonie would scratch back. Rescue herself. She wouldn't be caught, trapped in the darkness. She wasn't a Jew. It wasn't her fault. She was French.

Ripping off her Englishness and casting it aside was as easy as unfastening the collar of her dress. Her fingers glided over the frilled edges of daisy-shaped buttons. Shaking off the very idea of a German father was a wriggle of the shoulders, the thin cotton sleeves pushed down. Becoming French was taking Thérèse as her twin sister and then taking the boy she wanted. Baptiste, baptism, his tongue sliding into her mouth and blessing her, she wanted to laugh, she kissed him back. Thérèse could share in this, they were sisters weren't they, in the past they'd shared everything, food games beds a sack of skin.

Léonie wanted to be found out. She wanted to find out what it was like, Antoinette in the cellar, Madeleine in the big marital bed, their mysterious life in the arms of men, the embrace which the daughter was banned from and which they told endless lies about. Poor stupid ignorant lying Thérèse. Let her find out too. What it felt like to be left out.

Léonie opened one eye and cocked it ready. She opened her knees, drew Baptiste to her, held him firmly with both hands. She saw colours stir, a flash of black amongst the greenery. She waited for the shouts from beyond the trees, for the priest and the Bishop and Thérèse to come running, to come upon this sinful worship on the ground.

THE CIGARETTE

More than twenty years later Thérèse had returned home to lie in the bed in which she'd been born and to quarrel with Léonie. She longed for sleep, to muffle her ears against Léonie's shrill voice, implacable words that spattered her skin like pellets of ice. She crossed her arms and shivered. She cowered under the gaze of the woman opposite her, who threatened the air with a cigarette. As though the air had sinned, becoming too stale and smoke-filled. In this hectic mood Léonie was exactly like Antoinette. Did she berate her daughters in these accusing tones? No wonder, Thérèse thought: none of them is at home.

She said: you don't really mean that. You haven't really spent all this time pretending I didn't exist. You're exaggerating because you're upset.

Dead, I said, Léonie corrected her: I said that for twenty years I pretended you were *dead*.

In the convent Thérèse had died to the world. A white death. It had snowed on the day of her clothing. White fringes decorated the windowsills. The sky was a canopy of white held over a bride. When she walked along the cloister on her way to the chapel, the cloister garth spun with whiteness fine as the veil covering her hair. She wished now she could hold the white sheets of her bed in front of her face, to blot out Léonie's furious eyes. Cover her up. Bury her in snow.

You didn't come to my clothing did you, she remarked: you wrote saying you couldn't afford the fare.

Over by the door Léonie rubbed her nose between finger and thumb, then sighed explosively.

I suppose you offered it up for the holy souls, she said: that was what you always used to do.

Madeleine and Louis came, Thérèse said: so why not you?

Léonie paced to and fro in the shadowy space between the bed and the door. She clasped the ashtray in one hand and made half-stifled gestures with the other.

You weren't able to attend their funerals, of course, she said: the convent rules forbade it. I'm sure you offered *that* up as a sacrifice too didn't you? You didn't think about me having to cope completely on my own.

You had Baptiste, Thérèse reminded her: and you also had the house, and the farm.

Inside herself she was held together by knotted strings. One by one Léonie was slicing through them. Soon her arms and legs would fall off, plop, then her head. Léonie's precise strokes with the knife dismembered her. Like a chicken carcase on the kitchen table, joints and sinews severed, skin ripped off then bundled away. Hunched in her bed, Thérèse held on to a pillow and squinted at the other woman. Clearly Léonie found pleasure in this, letting out stored words in a rush, found a warmth and energy. Keen focus. Slapping cut after cut into the flesh of her target, then drawing breath to let fly again. Thérèse never got angry. She'd rooted it out of herself in the early days in the convent, along with all the other feelings that got in God's way. She'd made herself empty as the square plot encircled by the cloisters, open to the silence, the falling snow.

Léonie said: all those lies you tried to make me swallow about Madeleine not being my real mother. About being adopted. About Antoinette being raped by a German. Your big hysterical drama. Just because you were jealous. Your nose put out of joint.

Thérèse flinched. She wanted to protest. She ducked. Her hand flew to her mouth then down again.

Léonie hissed like a cross goose barring the path to the pond with flapping wings. She looked like one, too, plump in shining grey silk, her neck darting forward, all her hair rumpled and on end.

You haven't come back because you care about those dead Jews being dug up out of their burial place, Léonie accused her: you just want to put yourself in the limelight again with some new story. Saint Thérèse the little flower. Our Lady's pet.

I've come back to put the record straight about the past, Thérèse said: I want to admit I made a mistake when I described those visions. I want to tell everyone how sorry I am, for misleading them.

Not a mistake, Léonie said: a lie. You told a lie.

Thérèse glanced up.

You don't care about the dead Jews either, she said: in fact the opposite. All you care about is your position in the village, your nice quiet life. That's all that matters to you. You're a real *bourgeoise* aren't you.

You'll only cause fresh trouble, Léonie cried: nobody'll thank you, you wait and see. We can't live in the past. We've got to get on with our lives.

Hypocrite.

They said the word together. Léonie turned and went out.

THE STATUE

*T*hérèse pushed the window open to get rid of the cigarette smoke, then fell asleep. For the first time since Antoinette's death she dreamed of her.

She and her sisters from the convent, clad in their brown habits, their white coifs and black veils, stood about the table on which Antoinette's body had been put. They were preparing her for burial. They stitched up the torn skin, moulded the features of the face back into position, set the broken bones, then coaxed the limbs to lie straight. The body having been made whole again, they washed and dried it, then wrapped it in a linen sheet. Léonie had been a shadowy figure in the corner, watching in silence. Now she came forward. She made a fist, let sand trickle from it on to the white shroud. She made a pattern. She wrote something, a line of sand. Thérèse craned forward to read it.

She awoke. It was still dark. Her sisters in the convent would be singing the Office at this hour, some of them rubbing the green sleep crusts from the corners of their eyes and stifling yawns, just as she used to, in the choir lit by candles. Far from perfect she'd been, and she'd ended up saying to God: take me as I am, over to you now, I can't do it all by myself.

The practice of twenty years could not easily be abandoned. She sat up, leapt out of bed, fell to her knees. No prayer rose from her heart to her lips. In her mind's eye she saw the edge of a photograph, Léonie smiling behind the lens of a camera. She got up from the floor and dressed, then went quietly downstairs. In her pocket her fingers touched the cigarette lighter that Léonie had left behind on the bedside table.

She went via the main *salon,* so that she could glance at the new photographs tucked into the framed collection hanging above the bureau, the photographs she'd never seen. Léonie and Baptiste on their wedding-day, Léonie in layers of net frills, her hair backcombed into a beehive stuck with orchids, feet encased in shiny white stilettos. Baptiste grinned in a tight dark suit, his bristly hair slicked down and his ears sticking out. Thérèse shook her head over the pair of them. Léonie's teenage rebellion hadn't amounted to much: just a few years of black pullovers and pretending to be a beatnik. She hadn't bothered with higher education. All she cared about, obviously, was her family and her house. The photo of her three daughters was gold-rimmed. A bland portrait of three well-fed *jeunes filles* with ribboned plaits and polite smiles, neatly lined up like some of Léonie's well-dusted china ornaments.

There was no photograph displayed of Thérèse in her habit, on the day of her clothing. Though Madeleine had taken several with Louis's old camera while he looked on, calling out jokes to make Thérèse smile. He'd cried at one point and Thérèse had turned away. Léonie had wiped those memories from the house, like a smudge on a windowpane.

Thérèse found that her knees were shaking.

Get out of here, she thought: but where can I go?

She looked at the photograph of her mother and let the dream sweep back. The mending. The limbs stitched together, the torn pieces reassembled. She looked at Antoinette's calm pale face, at her buttoned high-heeled shoes. She remembered the dark red shoe she'd found in the cellar, behind the barrel, that day with Léonie.

Go back down to the cellar, of course. Antoinette had tried to tell her, as she lay dying, what was down there. But Thérèse had not been able to understand.

The cellar door was unlocked. Thérèse switched on the light and started down the wooden stairs.

The heap of sand, she discovered, was indeed still there. Right at the back, under a low vault, where the light hardly reached and the air was dank. Thérèse knelt down on the earthen floor, pushed her fingers in and moved them about. She made her hands into spades, shovelling the damp grit to one side, digging as deep as she could go.

Her fingernail scratched it. She forced herself to burrow around it, slowly, carefully. She brought it out piece by piece, she brushed the sand from it as best she could, she explored it with her fingertips, she caressed it. A stone leg, stone foot, the pleated dress falling down, exposing the long slender toes. A stone hand clasping a broken sheaf of corn. Half a stone head, one eye slanted, half a curled mouth, tender, amused.

Louis must have helped Antoinette gather up the broken statue and bring it here, Thérèse thought: after the priest smashed it, he must already have been in love with her, else why should he have bothered? He must have helped her to bury it next to the cider and the wine hidden under the sand. Antoinette had been a very pious woman. Just once in her life she'd defied the priest, the Church, acted in a way that allied her to the villagers not just to the big house. No wonder afterwards she'd tried to forget it ever happened. Just one more odd little episode in the war. But Rose and Victorine must also have known, Madeleine too. Perhaps at first they'd simply waited, biding their time until the Germans were gone, then realizing how long they'd have to wait until the hostile young priest retired or moved to another parish. Perhaps, as time went on, it all became less important. Hiding the broken statue had been some sort of daring and romantic game, part of cocking a snook, like cutting off bits of German soldiers' uniforms on trains without being seen, coming home to display scraps of epaulettes, braid. Probably the stone girl's significance had lessened. Local folklore, very nice, but they didn't want any more trouble.

So they'd waited, and they'd forgotten. And the trampled statue had remained in the airless dark, her mouth and nostrils and ears and eyes clogged with dirt and sand. Shut up like a cry in a box, the weight of the house pressing on her.

Yellow stars. A young woman with a dark gold face.

The communal grave. The boulders from the old shrine piled on top of it.

Thérèse thought: Léonie knows all about this, she must do, she'll have had it from Baptiste, who'll have had it from Rose. She's been deliberately not telling me. Her way of punishing me for stealing the lady she saw. But now I've found out.

She dusted off the palms of her hands against her dress and stood up. She left the bits of stone behind her, carefully arranged on the floor, and went back up the stairs into the kitchen. She lifted the latch of the outer door and left the house.

THE CIGARETTE LIGHTER

Somewhere a dog was asleep, must not be disturbed. Thérèse trod cautiously across the gravel and out of the gate. The light was a milky blue, a heel of white moon showing. The air was fresh and cold, dampening towards rain.

She met no one on the road. Too early, on a Sunday, for people to be about. The village slept behind closed shutters wet with dew. Shops too had their eyelids pulled down. The streets were empty and clean. As though everyone had been evacuated in the night and sent to live somewhere else.

Thérèse shivered in the light wind. While she was on the road night had changed imperceptibly into morning, a grey shift, the sky smudged pink. In this pearly light the façade of the church looked newly washed, yellow as a biscuit.

The door of the church was not locked. Its handle swung forwards, into her hand. She pulled the little door open, and went in.

She remembered the church as a vast, dark place. Edges lost in shadows. Gloomy side-aisles down which you crept on your way back to your pew after receiving communion. Now it seemed small and light, and she could see round it in one glance. Airy, cream-coloured. Stripped. Gone were all the old banners and draperies, the priestly thrones, the tiers of misericords, the carved screens. Gone was the old smell of damp stone and incense and decay. The church smelled of beeswax and grass and flowers.

Thérèse walked slowly up the centre aisle. The pews on either side had been left intact. Worn pine, dark and polished like the *buffet* at home. At the near end of each one, two wooden straw-seated chairs and prie-dieux. Beyond them, in the side aisles, the old statues still occupied their brackets on the wall in between the stained-glass windows. The high altar had been taken away. In its place was a simple stone slab half-covered by a strip of white linen and flanked by pots of yellow dahlias.

The church was all ready for the harvest festival Mass. Decked out as though for a party. The vaulted roof of the nave was crisscrossed by lines of string fringed with sprigs of barley and oats, feathery, fragile. Baskets of wheat and dried blue cornflowers were tied to the columns, to the ends of the pews. Wreaths of wheat were looped over the volutes of the capitals. Larger bouquets, pale gold sprouts, decorated the altar-rails. Bunches of grapes, black and green, dangled from the lectern. Near this stood a three-tiered bier, like a cake-stand, seemingly made entirely of twisted corn. On its lower shelves reposed loaves of bread, bundles of leeks and tomatoes and apples, a basket of corn sprigs tied up with multicoloured ribbons, a straw tray of *brioches*.

You could tell that the sun was starting up outside. The light inside the church strengthened. Light gathered itself up and streamed in. It fell through the coloured glass of the windows on to the columns, the floor. Bars of scarlet, green and blue. Little solid rainbows on the plaits and wreaths, on the statue of Our Lady of Blémont-la-Fontaine. Thérèse's statue. Fashioned according to Thérèse. Our Lady of would-be-saint Thérèse. Our Thérèse's Lady.

She stood ready to be taken around the village in procession after the harvest Mass. For this she had been brought out of her smart concrete chapel in the woods that the *curé* and the Bishop had built for her, for this she'd been set on top of the bier woven of corn. Wires twisted about her feet fastened her in place. When the bearers lifted her up she would not stagger or fall off.

She was child-sized. Plaster painted to look like polychromed wood. She was faintly gilded. Through the gold showed the greeny-blue streak of her dress, the yellow of her hair under her long veil. Cherubs and a cushiony cloud bore her up. Her hands were clasped and ecstatically raised, her bare feet rested on roses. Her eyes and sash were blue. She'd been redone since Thérèse saw her last. Her features were blurred under fresh paint and gilding, her fingers thickened. Her smile was unchangingly sweet.

She was the Virgin Mother of God. She was flat as a boy. She was the perfect mother who'd never had sex. To whom all earthly mothers had to aspire.

Easy to pretend your own mother never had sex. It meant you

didn't have to feel jealous when she went away to be with your father. It meant you could punish her by imagining you'd married your father yourself. You didn't need your mother, you told yourself. Anyway she wasn't there.

Antoinette had often prayed to the Mother of God. She'd aspired to her. But she hadn't been perfect. She'd had sex. She'd had a baby. She'd had red hair before it faded to gingery blonde. Her thick white skin had freckled and blotched in summer. She'd been broad-shouldered and tall. She'd often looked anxious. Her high voice called through the house. Listen! Where are you? Thérèse hadn't had time to get to know her.

Antoinette had gone away. Cancer had rubbed her out. Her voice got fainter. Her words stopped. She was off somewhere else where Thérèse couldn't follow her. Nothing left to live for, she'd mumbled right at the end. That's what death was. Just the next step. And the dying took it. Finally they desired it. They let go into it, stopped caring about anything, except death, they moved away and didn't miss you, they had their eyes fixed on death.

You won't die yet, Thérèse had whispered to her mother. And Antoinette's lips had twisted: oh yes I think I will.

She had gone off and abandoned Thérèse, she had expected her to grow up and manage on her own, in the end she'd forgotten her daughter, death was impatient and wouldn't wait.

Thérèse had been right behind her, forced to halt when Antoinette disappeared, left there alone on the brink, thrown back. Antoinette had pushed her away, hadn't let her come too, hadn't needed or wanted her company. She didn't care about how angry Thérèse felt, it didn't matter to her, she'd gone to a place where it wasn't important. Thérèse couldn't be angry with Antoinette: how could you be angry with someone who wasn't there, no mother to kiss the hurt better, take the pain away.

The pain was Thérèse's. It belonged to her. It stuck in her like stabbing swords. It lodged in her throat and made her choke. It burned, sour, in her stomach. It ticked in her blood, threatened to rip her apart and make her explode.

Thérèse had done the best she could. She'd found herself another mother, she'd been sold one ready-made by the priests of her Church. Perfect, that Mother of God, that pure Virgin, a holy doll who never felt angry or sexy and never went away. The convent was the only place where Thérèse could preserve that image intact.

Away from there it melted in the heat of her hands. It couldn't console her any more for Antoinette's loss. What she needed now was a funeral, a fire.

Around the bland-faced statue she piled the baskets of corn, the wreaths of wheat. All along the aisle, between the altar-rails and the door, she unrolled a narrow carpet, sprigs of barley and of oats.

After the harvest you set fire to the stubble and burned the fields. First ablaze then blackening. Scorched smell, red cinders flying in the wind, bald stalks. She took Léonie's lighter out of her pocket, flicked the silver top, pushed the leaping flame into the heap of harvest offerings.

Then, at last, after all these years, she saw her for the first time, that red and gold lady. The flames sliding up her forced her old clothes off, gave her new ones. With a red coat and slippers she flew. Skimmed into the air quick and bright as a rocket. She was outlined in gold, she held out her hands to her daughter, to pull her in, to teach her the steps of the dance.

Thérèse ran down the aisle of the church towards the door. After her ran the fire on red crackling feet. But Thérèse was too quick. Can't catch me. She flew with her one red wing. Her spine flared, one great red fin.

She jumped clear of those rags and tatters of flame. She cried *Maman*, and flung herself at the church door.

THE ALARM CLOCK

*L*éonie hadn't closed the shutters before getting into bed. She had pulled the heavy curtains of sprigged yellow chintz but had left a gap between them. As the darkness outside changed, and lightened, the space between the edges of the curtains became a streak of grey. A misty grey light leaked into the room, through Léonie's eyelashes that rubbed against her grey silk sleeve.

She focused on the chest of drawers next to the bed, the tin moon face of the alarm clock. It was an old one she liked because it had been Madeleine's. Green arrow hands, clear Roman numerals, stubby feet. It appeared to have stopped. No noisy tick. She stretched out an arm, picked it up, shook it. Was it really six o'clock? She wasn't sure.

Baptiste did not use alarm clocks. He liked to boast how he woke up exactly when he needed to, whatever the season. After his meeting last night, and his late session in the bar, he had slept deeply, snoring from time to time. But he was waking now. His head moved on the pillow, nearer to hers. Eyes firmly shut, he reached out and scooped her closer in, curved round her, both solid and soft. Still pretending to be fast asleep, he pulled the duvet almost over their heads to shut out the grey light.

Léonie whispered: I can't stop thinking about Thérèse. I've been dreaming about her all night.

Baptiste sighed and grunted.

Stupid girl.

Arms wrapped round her, he slid back into sleep. She lay inside his embrace, feet tangled up with his, mouth against his stubbly chin. He stroked her hip with one hand, muttered something.

She couldn't remember the dream exactly. About Thérèse. Something violent. The screech of bolts being drawn across, a red streamer tossed up into a vault of air.

Something was going to happen, to be upset. Léonie lay back, tried to reassert control over her world. She applied her usual formula for overcoming anxiety. She wandered in imagination through her house. She listed her numerous possessions one by one. She caressed her well-tended furniture. She chanted her triumphs of domestic organization, she recited her litany of solid objects, firm floors, walls that were not cracked and not defaced, foundations that did not crumble, a roof that was safe. She counted and inspected the contents of larders, china cabinets, cupboards. She raised her hand and set it all going. It clicked, whirred, chirruped. It was hers. It was her house. Her kingdom, firmly in her control. Peopled with daughters who looked like their mother and loved her comfortably and did as they were told.

No good. This morning the spell would not work.

Something glinted in the corner. Something started to shine. Léonie wriggled her chin over the edge of the duvet so that she could see. The room was emptying itself of darkness. The furniture was outlined with silvery light.

A light wind moved against the back of her neck. Not a draught but a wind. She'd shut the windows in here last night, she was certain. She lifted her head and looked.

The air seemed to ache and yawn. Like mayonnaise when it curdles, separates into tiny dots. The air was breaking up. Léonie fought against a memory which was coming too close. The shape of loss. The air hardening around what was lost for ever. Something might be disturbed, might break. She sucked in her breath.

She was inside what happened, and also outside. Her edges were of warm flesh, arms that held, contained. The world bent forwards, over her and into her, and she seized the world and leapt into it. Sweetness was her and it, her two hands grasping, her mouth demanding and receiving the lively flow. She was in a good place. Where the arms that held her would not let her drop, where her name was called over and over, where she was wanted, where she could stay and enjoy. The name of Léonie was the name of bliss. There was enough room for her. She did not have to go, to stop, to stand back. Rose sat easily, a baby on each arm. She looked from Léonie to Thérèse and she smiled. Of course I fed you both, silly. I had plenty of milk didn't I. Of course I fed you both. Rose, foster-mother, mother-in-law, second mother, fostering mother. Rose in her chair by the fire, feet up, blouse undone, a lapful of babies, a shout of joy, the smell of milk, there, my dears, there.

I took you both to the shrine in the woods, Rose said: what was left of it. It's what we always did with babies in the village. In the old days. I wanted you both to live, not to die like my child did. I dipped your feet in the spring. I popped you in, one after the other, just to make sure. That way you wouldn't come to any harm.

Rose leaned forward. She winked her green glass eye. Then she went away, quiet, abrupt, no goodbyes, nothing. Léonie jumped out of bed, to reach for her, call her back. She stood in the middle of the floor. She touched emptiness. Rose had gone.

THE WORDS

*L*éonie stood at the top of the kitchen stairs and faced the door of her old room. Her hand faltered out towards the china handle then slotted itself back into her woollen pocket. She fingered the roughly rolled edge of a seam, the cool cellophane wrapping her packet of cigarettes.

She wanted food inside her. Breakfast. To wall off the uncertain future. To shore her up. She wanted fresh white bread, salty and spongy under its crackling crust, spread with a slab of cold butter and a dollop of apricot jam scented with almonds. She wanted a cup of strong black coffee, well-sugared. More than this she wanted the best cigarette of the day: the first one. But she'd left her lighter in Thérèse's room last night and would have to go downstairs to find a match.

She was a coward who wanted to run away. The words she was frightened to say were fastened up inside this room. She thought she'd lost them, she'd forgotten she'd put them away in here. For twenty years. For thirty. Until Thérèse had arrived back and reminded her.

The grave in the cemetery had been forced open, made to give up its dead. At the same moment mouths had opened to shout words that Léonie had tried not to hear, tried to believe no one still spoke, would ever utter again. Death words shouted out in the night. Murderous red signs painted on the headstone of the grave. Léonie had to look steadily at what was rising up in her village, out of the grave of the war, the unburied and the undead arriving to lay hands upon them all, claim them for its own.

The murdered Jews had spent their last night in this room alongside the French peasant shot with them in the morning. They had chanted prayers in a language she could not understand. They had called out their own names and the name of the informer who had betrayed them. Léonie knew the names of the three members of the Jewish family, and she knew the name of the person who had led them to their deaths. She had heard them, night after night when she was ten years old. She had put the words away in here and left them because she was afraid.

She'd listed the contents of the room for her inventory simply as *bric à brac*. She'd kept the place as an extra store for junk overflowing from the attic. All the bits and pieces Madeleine had brought from the flat in suburban London she'd shared briefly with Maurice and then with Léonie. Maurice's books and army things, a clutter of valises and tin boxes and cracked leather bags. All the Englishness that Léonie had inherited from her father she had shut away in here. Then she had got on with marriage to a Frenchman, with becoming wholly French. She'd discouraged her daughters from ever entering this room. In case they heard those voices crying out and were frightened by them. In case they asked her to explain. But history was voices that came alive and shouted. Thérèse knew that, which was why she had come home.

Léonie had tried to cut Thérèse out of herself like the bad flesh from an apple. The rotten spot in her. Thérèse stood for the father, for God, for suffering. For everything that Léonie wanted to forget. But Thérèse had returned, she wouldn't be got rid of, she foretold a groan and heave of change.

Léonie would have to attend the enquiry into the desecrated grave, tell the lawyers the names of the slaughtered Jews buried there with Henri Taillé. She would have to confess that she had been silent all these years about the informer's identity. I had no real proof it was the priest, she'd argued with herself. So let them find it, she told herself now. Thérèse would have to accompany her, recount her side of the story. How much can either of us remember? Léonie thought: it's so long ago. Thérèse has begun writing it down. But I don't know that I'd want to rely on her memory alone. She'll have got half of it wrong.

She'd forgotten to put on her slippers. Her toes curled away from the cold floorboards. She took her hands out of her pockets and thrust them into her dressing-gown sleeves. Partly to warm herself. Partly to hold on to something, since she had no magic cigarette to stuff her mouth with, shut herself up with, defend herself with. Without a cigarette her sixth sense came back.

She had the idea that Thérèse was waiting for her on the other side of the door, along with the Jewish family and Henri Taillé. Her father Maurice was with them too. All she had to do was go in and join them, listen to what they had to say, unravel and reravel the different languages that they used.

She twisted the handle of the door. She opened it. She paused in the doorway, then went in.

The voices came from somewhere just ahead, the shadowy bit she couldn't see. She stepped forward, into the darkness, to find words.

PETER TEGTMEYER (20), Department of Molecular Genetics and Microbiology, State University of New York at Stony Brook, Stony Brook, New York 11794

C. UPTON (24), Department of Biochemistry, University of Alberta, Edmonton, Alberta, Canada T6G 2H7

YUN WANG (20), Department of Molecular Genetics and Microbiology, State University of New York at Stony Brook, Stony Brook, New York 11794

MARTIN WEBER (25), Zentrum für Molekulare Biologie, University of Heidelberg, 69120 Heidelberg, Germany

GEORG F. WEILLER (10), Research School of Biological Sciences, Australian National University, Canberra, ACT 0200, Australia

RICHARD A. YOUNG (1), Whitehead Institute for Biomedical Research, Cambridge, Massachusetts 02142, and Department of Biology, Massachusetts Institute of Technology, Cambridge, Massachusetts 02139

A. L. N. RAO (16), Department of Plant Pathology, University of California at Riverside, Riverside, California 92521

MICHAEL REED (20), Department of Molecular Genetics and Microbiology, State University of New York at Stony Brook, Stony Brook, New York 11794

HOLGER H. ROEHL (13), Department of Microbiology and Molecular Genetics, College of Medicine, University of California at Irvine, Irvine, California 92717

MARILYN J. ROOSSINCK (17), Plant Biology Division, The Samuel Roberts Noble Foundation, Ardmore, Oklahoma 73402

MICHAEL ROSHON (11), Department of Microbiology and Immunology, Vanderbilt University School of Medicine, Nashville, Tennessee 37232

H. EARL RULEY (11), Department of Microbiology and Immunology, Vanderbilt University School of Medicine, Nashville, Tennessee 37232

CHARLES E. SAMUEL (15), Department of Biological Sciences, and Graduate Program of Biochemistry and Molecular Biology, University of California at Santa Barbara, Santa Barbara, California 93106

HEINZ SCHALLER (25), Zentrum für Molekulare Biologie, University of Heidelberg, 69120 Heidelberg, Germany

CHRISTINA SCHERER (11), Department of Biology, and Center for Cancer Research, Massachusetts Institute of Technology, Cambridge, Massachusetts 02139

CARL L. SCHILDKRAUT (21), Department of Cell Biology, Albert Einstein College of Medicine, Bronx, New York 10461

BERT L. SEMLER (13), Department of Microbiology and Molecular Genetics, College of Medicine, University of California at Irvine, Irvine, California 92717

PETER R. SHANK (5), Division of Biology and Medicine, Brown University, Providence, Rhode Island 02912

ER-GANG SHI (11), Department of Microbiology and Immunology, Vanderbilt University School of Medicine, Nashville, Tennessee 37232

NAHUM SONENBERG (4), Department of Biochemistry, and McGill Cancer Center, McGill University, Montreal, Quebec, Canada H3G 1Y6

STEVEN L. SPITALNIK (8), Departments of Pathology and Laboratory Medicine, University of Pennsylvania Medical Center, Philadelphia, Pennsylvania 19104

KELLY STEFANO (9), Graduate Group in Immunology, University of Pennsylvania Medical Center, Philadelphia, Pennsylvania 19104

KUAN-TEH JEANG (2), Laboratory of Molecular Microbiology, National Institute of Allergy and Infectious Diseases, National Institutes of Health, Bethesda, Maryland 20892

MICHAEL G. KATZE (12), Department of Microbiology, University of Washington School of Medicine, Seattle, Washington 98195

DENNIS L. KOLSON (6), Departments of Neurology and Microbiology, University of Pennsylvania Medical Center, Philadelphia, Pennsylvania 19104

CARLA L. KUIKEN (10), Department of Virology, Academic Medical Center, University of Amsterdam, 1105 AZ Amsterdam, The Netherlands

F. C. LAHSER (16), Institute of Developmental and Molecular Biology, Texas A&M University, College Station, Texas 77843

RANDALL D. LITTLE (21), Department of Cell Biology, Albert Einstein College of Medicine, Bronx, New York 10461

YING LUO (3), Howard Hughes Medical Institute, Departments of Medicine, Microbiology, and Immunology, University of California at San Francisco, San Francisco, California 94143

G. MCFADDEN (24), Department of Biochemistry, University of Alberta, Edmonton, Alberta, Canada T6G 2H7

NEAL NATHANSON (7), Department of Microbiology, University of Pennsylvania School of Medicine, Philadelphia, Pennsylvania 19104

MICHAEL NEWSTEIN (5), Division of Biology and Medicine, Brown University, Providence, Rhode Island 02912

KATJA NIESELT-STRUWE (10), Max-Planck-Institut für Biophysikalische Chemie, 37077 Göttingen, Germany

MICHELLE A. OZBUN (19), Division of Molecular Virology, Baylor College of Medicine, Houston, Texas 77030

PETER PALUKAITIS (17), Department of Plant Pathology, Cornell University, Ithaca, New York 14853

B. MATIJA PETERLIN (3), Howard Hughes Medical Institute, Departments of Medicine, Microbiology, and Immunology, University of California at San Francisco, San Francisco, California 94143

ROBERT F. RAMIG (14), Division of Molecular Virology, Baylor College of Medicine, Houston, Texas 77030

Preface

The new series *Methods in Molecular Genetics* provides practical experimental procedures for use in the laboratory. Because the introduction of molecular genetic techniques and related methodology has revolutionized biological research, a wide range of methods are covered. The power and applicability of these techniques have led to detailed molecular answers to important biological questions and have changed the emphasis of biological research, including medical research, from the isolation and characterization of cellular material to studies of genes and their protein products.

Molecular genetics and related fields are concerned with genes: DNA sequences of genes, regulation of gene expression, and the proteins encoded by genes. The consequences of gene activity at the cellular and developmental levels are also investigated. In medical research, knowledge of the causes of human disease is reaching an increasingly sophisticated level now that disease genes and their products can be studied. The techniques of molecular genetics are also being widely applied to other biological systems, including viruses, bacteria, and plants, and the utilization of gene cloning methodology for the commercial production of proteins for medicine and industry is the foundation of biotechnology. The revolution in biology that began with the introduction of DNA sequencing and cloning techniques will continue as new procedures of increasing usefulness and convenience are developed.

In addition to the basic DNA methods, instrumentation and cell biology innovations are contributing to the advances in molecular genetics. Important examples include gel electrophoresis and DNA sequencing instrumentation, *in situ* hybridization, and transgenic animal technology. Such related methodology must be considered along with the DNA procedures.

In this volume, molecular biology and genetic techniques for the analysis of viral genes and chromosomes are presented. The viruses discussed include HIV (human immunodeficiency virus), other viruses with RNA genomes (of both animals and plants), and viruses with DNA genomes. The volume is therefore composed of three sections titled Human Immunodeficiency Virus; Retroviruses/RNA Viruses; DNA Viruses.

Methods in Molecular Genetics will be of value to researchers, as well as to students and technicians, in a number of biological disciplines because of the wide applicability of the procedures and the range of topics covered.

KENNETH W. ADOLPH

Methods in Molecular Genetics

Section I

Human Immunodeficiency Virus

[1] Construction and Analysis of Human and Simian Immunodeficiency Virus Mutants

Anna Aldovini and Richard A. Young

Introduction

The ability to introduce molecularly cloned human (HIV) and simian (SIV) immunodeficiency virus genomes into cells permits the production of defined virus stocks and allows the investigator to study the effects of specific mutations on retroviral biology. Virus produced from a molecularly cloned genome is better defined genetically and more homogeneous than that isolated from infected individuals. These features are essential for the interpretation of experiments involving molecular genetic manipulation of HIV and SIV.

We describe methods for the construction of mutant HIV and SIV genomes and for their analysis. The topics that are covered include selection of appropriate biologically active molecular clones, site-directed mutagenesis to produce the desired mutations, and analysis of the mutants in cell culture by transient transfection and through establishment of stable cell lines. In addition, we briefly describe approaches to investigate mutant viruses *in vivo*.

Construction of Mutations in Viral Genomes

Biologically Active Molecular Clones

Molecular genetic manipulation of HIV or SIV requires the use of well-characterized molecular clones of the virus. A variety of HIV clones has been described in some detail (1–3). These clones are considered biologically active when they are competent for virus production in a cell culture system. It should be noted that the virus generated from HIV clones may not be capable of inducing acquired immunodeficiency syndrome (AIDS) pathogenesis. An animal model for AIDS induced by HIV is not yet available, so it is not clear whether these biologically active clones are fully representative of the virus that is responsible for pathogenesis *in vivo*.

Multiple biologically active SIV molecular clones have also been established (4,5). Although all of these support viral replication, many do not produce virus that appears capable of producing disease in animals. One of them, $SIV_{mac}239$, induces AIDS in rhesus macaques in a manner that is typical of disease produced by uncloned SIV (6).

The choice of a specific clone for experimentation will depend on the biological issue under study. For example, SIV clones with mutations in accessory genes may be appropriate for certain experiments *in vitro,* but SIV clones that contain functional copies of the *nef* gene are likely to be important for studies of pathogenesis *in vivo.*

Site-Directed Mutagenesis and Plasmid Construction

The first step in the genetic manipulation of the HIV or SIV genome is to isolate a DNA fragment containing the region to be mutated from a biologically active viral clone and introduce it into M13 mp18 or M13 mp19. Oligonucleotide-mediated site-directed mutagenesis is performed on the single-stranded DNA produced by M13. Although a variety of protocols is available for this step, that of Kunkel *et al.* (7) offers a simple and efficient way to introduce mutations of interest into the desired DNA fragment. A commercial kit is available for this technique (Cat. No. 170-3571; Bio-Rad, Richmond, CA).

The design of oligonucleotides for site-directed mutagenesis depends on whether the objective is to obtain a point mutation or a deletion. In the case of point mutations, the desired nucleotide change is introduced into the oligonucleotide primer to modify the codon specific for the amino acid that will be altered. If a deletion is desired, an oligonucleotide that contains sequences upstream and downstream of the desired deletion is synthesized.

The length of the oligomer will have considerable influence on the stability of the hybrid molecules that are formed during mutagenesis. The successful introduction of deletion mutations into the Ψ region of HIV was obtained using 40-nucleotide-long oligonucleotides that contained approximately 20 nucleotides upstream and 20 nucleotides downstream of the DNA sequence to be deleted (8). The sizes of the deletions obtained with this approach were 21 and 39 nucleotides, but larger deletions can be obtained. Point mutations in the p7^{gag} protein of HIV were constructed using oligonucleotides of similar length (8).

For convenience of analysis of the mutated product, it is sometimes useful to introduce a restriction site into the genome through the oligonucleotide. For example, in the case of a deletion, the nucleotides for a restriction site can be introduced between upstream and downstream sequences and recombinant clones containing the deletions can be screened by restriction site. In the case of changes affecting an open reading frame, attention to conservation of the correct frame will be necessary.

The introduction of the desired mutations is verified by sequencing the product M13 recombinants, using primers that bind near the site of the mutation. Protocols for M13 sequencing can be found in Refs. 9 and 10. The shortest possible restriction fragment from the mutated M13 recombinant clones should be used to transfer mutations to the wild-type, biologically active clone. The entire region of DNA that is transferred from the M13 replicative form into the final viral genome should be sequenced using double-stranded DNA sequencing methods (9) to confirm the presence of the desired mutations and to ensure that no other alterations have been introduced into the clone.

Full-length molecular clones of HIV and SIV can be unstable when subcloned into plasmid vectors in *Escherichia coli*. Sequences at the 3' end of the viral genome appear to be particularly unstable. Investigators have attempted to minimize this problem with a variety of strategies. One approach is to use λ phage recombinants to carry the provirus, and to use DNA from these clones for transfection (4). The other approach has been to subclone two halves of the proviral DNA in two independent plasmids, which helps minimize deletions or rearrangements when the plasmid is grown in *E. coli*. To reconstitute the full-length viral genome, the two plasmids are digested with an enzyme that cuts the viral DNA once internally, and the DNA fragments are ligated together and used directly for transfection (11). Another strategy is to select *E. coli* hosts in which these large HIV- or SIV-containing plasmid molecules do not undergo frequent recombination. Plasmids derived from pHXB2gpt (1), which are extremely unstable in *E. coli* bacterial strains such as DH1 and DH5α, can be faithfully propagated in HB101 and in MC1061.

Introducing DNA into Mammalian Cells

Transfection of molecular clones of HIV and SIV into selected cells is useful for producing stocks of molecularly cloned viruses and for the analysis of the effects of specific mutations on virus biology. This section discusses features of the materials and methods that are important for successful transfection.

Preparation of DNA for Transfection

Plasmid DNA is the most common source of DNA for transfection of mammalian cells, but whether it is plasmid or phage DNA, the quality of the DNA preparation is extremely important for efficient transfection and for further experimental analysis. Plasmid DNA should be purified through cesium chloride density gradients and, ideally, it should be banded twice.

Cell Lines

Lymphocyte cell lines and monocyte/macrophage cell lines are frequently used in HIV-1 transfection experiments. H9, HUT78, MT4, JURKAT, CEMx174, and MOLT4 are among the lymphocyte lines, and U937, THP-1, and GCT are among the monocyte/macrophage lines most frequently used in transfection experiments. All of these cells are susceptible to HIV infection and, with the exception of GCT, all grow in suspension. Some adherent cell lines that are normally not susceptible to infection are nonetheless capable of expressing HIV genes and producing virus on transfection, and are regularly employed in studies of HIV gene regulation. These include COS-1, CV-1, HeLa, and HOS.

Normal peripheral blood mononuclear cells (PBMCs) can also be transfected and will support viral replication. However, PBMCs are sensitive to the presence of some of the chemical agents used to introduce DNA into cells and they are not very tolerant of the manipulations involved in transfection protocols. In addition, the cytopathic effect generated by HIV infection will deplete the susceptible target cell population, necessitating the timely addition of fresh cells to sustain virus production.

The lymphocytic and monocytic cells grow in 90% (v/v) RPMI-1640, 10% (v/v) fetal bovine serum, and antibiotics. COS-1, CV-1, HeLa, and HOS grow in 90% (v/v) Dulbecco's modified Eagle's medium with 10% (v/v) fetal bovine serum and antibiotics. Peripheral blood mononuclear cells can be grown in 80% (v/v) RPMI-1640, 20% (v/v) fetal bovine serum, and antibiotics. Lectin-free interleukin 2 (IL-2) (Electronucleonics, Inc., Fairfield, NJ) is added to the complete medium at a final concentration of 10% (v/v). Recombinant human IL-2 (Cat. No. 799068; Boehringer Mannheim, Indianapolis, IN) is normally used at a final concentration of 20–25 units (U)/ml. All of the cell lines described here can be obtained from the American Type Culture Collection (Rockville, MD) or from the AIDS Research and Reference Program (12).

Both the cell lines and the primary mononuclear cells should be in logarithmic growth prior to initiating transfection.

Transfection Techniques

A variety of techniques can be used to introduce the proviral form of an HIV or SIV genome into cells. An important consideration in selecting the transfection technique is the cell type to be transfected. Some cells are amenable to transfection by one procedure but not another. The other major consideration is whether there is a need for high transfection efficiencies.

If the goal is to generate molecularly defined virus stocks or to test mutant viruses that are replication competent, low-efficiency transient transfection methods can be employed. The limited number of virions produced in this manner will ultimately, through multiple rounds of infection, infect essentially all of the cells in the culture. In this case, DEAE dextran transfection or electroporation, which are the least time-consuming methods, can be employed.

Studies of replication-defective mutants and of viral gene regulation in which viral RNA and protein products need to be analyzed require higher transfection efficiency and stable integration of the transfected plasmid. In these cases, techniques such as calcium phosphate or electroporation should be employed. Calcium phosphate coprecipitation is a relatively efficient and reproducible method of introducing exogenous DNA into tissue culture cells, but its use is restricted to adherent cell lines. Lipofection, which is normally used for transient transfection assays, is also useful to consider because of its simplicity and because of the high efficiency of transfection that can be obtained in certain cell types. Protoplast fusion was used to provide the first experimental evidence of the biological activity of a cloned provirus (1), but this technique is laborious and does not offer any advantages relative to alternative methods. All of these techniques are described in detail elsewhere (11). A protocol for electroporation is provided below, as this method is generally useful with a wide variety of cell types (13–15); it can be applied to adherent or suspension cells and it can be used for transient studies or for establishing stable cell lines.

Electroporation

The exposure of cells to a high-voltage electric field leads to the transient formation of pores in the membrane that allow DNA to enter the cell. Electroporation can be used to transform a wide variety of cell types, and is especially useful for the transformation of normal peripheral blood lymphocyte and monocytic cell lines (16–18), which are very inefficiently transfected with other procedures. This technique is especially useful for transfections in which the goal is the establishment of stable cell lines. Electroporation is quite efficient for the stable transfection of both adherent and nonadherent cell lines, but for nonadherent cell lines it is the transfection technique of choice. The conditions that yield optimal transient transfection of a given cell line will usually also result in the most efficient stable transfection. Cells transfected by electroporation obtain a small number of plasmid DNA molecules, whereas cells transfected using a $CaPO_4$ technique are exposed to large amounts of incoming DNA, which often integrates in large tandem arrays and can be a source of instability.

The success of electroporation is dependent on two parameters: the voltage of the shock and the duration of the pulse. The voltage and capacitance parameters described below yield the greatest efficiency of transfection in primary cells and in most cell lines used for HIV–SIV studies. The time constant will be influenced by the composition of the transfection cocktail, but phosphate-buffered saline (PBS; including calcium and magnesium), 4-(2-hydroxyethyl)-1-piperazine-ethanesulfonic acid (HEPES)-buffered saline, and tissue culture media have all been successfully employed. Detailed discussions of these issues can be found in Refs. 14 and 16 for both adherent cell lines and human peripheral blood lymphocytes. Electroporation typically results in some cell death; the efficiency of the technique is frequently optimal when approximately half of the cells survive electroporation.

An important difference between the application of electroporation for transient or stable transfection is the state of the transfected DNA. Efficient stable transfection requires a linearized plasmid template (cleaved outside the gene of interest or selectable marker), whereas transient expression is usually higher from a supercoiled plasmid template.

Reagents and Solutions
>   Complete medium
>   PBS with Mg^{2+} and Ca^{2+}
>   Electroporation cuvettes, 0.4 cm (Cat. No. 165-2085; Bio-Rad)
>   Electroporation power source [Bio-Rad, Bethesda Research Laboratories (Gaithersburg, MD), Promega (Madison, WI)]

Method

1. Grow cells to exponential phase. Use approximately 2×10^7 cells for transient transfection or about 5×10^6 cells for permanent transfection.

2. Pellet the cells, wash twice with ice-cold PBS, and resuspend the pellet in 0.4 ml of ice-cold PBS. (Some investigators prefer to wash and resuspend cells in RPMI-1640 with 20% fetal bovine serum, especially if working with peripheral blood lymphocytes. Cell viability is reportedly better in this medium.)

3. Add $10-20$ μg of plasmid vector DNA (linearize the plasmid by restriction enzyme digestion if cell lines with stably integrated DNA are to be selected) and transfer the suspension to a 0.4-cm electroporation cuvette.

4. Place the cuvette in ice for 5 min.

5. Mix the DNA–cell suspension by flicking the bottom and place the cuvette in the holder in the electroporation apparatus.

6. Shock once with voltage set at 250 V and capacitance at 960 μF.

7. Return the cuvette to ice for 10 min.

8. Transfer to a 15-ml centrifuge tube and add complete medium.

9. Spin the cells for 3 min at 2100 rpm.

10. Resuspend the cells in 10 ml of medium and transfer to a T25 flask.

11. Transfer to a 37° C, 5% CO_2 incubator.

12. For transient expression, harvest the cells and assay for expression 48 hr after transfection. For selection of stably transfected cell lines, place the cells in selective medium $2-3$ days after transfection.

Gene Expression and Virus Production in Transiently Transfected Cells

Transient transfection studies have led to important insights into the biology of the human and simian immunodeficiency viruses. The major transient expression systems utilized in the study of HIV and SIV mutants include the simian virus 40 (SV40)-based transient amplification and expression system using the African Green monkey kidney COS cell line (18). The use of COS cells has provided significant insights into the function of viral regulatory gene products and has permitted the analysis of replication-incompetent proviral genomes. As described below, transient transfection analysis is particularly useful for investigating gene expression, for analyzing viral particles, and for investigating viral infectivity with specific HIV and SIV mutant genomes.

Viral Gene Expression

The expression of viral RNA and protein products can be examined in transiently transfected cells. RNA can be extracted from transfected cells and Northern blot analysis and RNA slot blot analysis are performed according to previously described techniques (9, 10). The expression of viral proteins can be investigated by radioimmunoprecipitation or Western blot analysis. For radioimmunoprecipitation analysis, cells transfected 48 hr earlier with 10 μg of plasmid DNA are metabolically labeled with radiolabeled amino acids (i.e., 500 μCi of [^{35}S]methionine) for 4 hr. Cell lysates are prepared and immunoprecipitations of viral proteins are performed using human or primate serum antibodies with demonstrated specificity for HIV-1 and SIV-1 antigens, as described (19). Immunoprecipitated proteins are resolved using sodium dodecyl sulfate-polyacrylamide gel electrophoresis (SDS-PAGE) (9, 20). If Western blotting is the assay of choice, cells transfected 48 hr earlier can be lysed in Laemmli buffer and protein lysates can be analyzed directly, after resolving them by SDS-PAGE and transferring them to a nitrocellulose membrane, using polyclonal sera or monoclonal antibodies specific for viral proteins (9).

Analysis of Viral Particles

The protein and nucleic acid composition of mutant viral particles produced from transfected cells can also be investigated. Virus in culture supernatants is pelleted by centrifugation for 2 hr at 27,000 rpm in an ultracentrifuge. The pellet is resuspended in dissociation buffer [0.01 M Tris-HCl (pH 7.3), 0.2% (v/v) Triton X-100, 0.001 M ethylenediaminetetraacetate (EDTA), 0.005 M dithiothreitol, 0.006 M KCl] if reverse transcriptase activity is to be measured (21), or in 0.2% Triton and Laemmli buffer (20) if protein analysis is to be performed. p24 (HIV) or p27 (SIV) analysis on tissue culture supernatants or on pelleted virus can be performed using a Du Pont (Wilmington, DE) p24 kit or a Coulter Immunologics (Hialeah, FL) p27 kit, according to the manufacturer. This assay is a very sensitive biochemical measure of the physical amount of virus present in the supernatant, but the physical amounts of virus do not necessarily correlate with the amount of infectious virus.

The amount of viral RNA present in virions can be investigated by extracting RNA directly from virions present in the supernatant, immobilizing the nucleic acid on nitrocellulose slot blots, and probing with virus-specific radiolabeled probes (11).

Virus particle morphology can be examined by electron microscopy, as reviewed in Ref. 22.

Infection of Target Cells with Mutant Viruses

The infectivity of the mutant viral particles produced by transient transfection can be investigated. To quantitate viral particle production by transfected cells, culture supernatants are harvested, filtered through a 0.45-μm pore size filter, and examined for the presence of reverse transcriptase activity and the Gag core protein p24 or p27. If these assays are positive, the infectivity of the viral particles in the supernatant is tested in an infection assay. Cells that are natural targets for these retroviruses (H9 and CEMx174) are exposed to the supernatant of transfected cells, and production of virus from these cells is monitored in a time course experiment (23). This assay indicates whether or not the mutant particle has retained the ability to replicate and be infectious. The 50% tissue culture infectious dose (TCID_{50})/ml can be measured to evaluate amounts of infectious virus (24).

Selection of Stable Cell Lines Expressing Mutant Viruses

The generation of cell lines stably expressing intact viral variants (such as replication-incompetent isolates or mutant genomes) can facilitate the study of biological defects due to genetic alterations in HIV and SIV genomes. These cell lines can also be used to produce antigens for candidate vaccines. Methods used to generate such stably transfected cell lines are described below.

The isolation of cell lines that stably express a transfected gene requires, in most instances, the simultaneous introduction of a selectable marker gene to permit the identification of the rare cells that have integrated the sequences of interest. These selectable markers generally encode an enzymatic activity that confers resistance to a specific antibiotic or the ability to grow in a metabolically selective medium. The transfection techniques used to introduce DNA into mammalian cells were described above. The following sections discuss selectable markers, vectors, and selection strategies that permit establishment of stable cell lines carrying HIV or SIV genomes.

Selectable Marker Genes

A variety of selectable marker systems have been developed to facilitate the generation of stably transfected cell lines. Some of these exploit cell lines selected for the absence or deficiency of a gene involved in host cell metabolism [i.e., thymidine kinase (TK), hypoxanthine–guanine phosphoribosyltransferase (HGPRT), or dihydrofolate reductase (DHFR)]. These cell lines can be maintained in supplemented media, but will die on transfer to selective media unless a gene able to complement the deficiency has been introduced. The metabolic selectable markers adenosine deaminase (ADA) and DHFR are of particular note, as they permit amplification of transfected genes through incremental increases in the selective pressure. An alternative approach employs dominant selectable markers. The dominant selectable markers, such as the *neo*^{r}, *hyg*^{r}, *his*^{r}, or *gpt* genes (11), are derived from prokaryotes. Thus, they have wider utility and are typically easier to use, as there is no endogenous enzymatic activity to be inhibited or overcome. The selectable markers may be either physically linked to the transfected gene (i.e., resident on the same plasmid) or unlinked (i.e., introduced by cotransfection of an independent plasmid).

HIV (SIV)-Based Retroviral Vectors for Generating Stable Cell Lines

The observation that the HIV *nef* gene is dispensable for viral replication *in vitro* led to the construction of replication-competent variants of HIV in which *nef* is replaced by a drug resistance gene (Fig. 1) (11,25). Introduction of these vectors into target cells confers drug resistance. An attractive feature of these vectors is that expression of the selectable marker depends on the HIV promoter, so drug resistance and viral production are linked in the cell lines produced with these vectors. The replication properties of these selectable marker-containing viruses parallel those of the wild-type virus.

Two different strategies can be used to introduce these vectors into cells to produce stable cell lines. Selection can be applied 48 hr after transfection. Alternatively, if the transfected genome can support viral replication, the virus generated by transient transfection can be harvested and used to infect a susceptible population.

For studies of mutants addressing SIV biology *in vitro* a similar approach can be taken and construction of vectors in which the *nef* gene is replaced by a selectable marker is possible.

Selection

The generation of stably transfected cell lines begins with the introduction of DNA into the target cell. Only a fraction of the cells is transfected with the introduced DNA, and in only a portion of these cells does the DNA become stably integrated and expressed. Selection is initiated 48 hr after transfection, regardless of the method of transfection. Stably transfected cells should emerge from selection in numbers sufficient to analyze in a period ranging from 2 to 4 weeks, depending on the target cell used and the transfection efficiency.

Following transfection, adherent cell lines are best split at various dilutions (i.e., 1:5, 1:10, 1:20), based on the expected transfection efficiency. Plates containing excessive numbers of cells will respond poorly to the selection. The ideal situation is

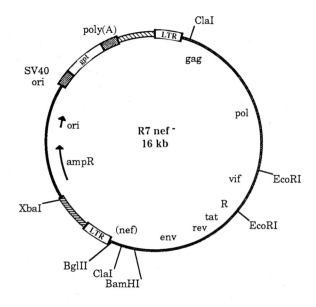

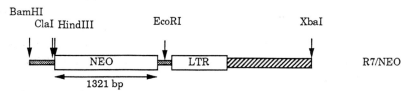

FIG. 1   The pR7nef⁻ plasmid is a modification of pHXB2gpt (1). In pHXB2gpt the *nef* ATG has been changed to a *Cla*I site (pHXB2Cla; M. Reitz, National Cancer Institute, Bethesda, MD) and, in this modified plasmid, the *Eco*RI site located 3′ of the 3′ HIV long terminal repeat (LTR) has been deleted (pR7nef⁻). The *neoʳ* selectable marker, previously cloned in the polylinker of a plasmid vector, has been inserted in pR7nef⁻. The *Cla*I–*Bgl*II fragment for the *NEO* gene has replaced the 255-bp *Cla*I–*Bgl*II fragment, located at the 5′ end of the *nef* gene in pR7nef⁻. The resulting vector is R7/NEO.

to obtain approximately 10–20 well-separated colonies per 10-cm plate. Selected adherent cells can frequently be harvested as clones from such plates and are best isolated and expanded using cloning cylinders (11). Nonadherent cells are most efficiently selected by seeding into 24- or 96-well plates (rather than in bulk culture). Optimal seeding density will vary according to cell type, but is usually between 10^5 and 10^6 cells/ml of selective medium.

   Aliquots of transfected cells should be frozen as soon after selection as possible, both as bulk populations and as clones, to avoid their loss due to genetic instability, overgrowth by cells not expressing the desired gene, or tissue culture contamination. Once a cell line is established, it is not always necessary to continue the drug or metabolic selection initially used. However, in some instances gene expression is lost following removal of selection, in particular when the nonselected gene encodes a product deleterious to the host cell. Whether selection must be continued needs to be empirically determined for each combination of selectable marker and nonselected gene studied.

Analysis of Mutant Viruses *in Vivo*

Infectious Mutants

The study of SIV biology and simian AIDS has benefited enormously from SIV variants that have been genetically engineered with specific mutations (4, 26). For example, the *nef* gene has been demonstrated to have an important role in the pathogenesis of simian AIDS, and an attenuated virus vaccine candidate has been constructed that exploits this observation (27). Knowledge that an intact *nef* gene is required for pathogenesis should influence the selection of appropriate molecular clones for further molecular genetic analysis of SIV.

Virus stocks produced as early as possible after transfection should be used to infect animals. Generally, 1 ml of tissue culture supernatant containing the virus of interest is injected into an animal (4). The supernatant is harvested approximately 2 to 3 weeks after the transfection, when reverse transcriptase values peak and when enough supernatant is available to freeze a homogeneous stock sufficient for an entire experiment. Cell-free virus stock for inoculation is prepared by centrifugation of an infected cell culture and filtration of viral supernatant through a 0.45-μm pore size filter (28). Virus stock aliquots (1 ml) are stored in liquid nitrogen and the TCID$_{50}$/ml for each is measured after thawing and at the time of injection. Virus stocks tend to lose approximately one log of infective titer in a freeze–thaw cycle. The TCID$_{50}$/ml ranges from 10^5 to 10^6 for both HIV and SIV before freezing. Blood samples are collected from the injected animals at intervals after virus inoculation to monitor seroconversion, to isolate and characterize virus, and to obtain a T cell count and measure immunocompetence. Onset of typical AIDS symptoms should also be monitored. This systematic analysis should offer a detailed picture of the *in vivo* phenotype of the engineered mutants.

Noninfectious Mutants

Cell lines that stably express mutated genomes that produce noninfectious particles can be established by the techniques described above. These replication-deficient viruses would not be capable of establishing a productive infection in experimental animals, but the material generated from these lines is of interest for vaccine purposes (29). Injection of animals with virus that does not exhibit infectivity *in vitro* should provide information on the safety of such genetically inactivated virus, its immunogenicity, and its therapeutic and prophylactic activity for immunotherapy and prevention of AIDS.

Characterization of Viral Mutants Occurring *in Vivo*

Isolation of Mutant Viruses from Peripheral Blood

RNA viruses are intrinsically prone to genetic variability, as no proofreading mechanism exists for viral polymerases. For retroviruses, and thus for HIV and SIV, errors can occur at three different stages: reverse transcription, plus-strand DNA synthesis, and transcription. As a result, every virus isolate from a biological specimen is likely to be, and usually is, a mixture of viral variants. Virus isolation in a tissue culture system can alter the representation of the different genetic species, or quasispecies, favoring the growth of a subset of viruses. In contrast, viral variants that are replication incompetent are lost during passage in cell culture. There are a variety of studies that have described the genetic variability that occurs *in vitro* and *in vivo* (30–39).

Virus isolation from biological specimens generally utilizes cell lines or PBMCs starting from a sample of peripheral blood that has been characterized as antibody positive for HIV or SIV. The following protocol summarizes virus isolation from blood of infected animals (36). Peripheral blood mononuclear cells are isolated from 5 ml of heparinized whole blood by Ficoll–Hypaque gradient centrifugation. Cells are washed, resuspended at 10^6/ml, and stimulated with phytohemagglutinin (PHA) at 5 μg/ml for 48 hr. Cultures are washed twice to remove any PHA and maintained for 5 days in complete medium with 10% (v/v) IL-2. At this point, cells are cocultivated with the cell line of interest. Supernatants are harvested every few days to monitor virus production. Once a productive cell line has been established, standard cloning techniques can be utilized to isolate viral DNA from extracted cellular and Hirt DNA or from PCR products (40). Similar protocols are available for virus isolation from human peripheral blood samples (41). This analysis should permit detection of different viruses that simultaneously infect the same host (30, 32, 34, 36, 37, 39). Spontaneously occurring mutants can be detected by continuing this analysis over a time course (31, 33, 35, 38).

Direct Characterization of Mutant Viruses from Pathological Specimens

Virus isolation from cell culture tends to bias the selection of the growing viruses toward those viral clones that are replication competent and lymphotropic. A major incentive for establishing chronically infected cell lines was the initial difficulty in cloning and characterizing viral isolates directly from biological specimens, as HIV and SIV tend to be present only in a small fraction of the cells of an infected tissue. The development of more sensitive technologies such as *in situ* and nested polymerase chain reaction (PCR) has allowed investigators to investigate more directly the complexity of the different viral quasi species present in a given animal or individual. Such analysis has revealed that there is a substantial level of viral variability *in vivo,* and the acquisition of new variants occurs continuously with time. Interestingly, the variability observed *in vivo* does not correlate with the variability observed with paralleled *in vitro* isolates or with the variability of these isolates with time *in vitro* (30–39).

Nested PCR has proven to be an excellent tool to investigate viral variability. A protocol for nested PCR using DNA from uncultured PBMCs, adapted from published protocols (36), is described below. This protocol should allow amplification of HIV or SIV sequences from any virus-containing sample, even when its sequence diverges from the primer sequence.

Nested Polymerase Chain Reaction Protocol

1. High molecular weight DNA is obtained from 20×10^6 PBMCs by standard procedures (9).

2. One to 3 μg of genomic DNA is amplified with an outer set of primers under low-stringency conditions (denaturing: 94°C, 1 min; annealing: 45°C, 1 min; extension: 68°C, 1 min; 30 cycles) to ensure amplification of divergent viral sequences. Amplification reactions are performed in a total volume of 100 μl containing 10 mM Tris–HCl (pH 8.3), 50 mM KCl, 1.5 mM MgCl$_2$, 0.01% (w/v) gelatin, 200 μM deoxynucleotide triphosphates, 20 pmol of each primer, and 2.5 U of *Taq* polymerase. Samples are overlaid with 100 μl of mineral oil.

3. The PCR products are analyzed by polyacrylamide gel electrophoresis [8% (w/v) nondenaturing gels] for the presence of specific DNA fragments. In the absence of a recognizable amplification product, an aliquot (3 μl) of the initial amplification mixture is reamplified under the original conditions with a second set of primers designed to amplify a DNA fragment that is internal to the first primer pair.

4. Following the second round of amplification, the PCR products are reexamined for appropriately sized DNA fragments by polyacrylamide gel electrophoresis.

5. All positive specimens are then amplified a third time under increased stringency conditions (denaturing: 94°C, 1 min; annealing: 60°C, 1 min; extension: 72°C, 1 min; 30 cycles) with modified second-round primer pairs that contain restriction enzyme sites to facilitate subsequent M13 cloning and sequence analysis.

The PCR analysis is carried out with the appropriate negative and positive control DNAs. The authenticity of the PCR products is confirmed by molecular cloning into

M13 and subsequent nucleotide sequence analysis according to standard procedures (9, 10).

Conclusion

Our present understanding of the biology of the primate lentiviruses has been greatly facilitated by our ability to manipulate viral genomes genetically and to analyze defective phenotypes using transfected cells. The techniques described provide an overview of the experimental approaches that are likely to continue to be used to further investigate the biology of these important viruses.

Acknowledgment

This work was supported by Public Health Service Grant AI26463 from the National Institutes of Health.

References

1. A. G. Fisher, E. Collalti, L. Ratner, R. C. Gallo, and F. Wong-Staal, *Nature (London)* **316,** 262 (1985).
2. M. Alizon, P. Sonigo, F. Barre-Sinoussi, J.-C. Chermann, P. Tiollais, L. Montagnier, and S. Wain-Hobson, *Nature (London)* **326,** 757 (1984).
3. P. A. Luciw, S. J. Potter, K. Steimer, D. Dina, and J. A. Levy, *Nature (London)* **312,** 760 (1984).
4. Y. M. Naidu, H. W. Kestler, III, Y. Li, C. V. Butler, D. P. Seilva, D. K. Schmidt, C. D. Troup, P. K. Sehgal, P. Sonigo, M. D. Daniel, and R. Desrosiers, *J. Virol.* **62,** 4691 (1988).
5. S. Dewhurst, J. E. Embretson, D. C. Anderson, J. I. Mullins, and P. N. Fultz, *Nature (London)* **345,** 636 (1990).
6. H. Kestler, T. Kodama, D. Ringler, M. Marthas, N. Pedersen, A. Lackner, D. Regier, P. Sehgal, M. Daniel, N. King, and R. Desrosiers, *Science* **248,** 1109 (1990).
7. T. A. Kunkel, J. D. Roberts, and R. A. Zakour, *in* "Methods in Enzymology," Vol. 154, p. 367. Academic Press, San Diego, 1987.
8. A. Aldovini and R. A. Young, *J. Virol.* **64,** 1920 (1990).
9. F. M. Ausubel, R. Brent, R. E. Kingston, D. D. Moore, J. G. Seidman, J. A. Smith, and K. Struhl, "Current Protocols in Molecular Biology 1987." Green Publishing Associates and Wiley Interscience, New York, 1987.
10. J. Sambrook, E. F. Fritsch, and T. Maniatis, "Molecular Cloning: A Laboratory Manual." Cold Spring Harbor Laboratory Press, Cold Spring Harbor, New York, 1989.
11. A. Aldovini and M. B. Feinberg, *in* "Techniques in HIV Research" (A. Aldovini and B. D. Walker, eds.), p. 147. Stockton Press, New York, 1990.
12. AIDS Research and Reference Reagent Program Catalog, 1992. NIH Publication No. 90-1536.

13. H. Potter, L. Weir, and P. Leder, *Proc. Natl. Acad. Sci. USA* **81,** 7161 (1984).
14. G. Chu, H. Hayakawa, and P. Berg, *Nucleic Acids Res.* **15,** 1311 (1987).
15. E. Eustis-Turf, K.-M. Wang, and L. B. Schook, *Bio-Rad Bull.* 1345 (1987).
16. A. J. Cann, Y. Koyanagi, and I. S. Y. Chen, *Oncogene* **3,** 123 (1988).
17. P. Barry, E. Pratt-Lowe, and P. A. Luciw, *Bio-Rad Bull.* 1349 (1989).
18. Y. Gluzman, *Cell (Cambridge, Mass.)* **23,** 175 (1981).
19. F. D. Veronese, M. G. Sarngadharan, R. Rahman, P. D. Markham, M. Popovic, A. J. Bodner, and R. C. Gallo, *Proc. Natl. Acad. Sci. USA* **82,** 5199 (1985).
20. U. K. Laemmli, *Nature (London)* **227,** 680 (1970).
21. M. D. Daniel, N. L. Letvin, N. W. King, P. M. Kannagi, P. K. Sehgal, R. D. Hunt, P. J. Kanki, M. Essex, and R. C. Desrosiers, *Science* **228,** 1201 (1985).
22. H. R. Gelderblom, *AIDS (USA)* **5,** 617 (1991).
23. M. Popovic, M. G. Sarngadharan, E. Read, and R. C. Gallo, *Science* **224,** 497 (1984).
24. V. A. Johnson, R. E. Byington, and P. I. Nara, *in* "Techniques in HIV Research" (A. Aldovini and B. D. Walker, eds.), p. 71, Stockton Press, New York, 1990.
25. M. B. Feinberg, D. Baltimore, and A. D. Frankel, *Proc. Natl. Acad. Sci. USA* **88,** 4045 (1991).
26. H. E. Kestler *et al., Cell (Cambridge, Mass.)* **65,** 651 (1991).
27. M. D. Daniel, F. Kirchhoff, S. C. Czajak, P. K. Sehgal, and R. C. Desrosiers, *Science* **258,** 1938 (1991).
28. R. C. Desrosiers, *in* "Techniques in HIV Research" (A. Aldovini and B. D. Walker, eds.), p. 121. Stockton Press, New York, 1990.
29. A. Aldovini and R. A. Young, *in* "Vaccine Research: A Series Advances" (W. Koff and H. Six, eds.), p. 43. Marcel Dekker, New York, 1991.
30. M. Goodenow *et al., J. Acquir. Immune Defic. Syndr.* **2,** 344 (1989).
31. A. Meyerhans *et al., Cell (Cambridge, Mass.)* **58,** 901 (1989).
32. Y. Li *et al., J. Virol.* **65,** 3973 (1991).
33. S. Delassus, R. Cheynier, and S. Wain-Hobson, *J. Virol.* **65,** 225 (1991).
34. J. P. Vartanian, A. Meyerhans, B. Asjo, and S. Wain-Hobson, *J. Virol.* **65,** 1779 (1991).
35. D. P. Burns and R. C. Desrosiers, *J. Virol.* **65,** 1843 (1991).
36. J. S. Allan *et al., J. Virol.* **65,** 2816 (1991).
37. Y. Li, Y. M. Naidu, M. D. Daniel, and R. C. Desrosiers, *J. Virol.* **63,** 1800 (1989).
38. L. P. Martins, N. Chenciner, B. Asjo, A. Meyerhans, and S. Wain-Hobson, *J. Virol.* **65,** 4502 (1991).
39. L. Pedroza Martins, N. Chenciner, and S. Wain-Hobson, *Virology* **191,** 837 (1992).
40. J. Overbaugh, S. Dewhurst, and J. I. Mullins, *in* "Techniques in HIV Research" (A. Aldovini and B. D. Walker, eds.), p. 131. Stockton Press, New York, 1990.
41. S. Gartner and M. Popovic, *in* "Techniques in HIV Research" (A. Aldovini and B. D. Walker, eds.), p. 53. Stockton Press, New York, 1990.

[2]   Expression Cloning of Genes Encoding RNA-Binding Proteins

Anne Gatignol and Kuan-Teh Jeang

Introduction

RNA-binding proteins are important regulatory factors in gene expression (1–6). Studies on the structure and function of these proteins are complicated by limitations in the technology for their isolation. One approach is to purify the protein, to perform microsequencing of amino acids, to construct degenerate oligonucleotides, and then to screen libraries. An alternative to this is to isolate expression cDNAs directly. Here, we describe a method to isolate cDNAs encoding RNA-binding proteins, using a direct screening assay for RNA–protein-binding interactions. For this purpose, we modified the procedure commonly used in isolating cDNAs encoding sequence-specific DNA-binding proteins (7–9). Using this protocol, we have isolated a 1.5-kb cDNA that codes for a human protein that binds to the human immunodeficiency type 1 (HIV-1) Tat-response element (TAR) RNA (6). Modifications and details of our technique are presented below.

Methods

The technique described here is designed to screen a cDNA expression library prepared in lambda (λ) ZAP using *Escherichia coli* XL1 Blue (10). It can be used for any expression libraries in λ using the appropriate *E. coli* recipient strain and selective agent. A simplified schematic of the overall procedure is presented in Fig. 1.

Materials

Tetracycline (or other selective agents)
LB broth
Bacto-agar (Difco, Detroit, MI), agarose
Maltose
$MgSO_4$
NZY broth
   NaCl   5 g
   $MgSO_4$ 2 g

Methods in Molecular Genetics, Volume 4

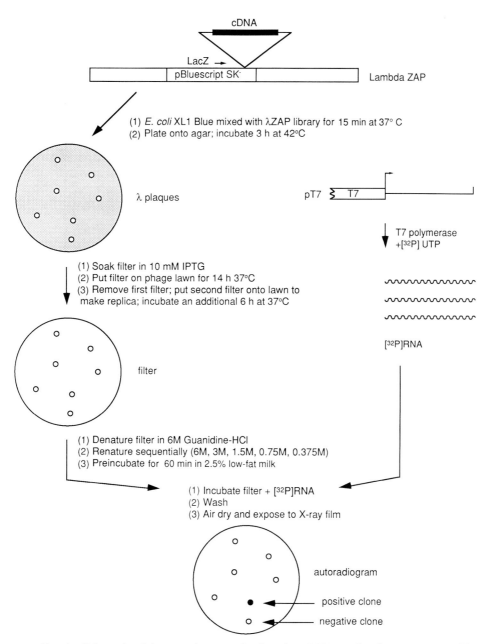

Fig. 1   Schematic of the screening steps for detecting cDNAs coding for sequence-specific RNA-binding proteins, using a ^{32}P-labeled recognition site RNA probe.

> Yeast extract   5 g
> NZ amine   10 g
> H$_2$O   1 liter
>   or NZY broth (21 g/liter)(Cat. No. M36350Q; GIBCO, Grand Island, NY)

Gelatin
Nitrocellulose filters, 132 mm
Petri or tissue culture dishes, 150 mm (Cat. No. 3025; Falcon; Becton Dickinson, Lincoln Park, NJ)
λ ZAP cDNA library (Stratagene, La Jolla, CA)
Escherichia coli XL1 Blue, grown in LB plus tetracycline (12.5 μg/ml) and stored on similar plates for a few months
[α-^{32}P]UTP (Cat. No. PB20383; Amersham, Arlington Heights, IL)
Sterile water (~4 liters)

Probe Preparation

DNA

We suggest using T7 (SP6 or T3) promoter to transcribe a downstream DNA sequence in order to generate an RNA probe with high specific activity. Digest DNA with the appropriate enzyme, extract once with saturated phenol, once with chloroform, precipitate with ethanol, and resuspend at a final concentration of 1 μg/μl.

In Vitro Transcription
This procedure is adapted from Promega (Madison, WI).

> Reaction buffer (5×)
>   Tris-HCl, pH7.5   200 m*M*
>   MgCl$_2$ 30 m*M*
>   Spermidine   10 m*M*
>   NaCl   50 m*M*
>
> NTPs mix
>   ATP   2.5 m*M*
>   CTP   2.5 m*M*
>   GTP   2.5 m*M*
>   UTP   0.25 m*M*

1. Add successively:

> Diethyl pyrocarbonate (DEPC)-treated H$_2$O   up to 25 μl
> Reaction buffer (5×)   5 μl

Dithiothreitol (DTT) 100 mM	2.5 μl
RNasin	1 μl
NTPs mix	5 μl
DNA	1 μl
[α-^{32}P]UTP	3 μl
T7 polymerase	1 μl

2. Incubate at 37–40°C for 1–2 hr.
3. Add RQ1 DNase   (1 μl).
4. Incubate at 37°C for 15–30 min.

If the probe has a high C or a low U content, [^{32}P]CTP should be used instead of [^{32}P]UTP and the CTP/UTP ratio inverted in the NTPs mix.

Gel Purification of Probe
See Ref. 11 (c.f., p. 6.46) for details.

1. Add 25 μl of formamide/bromophenol blue.
2. Heat for 5 min at 90°C.
3. Run the entire reaction on a polyacrylamide/urea gel (6 or 8% for an 80-nucleotide probe). We use Gel-mix from Bethesda Research Laboratories (Gaithersburg, MD).
4. Expose the gel to X-ray film for about 5 min, develop the X-ray film, and use it as a guide to cut the desired band from the gel covered with plastic wrap.
5. Transfer the polyacrylamide piece to a 15-ml tube.
6. Fragment it into small pieces.
7. Add 1 ml of elution buffer:

Ammonium acetate	0.5 M
Magnesium acetate	10 mM
Ethylenediaminetetraacetic acid (EDTA)	1 mM
Sodium dodecyl sulfate (SDS)	0.1%

8. Agitate for 2 hr (use this amount of time for short probes; a longer time may be necessary for longer RNAs) at 37°C.
9. Separate the buffer from the acrylamide using SpinX columns (Cat. No. 8160; Costar, Cambridge, MA).
10. Transfer the buffer with as few as possible contaminating gel slices to the column; spin in the microfuge for 2 min; wash the rest of the gel with 200 μl of buffer; add this buffer to the column and spin again.
11. Extract with phenol once, followed by chloroform once.
12. Transfer to two microtubes (400 to 500 μl each) and precipitate with 1 ml of ethanol.

The probe can be kept up to 2 weeks in ethanol or for 1 week when resuspended in DEPC-treated sterile water.

Plating the Library

The protocol is used to screen 5×10^5 plaques on 20 150-mm plates, which yields roughly 25,000 plaques per plate. We found this to be the optimal number in order to obtain a good signal.

Day 1

1. Sterilize maltose (20%, w/v) by filtration.
2. Sterilize $MgSO_4$ (1 *M*) by filtration.
3. Prepare 20 150-mm plates of NZY agar:

NZY broth (GIBCO)	21	g
Bacto-agar	15	g
H_2O	1	liter

Prepare no more than 700 ml in a 1-liter bottle and autoclave for 30 min at 120°C. Cool to 55°C and pour the plates.

4. Prepare top agarose:

NZY broth (GIBCO)	21	g
Agarose	8	g
H_2O	1	liter

Prepare 200 ml in a 500-ml bottle and autoclave. This preparation can be stored at room temperature indefinitely.

5. Prepare gelatin (2%, w/v).
6. Prepare SM buffer:

NaCl	5.8 g	
$MgSO_4$	2.0 g	
Tris-HCl, pH 7.5 (1 *M*)	50	ml
Gelatin (2%)	5	ml
H_2O	up to 1	liter

Autoclave for 30 min.

7. Grow XL1 Blue in 50 ml of LB supplemented with 0.2% maltose (0.5 ml), 10 mM MgSO$_4$ (0.5 ml) overnight at 30°C.

8. Set up a water bath at 48°C.

Day 2

1. Check the OD$_{600}$ of the bacteria. It should be between 1 and 1.5. If higher, dilute in LB supplemented as previously described and let grow for an additional 1 hr and repeat the OD$_{600}$ check.

2. Melt the top agarose (200 ml if screening 20 plates) and let it cool to 48°C.

3. Prewarm plates to 42°C.

4. Spin the cells down at 2000 rpm for 10 min at room temperature.

5. Dilute phage stock in SM buffer to have a convenient volume to add to the bacteria. For a library of 3 × 10^{10} plaque-forming units (pfu)/ml, dilute by 10^{-3} so that 17 μl contains 5 × 10^5 pfu.

6. Prepare 50 ml of 10 mM MgSO$_4$ in sterile water. Use about 15 ml to resuspend the bacterial pellet (do not vortex). Make serial 1:1 dilutions and check the OD$_{600}$. Keep the tube that has an OD$_{600}$ of 0.5. Take 12 ml of these cells and add 5 × 10^5 pfu of phage. Incubate for 15–20 min at 37°C.

7. Mix 200 ml of top agarose with the 12 ml of infected bacteria. Try to maintain the mix at 48°C. Spread 9 ml evenly onto a prewarmed 150-mm plate of bottom agar. The temperature of the top agarose is important: 55°C can kill the bacteria and 42°C can cause the top agarose to gel too fast.

8. Let the plates stand for 15 min at room temperature. Then place the plates at 42°C until tiny plaques are visible (3–4 hr).

9. Prepare 20 nitrocellulose filters by soaking them in 10 mM isopropyl-β-D-thiogalactopyranoside (IPTG) for 15–30 min. Allow the filters to air dry and number them with indelible ink.

10. As soon as tiny plaques are visible, overlay the plates with filters. Avoid trapping any air bubbles under the nitrocellulose filters. Incubate overnight at 37°C.

Day 3

1. Remove the plates from the incubator. They can be placed for 2 hr at 4°C if the filter is difficult to remove. This is generally not the case if top agarose is used instead of top agar.

2. Prepare 20 new IPTG-soaked filters as described in step 9 (day 2).

3. Mark the position of each filter on the plates (with a needle or ink) and on the back of the dish before lifting filters. Markings need to be asymmetrical. Allow all filters to air dry.

4. Overlay with a new set of filters and copy the asymmetrical markings onto each filter. Incubate for 6 hr at 37°C.

5. Lift the duplicate filters and allow them to air dry before proceeding.

If convenient, the first set of filters can be left for 6 hr and the second set for overnight. If this is done, the signal will be more intense on the second set than on the first set.

We recommend performing the affinity-screening protocol as soon as possible after plating (following day) because, over time, the phage in the top agarose tends to diffuse and the resolution of the plaques will decrease. However, for short-term storage, we do recommend keeping the plated phage at 4°C.

Affinity Screening Protocol

Prepare the following solutions (with sterile water).

Binding buffer (BB) ($1 \times$)
4-(2-hydroxyethyl)-1-piperazine-ethansulfonic acid
   (HEPES), pH 7.3   20 mM
KCl   40 mM
MgCl$_2$ 1.5 mM
DTT   1 mM
Guanidine-hydrochloride (6 M in $1 \times$ binding buffer)
Prebinding solution: 2.5% (w/v) low-fat milk in $1 \times$ binding buffer. Mix thoroughly. Prepare approximately 10 ml/filter

All of the following steps are performed at room temperature. Treat 40 filters as a batch.

1. Immerse dry filters in 6 M guanidine solution. Use approximately 10 ml/filter. Gently agitate on a rocking platform for 5 min.
2. Remove one-half the volume of the buffer and replace with an equal volume of fresh BB. Agitate the filters for 5 min.
3. Repeat step 2 four times.
4. Remove all the buffer and wash the filters with fresh BB, for 5 min with gentle agitation.
5. Incubate the filters in prebinding buffer for 60 min on a rocking platform.
6. Wash the filters twice (5 min each) with fresh BB.
7. Prepare binding cocktail (1 ml/filter). In $1 \times$ BB, add
   Calf thymus DNA (10 μg/ml)
   Competitor RNA[1] (10 μg/ml)
   Labeled RNA (10^6 cpm per filter)
8. Place the filters in a sealable bag and add binding cocktail. Place eight filters

[1] Various competitor RNAs can be used depending on the stringency of screening.

in each bag. Remove all air bubbles from between the bag and filters by rubbing a pipette across the bag.

9. Seal the bag completely.

10. Agitate for 1 hr on a rocking platform at room temperature.

11. Remove the filters from the binding cocktail and wash thoroughly with $1 \times$ BB until radioactivity no longer washes off. Do three to four washes with 15 to 20 ml/filter for durations of 10 min. The total washing time should be less than 1 hr.

12. Air dry the filters.

13. Expose the filters to X-ray film for 24 hr at $-70°C$ with an intensifying screen.

14. Pick clones that are positive on both the first and the second set of filters. Stray signals generally show up on either set but not simultaneously on both sets. Suspend the phage plaques into 1 ml of SM buffer and add one or two drops of chloroform.

Positive clones should be distinguished from false positives. A small, bright dot (Fig. 2, top left) is probably a false positive whereas a wide, diffuse spot, more clear in the middle (Fig. 2, middle and bottom left), is more likely to be a true positive clone.

15. Rescreen all positive clones through a second round and further rounds until 100% homogeneity is reached.

Successive Rounds of Screening

Each plaque in 1 ml of SM buffer will contain approximately $1-5 \times 10^8$ pfu/ml.

1. Dilute each clone 10^{-3} in SM buffer.

2. Prepare cells as previously described and resuspend in 10 mM MgSO$_4$ to an OD$_{600}$ of 0.5.

3. Prepare individual tubes with 600 μl of cells in each. Add either 100, 10, or 1 μl of diluted phage.

4. Incubate for 15–20 min at 37°C. Add 9 ml of top agarose and plate immediately.

5. Process the filters as described above.

Continue purification over as many rounds as necessary to obtain a homogeneous population that gives 100% binding (Fig. 2, bottom; compare left and right).

Designing Binding Cocktail

The competitor RNA in our original protocol for TAR RNA-binding proteins (6) was total yeast RNA (10 μg/ml). We have since found that RNA-binding proteins (e.g.,

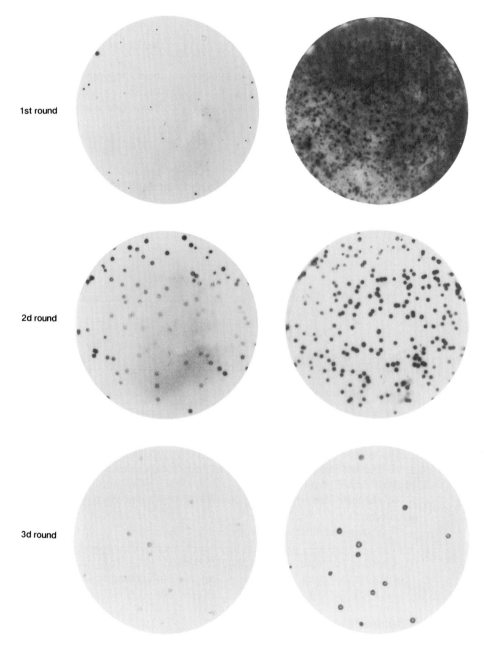

FIG. 2   An example of the isolation and purification of a λ phage plaque expressing a TAR RNA-binding protein (3). Filters (left) show three successive rounds of screening with [^{32}P]TAR RNA. Filters (right) represent hybridization of corresponding replica filters with ^{32}P-labeled λ phage DNA to demonstrate the total number of λ plaques on each filter.

TRBP) that prefer double-stranded RNA with G + C-rich characteristics can be competed away if the binding cocktail contains poly(I) · poly(C) (12). In a Northwestern assay performed under conditions that are identical to the screening procedure described here, we illustrate this point. In this experiment, TRBP–TAR RNA complex is partially sensitive to poly(I) · poly(C) competition (Fig. 3), while being much less sensitive to competition by total yeast RNA. These results suggest some RNA-binding proteins would not be identified if different competitor RNAs are used. Thus, we suggest a modification of the binding cocktail depending on the type of RNA-binding proteins to be isolated: For cloning cDNAs encoding RNA-binding proteins with double-stranded binding characteristics, total RNA or single-stranded RNA should be used as competitor. For cloning cDNAs for single-stranded RNA-binding proteins, poly(I) · poly(C) should be used in the binding cocktail.

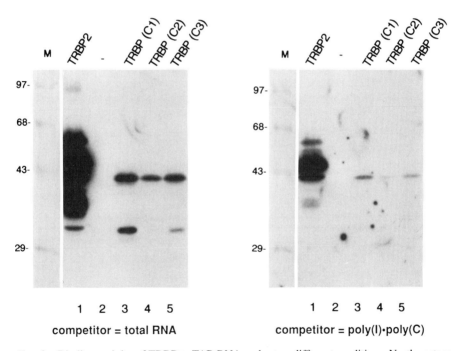

FIG. 3   Binding activity of TRBP to TAR RNA under two different conditions. Northwestern analyses were performed as described previously (9). Both filters, except for competitor RNA, were treated identically and were exposed for the same duration to X-ray film. Competitor RNA was yeast RNA (10 μg/ml) (left) or poly(I) · poly(C) (10 μg/ml) (right). TRBP C1–3 are identical, independently isolated cDNAs (3); TRBP2 is a separate cDNA that has a modified N-terminal amino acid sequence compared to TRBP.

This technique (or modified versions) has been used with success in isolating several cDNAs for RNA-binding proteins (6, 13, 14). Although not all RNA-binding proteins can be detected by this approach, a large number of RNA sequences and cognate binding proteins can be easily and rapidly assayed using this protocol.

Acknowledgments

This work was supported in part by the intramural AIDS antiviral targeted program from the office of the Director of the NIH and by the Council for Tobacco Research-USA, Inc.

References

1. I. W. Mattaj, *Cell (Cambridge, Mass.)* **57,** 1 (1989).
2. T. A. Rouault, C. D. Stout, S. Kaptain, J. B. Harford, and R. D. Klausner, *Cell (Cambridge, Mass.)* **64,** 881 (1991).
3. R. J. Bandziulius, M. S. Swanson, and G. Dreyfuss, *Genes Dev.* **3,** 431 (1989).
4. D. J. Kenan, C. C. Query, and J. D. Keene, *Trends Biochem. Sci.* **16,** 214 (1991).
5. P. D. Zamore, M. L. Zapp, and M. R. Green, *Nature (London)* **348,** 485 (1990).
6. A. Gatignol, A. Buckler-White, B. Berkhout, and K.-T. Jeang, *Science* **251,** 1597 (1991).
7. H. Singh, J. H. LeBowitz, A. S. Baldwin, and P. A. Sharp, *Cell (Cambridge, Mass.)* **52,** 415 (1988).
8. C. R. Vinson, K. L. LaMarco, P. F. Johnson, W. H. Landschulz, and S. L. McKnight, *Genes Dev.* **2,** 801 (1988).
9. H. Singh, R. G. Clerc, and J. H. LeBowitz, *BioTechniques* **7,** 252 (1989).
10. J. M. Short, J. M. Fernandez, J. A. Sorge, and W. D. Huse, *Nucleic Acids Res.* **16,** 7583 (1988).
11. J. Sambrook, E. F. Fritsch, and T. Maniatis, "Molecular Cloning: A Laboratory Manual," 2nd ed. Cold Spring Harbor Laboratory Press, Cold Spring Harbor, New York, 1989.
12. A. Gatignol, C. Buckler, and K.-T. Jeang, *Mol. Cell. Biol.* **13,** 2193 (1993).
13. P. Constantoulakis, M. Campbell, B. K. Felber, G. Nasioulas, E. Afonina, and G. N. Pavlakis, *Science* **259,** 1314 (1993).
14. D. St. Johnston, N. H. Brown, J. G. Gall, and M. Jantsch, *Proc. Natl. Acad. Sci. USA* **89,** 10979 (1992).

[3]   RNA-Binding *trans*-Activators in Cells

Ying Luo and B. Matija Peterlin

Introduction

Many transcription activators are composed of small structurally and functionally independent domains. The best known are eukaryotic transcription activators such as Gal4 and VPl6. For example, the Gal4 activator contains two distinct functional domains. Its N-terminal DNA-binding domain [amino acids (aa) 1–147] can be separated from its highly acidic C-terminal activation domain (aa 148–238 and 768–881) (1) and fused to other heterologous transcription activators such as the VP16 activation domain (aa 413–491) (2, 3). This hybrid protein can be targeted to any promoter containing Gal4 DNA-binding sites. Vice versa, the acidic activation domain of Gal4 can be fused to other heterologous DNA-binding proteins to activate transcription from promoters containing those heterologous DNA-binding sites. Moreover, it has been demonstrated that these DNA-binding transcription factors increase rates of initiation of transcription (4–6).

Tat and Rev, two essential viral proteins for the replication of the human and simian immunodeficiency viruses, represent a new class of RNA-binding activator proteins (7–10). Tat binds to an RNA element called TAR (*trans*-activation response element), which is located at the 5′ end of all viral transcripts, to activate transcription from the viral long terminal repeat (LTR) (11, 12). Because there is no evidence that unmodified Tat can also bind to DNA and because Tat can increase the processivity of RNA polymerase II (13–16), these differences are also reflected in its distinct mechanism of transcription activation.

Rev is involved in the posttranscriptional regulation of HIV by binding to another viral RNA element called the RRE (Rev response element) (17), which is located in the middle of the *env* gene. Rev is crucial for the transport of unspliced viral mRNA species out of the nucleus (7, 18). It is not clear whether Rev affects RNA splicing, RNA transport, RNA stability, or all of these. Although no cellular proteins have been found that function in ways similar to Tat and Rev, bacteriophage λ does regulate transcription elongation via RNA-binding proteins N and NusA (19, 20).

Heterologous RNA-Binding Proteins

To investigate the structure and function of these RNA-binding activators (Tat and Rev), we used heterologous targeting strategies similar to those described for DNA-binding transcription activators. However, unlike DNA-binding transcription activators, few well-studied heterologous RNA-binding proteins were available.

Prokaryotic RNA-Binding Regulatory Proteins

One of the best characterized prokaryotic regulatory proteins is the coat protein of bacteriophage R17 or MS2 (129 amino acids) (21). The coat protein binds to its operator (23 nucleotides) as a dimer and inhibits the translation of the bacteriophage R17 (MS2) replicase gene. The operator RNA forms a small stem–loop with a free energy (ΔG) of -21.8 kcal/mol^{-1} (Fig. 2) (22). The dissociation constant (K^d) of the binding of the coat protein to the operator is 30 nM (21).

Bacteriophage λ phage N protein is another prokaryotic transcription activator that is well characterized. N binds to a specific RNA stem–loop (NutB) on λ transcripts and interacts with the core RNA polymerase (19, 20). Interactions between N, NusA, and RNA polymerase increase the processivity of bacteriophage λ transcription through downstream termination sites. In addition to N and NusA protein, NusB, NusG, and S10 are also involved in this complex transcription regulation (19, 20).

Eukaryotic RNA-Binding Regulatory Proteins

HIV-1 trans-Activator Tat

Tat is a protein of 86–101 amino acids, depending on the viral isolate. It can be divided into five structural domains (Fig. 1), which include N-terminal proline-rich (aa 1–21), cysteine-rich (aa 22–37), conserved core (aa 38–47), basic arginine-rich (aa 48–57), and glutamine-rich C-terminal (aa 58–101) domains (23). Extensive *in vitro* studies suggested that the basic region of Tat, which consists primarily of arginines and lysines, is not only required, but is also sufficient for its binding to TAR RNA (Fig. 2) (24–26). However, the precise sequence of amino acids in the basic region of Tat is not essential. It is, rather, the number of arginines and lysines that determines whether or not Tat can bind to TAR RNA *in vitro*. Interestingly, basic regions of Rev and bacteriophage N protein can also functionally replace the basic domain of Tat (27).

Tat response element RNA forms a 59-nucleotide long stem–loop with a predicted free energy (ΔG) of -37.6 kcal/mol^{-1} (28) (Fig. 2). Previous *in vitro* studies determined that the 5' bulge in TAR RNA is required for the basic region of Tat to bind to TAR. Actually, a peptide containing only the basic region of Tat can bind to TAR as strongly as the full-length protein with a K_d in the nanomolar range (12 nM) (25, 29, 30). Although the central loop in TAR is absolutely required for Tat effects *in vivo,* it is not required for Tat to bind to TAR *in vitro.* Although structural studies demonstrated that Tat binds to TAR as a monomer, Tat can also form homodimers in yeast (31). The physiological significance of such dimers remains unclear.

HIV-1 Posttranscriptional Regular Rev

Rev can also be divided into several structural domains (Fig. 1), which include arginine-rich (aa 35–50), effector (aa 75–83), and multimerization domains (aa 18–

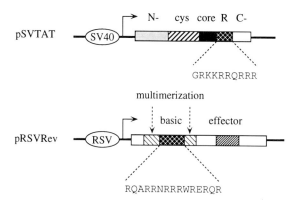

FIG. 1   Structure of Tat and Rev of human immunodeficiency virus type 1 (HIV-1). pSVTAT contains Tat from HIV-1$_{SF2}$, which was under the control of the simian virus 40 (SV40) promoter. Domains are labeled as follows: the N terminus (N$^-$) is proline rich (aa 1–21), cys represents the cysteine-rich region (aa 22–37), core is most conserved (aa 38–47), R is arginine rich (aa 48–57), and the C terminus (C$^-$) is glutamine rich (aa 58–101). N$^-$, cys, and core regions comprise the *trans*-activation domain of Tat. The basic region binds to TAR RNA *in vitro* and the C terminus is dispensable. pRSVRev contains the *rev* gene from HIV-1$_{SF2}$, which was under the control of the Rous sarcoma virus (RSV) promoter. The basic region (aa 35–50) is arginine rich and is involved in the binding of Rev to the RRE. It is flanked by multimerization domains (aa 18–34 and 51–56). The effector domain is also essential for Rev function (aa 75–83). Although the region separating effector and multimerization domains (aa 57–74) is also required, its precise sequence of amino acids is not important.

34 and 51–56) (7, 32–35). The arginine-rich region is involved in the binding of Rev to the RRE and represents the nuclear localization signal of Rev (35, 36). This region can form an α-helical structure with the help of flanking amino acids (37). In contrast to interactions between Tat and TAR, the precise amino acid sequence of this region is required and sufficient for Rev to bind to the RRE *in vivo* and *in vitro* (36).

The RRE RNA can also form a complex secondary structure containing several stem–loops larger than TAR (234 nt) (Fig. 2) (17). This RNA structure is very stable ($\Delta G = -115.1$ kcal/mol^{-1}). *In vivo* and *in vitro* experiments demonstrated that stem–loop II (SLII) is where Rev binds to the RRE most tightly. Full-length Rev or its basic domain binds to the RRE with a dissociation constant (K_d) of 10 nM (36). Although mutational analyses revealed that multimerization is not necessary for the binding of Rev to the RRE, it is essential for Rev function (38).

Other Lentiviral RNA-Binding Activators

Other lentiviruses such as HIV-2, simian immunodeficiency virus (SIV), and equine infectious anemia virus (EIAV) also have their own Tats, TARs, Revs, and RREs. These viral proteins can also be used as heterologous RNA-targeting proteins. Like

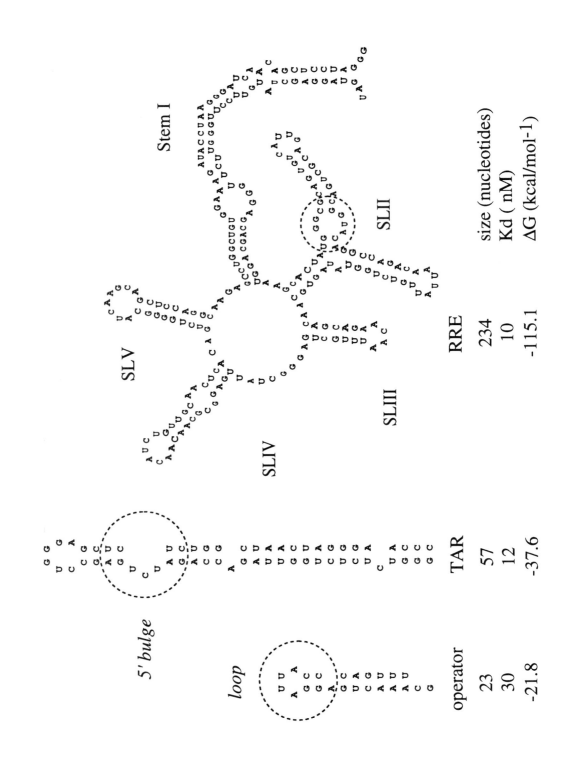

Stem I

SLV

SLIV

SLIII

SLII

RRE

	size (nucleotides)	Kd (nM)	ΔG (kcal/mol^{-1})
RRE	234	10	-115.1

5' bulge

loop

	TAR	operator
	57	23
	12	30
	-37.6	-21.8

Tat of HIV-1, Tat of EIAV/SIV can be subdivided into N-terminal, core, basic, and C-terminal domains (23, 39). However, most of its identity with Tat of HIV-1 lies within the core region. Correspondingly, the TAR of EIAV forms an RNA stem–loop, although its sequence is very different from that of the TAR of HIV-1 (Fig. 2) (40). However, hybrid proteins between Tats of HIV-1 and EIAV were used to define a minimal lentiviral Tat (23, 41).

Studies of Activation Domain of Tat Using Hybrid Tat–Coat Proteins

To separate the RNA-binding from its activation properties, we fused Tat to other heterologous RNA-binding proteins. We chose the coat protein as the fusion partner for Tat.

Hybrid Protein Construction

In contrast to the N-terminal half, the C-terminal sequences of Tat are not required for efficient *trans*-activation. Therefore, the coat protein (aa 2–129) was fused to the C terminus of Tat to form pSVTATCP (Fig. 3) (11). A proline bridge was inserted between these proteins to avoid any possible structural interference between the coat protein and Tat. Mutations were also introduced into different regions of Tat in pSVTATCP to define the activation domain of Tat. Because the binding of the coat protein to its operator replaced the binding of Tat to TAR, these mutations in Tat were designed to define the activation domain of Tat independent of its RNA-binding sequences.

Reporter Plasmid Construction

pHIVSCAT contains the *CAT* reporter gene under the control of the HIV-1 LTR (11, 12). CAT activities were used to measure *trans*-activation by the wildtype Tat. In pHIVSRCAT, the operator replaced TAR. There are several advantages to the use of the operator. First, it has no sequence homology with TAR. Second, coat protein is a

---

Fig. 2   RNA structures of the R17 (MS2) operator, TAR, and RRE. Stem–loop (SL) structures are shown with their sizes, stabilities, and binding affinities. Dashed circles represent the most important region involved in the binding of coat protein, Tat, and Rev, respectively. Transcription from the HIV-1 LTR starts at the 5′-most G of TAR. However, the lower stem is not required for effects of Tat. The RRE is divided into five stem–loop structures, although the SLII binds Rev most avidly.

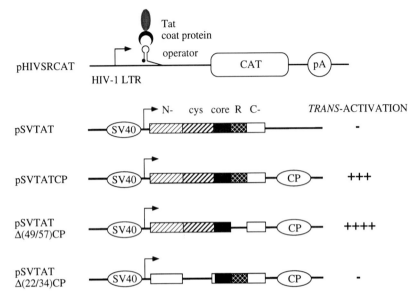

FIG. 3   Mapping the activation domain of Tat. pHIVSRCAT was the reporter plasmid, which contains the *CAT* gene under the control of the HIV-1 LTR. The TAR was replaced by the operator. Coat protein was linked to the C terminus of Tat. In HeLa cells, which were co-transfected with pHIVSRCAT and pSVTATCP, activation by pSVTATCP is represented by + and −. Whereas +++ represents more than 20-fold activation, − represents less than 2-fold activation. Mutations in different regions of Tat are presented.

prokaryotic RNA-binding protein with size and binding characteristics similar to Tat and Rev. Third, no eukaryotic cellular factor should bind to the operator. Fourth, the operator is not only very small [23 nucleotides (nt)], but interactions between the coat protein and the operator are specific and have been characterized extensively (21).

Unlike binding sites for many eukaryotic transcription activators, the location of TAR is extremely important for *trans*-activation (12). The TAR is naturally located at the extreme 5′ end of all viral transcripts. Increasing the distance between TAR and the cap site exponentially diminishes levels of *trans*-activation. Therefore, when replacing TAR, the operator must be placed next to the HIV-1 cap site (see pHIVSRCAT in Fig. 3).

The RRE has also been used to replace TAR. In this case, Tat was fused to Rev so that it could be targeted to the HIV-1 LTR (42). Although the RRE is much bigger than TAR, the second stem–loop of the RRE (SLII) is sufficient for the binding of Rev *in vivo*. Indeed, a hybrid Rev–VP16 fusion protein can activate the HIV-1 LTR via SLII in the position of TAR (38). However, a possible disadvantage to the use of the RRE to replace TAR is that other cellular proteins bind to the RRE and might affect the binding of Rev (43).

Cell Lines

HeLa cells are used in our studies. However, COS, CV-1, and D17 cells can also be used, except that higher basal promoter activities and lower levels of *trans*-activation by Tat are observed in these cells. Although others have also observed effects of Tat in L929 cells, we do not detect any activity of Tat in mouse Ltk⁻ and NIH 3T3 cells.

Transfections and CAT Assays

1. HeLa cells are grown overnight to one-third confluence on 10-cm tissue culture plates in Dulbecco's modified Eagle's medium (DMEM) with 10% (v/v) fetal calf serum.

2. The DEAE-dextran transfection method is used to cotransfect 5 μg of pSVTATCP and 5 μg of pHIVSRCAT (200-μg/ml final concentration of DEAE-dextran). We observe lower levels of *trans*-activation by the $CaPO_4$ precipitation method. Before transfection, the old medium is removed. The cells are washed twice with phosphate-buffered saline (PBS). Four milliliters of DMEM without serum is then added to the cells (44).

3. DNA/DEAE-dextran mix (120 μl) is added to the medium for a 4-hr incubation period.

4. Medium containing DNA is removed. Cells are washed twice with prewarmed PBS. Then 10 ml of fresh DMEM containing 10% fetal calf serum is added.

5. Cells are harvested 40–48 hr after transfection and lysed in Triton X-100 lysis buffer for CAT assays as previously described. Protein concentrations are measured by using a Bio-Rad (Richmond, CA) kit to normalize differences in CAT results. Amounts of pSVTATCP and pHIVSRCAT should not be decreased under our transfection conditions. Otherwise, much lower CAT activities will be observed.

A generalization of our results is presented in Fig. 3. They demonstrate that N-terminal, cysteine-rich, and core domains are sufficient for Tat *trans*-activation when fused to the coat protein. Moreover, the basic arginine-rich region could be deleted without negative effects. We therefore mapped the activation domain of Tat to its N-terminal 48 amino acids. It should be noted that the deletion of the basic domain in pSVTATΔ(49/57)CP actually increased levels of *trans*-activation. This may be due to interactions between activation and arginine-rich domains that block the full access of cellular proteins to the activation domain of Tat (11).

Analysis of RNA-Binding Domain of Tat

Although *in vitro* studies suggested that the arginine-rich domain is the RNA-binding domain of Tat (25, 26, 29, 45), several questions remained unresolved. First, if the arginine-rich sequence were sufficient for Tat to bind to TAR, than any cellular protein with an arginine stretch should be able to compete with Tat for TAR and inhibit *trans*-activation. Second, although it was demonstrated that the 5′ bulge in the stem–loop of TAR was sufficient for Tat to bind to TAR *in vitro,* the central loop in TAR was also required for efficient *trans*-activation *in vivo* (40, 46–48). Therefore, *in vitro* studies did not reveal all interactions that occur between Tat and TAR *in vivo*.

To solve the discrepancy between *in vitro* and *in vivo* results and to define the RNA-binding domain of Tat in cells, we fused Tat to other heterologous RNA-binding proteins such as Rev and the coat protein.

Indirect Analyses of TAR RNA-Binding Domain of HIV-1 Tat, Using Hybrid Tat–Coat Proteins and TAR Decoys

Plasmid Constructions

The RNA-binding domain of Tat should function independently when separated from other domains of Tat. To test whether the basic amino acids indeed represent an independent RNA-binding domain in cells, we inserted irrelevant or duplicated amino acids from Tat between core and basic amino acids in Tat (Fig. 4A). Introduced amino acids were flanked by a proline and a glycine to prevent the formation of new secondary structures between introduced amino acids and the wild-type Tat.

Transfection and CAT Assays

1. Plasmids (5 μg), which contained mutated Tats, are cotransfected together with 5 μg of pHIVCAT into one-third confluent HeLa cells on 10-cm plates.
2. The DEAE-dextran transfection procedure previously described is used.
3. Cells are harvested 40–48 hr after transfection and CAT activities are analyzed as before.

Our results demonstrated that insertions between core and basic regions completely abolished *trans*-activation by Tat (Fig. 4A).

Because this loss of *trans*-activation could be caused by the disruption of the activation or binding domains of Tat, hybrid Tat–CP and Rev–Tat proteins were con-

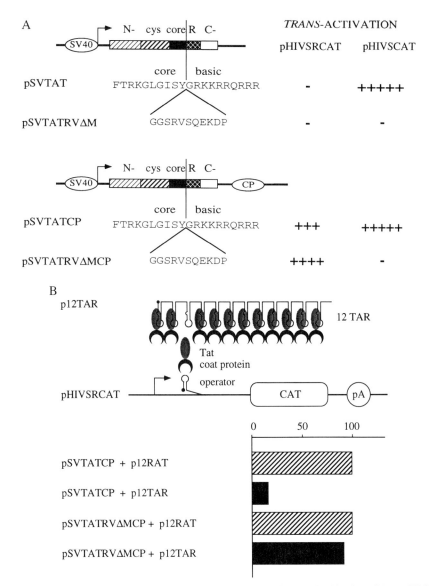

FIG. 4 Separation of activation and basic domains of Tat disrupts the binding of Tat to TAR. (A) Insertions between activation and basic domains did not interrupt the function of the activation domain. Inserted amino acids and the position of the insertion are presented. Effects of mutant Tats on pHIVSRCAT and pHIVSCAT are given by + or −. (B) Competition binding assays with TAR decoys. TAR decoys (p12TAR) were cotransfected into HeLa cells to compete with pHIVSRCAT for the binding of the hybrid Tat–coat protein. Striped bars represent levels of *trans*-activation by pSVTATCP in the absence of TAR decoys. Black bars represent levels of *trans*-activation by pSVTATCP in the presence of TAR decoys. Antisense TAR (p12RAT) was used as the control.

structed to distinguish between these two possibilities. First, we fused the coat protein to the C terminus of Tat (pSVTATRVΔMCP) and replaced TAR with the operator (pHIVSRCAT). Cotransfections demonstrated that pSVTATRVΔMCP could still activate the pHIVSRCAT but not pHIVSCAT (Fig. 4A). Thus, the activation domain was not affected by our insertions in Tat. We then used TAR decoys to demonstrate that the binding of Tat to TAR was disrupted by these insertions. This strategy is shown in Fig. 4B. For each 10-cm plate, 5 μg of pSVTATCP or pSVTATRVΔMCP was cotransfected with 5 μg of pHIVSRCAT into HeLa cells in the presence or absence of TAR decoys [10 μg of p12TAR (49) per plate]. If hybrid Tat–coat protein bound to the operator in pHIVSRCAT, we should have seen an increase in the CAT activity. However, if TAR decoys titrated the hybrid Tat–coat protein away, then we should have observed a much lower CAT activity. Our results demonstrated that although the hybrid Tat–coat protein was titrated away from pHIVSRCAT, pSVTATRVΔMCP was not affected. This provided direct evidence that insertions between core and basic domains disrupted interactions between Tat and TAR. Thus, we conclude that the basic amino acids do not form an independent RNA-binding domain.

When we used TAR decoys to compete directly with the operator in pHIVSRCAT to analyze the binding of Tat to TAR in cells, this strategy ruled out other possibilities for effects of our insertions. However, high levels of expression of TAR decoys are necessary to ensure that the competition can be observed. In our experiments, we used an actin promoter linked to the 12-TAR template (p12TAR). Antisense 12-TAR (p12RAT) was also used as control. There are other reports that a PolIII promoter driving a single TAR template can also act as a TAR decoy.

Direct Binding Assay Using Hybrid Tat–Rev Proteins

Hybrid Protein and Reporter Plasmid Constructions

Tat was also fused to the N terminus of Rev to minimize the effects on the structure of Tat (50, 51). This is because the C-terminus of Tat is dispensable for Tat function. In pRev/MS2, the bacteriophage coat protein was fused to the C terminus of Rev. It should be noted that C-terminal (aa 84–121) and N-terminal (aa 1–17) amino acids of Rev can be mutated without affecting Rev function (35, 52). In this experiment, Tat was used as the heterologous RNA targeting protein for Rev function. Mutations were introduced into Tat, and the activity of Rev was measured to analyze effects of these mutations on the binding of Tat to TAR. A plasmid, in which the *CAT* reporter gene was inserted into the *env* gene of HIV-1, served as the target (Fig. 5) (53–55). If Rev binds to the RRE located in this intron, unspliced mRNA is transported from the nucleus and leads to elevated levels of CAT expressed in the cytoplasm. In our Tat-binding studies, the RRE was replaced by multiple TARs (up to 12).

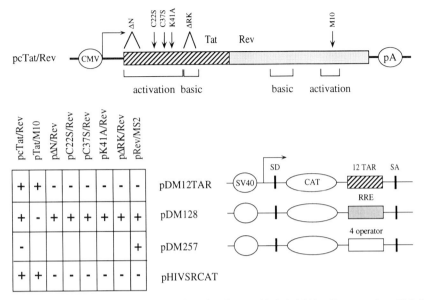

FIG. 5   Mapping of the RNA-binding domain of Tat with hybrid Tat–Rev proteins. Hybrid Tat–Rev proteins were under the control of the cytomegalovirus (CMV) promoter. Indicated are positions of introduced mutations. Target plasmid constructions are also presented. SD, Splice donor site; SA, splice acceptor site. The *CAT* reporter gene was located in the *env* intron. In pDM12TAR and pDM257, the RRE was replaced by either 12 TARs or four operators. Cotransfection results are presented as + or −.

Cells, Transfections, and CAT Assays

Although NIH 3T3, L929, QL3, COS, and HeLa cells can also be used (32), CV-1 cells were transfected in all of these experiments. Again, we observed lower Rev effects in COS cells, probably due to the ability of the reporter plasmid to replicate in these cells. Because the DEAE-dextran method gives lower CAT levels, $CaPO_4$ transfection protocols were used in these studies. The procedure is outlined below (44).

1. Cells are grown to one-third confluent on a 10-cm tissue culture plate in DMEM containing 10% fetal calf serum.
2. Medium is changed 2–4 hr before transfection to stimulate cell growth.
3. DNA mix is dissolved in 450 μl of water. Then 50 μl of 2.5 *M* $CaCl_2$ is added to the DNA solution to a final concentration of 0.25 *M*.
4. This DNA–$CaCl_2$ mix is added dropwise into a tube containing 0.5 ml of 2 × BBS, with vortexing at the same time.

5. The DNA–CaPO$_4$ mix is incubated at room temperature for 20 min.

6. The DNA–CaPO$_4$ mix is added dropwise to the cells. Plates were then gently swirled to disperse the precipitated DNA evenly.

7. Cells are incubated in a 37°C CO$_2$ incubator for 12–18 hr.

8. The medium is changed and cells are incubated for another 30 hr.

9. Cells are harvested for CAT assay as previously described (11, 12, 56).

All plasmids containing mutant hybrid Tat–Rev proteins were cotransfected with either pHIVSCAT for Tat assays, or pDM128 for Rev assays. These controls were done to ensure that our mutants were functionally expressed in CV-1 cells. In Tat assays, 5 μg of pHIVSCAT and 5 μg of various Tat/Rev plasmids were used in each transfection. In Rev assays, 20 μg of pTat/Rev and other Tat and Rev mutants over control plasmids and 1 μg of target plasmids (pDM257 or pDM12TAR) were used for each 10-cm tissue culture plate. For pDM128, only 1 μg of Tat/Rev plasmids was used in cotransfection experiments. The reason for using less pTat/Rev on pDM128 reporter is that Rev activity on the RRE reaches its highest levels at very low concentrations of the target plasmid (55). Decreasing the amount of pTat/Rev used in our transfections decreased the activity of Rev on pDM12TAR, which is probably due to the weaker interactions between Tat and TAR than between Rev and RRE. Additionally, when Rev was fused to the coat protein (pDM191), much more pRev/CP (up to 10 μg/10-cm plate) was required to obtain Rev function via the operator.

CAT levels were increased when pTat/Rev was cotransfected with pDM12TAR, which means that Tat binds to multiple TARs and brings Rev to the intron of pDM12TAR to rescue unspliced transcripts that contain CAT. However, mutations in the activation domain disrupted the binding of Tat to TAR (pΔN/Rev, pC22S/Rev, pC37S/Rev, pK41A/Rev, and pΔRKII/Rev) (Fig. 5). pDM257 was used to demonstrate that once TARs were replaced by operators, no activation of CAT by pTat/Rev was observed. This was used as a specificity control for the binding of Tat to TAR in our studies. These results demonstrate that the binding of Tat to TAR in cells requires more than the basic region of Tat.

Discussion

It should be noted that the number of TARs used to replace RRE in our assay is important. Single or double TARs are not sufficient to replace the RRE. Similar data were obtained with experiments in which the RRE was replaced by four operators and in which Rev was fused to the coat protein (53). Although the coat protein binds to the operator as a dimer, a single operator was not sufficient for the hybrid Rev/CP to bind to the intron. Up to four MS2 operators were also needed to observe full Rev effects (53). Because Rev must multimerize on the RRE for function, more heterologous RNA-binding sites are needed to bring enough hybrid Rev to the intron. An alternative explanation is that the operator and TAR are not able to form stable secondary structures in the intron as easily as the RRE, due to their different free energies.

Concluding Remarks

We used hybrid proteins to define binding and activation domains of Tat *in vivo*. Unlike similar methods used in the analysis of domains of DNA-bound transcription factors, there is only a limited number of RNA-binding proteins that are suitable for these studies. This limits the choice of heterologous proteins that can be used to make hybrid proteins. Because the mechanisms of Tat and Rev function are not yet clear, this increases the difficulty of constructing hybrid proteins and interpreting results.

It should also be noted that there are some special characteristics associated with these RNA-binding proteins. For example, it is proposed that RNA polymerase II may pause shortly after initiation of transcription of HIV-1.

The short transcripts produced are sufficient to form the TAR stem–loop, which is only 59 nt long (13). Then Tat binds to TAR and activates the paused RNA polymerase II to produce full-length transcripts. Therefore, the heterologous RNA-binding site used to replace TAR must be short enough so that it can form the right secondary structure for heterologous RNA-binding protein to bind.

Multimerization is another characteristic that should be kept in mind. Reports show that up to more than nine Revs can bind to a single RRE *in vitro* (33). Tat is also able to multimerize in the eukaryotic nucleus (31). However, the importance of multimerization is poorly understood. Therefore, we should be cautious in deciding how many heterologous RNA-binding sites should be introduced to replace the function of Rev.

Acknowledgments

The authors thank Dr. Xiaobin Lu, Dr. Subir Ghosh, and Dr. Alicia Alonso for helpful discussions. We also appreciate the pTat/Rev plasmids kindly given to us by Dr. Bryan Cullen and the pDM128 plasmid from Dr. Tristram Parslow. The authors also thank Michael Armanini for expert secretarial assistance. Dr. Ying Luo is a recipient of a Universitywide AIDS Research Postdoctoral Fellowship Award.

References

1. M. Ptashne, *Nature (London)* **335,** 683 (1988).
2. H. J. Himmelfarb, J. Pearlberg, D. H. Last, and M. Ptashne, *Cell (Cambridge, Mass.)* **63,** 1299 (1990).

3. M. Carey, J. Leatherwood, and M. Ptashne, *Science* **247,** 710 (1990).

4. R. G. Roeder, *Trends Biochem. Sci.* **16,** 402 (1991).

5. Y. S. Lin and M. R. Green, *Cell (Cambridge, Mass.)* **64,** 971 (1991).

6. M. Ptashne and A. A. F. Gann, *Nature (London)* **346,** 329 (1990).

7. B. R. Cullen and M. H. Malim, *Trends Biochem. Sci.* **16,** 33 (1991).

8. B. R. Cullen, *Cell (Cambridge, Mass.)* **63,** 655 (1990).

9. C. A. Rosen and G. N. Pavlakis, *AIDS* **4,** 499 (1990).

10. B. M. Peterlin, M. Adams, A. Alonso, A. Baur, S. Ghosh, X. Lu, and Y. Luo, *in "Human Retroviruses"* (B. R. Cullen, ed.), p. 75. Oxford University Press, Oxford, 1993.

11. M. J. Selby and B. M. Peterlin, *Cell (Cambridge, Mass.)* **62,** 769 (1990).

12. M. J. Selby, E. S. Bain, P. A. Luciw, and B. M. Peterlin, *Genes Dev.* **3,** 547 (1989).

13. S. Y. Kao, A. F. Calman, P. A. Luciw, and B. M. Peterlin, *Nature (London)* **330,** 489 (1987).

14. R. A. Marciniak and P. A. Sharp, *EMBO J.* **10,** 4189 (1991).

15. X. Lu, T. M. Welsh, and B. M. Peterlin, *J. Virol.* **67,** 1752 (1993).

16. H. Kato, H. Sumimoto, P. Pognonec, C. H. Chen, C. A. Rosen, and R. G. Roeder, *Genes Dev.* **6,** 655 (1992).

17. M. H. Malim, J. Hauber, S. Y. Le, J. V. Maizel, and B. R. Cullen, *Nature (London)* **338,** 254 (1989).

18. B. R. Cullen, *FASEB J.* **5,** 2361 (1991).

19. D. Lazinski, E. Grzadzielska, and A. Das, *Cell (Cambridge, Mass.)* **59,** 207 (1989).

20. S. Mason and J. Greenblatt, *Genes Dev.* **5,** 1504 (1991).

21. P. J. Romaniuk, P. Lowary, H. N. Wu, G. Stormo, and O. C. Uhlenbeck, *Biochemistry* **26,** 1563 (1987).

22. I. Tinoco, P. N. Borer, B. Dengler, M. Levine, O. C. Uhlenbeck, D. M. Crothers, and J. Gralla, *Nature New Biol.* **246,** 40 (1973).

23. D. Derse, M. Carvalho, R. Carroll, and B. M. Peterlin, *J. Virol.* **65,** 7012 (1991).

24. M. G. Cordingley, R. L. LaFemina, P. L. Callahan, J. H. Condra, V. V. Sardana, D. J. Graham, T. M. Nguyen, K. LeGrow, L. Gotlib, A. J. Schlabach, and R. J. Colonno, *Proc. Natl. Acad. Sci. USA* **87,** 8985 (1990).

25. B. J. Calnan, B. Tidor, S. Biancalana, D. Hudson, and A. D. Frankel, *Science* **252,** 1167 (1991).

26. C. Dingwall, I. Ernberg, M. J. Gait, S. M. Green, S. Heaphy, J. Karn, A. D. Lowe, M. Singh, and M. A. Skinner, *EMBO J.* **9,** 4145 (1990).

27. T. Subramanian, R. Govindarajan, and G. Chinnadurai, *EMBO J.* **10,** 2311 (1991).

28. M. A. Muesing, D. H. Smith, and D. J. Capon, *Cell (Cambridge, Mass.)* **48,** 691 (1987).

29. B. J. Calnan, S. Biancalana, D. Hudson, and A. D. Frankel, *Genes Dev.* **5,** 201 (1991).

30. J. Tao and A. D. Frankel, *Proc. Natl. Acad. Sci. USA* **89,** 2723 (1992).

31. H. P. Bogerd, R. A. Fridell, W. S. Blair, and B. R. Cullen, *J. Virol.* **67,** 5030 (1993).

32. L. S. Tiley, M. H. Malim, and B. R. Cullen, *J. Virol.* **65,** 3877 (1991).

33. M. H. Malim and B. R. Cullen, *Cell (Cambridge, Mass.)* **65,** 241 (1991).

34. S. Kubota, T. Nosaka, B. R. Cullen, M. Maki, and M. Hatanaka, *J. Virol.* **65,** 2452 (1991).

35. M. H. Malim, S. Bohnlein, J. Hauber, and B. R. Cullen, *Cell (Cambridge, Mass.)* **58,** 205 (1989).

36. J. Kjems, B. J. Calnan, A. D. Frankel, and P. A. Sharp, *EMBO J.* **11,** 1119 (1992).

37. R. Tan, L. Chen, J. A. Buettner, D. Hudson, and A. D. Frankel, *Cell (Cambridge, Mass.)* **73,** 1031 (1993).
38. L. S. Tiley, S. J. Madore, M. H. Malim, and B. R. Cullen, *Genes Dev.* **6,** 2077 (1992).
39. Z. Q. Liu, D. Sheridan, and C. Wood, *J. Virol.* **66,** 5137 (1992).
40. M. Carvalho and D. Derse, *J. Virol.* **65,** 3468 (1991).
41. R. Carroll, L. Martarano, and D. Derse, *J. Virol.* **65,** 3460 (1991).
42. C. Southgate, M. L. Zapp, and M. R. Green, *Nature (London)* **345,** 640 (1990).
43. P. Constantoulakis, M. Campbell, B. K. Felber, G. Nasioulas, E. Afonina, and G. N. Pavlakis, *Science* **259,** 1314 (1993).
44. F. M. Ausubel, R. Brent, R. E. Kingston, D. D. Moore, J. G. Seidman, J. A. Smith, and K. Struhl, "Current Protocols in Molecular Biology." Greene Publishing Associates and Wiley-Interscience, New York, 1989.
45. S. Roy, U. Delling, C. H. Chen, C. A. Rosen, and N. Sonenberg, *Genes Dev.* **4,** 1365 (1990).
46. B. Berkhout and K. T. Jeang, *Nucleic Acids Res.* **19,** 6169 (1991).
47. S. Feng and E. C. Holland, *Nature (London)* **334,** 165 (1988).
48. J. A. Garcia, D. Harrich, E. Soultanakis, F. Wu, R. Mitsuyasu, and R. B. Gaynor, *EMBO J.* **8,** 765 (1989).
49. G. J. Graham and J. J. Maio, *Proc. Natl. Acad. Sci. USA* **87,** 5817 (1990).
50. S. J. Madore and B. R. Cullen, *J. Virol.* **67,** 3703 (1993).
51. Y. Luo, S. J. Madore, T. G. Parslow, B. R. Cullen, and B. M. Peterlin, *J. Virol.* **67,** in press (1993).
52. M. H. Malim, D. F. McCarn, L. S. Tiley, and B. R. Cullen, *J. Virol.* **65,** 4248 (1991).
53. D. McDonald, T. J. Hope, and T. G. Parslow, *J. Virol.* **66,** 7232 (1992).
54. T. J. Hope, N. P. Klein, M. E. Elder, and T. G. Parslow, *J. Virol.* **66,** 1849 (1992).
55. T. J. Hope, B. L. Bond, D. McDonald, N. P. Klein, and T. G. Parslow, *J. Virol.* **65,** 6001 (1991).
56. Y. Luo and B. M. Peterlin, *J. Virol.* **67,** 3441 (1993).

[4] Characterization of Interaction between Human Immunodeficiency Virus Type 1 *trans*-Activator Tat and TAR RNA by RNA Mobility Shift and Transient Transfection Assays

Ulrike Delling and Nahum Sonenberg

Introduction

The initial phase of human immunodeficiency virus type 1 (HIV-1) replication is regulated by the combined action of two essential viral proteins: Tat and Rev. The action of the *trans*-activating protein Tat on the viral promoter stimulates viral gene expression by more than 100-fold. In the presence of Tat only small, multiply spliced mRNAs accumulate in the cytoplasm of HIV-infected cells. These mRNAs code for Tat itself and for the other essential viral *trans*-activator, Rev. When Rev reaches a critical level, it inhibits further synthesis of the regulatory mRNAs and promotes the expression of mRNAs encoding the structural proteins Gag, Pol, and Env. The synthesis and processing of these proteins is the first step in the replication cycle leading to virion assembly.

Since the discovery in 1985 that the HIV-1 genome encodes Tat, a great amount of information has been accumulated about its function and mechanism of action. Tat is a 14-kDa nuclear protein that activates viral gene expression through interaction with an RNA stem–loop structure termed TAR (*trans*-activation response region), which is positioned immediately downstream of the viral transcriptional start site. Extensive mutational analysis has identified critical determinants for the interaction between Tat and TAR. A highly basic stretch comprising 2 lysines and 6 arginines within 12 amino acids close to the carboxy terminus of Tat mediates its RNA-binding activity. Tat has been proposed to recognize a precise structural conformation adopted by the TAR stem–loop. A bulge of two to three nucleotides in the TAR stem structure is the contact site for arginine residues of the basic domain of Tat (Fig. 1A).

Increasing evidence suggests that the interaction of Tat with TAR RNA localizes Tat in close proximity to the viral promoter region. This would allow an interaction between Tat and some not yet identified factor of the transcriptional machinery. Such an interaction has been proposed to increase the processivity of the RNA polymerase II and lead to a more efficient transcriptional elongation. This model is supported by results obtained in several laboratories; however, the exact mechanism of gene activation by Tat remains to be elucidated. For reviews on Tat-mediated *trans*-activation see Refs. 1 and 2.

Methods in Molecular Genetics, Volume 4

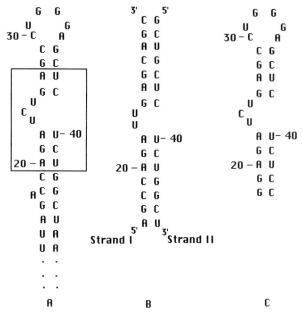

Fɪɢ. 1   Schematic representation of TAR RNA structures used in the RNA gel mobility shift assay. (A) Predicted stem–loop structure of TAR RNA. The Tat recognition site is boxed. (B) TAR duplex structure formed by hybridization of two single-stranded oligoribonucleotides. (C) Synthetic 27-mer TAR RNA sufficient for binding of Tat.

The motivation for studying the interaction between Tat and TAR RNA is twofold. The detailed analysis of Tat function is of great significance because of its unique manner of activating HIV gene expression, and also because of its essential role in the life cycle of HIV as it is implicated in latency in HIV infection. Tat is the first transcriptional activator described in a eukaryotic system that exerts its activity via an RNA responsive element. There appear to be some similarities to the antitermination mechanism of the *Escherichia coli* phage λ, in which the λ N protein interacts through an arginine-rich domain with a viral RNA stem–loop termed the λ Nut site. This interaction overcomes the transcriptional termination activity of rho (ρ) factor. However, no such termination activity has been identified in the Tat–TAR system.

The activation of HIV gene expression through the Tat–TAR interaction is highly specific and limited to two viral components. Therefore, this mechanism is an ideal target for therapeutic intervention against acquired immunodeficiency syndrome (AIDS). Indeed, several studies showed that overexpression of TAR RNA in cells renders them less susceptible to killing by HIV. In another approach, small basic peptides comprising the RNA-binding domain of Tat have been shown to inhibit viral replication in cultured cells. Clinical trials should determine the therapeutic potential of these peptides (reviewed in Refs. 3 and 4).

The analysis of the physical interaction between Tat and TAR RNA and the study of their specificity has been greatly advanced by the use of the gel mobility shift assay. This assay is an efficient tool to study interactions between proteins and polynucleotides. It is an elegant experimental approach that was first established to analyze DNA–protein interactions, but is now widely used in a slightly modified manner to investigate RNA-binding proteins and their targets (for a detailed review see Ref. 5). The fact that the gel mobility shift assay allows a qualitative as well as quantitative analysis of nucleoprotein complexes, the possibility to actually visualize the complex as a discrete band in the gel, and the rather uncomplicated experimental procedure, make it a popular method in molecular biology.

To elucidate the function of the protein or nucleoprotein complex of interest, it is important to perform *in vivo* studies. This is commonly done by examining the expression of a reporter gene after transient transfection into cultured cells. A wide range of literature describing this method and its multiple modifications is available, allowing the individual researchers to choose the most appropriate application for their purposes (6, 7). Here we present protocols used for the *in vitro* and *in vivo* analysis of the interaction between the HIV-1 *trans*-activating protein Tat and its responsive RNA element, termed TAR. These protocols have been used successfully in our laboratory for several years (8, 9). We also discuss modifications introduced by other groups to study HIV-1 gene regulation by Tat.

Protocols

TAR RNA Used in RNA Mobility Shift Assay

The RNA used in the mobility shift assay can be generated by *in vitro* transcription from plasmid or oligodeoxynucleotide templates with SP6 or T7 RNA polymerases. Plasmids must be linearized by restriction endonucleases prior to *in vitro* transcription. The transcripts are internally labeled by adding radiolabeled nucleotides to the transcription reaction. Several *in vitro* transcription systems are commercially available and contain established protocols (e.g., Promega, Madison, WI). Alternatively, chemically synthesized oligoribonucleotides can be used. These oligonucleotides are 5′ end labeled using T4 polynucleotide kinase and [γ-^{32}P]ATP. The use of synthetic RNA has the advantage that single residues can be substituted by nucleoside analogs to identify specific functional groups involved in complex formation (10–12). Moreover, the role of base or sugar moieties of individual nucleotides can be analyzed by substituting nonnucleotide linkers (13, 14). The oligoribonucleotides used in our studies were synthesized by the method of Ogilvie *et al.* (15), employing 5'-dimethoxytrityl-2'-*tert*-butyldimethylsilylribonucleoside 3'-CE-phosphoramidites (ChemGenes Corp., MA, or Peninsula Laboratories, Belmont, CA).

5' End Labeling of Synthetic Oligoribonucleotides

Chemically synthesized oligoribonucleotide	50 pmol
[γ-^{32}P]ATP (3000 Ci/mmol)	1 μl
T4 polynucleotide kinase	1–5 units
ATP	12.5 μM
Tris-HCl (pH 8.0)	100 mM
MgCl$_2$	10 mM
Dithiothreitol (DTT)	10 mM

The total volume should be 20 μl.

1. Incubate 30–45 min at 37°C.
2. Add 180 μl of H$_2$O.
3. Extract with phenol–chloroform.
4. Purify over a Sephadex G-50 spin column.
5. Determine the counts per minute (cpm) by scintillation counting.
6. Use 2000–4000 cpm in an RNA gel mobility shift assay.

The essential region of TAR for interaction with Tat has been limited to a segment within the Tar stem–loop comprising three nucleotide pairs above and below an essential three-nucleotide bulge. A synthetic 27mer oligoribonucleotide (oligo) comprising nucleotides 18 to 44 of TAR and including the essential contact region is sufficient for Tat binding (Fig. 1C). Moreover, as the loop sequence is not required for the interaction with Tat, the minimal binding region can be substituted by two single-stranded oligoribonucleotides that are hybridized to form a TAR duplex (Fig. 1B). The use of these synthetic oligonucleotides allows the analysis of a large number of nucleotide substitutions, as many combinations of single strands with different modifications can be hybridized to form the TAR duplex. Using single-strand-based TAR duplexes also has the advantage of reducing the length of oligonucleotides that must be synthesized. The oligos are heated for 5 min at 90°C in 20 mM Tris-HCl (pH 7.6), 3 mM magnesium acetate, 400 mM NaCl, 1 mM DTT and cooled slowly to room temperature to allow the formation of the double-stranded TAR stem structure (10, 11). Heat denaturation and slow cooling are used routinely for refolding the functionally important secondary structure of the RNAs.

RNA Gel Mobility Shift Assay with Pure Tat Protein

This protocol is based on a publication by Konarska and Sharp (16) and has been modified to analyze the interaction between Tat and TAR RNA (8). The Tat protein is overexpressed in *E. coli* and purified with a nickel-affinity column (18). Purified Tat protein was kindly provided to us by C. Rosen (Roche Institute, Nutley, NJ). This protocol is also used for binding studies using small synthetic peptides that comprise only the basic domain of Tat (9, 11).

Binding buffer (4×)

Tris-HCl (pH 7.5)	40 mM
DTT	4 mM
NaCl	200 mM
Ethylenediaminetetraacetic acid (EDTA)	4 mM
Bovine serum albumin (BSA)	0.36 μg/μl
RNasin	2.0 U/μl
Glycerol	20% (v/v)

Binding Reaction

Binding buffer (4×)	5 μl
Tat (20 ng/μl)	5 μl
TAR RNA (400 cpm/μl)	5 μl
H$_2$O (double-distilled, autoclaved)	5 μl

1. Prepare the binding reaction by mixing 50–100 ng of Tat protein with 0.7 ng of labeled TAR RNA (2000 cpm, 2.94 × 10⁶ cpm/μg) in a total volume of 20 μl of binding buffer. A 4×-concentrated binding buffer is made fresh before each experiment.

2. Incubate for 10 min at 30°C.

3. Load the complete binding reaction slowly on a 5% (w/v) nondenaturing polyacrylamide gel. It is not necessary to add any loading buffer as there is enough glycerol in the binding buffer to give the required density to the sample. We found that the dyes commonly used in electrophoresis loading buffer (bromophenol blue and xylene cyanol) tend to interfere with Tat–TAR complex formation.

Gel Electrophoresis

Gel recipe (for 75 ml)

Acrylamide (30%, w/v)	12.2 ml
Bisacrylamide (2%, w/v)	4.7 ml
H$_2$O	50.0 ml
TBE (10×): 0.9 M Tris–borate, 0.02 M EDTA (see Ref. 17)	3.75 ml

Glycerol	3.75 ml
Ammonium persulfate (10%, w/v)	540 μl
TEMED	26.3 μl

1. Pour a 1.5-mm thick 5% polyacrylamide gel and allow to cool to 4°C overnight.

2. Prepare 0.5× TBE as the running buffer. Keep at 4°C.

3. Prerun the gel for 30 min at 30 mA. After that time the voltage should be at 420–430 V. Rinse the wells carefully with running buffer before loading the samples.

4. Separate the RNA protein complexes for 2.5 hr at a constant current of 30 mA at 4°C. The voltage should not exceed 480–500 V; otherwise the gel warms up, which leads to diffused, smeary bands.

5. Transfer the gel onto gel blot paper (Cat. No. 31540; Schleicher & Schuell, Keene, NH) and dry for 1 hr at 80°C. Expose overnight at −70°C to X-ray film (Fuji RX; Fuji, Tokyo, Japan).

For quantitative studies it has been shown to be useful to dry the gel on DEAE paper (DE81; Whatman, Clifton, NJ) to prevent fluctuations in the amount of radioactive label remaining in the gel (14).

Modifications

Weeks and Crothers (19) found that addition of Nonidet P-40 (0.01%, v/v) to the binding reaction improved reproducibility and increased the binding constant of small, basic Tat peptides to TAR RNA. This was assumed to be due to the elimination of aggregation of peptides and therefore the increase of the concentrations of active peptide. Churcher and co-workers (20) also included a nonionic detergent [0.1% (v/v) Triton X-100] in the binding reaction, binding gel, and running buffer to minimize Tat protein and peptide aggregation.

Troubleshooting

After autoradiography of the dried gel, two bands should be visible that correspond to the free TAR RNA and to the portion complexed to Tat (Fig. 2C). If there is no signal at all, it might be necessary to check for ribonuclease activity at all possible steps throughout the experiment. All reaction buffers should be autoclaved and only ribonuclease-free reagents (e.g., BSA) should be used. However, the exposure to ribonuclease activity in these *in vitro* studies with pure Tat protein or synthetic Tat peptides is not as great as when cellular extracts are used, and the special treatment of the water with diethylpyrocarbonate (DEPC) is not necessary. If no complex formation is observed in the presence of Tat, the protein might not be fully active and/or the amount added might not be sufficient. Complexes that appear as a smeared streak on the autoradiogram can be focused by varying the concentration of BSA or

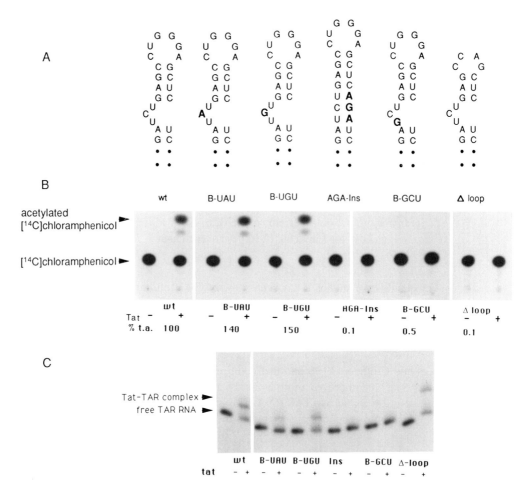

FIG. 2   Effect of mutations in the HIV-1 TAR region on Tat-mediated *trans*-activation *in vivo* and on Tat binding to TAR RNA *in vitro*. (A) Schematic representation of wild-type and mutant TAR RNAs. The mutated nucleotides are printed in bold. (B) Representative CAT assay after transfection of wild-type and mutant TAR expression vectors into HeLa or Tat-expressing HeLa cells. (C) RNA gel mobility shift assay with wild-type or mutant TAR RNAs. Mutation of the first uridine residue at position 23 of TAR abolishes Tat binding to TAR RNA and *trans*-activation *in vivo* (B-GCU; 8). Mutation of the second nucleotide in the pyrimidine bulge does not affect Tat binding or Tat-mediated *trans*-activation *in vivo* (B-UAU, B-UGU; 10). The nucleotides in the bulge must be unpaired, as insertion of three complementary nucleotides in the opposite strand abolishes Tat binding and *trans*-activation (AGA-Ins; 10). The hexanucleotide loop of TAR is essential for *trans*-activation *in vivo*, but dispensable for Tat binding to TAR RNA *in vitro* (Δ-loop, 10).

glycerol, or by increasing the acrylamide concentration. These smears can also be caused by overheating of the gel during electrophoresis. Therefore, gels should be precooled and run at 4°C.

In Vivo Expression Studies

Transient transfection assays have been used with a variety of cell lines to identify important functional determinants in Tat protein and in TAR RNA. However, the *trans*-activation response is not efficient in some mouse cell lines (e.g., NIH 3T3), probably due to an essential factor limiting in these cells. The cell lines used for the transfection experiments should be chosen on the basis of the expression vector used (or vice versa), as different promoters have different activities depending on the cell line. The *trans*-activation capacity of Tat *in vivo* is determined by measuring the activity of a reporter gene fused downstream of the viral long terminal repeat that contains the TAR element. Several reporter genes are available: bacterial chloramphenicol acetyltransferase (CAT), firefly luciferase (luc), human growth hormone (hGH), and β-glucuronidase (GUS). For a detailed review on reporter genes and transfection methods, see Ref. 6.

We transfected HeLa S3 cells (which were grown as a monolayer culture) or HeLa cells (which stably expressed the *tat* gene) with a TAR–CAT reporter plasmid (21). This plasmid (pU3RIII) contains the reporter gene *CAT* under the control of the HIV-1 long terminal repeat (LTR) (22). Alternatively, HeLa, COS, or QT6 (quail fibroblast) cells were cotransfected with pU3RIII and a Tat expression vector, pSVEXtat (23). We routinely used HeLa cells for these experiments, but the *trans*-activation assay in COS or QT6 cells was also efficient. The advantage of using QT6 cells is their high efficiency of transfection. Eukaryotic expression vectors for the expression of Tat have been constructed in several laboratories and some of them are now available commercially. A good overview on plasmids, cell lines, and peptides related to HIV is given in Ref. 24.

Transient Transfection of HeLa S3 Monolayer Cells
Solutions
    Transfection buffer
        NaCl   137 mM
        KCl   5 mM
        $Na_2HPO_4 \cdot 7H_2O$ 0.37 mM
        Tris-HCl   25 mM

    Adjust pH to 7.4. Add H_2O to 1 liter and autoclave

    TS
        Transfection buffer   100 ml
        $MgCl_2/CaCl_2$ (stock solution: 10 mg/ml of each)   1 ml

Make fresh and sterile filter

DEAE-dextran solution: Prepare a solution of 10 mg/ml in H_2O. Sterile filter and keep aliquots at $-20°C$

Chloroquine: Prepare a stock solution of 10 mM in H_2O. Sterile filter and keep aliquots at $-20°C$

Tris–EDTA–NaCl solution (TEN)

Tris-HCl (pH 7.5)	150 mM
NaCl	40 mM
EDTA	1 mM

Transfection Protocol

1. Four to 6 hr prior to transfection, trypsinize HeLa cells and seed 1×10^6 cells in 10-cm petri dishes

2. Prepare DNA mix:

TS	0.9 ml
DEAE-dextran (10 mg/ml)	0.05 ml
Plasmid	0.5–10 μg

The optimal amount of plasmid varies with cell line and expression vector used and should be determined by titration.

3. Rinse the cells once with TS.

4. Add DNA mix to the cells. Incubate for 20 min at room temperature (in the tissue culture hood).

5. Add 10 ml of complete medium plus 0.1 ml of 10 mM chloroquine (final concentration, 100 μM). Incubate for 45 min at 37°C (in tissue culture incubator).

6. Remove the chloroquine-containing medium. Add fresh medium. Incubate for 40–48 hr at 37°C.

7. Harvest the cells. Place the petri dishes on ice. Rinse the cells twice with phosphate-buffered saline (PBS). Add 1 ml of TEN solution. Incubate for 10 min on ice. Scrape the cells with a spatula or rubber policeman and transfer into a 1.5-ml Eppendorf tube. The cells are now ready for the preparation of extracts to determine the activity of the reporter gene *CAT* (25).

Preparation of Extracts for CAT Assays

1. Centrifuge the cells for 30 sec in a microfuge and remove the supernatant carefully by vacuum suction.

2. Resuspend the cell pellet in 100 μl of 0.25 M Tris-HCl (pH 7.5) by vortexing.

3. Freeze the cells in a dry ice–methanol bath for 2 min. Quick-thaw the cells in a 37°C water bath for 2 min. Vortex vigorously to lyse the cells. Repeat the freeze–thaw cycle for a total of three times.

4. Incubate for 5 min at 60°C. This step denatures cellular acetyltransferases.

5. Centrifuge for 5 min at 4°C.

6. Transfer the supernatant to a fresh Eppendorf tube.

7. Determine the protein content of the extract by Bradford assay (Cat. No. 500-600; Bio-Rad, Richmond, CA).

CAT Assay

1. Mix 5–50 μg of protein extract with 0.25 M Tris (pH 7.5) in a total volume of 143 μl.

2. Prepare an acetyl-CoA stock solution of 14 mg/ml (16 mM) (Cat. No. 27-6200-03; Pharmacia, Piscataway, NJ).

3. Make a master mix of 1 μl of [^{14}C]chloramphenicol (68 mCi/mmol, 0.1 μCi/μl) (Cat. No. 1206083; ICN, Costa Mesa, CA) and 5 μl of acetyl-CoA (14 mg/ml) per sample. Add 6 μl of this master mix to each assay.

4. Incubate for 1 hr at 37°C.

5. Briefly centrifuge the samples to collect condensate from the lids.

6. Add 1 ml of ethyl acetate to each sample. This extracts chloramphenicol and acetylated chloramphenicol into the organic phase. Vortex for 30 sec. Centrifuge for 30 sec and carefully transfer the upper phase to a fresh Eppendorf tube.

7. Dry the ethyl acetate in a Speed-Vac (Savant, Hicksville, NY) evaporator. Turning on the heat reduces drying time, and we did not observe any negative effects on the CAT assay.

8. Add 10 μl of ethyl acetate to each sample. Vortex briefly.

9. Prepare the running solution of chloroform–methanol (19 : 1). Equilibrate a chromatography tank with 190 ml of chloroform, 10 ml of methanol, and a piece of filter paper (about the size of the tank) for 30 min. Seal the lid well with grease. Do not use the tank after a period of 24 hr as methanol evaporates easily, which alters the chloroform-to-methanol ratio.

10. Spot 0.5–1 μl at a time on a thin-layer chromatography (TLC) sheet at distances of 1.5 and 2 cm from the edge. Be careful not to scratch the surface of the TLC sheet (Bakerflex silica gel1B, Cat. No. 4462-04; J. T. Baker, Phillipsburg, NJ).

11. Carefully place the TLC sheet in the equilibrated chromatography tank. Make sure not to drop the TLC in at an angle. This causes distorted signals on the autoradiograph, as the solvent mixture starts migrating right away.

12. Let the chromatograph develop until the solvent front is about 1.5 cm from the upper edge of the TLC sheet. Remove and air dry.

13. Wrap the TLC sheet in plastic film to avoid contamination of cassettes and expose overnight at room temperature to X-ray film.

14. Determine the activity of the reporter gene *CAT* by quantifying the percentage of acetylated chloramphenicol. This is done by aligning the autoradiogram with the TLC sheets and cutting out the regions corresponding to the chloramphenicol and monoacetylated chloramphenicol. The pieces are then counted in a scintillation counter. Alternatively, the autoradiogram can be quantitated using a Phospho-Imager.

trans-Activation is determined as the ratio of CAT activity obtained in cells expressing Tat and TAR–CAT to that obtained in cells expressing TAR–CAT alone.

Linearity of CAT Assay

To interpret the results correctly, the CAT assay must be in the linear range, which is between 1 and 20% conversion. The appearance of a fourth signal on the autoradiogram representing the faster migrating diacetylated form of chloramphenicol is an indication that the assay is not linear. Nonlinear CAT assays must be repeated, using less protein extract or reducing the time of incubation.

Troubleshooting

Low levels of CAT activity can be caused either by problems in the transient transfection assay or in the CAT assay itself. To assure that the CAT assay is performed correctly, pure CAT enzyme should be used as a positive control (Cat. No. 27-0847-01; Pharmacia). If the signal obtained with cell extract is weak, and remains weak even with higher amounts of extract, the promoter in the construct used to drive CAT expression might be poorly active in the transfected cell line. Low levels of CAT activity could also indicate a low transfection efficiency. The optimal transfection protocol must be established for the individual cell line used (6). We observed that the DEAE method gives good results with HeLa cells, and the calcium phosphate method is efficient for COS, QT6, and P19 (mouse embryonic carcinoma) cells. Introduction of plasmids by lipofection increased the transfection efficiency by 10-fold in COS cells, whereas the efficiency in P19 cells was comparable to the calcium phosphate method. The HIV promoter is generally weak in HeLa cells, but cotransfection with a Tat-expressing vector results in a strong stimulation of CAT expression; therefore, low transfection efficiency is not that much of a problem.

However, if no *trans*-activation at all is observed in the presence of Tat, it must be determined if the cell line used is competent for the *trans*-activation response. As cellular factors are implicated in this mechanism, the Tat–TAR system will not function in cell lines lacking these factors. It should also be verified if the Tat protein is properly expressed (stably or transiently) in the cell. This can be done by Western blotting or immunoprecipitation with an antibody specific for Tat. Errors during the construction of the expression vectors can also be a reason for lack of Tat activity. Mutations in the TAR region or in the *tat* gene will prevent *trans*-activation. If this appears to be the case, the expression vectors should be checked by sequencing for possible mutations.

Acknowledgments

This work was supported by grants (to N.S.) from the Health and Welfare Canada-National Health Research and Development Program and the Cancer Research Society (Montreal). U.D. is the recipient of a studentship from the Medical Research Council of Canada.

References

1. J. Karn and M. A. Graeble, *Trends Genet.* **8,** 365 (1992).
2. B. R. Cullen, *Microbiol. Rev.* **56,** 375 (1992).
3. K. Steffy and F. Wong-Staal, *Microbiol. Rev.* **55,** 193 (1991).
4. C. A. Rosen, *AIDS Res. Hum. Retroviruses* **8,** 175 (1992).
5. J. Carey, *in* "Protein–DNA Interactions" (R. T. Sauer, ed.), p. 103. Academic Press, Orlando, Florida, 1991.
6. F. M. Ausubel *et al.* (eds.), "Current Protocols in Molecular Biology" Chapter 9. Wiley Interscience, New York, 1989.
7. B. R. Cullen, *in* "Guide to Molecular Cloning Techniques" (S. L. Berger and A. R. Kimmel, eds.), p. 684. Academic Press, Orlando, Florida, 1987.
8. S. Roy, U. Delling, C.-H. Chen, C. A. Rosen, and N. Sonenberg, *Genes Dev.* **4,** 1365 (1990).
9. U. Delling, S. Roy, M. Sumner-Smith, R. Barnett, L. S. Reid, C. A. Rosen, and N. Sonenberg, *Proc. Natl. Acad. Sci. USA* **88,** 6234 (1991).
10. M. Sumner-Smith, S. Roy, R. Barnett, L. S. Reid, R. Kuperman, U. Delling, and N. Sonenberg, *J. Virol.* **65,** 5196 (1991).
11. R. W. Barnett, U. Delling, R. Kuperman, N. Sonenberg, and M. Sumner-Smith, *Nucleic Acids Res.* **21,** 151 (1993).
12. F. Hamy, U. Asseline, J. Grasby, S. Iwai, C. Pritchard, S. Slim, P. J. G. Butler, J. Karn, and M. J. Gait, *J. Mol. Biol.* **230,** 111 (1993).
13. U. Delling, L. S. Reid, R. W. Barnett, M. Y.-X. Ma, S. Climie, M. Sumner-Smith, and N. Sonenberg, *J. Virol.* **66,** 3018 (1992).
14. M. Y.-X. Ma, L. S. Reid, S. C. Climie, W. C. Lin, R. Kuperman, M. Sumner-Smith, and R. W. Barnett, *Biochemistry* **32,** 1751 (1993).
15. K. K. Ogilvie, N. Usman, K. Nicoghosian, and R. Cedergren, *Proc. Natl. Acad. Sci. USA* **85,** 5764 (1988).
16. M. M. Konarska and P. A. Sharp, *Cell (Cambridge, Mass.)* **46,** 845 (1986).
17. J. Sambrook, E. F. Fritsch, and T. Maniatis, "Molecular Cloning: A Laboratory Manual," 2nd ed. Cold Spring Harbor Laboratory Press, Cold Spring Harbor, New York, 1989.
18. R. Gentz, C.-H. Chen, and C. A. Rosen, *Proc. Natl. Acad. Sci. USA* **86,** 821 (1989).
19. K. W. Weeks and D. M. Crothers, *Cell (Cambridge, Mass.)* **66,** 577 (1991).
20. M. J. Churcher, C. Lamont, F. Hamy, C. Dingwall, S. M. Green, A. D. Lowe, P. J. G. Butler, M. J. Gait, and J. Karn, *J. Mol. Biol.* **230,** 90 (1993).
21. C. A. Rosen, J. G. Sodroski, K. Campbell, and W. A. Haseltine, *J. Virol.* **57,** 379 (1986).
22. J. G. Sodroski, C. A. Rosen, F. Wong-Staal, S. Z. Salahuddin, M. Popovic, S. Arya, R. C. Gallo, and W. A. Haseltine, *Science* **227,** 171 (1985).
23. S. Ruben, A. Perkins, R. Purcell, K. Joung, R. Sia, R. Burghoff, W. A. Haseltine, and C. A. Rosen, *J. Virol.* **63,** 1 (1989).
24. AIDS Research and Reference Reagent Program Catalog, National Institutes of Health, Rockville, Maryland (1993).
25. C. M. Gorman, L. F. Moffat, and B. H. Howard, *Mol. Cell. Biol.* **3,** 1044 (1982).

[5]   Transient Transfection Studies to Characterize Human Immunodeficiency Virus Tat *trans*-Activation

Michael Newstein and Peter R. Shank

Introduction

Human immunodeficiency virus (HIV), the etiological agent of acquired immunodeficiency syndrome (AIDS), differs from many retroviruses in the complexity of its genetic organization. As with other members of the lentivirus family (1) the HIV-1 genome includes sequences coding for Gag (virion core proteins), Pol (viral reverse transcriptase, ribonuclease H, protease, and endonuclease), and Env (virion membrane glycoproteins). In addition, several additional open reading frames (ORFs) specify viral proteins of incompletely characterized function, such as the *tat, rev, vif, vpr, nef, vpu,* and *tev* genes (2–4) (Fig. 1). HIV-1 therefore represents the most complex retrovirus described to date. Much of our understanding of the function of these unique regulatory and accessory genes has come through elegant molecular biological studies that have allowed the examination of individual viral gene products. This type of analysis is particularly important when genes overlap one another. We will briefly describe the Tat-mediated *trans*-activation system of HIV-1 and describe how molecular biological techniques can be used to begin to dissect the mechanism of action of this unique regulatory protein.

HIV Tat Protein

The *tat* gene encodes a powerful *trans*-activator of viral gene expression that plays an essential role in the positive regulation of HIV gene expression (5). The 14-kDa Tat protein is translated from a doubly spliced mRNA and is composed of 86–101 amino acids in HIV-1 and 130 amino acids in HIV-2. Surprisingly, deletion of the second coding exons of both HIV-1 and HIV-2 appears to result only in a modest decrease in *trans*-activation (6, 7). In the case of HIV-2 results suggest that this may be due to inefficient splicing of the first and second coding exons (8).

HIV-1 Tat protein is localized in the nucleus of cells and is concentrated predominantly in the nucleolus (9, 10). The biological significance of this nucleolar localization of Tat remains controversial. One possibility is that RNA-binding factors that function in ribosomal RNA synthesis and transport may also participate in binding Tat to the *trans*-activation response element termed TAR. Alternatively, Tat localization to the nucleolus may just be an artifact of a nonspecific Tat–rRNA interaction.

Methods in Molecular Genetics, Volume 4

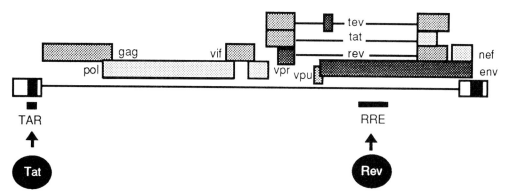

FIG. 1   Genetic organization of HIV-1. Tat and Rev are virus-encoded *trans*-acting factors that act on the TAR (Tat-responsive element) and RRE (Rev-responsive element), respectively.

Several functional domains have been characterized in HIV-1 Tat (Fig. 2). A highly conserved domain containing seven cysteine residues is found in both HIV-1 and HIV-2. Mutation of all but one of these cysteines results in markedly decreased Tat *trans*-activation (11). The Tat-1 and Tat-2 proteins have been reported to form a zinc finger-like motif that binds heavy metals. Metal-linked dimer formation of HIV-1 Tat expressed in *Escherichia coli* has been reported, with the cysteine-rich motifs of each monomer interacting with a bridging metal ion (9, 12). However, it is not clear whether metal binding or the predicted Tat dimer formation is relevant to *in vivo trans*-activation.

Several lines of evidence suggest that a basic amino acid-rich region near the carboxyl terminus of the first coding exon represents a nuclear targeting domain in HIV-1 and HIV-2 Tat. Mutations in this basic region result in cytoplasmic localization of Tat and elimination of *trans*-activation (9, 10). Moreover, when amino acids 48 to 52 (GRKKR) of HIV-1 Tat are linked to the amino terminus of the normally cytoplasmic β-galactosidase enzyme, the β-galactosidase accumulates in the nucleus (9). No Tat sequence conferring nucleolar localization has been identified to date. The basic region of Tat also functions as the binding motif for the *trans*-activation response element (TAR). Small fragments of Tat containing the basic sequence bind specifically to TAR RNA (13). Tat may be considered a member of the arginine-rich class of sequence-specific RNA-binding proteins, which also includes the HIV Rev protein.

Structure of HIV-1 and HIV-2 TAR

The *cis*-acting response element for HIV-1 and HIV-2 Tat is located in the R region of the long terminal repeat (LTR). This region has been termed the *trans*-activation

A

B

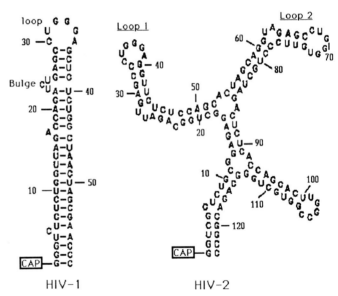

FIG. 2   (A) Comparison of HIV-1 and HIV-2 Tat proteins. Gray boxes represent the cysteine-rich domains; black boxes represent the arginine/lysine-rich domains; striped boxes are regions unique to HIV-2 Tat. The arrowheads represent the junction between exon 1 and exon 2. (B) *trans*-Activation response regions (TARs) of HIV-1 and HIV-2. These secondary structures are based on computer analysis. (Modified from Ref. 21.)

response element, or TAR. The TAR region is transcribed as part of the leader sequence present on all viral encoded RNAs. The location of a *cis*-acting sequence for
a viral activator within the transcriptional unit was quite unusual. The TAR sequence
of HIV-1 was initially localized to nucleotides +1 to +80 of the LTR by deletion
analysis and transient transfection assays (14). Subsequent reports have further localized the 3' region of the TAR to +14 to +44 (15, 16). The TAR sequence forms
a thermodynamically stable, secondary structure RNA transcript in both HIV-1 and
HIV-2 (17) (Fig. 2). These secondary structures are based on both computer predictions of RNA folding and nuclease mapping (18). The HIV-1 TAR RNA structure
consists of a stem and a six-nucleotide single-stranded loop. In addition, a three-
nucleotide single-stranded "bulge" is present four nucleotides below the loop. The
function of the HIV-1 TAR is strictly orientation and position dependent (18, 19).
An intact TAR region is also required for viral infectivity of cultured lymphocytes,
as determined by mutational analysis of proviral constructs (20).

The HIV-2 TAR region is larger (+1 to +125) and has a more complex secondary
structure than the HIV-1 TAR. The TAR region of HIV-2 contains two stem–loop
structures that share the six-nucleotide loop and similar bulge regions with the
HIV-1 TAR RNA. In addition, the HIV-2 TAR RNA contains a third stem–loop that
shares less homology with the HIV-1 TAR. Mutational analysis of HIV-2 LTR–
reporter gene constructs has revealed that the third loop of HIV-2 TAR is dispensable
for efficient HIV-2 Tat-mediated *trans*-activation, whereas mutation of either the first
or second loop resulted in reduced responsiveness to Tat-1 and Tat-2 (7, 21).

Several lines of evidence indicate that the functional form of the TAR is RNA.
First, mutations that disrupt the stem structure of TAR dramatically reduced Tat responsiveness. In contrast, many compensatory mutations that change the primary
sequence of the TAR stem region and preserve its secondary structure can be efficiently *trans*-activated by Tat (15, 22). Furthermore, through a novel mutagenesis
strategy, Berkhout *et al.* (23) also inhibited Tat responsiveness by introducing a complementary sequence that prevented formation of the stem–loop in the newly transcribed TAR RNA. Thus Tat must recognize the stem–loop structure in the nascent
transcript to stimulate LTR expression (23).

The role of the RNA form of TAR in Tat-mediated *trans*-activation has been further strengthened by the identification of specific protein–TAR RNA interactions.
The six-nucleotide loop of HIV-1 TAR RNA binds to a 68-kDa cellular protein, and
mutations that disrupt this binding dramatically inhibit Tat responsiveness (24). In
addition, this 68-kDa protein has been reported to stimulate Tat-mediated *trans*-
activation in an *in vitro* transcription system (25). Besides this 68-kDa protein, several other cellular proteins have been identified that bind to different regions of the
TAR (26–28). Some of these TAR-binding proteins appear to require specific complementary RNA sequences in the stem for binding (29).

The HIV-1 Tat protein binds to the HIV-1 TAR *in vitro*. Tat–TAR binding requires
the three-nucleotide bulge. When the bulge is deleted from the TAR, Tat fails to bind,

and Tat responsiveness of the mutated TAR is significantly decreased (30, 31). In contrast to the base of the TAR stem, the primary sequence of the stem region immediately adjacent to the bulge region appears to be important for Tat responsiveness (32, 33).

Methods

Our approach for analyzing HIV gene regulation involves transient transfection analysis with single-gene products. Other workers have approached this problem by using infectious viral molecular clones that contain specific mutations in regulatory regions (20, 34). An important advantage of transient transfection analysis is the effective dissection of *cis*-acting elements and *trans*-acting factors. This is an important consideration with HIV, given the complexity of its genome, and the presence of multiple *trans*-acting regulatory factors. Furthermore, this technique allows the manipulation of the quantity and stoichiometry of transfected constructs, use of different cell types, and mutational analysis of HIV regulatory sequences and/or effector genes. One disadvantage of transient transfection analysis of Tat function is that levels of Tat protein produced in these assays may not reflect levels found *in vivo* in HIV replication.

Transient transfection analysis involves the use of *cis*-acting regulatory elements linked to a reporter gene which can be introduced into cultured cells in the presence or absence of expression plasmids for *trans*-regulatory factors. Following the transfection of plasmid DNA and after allowing sufficient time for the expression of transfected genes, the reporter gene products are assayed and/or mRNA of reporter or effector genes are characterized and quantified (Fig. 3). In this study, the HIV-1 and HIV-2 LTRs were linked to reporter genes and transiently transfected into various cell types in the presence and absence of Tat expression vectors.

Cell Culture

Cell lines used in this study included human HeLa cells, mouse A9 cells, and a series of microcell hybrids based on A9 cells containing individual human chromosomes. All cell lines were maintained in Dulbecco's modified Eagle's medium (DMEM) supplemented with 10% (v/v) fetal calf serum in a humidified 5.0% CO_2 incubator at 37°C. The A9-based microcell hybrids were maintained in medium supplemented with G418 antibiotic (600 μg/ml) to maintain selection for the human chromosome.

DNA Constructs

Two reporter genes were used to investigate HIV LTR-directed expression. The first, the bacterial chloramphenicol acetyltransferase gene (*CAT*), is one of the most fre-

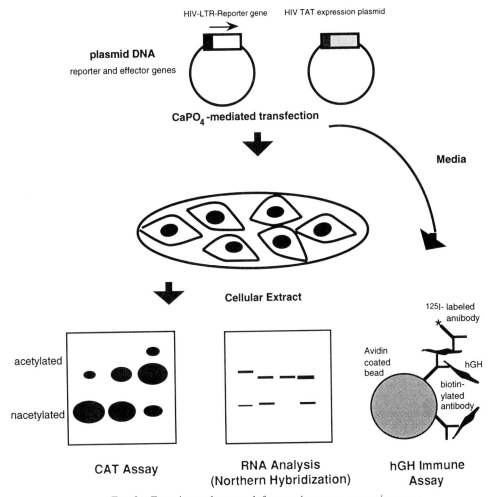

FIG. 3   Experimental approach for transient gene expression assay.

quently used reporter genes to study eukaryotic gene regulation (35). The *CAT* reporter gene system has the advantages of the absence of any homologous eukaryotic enzymatic activity, a relatively simple enzymatic assay, and the stability of the CAT enzyme, which allows for prolonged assay times, thereby amplifying the signal. The expression vectors used in this study contain the *CAT* gene under the control of HIV LTR sequences necessary for transcription and Tat-mediated *trans*-activation (Fig. 4). We have also created constructs in which the HIV LTR drives expression of the human growth hormone gene. The level of HIV LTR-directed expression is quantified by measuring CAT enzymatic activity expressed as percent conversion of

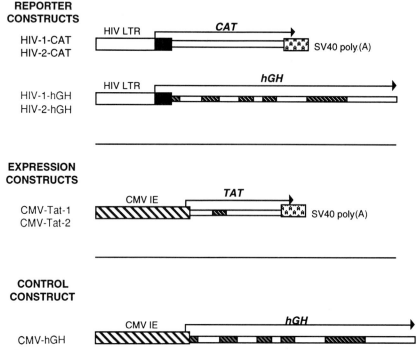

FIG. 4   Plasmid vectors. HIV reporter constructs contain the HIV-1 and HIV-2 LTRs linked to the bacterial *CAT* gene or the pituitary-specific human growth hormone (hGH) gene. The Tat expression vectors CMV-Tat-1 and CMV-Tat-2 were obtained from B. Cullen (Duke University Medical School, Durham, NC). These plasmids contain genomic sequences encoding HIV-1 and HIV-2 Tat linked to the immediate–early promoter of cytomegalovirus (CMV). The transfection control plasmid CMV-hGH contained the CMV promoter linked to the hGH gene. The solid black portions of the HIV LTRs represent the R and U5 regions. The striped portions present in the hGH and Tat constructs represent introns.

[^{14}C]chloramphenicol to its acetylated forms. The acetylated forms of chloramphenicol were separated from the unacetylated form by thin-layer chromatography. *trans*-Activation ratios were then calculated by dividing the level of CAT activity in the presence of TAT by the level of CAT activity in the absence of Tat.

The second transient reporter vector used was the gene encoding the human growth hormone (hGH). A genomic copy of the hGH gene, containing several introns, has been engineered behind the human cytomegalovirus (CMV) or mouse metallothionein promoter (Fig. 4). Promoter-directed gene expression was measured as secreted hGH by an immune assay using ^{125}I-labeled anti-hGH antibodies (36). The hGH gene product is not expressed in the vast majority of cultured cells, because the native hGH promoter is tissue specific and is active only in pituitary derived cells.

The hGH reporter system has two advantages: it can be analyzed rapidly, and it does not require the destruction of the transfected cells. Human growth hormone levels can be measured for one transfection at multiple time intervals or hGH can be easily measured in conjunction with CAT assays or RNA analysis. Thus hGH assays can be used as internal controls for transfection efficiency during CAT assays. Both CAT and hGH transient expression systems are linear over a large range of reporter gene product concentrations.

HIV Tat expression vectors utilized in these studies are shown in Fig. 4. HIV-1 and HIV-2 Tat expression vectors containing two coding exons of Tat under transcriptional control of the cytomegalovirus (CMV) early promoter were used. The CMV-based Tat expression vectors CMV-HIV1-Tat and CMV-HIV2-Tat were generously provided by B. Cullen (Duke University Medical School, Durham, NC) and described elsewhere (21, 37).

Calcium Phosphate Transfection

Calcium phosphate–DNA coprecipitation has been used to introduce DNA into many cell types (38). Cells are plated 24 hr before transfection at a density of 3×10^5 cells/60-mm dish. Cells are fed 4 hr prior to transfection with fresh medium. The $CaPO_4$–DNA mixture is prepared by combining in order: H_2O (to a final volume of 500 μl), plasmid DNA (1–10 μg), 31 μl of 2 M $CaCl_2$, and 250 μl of 2× HBSP [1.5 mM Na_2HPO_4, 10 mM KCl, 280 mM NaCl, 12 mM glucose, 50 mM 4-(2-hydroxyethyl)-1-piperazine-ethanesulfonic acid (HEPES), pH 7]. We typically use 2 μg of reporter plasmid, 1 μg of Tat expression plasmid, 0.1 μg of CMV–hGH as an internal control, and salmon sperm DNA to a total of 5 μg/plate. Individual transfections are typically run in triplicate. The precipitate is allowed to form for 30 min at room temperature, mixed, slowly added to the cells, and incubated for 20 hr at 37° C in a humidified CO_2 incubator. After 16–20 hr the precipitate is removed, cells are refed with 5 μl of fresh medium, and incubated for an additional 48 hr.

Chloramphenicol Acetyltransferase Assay

Following transfection, medium is removed for hGH assay, cells are harvested by trypsinization, washed three times with phosphate-buffered saline (PBS), recovered by centrifugation at 12,000 g for 10 sec in a microcentrifuge, and resuspended in 100 μl of 0.25 M Tris, pH 7.8. Cells are vortexed and sonicated and cellular debris is removed by centrifugation at 12,000 g in a microcentrifuge for 5 min at 4° C. The cellular extract is incubated at 60° C for 10 min to inactivate deacetylases. The particulate material is then removed by a 2-min centrifugation in a microcentrifuge at 4° C. The reaction mixture for each sample contains the following: 70 μl of 1 M Tris-HCl (pH 7.8), 2 μl of [^{14}C]chloramphenicol (60 mCi/mmol), 20 μl acetyl coenzyme A (3.5 μg/μl in H_2O), 5–50 μl of cellular extract, and H_2O to make a 150-μl total reaction volume. Reagents are premixed and stored in frozen aliquots to minimize variation of assay conditions between experiments. Incubation times vary depending on the amount of extract used and the concentration of CAT in the cell extract. In most cases samples are incubated from 30 min to 2 hr. To be within the linear range of the assay, levels of acetylation must be maintained below 50%. Chloramphenicol is extracted with 1 μl of ethyl acetate by vortexing for 30 sec and the organic phase is separated by centrifugation at 12,000 g for 3 min at room temperature. The organic phase is evaporated under a stream of air, reaction products are resuspended in 30 μl of ethyl acetate and spotted on a 25-mm silica gel thin-layer chromatography (TLC) plate. The TLC plates are placed in a tank preequilibrated for 1 hr with 500 μl of chloroform–methanol (19:1). Ascending chromatography is performed until the solvent front migrates approximately 80% of the distance. Plates are air dried and exposed to X-ray film for 16–24 hr. The CAT activity is quantified by excising the radioactive spots and measuring radioactivity by liquid scintillation counting.

Human Growth Hormone Assay

The human growth hormone (hGH) transient assay system is based on immunological detection of hGH secreted by transfected cells (36) (Allégro hGH transient gene expression assay system, 1991, Nichols Institute Diagnostic, San Juan Capistrano, CA). Media samples of 100 μl are mixed with 100 μl of ^{125}I-labeled antibody solution {^{125}I-labeled monoclonal hGH antibody [mouse Ab(1)] and biotin-coupled monoclonal hGH antibody [mouse Ab(2)]} and one avidin-coated polystyrene bead in a test tube. The tube is incubated on a horizontal rotator (170 rpm) at room temperature for 90 min. Beads are then washed twice with 2 μl of wash solution (surfactant in phosphate-buffered saline with 0.3% sodium azide). Beads are then counted for radioactivity in a γ counter for 1 min. Human growth hormone standard solutions (0, 0.5, 1.5, 5, 15, and 50 ng/μl) and 100 μl of media from mock-transfected cells are also included as controls. The counts per minute (cpm) of ^{125}I-labeled anti-hGH antibody is corrected by subtracting the background from the mock-transfected media and the generated standard curve is used to calculate the amount of hGH secreted. This amount of hGH is then used to normalize the CAT assay for plate-to-plate variation. Variation between plates within a cell type is typically less than 20%.

Results

Species Specificity of HIV-2 Tat and TAR

As discussed earlier, the HIV-2 TAR sequence is larger and the RNA secondary structure is more complex than HIV-1 TAR (Fig. 2). Likewise, HIV-2 Tat is larger and contains regions that are not homologous to HIV-1 Tat. Because HIV-1 *trans*-activation is restricted in rodent cells, we have investigated whether the divergence of the HIV-1 and HIV-2 Tat and TAR sequences was associated with a differing requirement for species-specific factors.

Figure 5 shows the results of an experiment in which we measured *trans*-activation of the HIV-2 LTR in human cells (HeLa), murine cells (A9), and A9-based micro-cell hybrids containing individual human chromosomes (39). As is the case with the HIV-1 Tat–TAR system, HIV-2 Tat *trans*-activates the HIV-2 LTR much more efficiently in HeLa cells compared to the murine A9 cells. The presence of chromosome 12 encodes a major factor that enhances HIV-2 Tat *trans*-activation of the HIV-2 LTR. However, human chromosome 8 appears to encode an additional factor that increases Tat-2 *trans*-activation of the HIV-2 LTR in rodent cells.

Interestingly, the data of Fig. 5 indicate that heterologous *trans*-activation of the HIV-2 LTR by HIV-1 Tat functions at relatively high levels in all the cells that were examined. For example, in A9 cells Tat-1 *trans*-activates the HIV-2 LTR 20-fold, as compared to 5-fold *trans*-activation of the HIV-2 LTR by Tat-2. Moreover, Tat-1 mediated *trans*-activation of the HIV-2 occurs at high levels in all microcell hybrids examined, including those that do not enhance homologous *trans*-activation of the HIV-1 and HIV-2 LTRs. These results indicate that heterologous *trans*-activation of the HIV-2 LTR by HIV-1 Tat may have a less stringent requirement for human-specific cofactors than the *trans*-activation of the HIV-1 or HIV-2 LTRs by their homologous Tat proteins. Similar data using different cell hybrids have been reported by Hart *et al.* (40).

Previous reports have demonstrated that in human cells heterologous *trans*-activation of the HIV-1 LTR by HIV-2 Tat is 10 to 30% as efficient as *trans*-activation with HIV-1 Tat (17, 41). We have observed similar results in both HeLa and A9 cells as well as the microcell hybrids (Fig. 5).

Discussion

The use of the transient transfection assay has revealed similarities and differences in the *trans*-activation system of HIV-1 and HIV-2. The work presented here focuses on the *trans*-activation of the HIV-2 Tat protein. Like the HIV-1 Tat protein, HIV-2 Tat does not *trans*-activate its homologous LTR efficiently in rodent cells. This restriction can be partially overcome by factors encoded on human chromosome 12 and on chromosome 8. These results suggest that like HIV-1 Tat, species-specific cellular proteins may play a critical role in *trans*-activation by HIV-2. Although Tat can interact directly with TAR RNA *in vitro,* several lines of evidence suggest that cellular TAR-binding factors are necessary for efficient *trans*-activation. First, mutations in the TAR loop that abrogate Tat responsiveness have no effect on Tat binding to TAR. Furthermore, *trans*-activation requires the presence of species-specific factors (42–44).

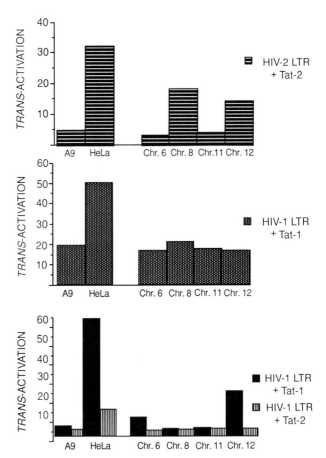

FIG. 5 *trans*-Activation of HIV-1 and HIV-2 LTRs by Tat-1 and Tat-2 in HeLa, A9, and microcell hybrids. HIV-1 CAT and HIV-2 CAT reporter constructs and CMV-Tat-1 and CMV-Tat-2 expression vectors were used in transient transfection experiments. The data presented are *trans*-activation ratios obtained by dividing CAT activity in the presence of Tat by CAT activity in the absence of Tat. CMV-hGH was used as an internal transfection control. The results represent the mean of three independent transfections with standard errors of less than 20%.

The most striking feature of these experiments is that the heterologous *trans*-activation of the HIV-2 LTR by HIV-1 Tat functions at relatively high levels in all cell types examined. For example, in mouse A9 cells Tat-1 *trans*-activates the HIV-2 LTR 20-fold as compared with 5-fold *trans*-activation of the HIV-2 LTR by Tat-2. Consistent with these results, heterologous *trans*-activation of the HIV-2 LTR by HIV-1 Tat is efficient in all hybrids examined. These results suggest that heterologous *trans*-activation of the HIV-2 LTR by HIV-1 Tat has a less stringent requirement for human-specific cofactors than the *trans*-activation of the HIV-1 or HIV-2 LTRs by the homologous Tat proteins. One interpretation of these results is that rodent cellular cofactors that bind to the HIV-2 TAR may interact in a more productive way with HIV-1 Tat than the HIV-1 TAR-binding factors.

References

1. O. Narayan and J. E. Clements, *J. Gen. Virol.* **70,** 1716 (1989).
2. Y. N. Vaishnav and F. Wong-Staal, *Annu. Rev. Biochem.* **60,** 577 (1991).
3. B. R. Cullen, *J. Virol.* **65,** 1053 (1991).
4. B. R. Cullen, *Microbiol. Rev.* **56,** 375 (1992).
5. A. G. Fisher, M. B. Feinberg, S. F. Josephs, M. E. Harper, L. M. Marselle, G. Reyes, M. A. Gonda, A. Aldovini, C. Debouk, R. C. Gallo, and F. Wong-Staal, *Nature (London)* **320,** 367 (1986).
6. L. J. Seigel, L. Ratner, S. F. Josephs, D. Derse, M. B. Feinberg, G. R. Reyes, S. J. O'Brien, and F. Wong-Staal, *Virology* **148,** 226 (1986).
7. B. Berkhout, A. Gatignol, J. Silver, and K.-T. Jeang, *Nucleic Acids Res.* **18,** 1839 (1990).
8. S. E. Tong-Starksen, A. Baur, X.-b. Lu, E. Peck, and B. M. Peterlin, *Virology* **195,** 826 (1993).
9. S. Ruben, A. Perkins, R. Purcell, K. Joung, R. Sia, R. Burghoff, and W. Haseltine, *J. Virol.* **63,** 1 (1989).
10. J. Hauber, A. Perkins, E. P. Heimer, and B. R. Cullen, *Proc. Natl. Acad. Sci. USA* **84,** 6364 (1987).
11. M. R. Sadaie, T. Benter, and F. Wong-Staal, *Science* **239,** 910 (1988).
12. A. D. Frankel, D. S. Bredt, and C. O. Pabo, *Science* **240,** 70 (1988).
13. K. M. Weeks, C. Ampe, S. C. Schultz, T. A. Steitz, and D. M. Crothers, *Science* **249,** 1281 (1990).
14. C. A. Rosen, J. G. Sodroski, and W. A. Haseltine, *Cell (Cambridge, Mass.)* **41,** 813 (1985).
15. J. Hauber and B. R. Cullen, *J. Virol.* **62,** 673 (1988).
16. A. Jakobovits, D. H. Smith, E. B. Jakobovits, and D. J. Capon, *Mol. Cell. Biol.* **8,** 2555 (1988).
17. M. Emerman, L. Guyader, L. Montagnier, D. Baltimore, and M. Muesing, *EMBO J.* **6,** 3755 (1987).
18. M. A. Muesing, D. H. Smith, and D. J. Capon, *Cell (Cambridge, Mass.)* **48,** 691 (1987).
19. B. M. Peterlin, P. A. Luciw, P. J. Barr, and M. D. Walker, *Proc. Natl. Acad. Sci. USA* **83,** 9734 (1987).
20. J. Leonard, C. Parrott, A. J. Buckler-White, W. Turner, E. K. Ross, M. A. Martin, and A. B. Rabson, *J. Virol.* **63,** 4919 (1989).
21. R. Fenrick, M. H. Malim, J. Hauber, S.-Y. Le, J. Maizel, and B. R. Cullen, *J. Virol.* **63,** 5006 (1989).
22. B. Berkhout and K.-T. Jeang, *J. Virol.* **63,** 5501 (1989).
23. B. Berkhout, R. H. Silverman, and K.-T. Jeang, *Cell (Cambridge, Mass.)* **59,** 273 (1989).

24. R. A. Marciniak, M. A. Garcia-Blanco, and P. A. Sharp, *Proc. Natl. Acad. Sci. USA* **87,** 3624 (1990).

25. R. A. Marciniak, B. J. Calnan, A. D. Frankel, and P. A. Sharp, *Cell (Cambridge, Mass.)* **63,** 791 (1990).

26. R. Gaynor, E. Soultanakis, M. Kuwabara, J. Garcia, and D. S. Sigman, *Proc. Natl. Acad. Sci. USA* **86,** 4858 (1989).

27. A. Gatignol, A. Kumar, A. Rabson, and K.-T. Jeang, *Proc. Natl. Acad. Sci. USA* **86,** 7828 (1989).

28. A. Gatignol, A. Buckler-White, B. Berkhout, and K.-T. Jeang, *Science* **251,** 1597 (1991).

29. M. P. Rounseville and A. Kumar, *J. Virol.* **66,** 1688 (1992).

30. C. Dingwall, I. Ernberg, M. J. Gait, S. M. Green, S. Heaphy, J. Karn, A. D. Lowe, M. Singh, M. A. Skinner, and R. Valerio, *Proc. Natl. Acad. Sci. USA* **86,** 6925 (1989).

31. C. Dingwall, I. Ernberg, M. J. Gait, S. M. Green, S. Heaphy, J. Karn, A. D. Lowe, M. Singh, and M. A. Skinner, *EMBO J.* **9,** 4145 (1990).

32. B. Berkhout and K.-T. Jeang, *Nucleic Acids Res.* **19,** 6169 (1991).

33. X.-M. Han, A. Laras, M. P. Rounseville, A. Kumar, and P. R. Shank, *J. Virol.* **66,** 4065 (1992).

34. C. Parrott, T. Seidner, E. Duil, J. Leonard, S. T. Theodore, A. Buckler-White, M. A. Martin, and A. B. Rabson, *J. Virol.* **65,** 1414 (1991).

35. C. M. Gorman, L. F. Moffat, and B. H. Howard, *Mol. Cell. Biol.* **2,** 1044 (1982).

36. R. F. Selden, K. Burke, M. E. Rowe, H. M. Goodman, and D. D. Moore, *Mol. Cell. Biol.* **6,** 3173 (1986).

37. B. R. Cullen, *Cell (Cambridge, Mass.)* **46,** 973 (1986).

38. L. G. Davis, M. D. Dibner, and J. F. Batty, "Basic Methods in Molecular Biology." Elsevier, New York, 1986.

39. P. J. Saxon, E. S. Srivatsan, G. V. Leipzig, J. H. Sameshima, and E. J. Stanbridge, *Mol. Cell. Biol.* **5,** 140 (1985).

40. C. Hart, M. A. Westhafer, J. C. Galphin, C.-Y. Ou, L. T. Bacheler, J. J. Wasmuth, S. R. J. Petteway, J. J. Wasmuth, I. S. Y. Chen, and G. Schochetman, *AIDS Res. Hum. Retroviruses* **7,** 877 (1991).

41. G. A. Viglianti and J. I. Mullins, *J. Virol.* **62,** 4523 (1988).

42. C. A. Hart, C.-Y. Oy, J. C. Galphin, J. Moore, L. T. Bachler, J. J. Wasmuth, S. R. Petteway, and G. Schochetman, *Science* **246,** 488 (1989).

43. M. Newstein, E. J. Stanbridge, G. Casey, and P. R. Shank, *J. Virol.* **64,** 4565 (1990).

44. C. E. Hart, J. C. Galphin, M. A. Westhafer, and G. Schochetman, *J. Virol.* **67,** 5020 (1993).

[6]   Functional Assay for Tat: *trans*-Activator Protein of Human Immunodeficiency Virus Type 1

Dennis L. Kolson, Christine Debouck, Ronald Collman, and Francisco Gonzalez-Scarano

Introduction and Background

Tat protein is a unique and potent regulator of human immunodeficiency virus type 1 (HIV-1) gene expression and is essential for virus replication (1). Although the precise mechanism of *trans*-activation is not clear, evidence suggests that Tat enhances the initiation of transcription and stabilizes elongation of mRNA (2,3) by direct interaction with nascent noncoding RNA transcribed from the TAR (*trans*-acting responsive) region just downstream of the transcriptional start site (4). Although it is necessary for virus replication, other possible roles for Tat in the pathogenesis of acquired immunodeficiency syndrome (AIDS) have been postulated. *In vitro* studies have suggested that soluble Tat protein may induce immune dysfunction (5), neurotoxicity (6), and tumorigenesis (7) through interaction with a variety of cell types, and that such effects may occur *in vivo* by release of Tat from infected cells and uptake by neighboring cells (7–10). It is thought to bind to cells by at least two mechanisms: (a) nonspecifically at its highly charged basic sequence (10), and (b) specifically at its RGD sequence, possibly by binding to integrin receptors (11). Other studies have demonstrated *trans*-activation of host cellular genes [tumor necrosis factor β (TNF-β) (12), fibronectin (13), and type I collagen (13)] in cells endogenously producing Tat.

Among the questions that such observations have raised are the following: (a) Which HIV-infected cells *in vivo* can release biologically active Tat and in which cells and under what conditions is there efficient uptake and transport of Tat? (b) Which regions of the Tat protein mediate *trans*-activation of host cellular genes? (c) Is transcellular Tat *trans*-activation/transport dependent on cell type and on Tat sequences other than those necessary for long terminal repeat (LTR) *trans*-activation? Utilization of a rapid, simple, functional Tat assay is critical for these evaluations.

Methods

The primary functional assay for Tat protein requires the demonstration of *trans*-activation of the HIV-1 LTR, typically by assay of a reporter gene, such as chloramphenicol acetyltransferase (CAT), linked to the TAR sequence at the 3' end of the LTR. Creation of stably transfected LTR/CAT-expressing cell lines allows one to assay Tat protein introduced by exogenous application, transcellular transport, or by transfection of Tat-expressing plasmids into the host cell. *trans*-Activation by exogenous Tat protein requires the presence of the lysosomotropic agent chloroquine in the medium (8–10), whereas *trans*-activation by Tat expressed within cells or presented transcellularly does not (8, 14–16). *trans*-Activation studies utilizing exogenous and endogenous Tat have allowed functional mapping of Tat through creation of mutations in its coding sequence (17, 18), as well as determination of the kinetics and mechanism of uptake of exogenous Tat by bystander cells (8).

Construction of LTR/CAT Reporter Cell Line

Human cell lines susceptible to transfection are used for functional Tat assays, as cellular Tat- and TAR-binding proteins encoded by the human genome are essential for optimal interaction between Tat and the TAR element and efficient *trans*-activation (19–22). The SK-N-MC/CAT cell line was derived by stable transfection of a human neuroepithelioma line, SK-N-MC (23), with a plasmid expressing both a *CAT* gene driven by the HIV-1 (IIIB) LTR and a neomycin resistance gene (*NEO*) driven by the simian virus 40 (SV40) early promoter (pHIV CAT-SVNEO; gift of K. Valerie, SmithKline Beecham Pharmaceuticals, King of Prussia, PA). SK-N-MC/ CAT is phenotypically stable and grows to confluent monolayers like the parental SK-N-MC line (23).

SK-N-MC cells (1.5×10^5) are washed in serum-free medium (Optimem; GIBCO, Inc., Grand Island, NY) prior to transfection. Twenty to 40 μl of lipofectin reagent (GIBCO, Inc.) is combined with 10–20 μg of pHIV CAT-SVNEO in a volume of 3 ml of Optimem, and layered over the washed cells (24). Twenty hours later, the medium is replaced with RPMI-1640 containing 10% (v/v) fetal bovine serum, which inactivates the lipofectin. Neomycin selection begins 2 days later with medium containing G418 (neomycin) (400 μg/ml) until confluent growth of neomycin-resistant populations (3–4 weeks). Cells are maintained in medium without G418 for subsequent experiments, and recycled through G418 selection at approximate 2-month intervals to assure maintenance of stable transfectants.

Tat Protein

Tat protein is 86–101 amino acids in length, depending on the HIV-1 strain of origin (25). Recombinant Tat protein (HIV-1 IIIB; 86 amino acids) is prepared as described (26). Aliquots of purified protein are stored at $-80°$C in Tat buffer [50 mM Tris (pH 8.0), 1 M NaCl, adjusted to 20% (v/v) glycerol] until use. Alternatively, native full-length Tat protein and Tat peptides may be purchased lyophilized and reconstituted in Tat buffer (ABT, Cambridge, MA; Repligen, Cambridge, MA).

trans-Activations

For assay of exogenous Tat protein, $\sim 5 \times 10^5$ SK-N-MC/CAT cells/well in a six-well culture plate ($\sim 50-75\%$ confluency) are fed fresh medium (e.g., RPMI-1640 with 10% fetal bovine serum) containing 100 μM chloroquine diphosphate (Sigma, St. Louis, MO) prepared immediately prior to use as a $1000 \times$ stock in water and filter sterilized. As chloroquine is light sensitive, solutions must be protected from light. Recombinant Tat protein or peptide in Tat buffer is added to a final concentration of $0.01-10$ μg/ml, and cultures are incubated for 24 hr at 37° C, followed by CAT assay.

For assay of endogenously produced Tat, $1-10$ μg of the Tat-expressing plasmid to be tested [e.g., pSV$_2$ Tat (27)], is transfected as described above into $\sim 5 \times 10^5$ cells, followed by CAT assay 48 hr later. Direct cocultivation of cells stably transfected with Tat-expressing plasmids ($\sim 0.5-1.0 \times 10^6$ cells) with LTR/CAT reporter cells may also result in a *trans*-activation response in $24-48$ hr (12–14).

CAT Assay

The CAT assay is modified after Neumann *et al.* (28). Confluent cultures in six-well plates ($\sim 10^6$ cells) are washed once in phosphate-buffered saline (PBS) and scraped into 300 μl of 250 mM Tris, pH 7.8 (sample buffer), followed by three rapid freeze–thaw cycles. Supernatant is clarified at 10,000 g for 5 min at 4° C and incubated at 65° C for 10 min, and a sample is assayed for total protein. Fifty microliters of supernatant is placed into a glass scintillation vial, and 200 μl of reaction mix containing 115 μl of H$_2$O, 25 μl of 1 M Tris (pH 7.8), 50 μl of 5 mM chloramphenicol, and 10 μl of ^{14}C-labeled acetyl coenzyme A (10 μCi/ml; Du Pont–New England Nuclear, Wilmington, DE) is added. The CAT standards are simultaneously prepared by serial dilution of purified CAT (5 Prime → 3 Prime, Inc., West Chester, PA) in sample buffer, through a range of $0.3-40$ pg/ml, and 50-μl aliquots are incubated as above. Four milliliters of Econofluor (Du Pont–New England Nuclear) or other organic scintillation cocktail is immediately added, and the mixture separates into two phases, an upper organic phase and a lower aqueous phase containing reactants. As this is a biphasic liquid extraction assay, use of an organic scintillation cocktail is essential. Acetylation proceeds in the aqueous phase and the reaction product, [*acetyl*-^{14}C]chloramphenicol, partitions into the organic phase, where β emissions generate a photic signal. Samples are counted in a β-scintillation counter over the next 24 hr, and CAT activity is calculated either as picograms of CAT (as calculated from standard curve) or total counts per minute (cpm) above blank, per unit protein in sample. Activity is expressed as a ratio of CAT activity in Tat-treated cells to activity in untreated control cells.

Applications and Representative Results

Exogenous Tat Assay

Determining the ability of a specific cell type to take up and internalize biologically active Tat requires that the cell be amenable to stable transfection and cultivation in the presence of chloroquine, because efficient *trans*-activation by exogenous Tat requires chloroquine in the medium. For these reasons, transformed cell lines and not primary cells are generally suited for functional Tat assay. Chloroquine alters intra-lysosomal pH, causes vacuole formation, and can be toxic to certain cells (29, 30); each cell type used, therefore, must be tested for viability during chloroquine exposure. Table I illustrates the Tat *trans*-activation response in the neuronal SK-N-MC/CAT cell line over a range of chloroquine concentrations. Maximal Tat-induced *trans*-activation is seen at a chloroquine concentration of 100 μM, which is typically used for a variety of cell types (8–10, 17, 18). This type of assay may be applied to a number of transfection-competent cell types, including monocytoid (31), cervical epithelial (8–10), embryonal carcinoma (D. L. Kolson *et al.*, unpublished observation, 1993), and neuronal (Fig. 1A and B).

TABLE I   Chloroquine Dependence of *trans*-Activation by Exogenous Tat[a]

Chloroquine (μM)	(−) Tat			(+) Tat			
	CAT (cpm)[b]	Protein (μg/ml)	Specific activity[c]	CAT (cpm)[b]	Protein (μg/ml)	Specific activity[c]	Ratio[d]
0	1.7×10^3	1480	1.1	—	—	—	—
25	2.8×10^3	660	4.3	4.6×10^2	1220	3.8	0.9
50	2.1×10^3	540	4.0	2.2×10^3	2260	9.9	2.5
75	2.3×10^3	960	2.4	9.4×10^3	1280	75.3	31.0
100	1.8×10^3	580	3.1	2.5×10^5	610	416.3	133.0
150	2.2×10^3	500	4.5	3.6×10^5	1220	292.3	66.0

[a] SK-N-MC/CAT cells were exposed to 1 μg of recombinant Tat protein (+Tat) per milliliter or Tat buffer alone (− Tat) in medium containing chloroquine diphosphate, and assayed for CAT activity 24 hr later, as described in Methods.
[b] Represents counts above background (2.2×10^3 cpm).
[c] Specific activity for each chloroquine treatment group was calculated as CAT activity divided by protein concentration in the cell lysate.
[d] Refers to the ratio of specific activity in the +Tat sample divided by that found in the − Tat sample for each chloroquine concentration.

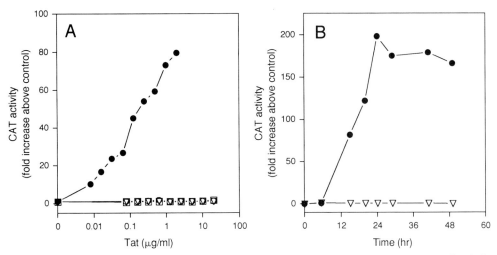

FIG. 1   Dose and time dependence for Tat *trans*-activation in SK-N-MC/CAT cells. Cells were exposed to Tat protein or peptide as described in Methods. (A) All incubations included 100 μM chloroquine diphosphate, with increasing amounts of full-length Tat protein (●, Tat 1–86) or peptide (▽, Tat 46–60; ○, Tat 56–70; □, Tat 73–86), and a CAT assay was performed 24 hr after exposure. (B) Parallel incubations included Tat protein (1 μg/ml) with (●) or without (▽) chloroquine (100 μM) as indicated. Cells were harvested at the indicated times after addition of the Tat and assayed. CAT activity is expressed as a -fold increase in specific activity in treated cells versus untreated controls.

The CAT assay predictably detects nanomolar concentrations of biologically active Tat in tissue culture medium, and CAT activity accurately reflects Tat concentration. Figure 1A illustrates a dose–response curve for SK-N-MC/CAT cells exposed to full-length Tat, as well as Tat peptides. A log–linear response is seen over a concentration range from ~10 ng/ml (1 nM full-length Tat) through 2 μg/ml (200 nM) for SK-N-MC/CAT cells. Only full-length or nearly full-length Tat protein is detected, as sequences responsible for *trans*-activation lie within the amino-terminal two-thirds of Tat (8, 9, 17, 18). Exogenous Tat will increase CAT activity maximally several 100-fold over untreated cells for most cell types (2–4, 8–10, 17, 18), and the response is maximal at approximately 24–30 hr (Fig. 1B).

Endogenous Tat Assay

Although the assay for exogenous Tat as outlined above requires chloroquine, the assay of endogenously produced Tat (as in Tat-transfected or HIV-1-infected cells) does not. Transient transfection of Tat-expressing plasmids into these cells results in

TABLE II   *trans*-Activation by Endogenous Tat[a]

pSV$_2$Tat transfected (μg)	CAT (cpm)[b]	Protein (μg/ml)	Specific activity	Ratio[c]
0	5.6×10^2	711	0.8	—
5	3.8×10^4	745	50.4	63.0
10	6.6×10^4	597	110.3	123.0

[a] SK-N-MC/CAT cells were transfected as described in Methods with different amounts of the Tat-expressing plasmid pSV$_2$Tat and assayed 24 hr later for CAT activity. CAT-specific activity was determined as the ratio of CAT (cpm) to total protein concentration of each lysate.

[b] Represents counts above background (2.3×10^3 cpm).

[c] Refers to the ratio of specific activity determined for each level of transfected pSV$_2$Tat to that seen in the control transfection (0 μg).

a dose–response effect with respect to amount of input plasmid (Table II), and thus allows one to study *trans*-activational effects of mutations in the Tat coding sequence. Furthermore, cocultivation of Tat-expressing cells (either stably transfected with Tat plasmids, or with replication-defective HIV-1) allows study of transcellular *trans*-activation/transport of Tat in relevant cell types, using this assay (12–14).

Acknowledgments

Supported by PHS Awards NS 27405, HL 02358, and A 1-24845, NIH Training Grant NS 07180, Clinical Investigator Development Award 1 KO8 NS01581-01, and grants from the W. W. Smith Charitable Trust and the Life and Health Insurance Medical Research Fund. We would like to thank members of the Neal Nathanson laboratory for many useful discussions, technical assistance, and reagents.

References

1. A. G. Fisher, M. B. Feinberg, S. F. Josephs, M. E. Harper, L. M. Marselle, G. Reyes, M. A. Gonda, A. Aldovini, C. Debouck, R. C. Gallo, and F. Wong-Staal, *Science* **320,** 367 (1986).
2. M. A. Muesing, D. H. Smith, and D. J. Capon, *Cell (Cambridge, Mass.)* **48,** 691 (1987).
3. M. F. Lapsia, A. P. Rice, and M. B. Mathews, *Cell (Cambridge, Mass.)* **59,** 283 (1989).
4. B. Berkhout, R. H. Silverman, and K.-T. Jeang, *Cell (Cambridge, Mass.)* **59,** 273 (1989).
5. R. P. Viscidi, K. Mayur, H. M. Lederman, and A. D. Frankel, *Science* **246,** 1606 (1989).
6. J.-N. Sabatier, E. Vives, K. Mabrouk, A. Benjouad, H. Rochat, A. Duval, B. Hue, and E. Hahraout, *J. Virol.* **65,** 961 (1991).
7. B. Ensoli, G. Barillari, S. Z. Salahuddin, R. C. Gallo, and F. Wong-Staal, *Nature (London)* **345,** 84 (1990).

8. A. D. Frankel and C. O. Pabo, *Cell (Cambridge, Mass.)* **55,** 1189 (1988).
9. A. D. Frankel, S. Biancalana, and D. Hudson, *Proc. Natl. Acad. Sci. USA* **86,** 7397 (1989).
10. D. A. Mann and A. D. Frankel, *EMBO J.* **10,** 1733 (1991).
11. D. A. Brake, C. Debouck, and G. Biesecker, *J. Cell Biol.* **111,** 1275 (1990).
12. L. Buonaguro, B. Giovanni, H. K. Chang, C. A. Bohan, V. Kao, R. Morgan, R. C. Gallo, and B. Ensoli, *J. Virol.* **66,** 7159 (1992).
13. J. P. Taylor, C. Cupp, A. Diaz, M. Chowdhury, K. Khalili, S. A. Jimenez, and S. Amini, *Proc. Natl. Acad. Sci. USA* **89,** 9617 (1992).
14. D. Helland, J. Welles, A. Caputo, and W. A. Haseltine, *J. Virol.* **65,** 4547 (1991).
15. A. Marcuzzi, J. Weinberger, and O. K. Weinberger, *J. Virol.* **66,** 4228 (1992).
16. A. Marcuzzi, J. Weinberger, and O. K. Weinberger, *J. Virol.* **66,** 4536 (1992).
17. M. Green and P. M. Loewenstein, *Cell (Cambridge, Mass.)* **55,** 1179 (1988).
18. S. Ruben, A. Perkins, R. Purcell, K. Joung, R. Sia, R. Burghoff, W. A. Haseltine, and C. Rosen, *J. Virol.* **63,** 1 (1989).
19. A. P. Rice and F. Carlotti, *J. Virol.* **64,** 6018 (1990).
20. A. Gatignol, A. Buckler-White, B. Berkhout, and K.-T. Jeang, *Science* **251,** 1597 (1991).
21. M. P. Rounseville and A. Kumar, *J. Virol.* **66,** 1688 (1992).
22. P. Nelbock, P. J. Dillon, A. Perkins, and C. A. Rosen, *Science* **248,** 1650 (1990).
23. J. L. Beidler, L. Helson, and B. A. Spengler, *Cancer Res.* **33,** 2643 (1973).
24. P. Felgner, T. Gadek, and M. Holm, *Proc. Natl. Acad. Sci. USA* **84,** 7413 (1987).
25. G. Meyers, A. B. Rabson, J. A. Bertofsky, and T. F. Smith, "Human Retroviruses and AIDS." Los Alamos National Laboratory, Los Alamos, New Mexico.
26. A. Aldovini, C. Debouck, M. B. Feinberg, M. Rosenberg, S. K. Arya, and F. Wong-Staal, *Proc. Natl. Acad. Sci. USA* **83,** 6672 (1986).
27. L. T. Bacheler, L. L. Strehl, R. H. Neubauer, S. R. Petteway, and B. Q. Ferguson, *AIDS Res. Hum. Retroviruses* **5,** 275 (1989).
28. J. R. Neumann, C. A. Morency, and K. O. Russian, *BioTechniques* **5,** 444 (1987).
29. P. O. Seglen, *in* "Methods in Enzymology," Vol. 96, p. 737. Academic Press, San Diego, 1983.
30. R. T. Dean, W. Jessup, and C. R. Roberts, *Biochem. J.* **217,** 27 (1984).
31. B. K. Felber and G. N. Pavlakis, *Science* **239,** 184 (1988).

[7]   Construction, Assay, and Use of Pseudotype
       Viruses between Human Immunodeficiency Virus
       and Cocal, a Vesiculovirus

Susan Gregory, William James, and Neal Nathanson

Introduction and Background

The cellular entry phase of the human immunodeficiency virus (HIV) replication cycle holds the key to many important aspects of HIV infection and disease pathogenesis. Among the questions that remain to be elucidated are the following: (a) What is the mechanism of HIV entry into CD4-negative cells? (b) Why do lymphocyte-tropic and macrophage-tropic HIV strains, all of which utilize the CD4 receptor, have a different host range? (c) Are there accessory cellular molecules that, in addition to CD4, are necessary for productive entry of HIV? (d) What is the role of Fc and complement receptors in the enhancement of infection?

The quantitation of pseudotype viruses is one methodology that can be used to elucidate questions regarding entry of HIV-1. Pseudotype virions are phenotypically mixed particles in which one virus contributes the genome and associated proteins, while another virus contributes the envelope (Fig. 1). Pseudotype viruses can be constructed between retroviruses and rhabdoviruses, and a large body of earlier work employed the combination of Rous sarcoma virus (RSV) and vesicular stomatitis virus (VSV) (1,2).

Pseudotype viruses in which VSV is pseudotyped on HIV, VSV(HIV), have been used by Weiss and colleagues to address several important questions in the biology of HIV, particularly to identify CD4 as a major receptor for HIV, and for the assay of neutralizing activity (3–5). Because pseudotype virus disconnects entry from the subsequent steps in replication and VSV produces plaques within 2 days, it is possible to obtain a rapid and quantitative assessment of the number of virions that have initiated productive interaction with cells. This facilitates certain experiments that are difficult to perform with a virus like HIV, which replicates relatively slowly and does not produce lytic plaques. We have developed a pseudotype system that is based on the assay of Weiss and collaborators (3–5) but differs from it in a number of ways, including the use of Cocal virus (COV) instead of VSV. The method, described in more detail elsewhere (6), has given us reproducible results when applied to several questions regarding cellular entry of HIV.

Methods in Molecular Genetics, Volume 4

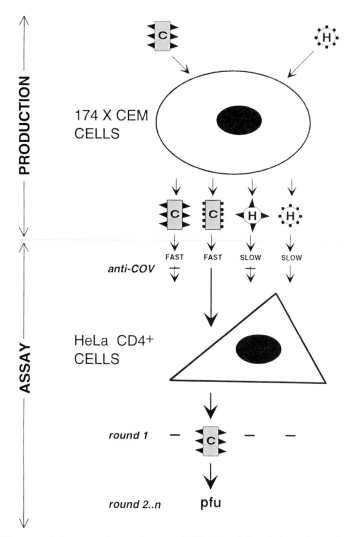

FIG. 1 Diagram of virus pseudotypes between HIV (H) and Cocal virus (C). Both viruses are used to coinfect a permissive cell line, and four possible types of progeny are produced: the two parent viruses and two pseudotype viruses, HIV(COV) and COV(HIV). For assay purposes, the two viruses with an HIV genome [HIV and HIV(COV)] replicate slowly and may be disregarded. Of the two viruses with a COV genome [COV and COV(HIV)], COV may be neutralized by specific antiserum, leaving only the pseudotype COV(HIV) to produce plaques. The authenticity of plaques due to putative COV(HIV) may be confirmed by several tests, including their failure to plaque on a cell line lacking CD4 receptors.

Cell Cultures

The 174XCEM T/B hybrid cell line (7) is maintained at 5×10^5 cells/ml in RPMI-1640 with 10% (v/v) fetal calf serum (FCS), penicillin (100 U/ml), streptomycin (100 μg/ml), and glutamine (300 μg/ml), and is split 1:4 twice weekly. HeLa CD4-positive clone 1022 and the congenic CD4-negative HeLa (8,9), CD4-expressing HeLa T4$^+$ and congenic CD4-negative HeLa cells (10), and baby hamster kidney clone 21 (BHK-21) cell lines are maintained in Dulbecco's modified Eagle's medium (DMEM) with 10% FCS, penicillin (100 U/ml), streptomycin (100 μg/ml), and glutamine (300 μg/ml), and are split 1:4 twice weekly.

Human Immunodeficiency Viruses

HIV-1 strain IIIB (11) was isolated from peripheral blood lymphocytes (PBLs) and is lymphocyte tropic, that is, will replicate in T cell lines and in PBLs but not in monocyte-derived macrophages. Virus stocks are grown in SUP-T1 cells (12), 174XCEM cells (7), or phytohemagglutinin/interleukin 2 (PHA/IL-2) stimulated PBLs (13), and clarified supernatant is frozen at $-80°$C. Virus stocks have titers of 500–2000 ng/ml or 10^5–10^6 TCID$_{50}$ (50% tissue culture infective dose) per milliliter.

Vesiculoviruses

Cocal virus (COV) (14) and vesicular stomatitis virus, Indiana strain (VSV), are members of the vesiculovirus group of the rhabdoviruses and are closely related antigenically. Vesiculoviruses are provided by the Yale Arbovirus Research Unit (New Haven, CT) and are grown in BHK-21 cells. Stocks with a titer of about 10^8 plaque-forming units (pfu)/ml are frozen at $-80°$C.

Plaque Assay for Cocal Virus

Assays are done in 24-well plates [Costar (Cambridge, MA), Nunc (Roskilde, Denmark), or Falcon] using BHK-21, HeLa, or other adherent cell lines, and growth medium for all dilutions. Virus dilutions are added, at 0.1 ml/well. A cell suspension is freshly made by trypsinizing a flask of proliferating cells and adjusting to a concentration of 10^6 cells/ml. To each well is added 0.5 ml of cell suspension (5×10^5 cells) and the plate is incubated at $37°$C for 0.5–1 hr to allow the cells to attach. (Alternatively, plaquing may be done with preformed monolayers, as in the pseudotype assay described below.) Carboxymethylcellulose [3% (w/v) CMC] overlay is added at 0.5 ml/well. The CMC overlay is made by combining equal volumes of two stocks: CMC, 6 g/100 ml of distilled water; and $2\times$ medium with 20% FCS. Cultures are incubated for 1–2 days and examined for plaque formation using an inverted microscope.

Once plaques are visible, cultures are fixed and stained: the overlay is removed by inversion over a disposal container, the plates are washed once by gentle application of isotonic NaCl, and are then fixed and stained with naphthalene black (naphthalene black, 1 g; glacial acetic acid, 60 ml; anhydrous sodium acetate, 13.6 g; and deionized water to 1 liter) for 5 min, followed by washing with tap water and drying. Stained plates can be stored indefinitely.

Rabbit Anti-cocal Virus Antiserum

Rabbits are immunized by infection followed by boosting with a virus–adjuvant mixture. The animals are infected with 0.5 ml of COV (10^6 pfu) injected intramuscularly, and undergo a silent infection. At 1 and 2 months after infection, the rabbits are immunized with concentrated virus (6) homogenized with an equal volume of Freund's complete adjuvant and administered in 10 separate intradermal sites (0.1 ml/site). Rabbits are bled before infection and 2–4 weeks after each immunization. After the second immunization, the serum has a neutralizing index of $>10^7$ at a 1:100 concentration and a neutralizing titer of about 10^4 against 100 pfu.

Production of Pseudotype Virus

Pseudotype virus is produced in a two-step procedure: HIV infection followed by superinfection with COV as shown in Fig. 2. A cell line, usually 174XCEM, at 2×10^5 cells/ml in growth medium is infected with an HIV stock diluted to give a final concentration of 10 ng of p24/ml cell suspension. The titer of the supernatant is followed until maximal levels are reached, about 1000–2000 ng/ml, which usually requires 7–14 days depending on the virus strain (13). Cocal virus is then added, at an input of 10^2–10^3 pfu/ml cell culture, and the culture is harvested 24–48 hr later, clarified by centrifugation at 5000 rpm for 5 min, and aliquots are frozen at $-80°$ C.

Assay of Pseudotype Virus

A standard protocol consists of four mixes, shown in Table I. Growth medium is used as diluent throughout. Mixes 1 and 2 contain a reference stock of COV, and are required to show that the rabbit anti-COV as used in the assay can totally neutralize a high-titer stock of COV. Mixes 3 and 4 contain the putative pseudotype stock that

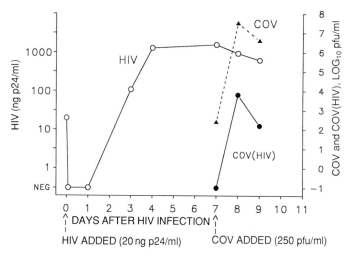

FIG. 2   Production of pseudotype virus by sequential infection of 174XCEM cells with HIV-1 strain IIIB (input, 20 ng/ml) followed by superinfection with COV (input, $10^{2.4}$ pfu/ml). The titers of HIV, COV, and COV(HIV) are shown. See text for details.

contains both COV and COV(HIV); mix 3 shows the titer of COV, and mix 4 is designed to determine whether the stock also contains authentic pseudotype virions, that is, virions that will escape neutralization with rabbit anti-COV antiserum.

All mixes are incubated at 37° C for 30 min and are then diluted in 10-fold steps through 10^6. The dilutions are added, 0.1 ml/well, to preformed monolayers of HeLa CD4 or other indicator cells and are incubated at 37° C for 4–6 hr to allow time for

TABLE I   Protocol for Assay of Pseudotype Virus[a]

| Mix | Virus stock | Virus present | | Antiserum treatment | Virus-producing plaques | Expected titer (pfu/ml) |
		COV	COV(HIV)			
1	COV	Yes	-	Normal rabbit serum	COV	$10^{7.6}$
2	COV	Yes	-	Rabbit anti-COV	None	$<10^1$
3	Putative COV(HIV)	Yes	Yes	Normal rabbit serum	COV	$10^{7.8}$
4	Putative COV(HIV)	Yes	Yes	Rabbit anti-COV	COV(HIV)	$10^{4.1}$

[a] To show composition of four mixes, each of which contains 0.05 ml of virus, 0.05 ml of serum, and 0.4 ml of diluent. Mixes 1 and 2 are designed to demonstrate that the rabbit anti-COV antiserum completely neutralizes any COV present, whereas mixes 3 and 4 are designed to demonstrate the titers of COV and of COV(HIV) present in the putative pseudotype stock. HIV and HIV(COV) present in mixes 3 and 4 can be disregarded because replication is slow and does not result in plaques.

virus entry. To remove the anti-COV in the mixes, each well is carefully aspirated, 0.75 ml of medium is added, aspirated, and the cultures are refed with 0.5 ml of medium. Overlay, 0.5 ml of 3% CMC, is added, and cultures are incubated until plaques develop (usually 2 days), followed by fixation and staining. For optimal results it is important to follow the steps in this protocol with precision.

Illustrative Pseudotype Assays

Figure 3 demonstrates an assay of a pseudotype stock (produced in 174XCEM cells with HIV-IIIB, a lymphocyte-tropic strain) which shows typical results. The COV titer is $10^{7.6}$ pfu/ml and the pseudotype stock has a titer of $10^{4.1}$ pfu/ml, similar to that reported by Weiss and collaborators (3–5).

To confirm that the residual titer in the assay protocol represents authentic COV(HIV), several controls can be used, including (a) repetition of the assay using CD4-negative congenic cells, (b) the demonstration that either anti-CD4 antibodies or recombinant soluble CD4 can reduce the pseudotype titer, and (c) the demonstration that anti-HIV antisera can reduce the pseudotype titer. We have relied mainly on the use of congenic CD4-negative HeLa cells because this control involves no changes in the assay protocol and because it demonstrates that the COV(HIV) infection is receptor dependent and therefore appropriate for studies of HIV entry. Figure 3B is a control using HeLa CD4-negative cells and shows that the COV(HIV) fails to register in the absence of an HIV receptor.

Requirements for Production and Assay of Pseudotype Virus Expressing HIV Envelope

Our experience has indicated that there are a number of critical requirements for the successful production of pseudotype virus that expresses HIV glycoproteins and can be quantified in a plaque assay: (a) A virus partner must be selected that will replicate rapidly to high titer in human lymphoid and monocytoid cells, including terminally differentiated monocytes; these attributes are necessary for the broadest application of HIV pseudotype viruses. Few animal viruses meet these criteria and, even within the *Vesiculovirus* genus, only selected viruses will replicate in human monocytes (6); (b) human lymphoid cell lines, such as 174XCEM and Molt 4 clone 8, which are permissive for selected macrophage-tropic strains of HIV, are essential for the production of pseudotype viruses expressing the envelope proteins of these strains of HIV; (c) the development of HeLa cell lines that express CD4 together with congenic CD4-negative HeLa cells provides a powerful system for the assay of putative pseudotype virus, although these cells can be used only for lymphocyte-tropic strains of HIV; (d) the somewhat unexpected variation of greater than 1000-fold in the

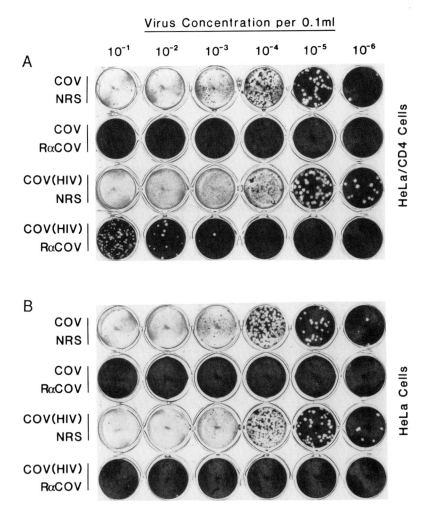

FIG. 3   Assay of pseudotype virus COV(HIV-1) on HeLa/CD4 (A) and HeLa cells (B). (A) COV and NRS, Titration of COV stock plus normal rabbit serum; COV and RαCOV, titration of the same COV stock plus immune rabbit antiserum against COV to show total neutralization of COV; COV(HIV) and NRS, titration of a putative stock of COV(HIV-1) plus normal rabbit serum to show titer of COV; COV(HIV) and RαCOV, titration of the same stock of COV(HIV-1) plus immune rabbit antiserum against COV to show that the pseudotype virus escapes neutralization and produces plaques. In this stock the titer of COV is $10^{7.6}$ pfu/ml and the titer of COV(HIV-1) is $10^{4.1}$ pfu/ml. Each well contains virus in a final volume of 0.1 ml at the concentration indicated. (B) The same stocks of virus are titrated on HeLa cells to show that the results are similar except for the bottom row [COV(HIV), RαCOV], where no plaques are seen. This negative control indicates that the putative pseudotype stock would not infect HeLa cells, which do not express CD4.

plaque titer of COV on different COV-permissive cell lines makes this parameter a critical one for assay of pseudotype virus, because COV(HIV) pseudotype stocks reached a titer about $10^3 – 10^4$ lower than that of COV. Thus, among CD4-expressing HeLa cells, the 1022 cell line is much more useful than the T4$^+$ cell line because of its greater sensitivity to COV; (e) among several alternate protocols for assay of COV(HIV), only one has been found to register pseudotype virus plaques consistently. These constraints on the production and application of pseudotype viruses may explain why the pseudotype method has not been more widely used in HIV research.

Applications of HIV Pseudotype Virus

Analysis of CD4 Receptor

CD4 has been established as the principal receptor for HIV, and an ongoing analysis of the interaction between the receptor and gp120, the virus attachment protein, includes a definition of the domains that are crucial for receptor binding. One study (15) utilized COV(HIV) pseudotype viruses to assess the relative efficiency of entry mediated by different human–rat chimeric CD4 molecules, as a technique to map the HIV-binding site.

Alternative Cellular Receptors for HIV

It is now well established that certain cell lines can be infected with HIV even though they do not express the CD4 receptor (16–19). Furthermore, there is at least one molecule that has been proposed as a candidate alternative receptor, that is, galactosylceramide (20,21), and there may be other receptors, because galactosylceramide is not expressed on some CD4-negative cells that can be infected with HIV (16,19). Because many of these CD4-negative HIV-permissive cell lines produce only low levels of infectious virus, the HIV pseudotype system may provide a useful bioassay for detailed studies of CD4-independent entry. We have found that COV(HIV) pseudotype virus will produce plaques on HT29 cells, a colonic carcinoma cell line (21) that is CD4 negative but expresses galactosylceramide (Peterson *et al.,* unpublished, 1992).

Host Range Differences in Strains of HIV

One of the major enigmas in the biology of HIV is the mechanism of differences in the cellular host range of different virus isolates, particularly the difference between lymphocyte-tropic strains, which fail to replicate in macrophages, and macrophage-tropic strains, which fail to replicate in most human T lymphocytoid cell lines (13). All of these strains utilize the CD4 receptor to gain entry to lymphocytes and macrophages (22), suggesting that they utilize a similar entry pathway. Yet genetic studies with recombinant viruses indicate that many of the virus determinants of tropism map to the envelope gene (23–26), suggesting that tropism is determined at the level of entry. This apparent paradox is under investigation in many laboratories, and studies with HIV pseudotype virus may help to dissect the role of envelope genes and accessory cellular molecules in entry and tropism.

Acknowledgments

Supported in part by Grants NS 27405, HL 02358, and HL 07586 from the United States Public Health Service, and by grants from the AIDS Directed Program, Medical Research Council. We would like to thank Laurence Turley and Vedapuri Shanmugam for expert technical support, Robert Shope for vesiculoviruses, Paul Clapham and Robin Weiss for advice, assistance, and reagents, and Bruce Chesebro for CD4-expressing (1022) cells.

References

1. D. Boettiger, *Prog. Med. Virol.* **25,** 37 (1979).
2. R. A. Weiss, *in* "Rhabdoviruses" (D. H. L. Bishop, ed.). CRC Press, Boca Raton, Florida, 1977.
3. P. R. Clapham, K. Nagy, and R. A. Weiss, *Proc. Natl. Acad. Sci. USA* **81,** 2886 (1984).
4. A. G. Dalgleish, P. C. L. Beverly, P. C. Clapham, D. H. Crawford, M. F. Greaves, and R. A. Weiss, *Nature (London)* **312,** 763 (1984).
5. R. A. Weiss, P. R. Clapham, R. Cheingsong-Popov, A. G. Dalgleish, C. A. Carne, I. V. D. Weller, and R. S. Tedder, *Nature (London)* **316,** 69 (1985).
6. S. A. Gregory, R. Collman, W. James, S. Gordon, F. Gonzalez-Scarano, and N. Nathanson, submitted (1993).
7. R. D. Salter, D. N. Howell, and P. Cresswell, *Immunogenetics* **21,** 235 (1985).
8. B. Chesebro and K. Wehrly, *J. Virol.* **62,** 3779 (1988).
9. B. Chesebro, K. Wehrly, J. Metcalf, and D. E. Griffin, *J. Infect. Dis.* **163,** 64 (1991).
10. P. J. Maddon, A. G. Dalgleish, J. S. McDougal, P. R. Clapham, R. A. Weiss, and R. Axel, *Cell (Cambridge, Mass.)* **47,** 333 (1986).
11. M. Popovic, M. G. Sarngadharan, E. Read, and R. C. Gallo, *Science* **224,** 297 (1984).
12. S. D. Smith, M. Shatsky, P. S. Cohen, R. Warnke, M. P. Link, and B. E. Glader, *Cancer Res.* **44,** 5657 (1984).
13. R. Collman, N. F. Hassan, R. Walker, B. Godfrey, J. Cutilli, J. Hastings, H. Friedman, S. D. Douglas, and N. Nathanson, *J. Exp. Med.* **170,** 1149 (1989).
14. A. H. Jonkers, R. E. Shope, T. H. G. Aitken, and L. Spence, *Am. J. Vet. Res.* **25,** 236 (1964).

15. J. H. M. Simon, C. Somoza, F. A. Schockmel, M. Collin, S. J. Davis, A. F. Williams, and W. James, submitted (1993).

16. Y. Cao, A. E. Friedman-Kien, Y. Huang, X. L. Li, M. Mirabile, T. Moudgil, D. Zucker-Franklin, and D. D. Ho, *J. Virol.* **64,** 2553 (1990).

17. J. M. Harouse, C. Kunsch, H. T. Hartle, M. A. Laughlin, J. A. Hoxie, B. Wigdahl, and F. Gonzalez-Scarano, *J. Virol.* **63,** 2537 (1989).

18. X. L. Li, T. Moudgil, H. V. Vinters, and D. D. Ho, *J. Virol.* **64,** 1383 (1990).

19. M. Tateno, F. Gonzalez-Scarano, and J. A. Levy, *Proc. Natl. Acad. Sci. USA* **86,** 4287 (1989).

20. J. M. Harouse, S. Bhat, S. L. Spitalnik, M. Laughlin, K. Stefano, D. H. Silberberg, and F. Gonzalez-Scarano, *Science* **253,** 320 (1991).

21. N. Yahi, S. Baghdiguian, H. Moreau, and J. Fantini, *J. Virol.* **66,** 4448 (1992).

22. R. Collman, B. Godfrey, J. Cutilli, A. Rhodes, N. F. Hassan, R. Sweet, S. D. Douglas, H. Friedman, N. Nathanson, and F. Gonzalez-Scarano, *J. Virol.* **64,** 4468 (1990).

23. A. J. Cann, M. J. Churcher, M. Body, W. O'Brien, J. Q. Zhao, J. Zack, and I. S. Y. Chen, *J. Virol.* **66,** 305 (1992).

24. J. M. Hwang, T. J. Boyle, H. K. Lyerly, and B. R. Curren, *Science* **253,** 71 (1991).

25. W. A. O'Brien, Y. Koyanagi, A. Namazie, J. Q. Zhao, A. Diagne, K. Idler, J. A. Zack, and I. S. Y. Chen, *Nature (London)* **348,** 69 (1990).

26. P. Westervelt, H. E. Gendelman, and L. Ratner, *Proc. Natl. Acad. Sci. USA* **88,** 3097 (1991).

[8] Application of High-Performance Thin-Layer Chromatography to Investigate Interactions between Human Immunodeficiency Virus Type 1 Glycoprotein gp120 and Galactosylceramide

David G. Cook, Jacques Fantini, Steven L. Spitalnik, and Francisco Gonzalez-Scarano

Introduction and Background

It is well established that the primary cellular receptor for human immunodeficiency virus type 1 (HIV-1) is the CD4 molecule and that CD4 binds the HIV-1 surface glycoprotein, gp120 (1–3). However, HIV-1 is also capable of infecting a diverse range of human cell types that do not express CD4 (4–7). Evidence has emerged indicating that the neutral glycolipid, galactosylceramide (Gal-Cer, or other closely related glycolipids such as galactosylsulfatide) may serve as an alternative receptor for several CD4-negative cell lines (8–12). Thus, as has been shown for several other viruses, infection by HIV-1 may involve interactions between viral envelope proteins and carbohydrate moieties expressed on the cell surface (13–15).

These findings give rise to a number of important questions regarding the role that glycolipids may play in HIV-1 infection and pathogenesis, including the following: (a) Does gp120, the HIV-1 envelope glycoprotein, bind to Gal-Cer? (b) What cell types contain glycolipids that bind gp120? (c) Do the envelope proteins of distinct HIV-1 isolates display differing affinities for Gal-Cer? (d) When it binds Gal-Cer, does gp120 undergo conformational changes that may be important for virus entry? Because Gal-Cer is a lipid, many of the techniques customarily employed to study protein–protein interactions are inappropriate to address such questions. High-performance thin–layer chromatography (HPTLC), on the other hand, is well suited for the study of protein–glycolipid interactions and has found wide application in the study of both bacterial and viral protein binding to glycolipids (16, 17).

We describe here an overlay technique adapted from Magnani *et al.* (18) to assess the binding of gp120 to Gal-Cer by indirectly probing gp120 with anti-gp120 antibodies. This technique has yielded reliable results that are easy to quantify and uses small amounts of gp120. Moreover, this method avoids many of the difficulties and biohazards associated with HPTLC techniques that require protein iodination.

Methods in Molecular Genetics, Volume 4

Methods

Preparation of HPTLC Plates

The general procedure for preparing HPTLC plates with immobilized bands of Gal-Cer is outlined in Fig. 1. Galactosylceramide (C-4905 or C-8752; Sigma, St. Louis, MO) is dissolved at 1 mg/ml in chloroform–methanol (2:1) and 5 μl is spotted in 5-mm lanes, using a Hamilton syringe. A 20×10 cm HPTLC plate will accommodate several lanes spaced 2 cm apart and 1.5 cm from either edge of the plate. The Gal-Cer is next chromatographed by placing the HPTLC plate in a TLC tank in which the solvent system (chloroform–methanol–water, 60:35:8, v/v) has

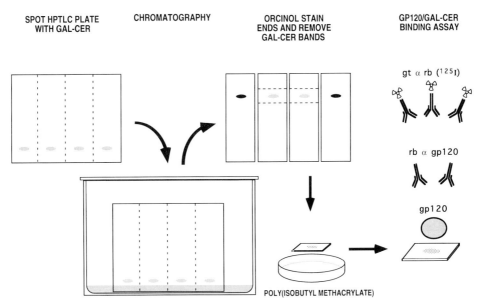

FIG. 1   Preparation of HPTLC plates and binding of gp120 to galactosylceramide. Galactosylceramide is spotted in lanes near the bottom of HPTLC plates, which are then chromatographed in a TLC tank containing the solvent system (chloroform–methanol–water, 60:35:8). Far left and right lanes are cut from the plate and stained with orcinol to reveal location of resolved Gal-Cer bands. Using orcinol-stained lanes as a guide, remaining Gal-Cer bands are cut from the HPTLC plate and treated with poly(isobutyl methacrylate) to reduce nonspecific background binding. Next, plates with immobilized Gal-Cer are serially washed and incubated with gp120, rabbit anti-gp120 antiserum, and iodinated goat anti-rabbit antiserum. gp120 Binding to Gal-Cer is revealed by autoradiography or direct phosphoimaging analysis of HPTLC plates.

equilibrated for at least 2 hr. When the solvent front migrates to the top of the HPTLC plate, it is removed from the tank and allowed to air dry.

To minimize the reagents used in subsequent incubations, small rectangles containing the Gal-Cer bands are cut from the plate with a razor blade. To determine the location of the resolved Gal-Cer bands, the far left and far right lanes are cut from the plate and stained with an orcinol spray (O-1875; Sigma), which stains all glycolipids. The orcinol solution is prepared as follows: 10 ml of orcinol stock [1% (v/v) in H_2O] stored in a brown bottle; 87 ml of H_2O; and 3 ml of concentrated H_2SO_4. The orcinol solution is sprayed on the plates, allowed to dry, and then heated for a few seconds over a hot plate to reveal the location of the glycolipid bands. Next, using the orcinol-stained lanes as a guide, the remaining Gal-Cer bands are cut from the plate centered on 2.5 $\times$ 1 cm rectangles. Finally, each plate is soaked for 1–3 min in a hexane solution containing dissolved poly(isobutyl methacrylate) beads (1 mg/ml) (Polysciences, Warrington, PA) to reduce nonspecific background staining. In addition, it has been speculated that the plastic may cause the glycolipids to orient their polar and nonpolar domains in the silica as they would in a lipid membrane (19). After the plates have dried at room temperature they are ready to be used.

It is tempting to speed up the preparation of plates by omitting the chromatography steps and simply to spot glycolipids onto small HPTLC plates. However, it has been our experience that glycolipids are unevenly deposited on the HPTLC plates, resulting in binding by gp120 that is irregular and difficult to quantify. The process of chromatography produces glycolipid bands of uniform density and vastly improves the results.

Binding of gp120 to Galactosylceramide

Immediately prior to immersing the HPTLC plates in a blocking solution, they should be lightly misted with phosphate-buffered saline (PBS). The plates are blocked for 1 hr at room temperature with 1% gelatin Tris-HCl (0.05 M Tris-HCl, 0.15 M NaCl; pH 7.3) and all dilutions are made in this solution. After blocking, the HPTLC plates are placed horizontally in a humidity chamber of individual pedestals (for the small plates described above, the snap tops from two 500-ml centrifuge tubes placed side by side in the bottom of the chamber work well). This procedure allows the surface of each plate to be covered by a minimal volume of the reagent mixture. To calculate the required volume in microliters, multiply the area of the plate (in cm^2) by 60.

Next, gp120 (1–4 μg/ml) (BH10 clone for the experiments described below) is overlaid on the plate and incubated for 30–60 min at room temperature (Fig. 1). Following this incubation, the plates are washed and then serially incubated and

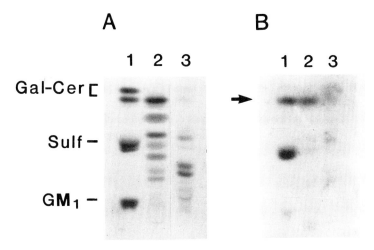

FIG. 2   (A) Orcinol stain of alkali-stable glycolipids. Neutral glycolipids extracted from HT29 and Caco-2 (clone 2) cells (200 mg derived from approximately 10^7 cells each) were chromatographed by HPTLC with a chloroform–methol–water (60:35:8) solvent system and sprayed with orcinol. Lane 1, Gal-Cer (upper and lower bands correspond to nonhydroxyl and α-hydroxyl species, respectively), galactosylsulfatide (Sulf), and GM_1 standards; lane 2, HT29 cells; lane 3, Caco-2 (clone 2) cells. In the HT29 lane but not the Caco-2 (clone 2) lane, a prominent band comigrated with α-hydroxyl-Gal-Cer standard. (B) Autoradiography of gp120 binding to glycolipid extracts. Instead of orcinol staining, a plate prepared identically to (A) was serially incubated with recombinant gp120 (1 μg/ml) (BH10 strain), polyclonal rabbit anti-gp120 antiserum (1:200) (both antibody and gp120 provided by R. Sweet of SmithKline Beecham), and iodinated goat anti-rabbit antiserum (1400 cpm/ml) (NEX 155; Du Pont). gp120 bound only to Gal-Cer standard, galactosylsulfatide standard, and the HT29 band comigrating with Gal-Cer. [Adapted from (12).]

washed (five times in PBS) with sheep anti-gp120 (1.5 mg/ml) (6205, International Enzymes, Fallbrook, CA), rabbit anti-sheep antiserum (1:500) (S-1265; Sigma), and iodinated goat anti-rabbit antiserum (6–10 μCi/μg, diluted to 1500 cpm/μl) (NEX 155; New England Nuclear–Du Pont, Wilmington, DE). We routinely use variations of this procedure that use fewer incubation steps (see Figs. 1 and 2B). This three-step labeling procedure is recommended because it uses antibody reagents that are commercially available and are relatively inexpensive.

Following the last washing steps the HPTLC plates are dried under a heat lamp and autoradiographed with an intensifying screen overnight at $-80°$ C (X-Omat AR; Kodak, Rochester, NY). The binding of gp120 to Gal-Cer can be quantified by scanning densitometry of the autoradiographs or by direct phosphoimaging analysis of the HPTLC plates (e.g., Molecular Dynamics, Sunnyvale, CA).

Applications

Cell-Specific Expression of Galactosylceramide Binding to gp120

In addition to binding gp120 to purified Gal-Cer, the procedures described above can be directly applied to examine the binding of gp120 to cellular glycolipid extracts [glycolipid extraction methods are described elsewhere (18, 20)].

An example of the application of this technique follows from the observation that the human colonic adenocarcinoma cell line HT29 can be productively infected by both HIV-1 and HIV-2 in a CD4-independent fashion (21–25). Moreover, antibodies raised against Gal-Cer block HIV infection of these cells, suggesting that HIV may use Gal-Cer as a receptor for HT29 cells (11, 12). In contrast to HT29, another colonic-derived cell line, Caco-2 (clone 2), is much less susceptible to infection by HIV (21). Can these differences in the level of infection be explained by differing levels of Gal-Cer expression? Does gp120 bind glycolipid extracts from HT29 but not Caco-2 (clone 2)?

Figure 2A shows orcinol staining of the alkali-stable lipids expressed in HT29 and Caco-2 (clone 2) cells, indicating that for each cell line a large number of neutral glycolipids are expressed to different levels (12). However, only the band in the HT29 lane comigrating with the lower α-hydroxyl-Gal-Cer standard binds gp120 (Fig. 2B). Additional analyses have verified that this comigrating band is authentic Gal-Cer (12). In contrast, there were no detectable levels of Gal-Cer expression or gp120/glycolipid binding to the Caco-2 (clone 2) lipid extract (12).

Additional Applications

Many of the remaining questions concerning the role glycolipid–gp120 interactions play in HIV-1 infection can be approached using the techniques described above.

Do glycolipids other than Gal-Cer bind gp120? Currently, gp120 has been reported to bind to galactosylsulfatide, psychosine, and G_{M_4}; all of these, like Gal-Cer, contain a proximal galactose (26). The related glycolipids glucosylceramide, lactosylceramide, and G_{M_1} all lack a proximal galactose and exhibit poor binding to gp120 (26).

What domains of Gal-Cer and gp120 mediate binding? The selectivity of gp120 for glycolipids with a proximal galactose indicates that the Gal-β-Cer linkage is essential for binding. Thus, in addition to galactose a lipid moiety is necessary for binding to Gal-Cer, but in most cases neither the structure of the ceramide nor the fatty acid chain of Gal-Cer has an impact on the binding of gp120 (26). However, in our hands gp120 binds only the α-hydroxyl form of Gal-Cer [Gal-Cer with an α-hydroxyl fatty acid chain (see Fig. 2)]. This may be due to proposed conformational differences between α-hydroxyl and nonhydroxyl Gal-Cer (27). The V3 domain of gp120 appears to be critical for binding to Gal-Cer because anti-V3 mono-clonal antibodies interfere with binding, whereas site-specific antibodies against other domains of gp120 do not block binding to Gal-Cer (28).

Work is now underway to investigate the Gal-Cer-binding properties of gp120 derived from different HIV-1 strains. In addition, HPTLC binding assays are being employed to address questions regarding both the conformational requirements for gp120–Gal-Cer interactions and the impact of gp120 multimerization on binding to Gal-Cer.

Acknowledgments

Supported in part by Grants NS27405, NS31067, and NS30606 from the U.S. Public Health Service, and by grants from the North Atlantic Treaty Organization (NATO) and the Fondation pour la Recherche Médicale to J.F. We thank Dr. Raymond Sweet (SmithKline Beecham, King of Prussia, PA) for the gift of recombinant gp120, and Andrew Albright for technical assistance.

References

1. A. G. Dalgleish, P. C. L. Beverly, P. R. Clapham, D. H. Crawford, M. F. Greaves, and R. A. Weiss, *Nature (London)* **312,** 763 (1984).
2. D. Klatzmann, E. Champagne, S. Charmaret, J. Gruest, D. Guetard, T. Hercend, J. C. Gluckman, and L. Montagnier, *Nature (London)* **312,** 767 (1984).
3. P. J. Maddon, A. G. Dalgleish, J. S. McDougal, P. R. Clapham, R. A. Weiss, and R. Axel, *Cell (Cambridge, Mass.)* **47,** 33 (1986).
4. F. Chiodi, S. Fuerstenber, M. Gidlund, B. Asjo, and E. M. Fenyo, *J. Virol.* **61,** 1244 (1987).
5. M. Tateno, F. Gonzalez-Scarano, and J. A. Levy, *Proc. Natl. Acad. Sci. USA* **86,** 4287 (1989).
6. J. M. Harouse, C. Kunsch, H. T. Hartle, M. A. Laughlin, J. A. Hoxie, B. Wigdahl, and F. Gonzalez-Scarano, *J. Virol.* **63,** 2527 (1989).
7. X. L. Li, T. Moudgil, H. V. Vinters, and D. D. Ho, *J. Virol.* **64,** 1383 (1990).
8. L. H. va den Berg, S. A. Sadiq, S. Lederman, and N. Latov, *J. Neurosci. Res.* **33,** 513 (1992).
9. J. Schneider-Schaulies, S. Schneider-Schaulies, R. Brinkmann, P. Tas, M. Halbrugge, U. Walter, H. C. Holmes, and V. Ter Meulen, *Virology* **191,** 765 (1992).
10. J. M. Harouse, S. Bhat, S. L. Spitalnik, M. Laughlin, K. Stefano, D. H. Silberberg, and F. Gonzalez-Scarano, *Science* **253,** 320 (1991).
11. N. Yahi, S. Baghdiguian, H. Moreau, and J. Fantini, *J. Virol.* **66,** 4848 (1992).
12. J. Fantini, D. G. Cook, N. Nathanson, S. L. Spitalnik, and F. Gonzalez-Scarano, *Proc. Natl. Acad. Sci. USA* **90,** 2700 (1993).

13. G. C. Hansson, K.-A. Karlsson, G. Larson, N. Stromberg, J. Thurin, C. Orvell, and E. Norrby, *FEBS Lett.* **170,** 15 (1984).

14. D. C. Wiley and J. J. Skehel, *Annu. Rev. Biochem.* **56,** 365 (1987).

15. E. Superti and G. Donelli, *J. Gen. Virol.* **72,** 2467 (1991).

16. G. C. Hansson, K.-A. Karlsson, G. Larson, A. A. Lindberg, N. Stromberg, and J. Thurin, *Proc. Int. Symp. Glycoconjugates,* 7th, 631 (1983).

17. G. C. Hansson, K.-A. Karlsson, G. Larson, N. Stromberg, and J. Thurin, *Anal. Biochem.* **146,** 158 (1985).

18. J. L. Magnani, S. L. Spitalnik, and V. Ginsburg, *in* "Methods in Enzymology," Vol. 138, p. 195. Academic Press, Orlando, Florida, 1987.

19. K.-A. Karlsson and N. Stromberg, *in* "Methods in Enzymology," Vol. 138, p. 220. Academic Press, Orlando, Florida, 1987.

20. S. Hakomori, *Handbook Lipid Res.* **3,** 1 (1983).

21. A. Adachi, S. Koenig, H. E. Gendelman, D. Daugherty, D. Gattoni-Celli, A. S. Fauci, and M. A. Martin, *J. Virol.* **61,** 209 (1987).

22. M. B. Omary, D. A. Brenner, L. Y. De Gradpre, K. A. Roebuck, D. D. Richman, and M. F. Kagnoff, *AIDS* **5,** 275 (1991).

23. S. W. Barnett, A. Barboza, C. M. Wilcox, C. E. Forsmark, and J. A. Levy, *Virology* **182,** 802 (1991).

24. J. Fantini, N. Yahi, and J. C. Chermann, *Proc. Natl. Acad. Sci. USA* **88,** 9297 (1991).

25. J. Fantini, N. Yahi, S. Baghdiguian, and J. C. Chermann, *J. Virol.* **66,** 580 (1992).

26. S. Bhat, S. L. Spitalnik, F. Gonzalez-Scarano, and D. H. Silberberg, *Proc. Natl. Acad. Sci. USA* **88,** 7131 (1991).

27. K.-A. Karlsson, *Annu. Rev. Biochem.* **58,** 309 (1989).

28. D. G. Cook, J. Fantini, S. L. Spitalnik, and F. Gonzalez-Scarano, *Virology,* in press (1994).

[9] Quantitative Analysis of Human Immunodeficiency Virus Internalization

Janet M. Harouse, Kelly Stefano, and Francisco Gonzalez-Scarano

Background

The initial event in the life cycle of a virus is attachment to specific receptors on the host cell membrane, and this step is often a major determinant of virus tropism and pathogenesis (1). Viruses usually use a host cell intrinsic membrane component as a specific receptor, and attachment to the receptor allows for the subsequent efficient entry of a virus into the host cell. It is well established that the CD4 molecule is the major cellular receptor for human immunodeficiency virus type 1 (HIV-1) in cells of lymphocytic and monocytic origin (2–4). The CD4 molecule binds the surface glycoprotein of HIV-1, gp120, which eventually leads to fusion of the viral and cellular membranes (5). A number of investigators have also demonstrated HIV-1 infection in CD4-negative cell lines *in vitro,* although the infection is less permissive than in CD4-positive cell lines (6–9). To ascertain whether the nonpermissive nature of these HIV-1 infections reflects a block at the level of viral binding and entry into the host cell, or whether an alternative pathway of entry exists for HIV-1, we utilized a viral internalization assay that allowed us to examine entry events (10). This assay has been successfully used to investigate a putative CD4-independent entry pathway in neurally derived cell lines. The conditions of the assay are such that we are measuring only internalization of HIV-1 into the host cell, and the assay is terminated prior to the initiation of reverse transcription. This technique has yielded reproducible results and avoids the biohazard of using purified HIV-1.

Methods

The method outlined below has been used with a number of CD4-positive and CD4-negative cell lines with excellent results. However, the parameters of incubation, washing, and particularly the trypsin treatment conditions, must be individualized for each cell line. A 0–4°C control must be performed with each assay. Furthermore, because of varying background levels, the assay must be performed a sufficient number of times to allow statistical analysis. We have not attempted to use this assay with primary lymphocytes, monocytes, or macrophages.

Cells and Virus

The CEMx174 cell line is a CD4-positive T cell/B cell hybrid line, and is highly susceptible to infection by many HIV-1 isolates (11). The control, nonsusceptible

721.174, is the CD4-negative parental B cell line of the hybrid CEMx174 (12). The cells are maintained in RPMI-1640 supplemented with 10% (v/v) fetal calf serum. The U373-MG cell line is a central nervous system (CNS)-derived, CD4-negative glioma cell line (obtained from D. Herlyn, Wistar Institute, Philadelphia, PA). The HTB-138 is also a CD4-negative glioma cell line, but is not infectable with HIV-1 (8). The glioma cell lines are maintained in Dulbecco's modified Eagle's medium (DMEM) containing 10% FCS. The HeLa T4 cell line is a stable CD4-positive transfectant of the cervical carcinoma cell line, HeLa (gifts of R. Axel, Columbia University, NY, NY). The experiments described are performed with the IIIB strain of HIV-1 (13).

Antibodies

Monoclonal antibody Leu3a directed against the CD4 molecule is obtained from Becton Dickinson (San Jose, CA) (14,15). Monoclonal antibody B33.1.1 recognizes antigenic determinants on the HLA II molecule (a gift from B. Perussia, Thomas Jefferson University, Philadelphia, PA). Rabbit polyclonal antiserum against galactosylceramide (anti-Gal-Cer) has been described previously (16–18). Control polyclonal rabbit sera is heat-inactivated rabbit sera.

Internalization Assay

In the viral internalization assay, 2×10^6 CEMx174 or 721.174 suspension cells are centrifuged for 10 min at a relative centrifugal force (RFC) of 400 g at 4°C. Cells are resuspended in 10 ml of phosphate-buffered saline (PBS) and again centrifuged as above. The cell pellet is then resuspended in 1 ml of RPMI-1640 containing HIV-1$_{\text{IIIB}}$ (final concentration, 500 ng of p24^{gag}/ml) and transferred to a sterile microfuge tube. The virus–cell suspension mixture is incubated for 60 min at either 0°C to allow for viral binding only, or at 37°C to allow for binding and internalization. After 1 hr, the cells are centrifuged for 2 min at 1000 g and all samples are transferred to ice. The inoculum is removed and the cell pellet is washed in 1 ml of cold PBS, and centrifuged again for 2 min at 1000 g. The entire volume of PBS is gently removed without disrupting the cell pellet. This wash is repeated three times. The residual, noninternalized inoculum is removed by the addition of 500 μl of trypsin (T-0646; Sigma Chemicals, St. Louis, MO), at a final concentration of 2.5 mg/ml, for 4 min at room temperature. The trypsin is then inactivated by the addition of 500 μl of fetal calf serum (FCS). Cells are again pelleted in microfuge tubes for 2 min at 1000 g at room temperature. The supernatant is gently removed and the wash step repeated three times. The cell pellet is resuspended in 200 μl of RPMI-1640. The cells are lysed by the addition of 20 μl of 10% (v/v) Triton X-100 for a final concentration of 1% Triton X-100. The concentration of p24^{gag} in the cell lysate is measured using a commercial p24^{gag} antigen capture assay.

Inhibition of Internalization

To assay the specificity of the internalization assay in a CD4-dependent infection, cells are first incubated with monoclonal antibody (MAb) (25 μg/ml in PBS) or PBS alone for 60 min at room temperature. Cells are centrifuged for 2 min at 1000 g at room temperature and resuspended in 1 ml of RPMI-1640 containing HIV-1 (500 ng of p24^{gag}/ml) prior to use in an internalization assay as above.

To screen antibodies for the ability to inhibit entry of HIV-1 into cells using a CD4-independent pathway, U373-MG or HTB-138 are removed from a monolayer by incubation with 0.01% (v/v) ethylenediaminetetraacetic acid (EDTA) in PBS not containing Ca^{2+}/Mg^{2+}, and pelleted for 10 min at 400 g at 4°C. Cells are then resuspended in 10 ml of PBS and the wash step repeated. Cells are suspended in 100 μl of serum-free medium and transferred to a microfuge tube. The cells are preincubated with 100 μl of antiserum for 30 min on ice. Cells are exposed to 100 μl of HIV-1$_{IIIB}$ (final concentration of p24^{gag}, 1.55 μg/ml) for 30 min on ice, and then transferred to 37°C for an additional 60 min. The virus–cell suspension is pelleted for 2 min at 1000 g at room temperature, and the inoculum is removed carefully so as not to disturb the cell pellet. Cells are washed in 1 ml of PBS and pelleted again at 1000 g for 2 min. The wash step is repeated twice. The cells are exposed to 100 μl of trypsin (100 μg/ml) for 30 min at room temperature, and the trypsin is quenched in 1 ml of PBS containing 10% (v/v) human serum. The cells are again pelleted at 1000 g for 2 min at room temperature and the supernatant is removed. The cells are lysed with 20 μl of 10% Triton in PBS. The concentration of p24^{gag} antigen is determined by antigen-capture enzyme-linked immunosorbent assay (ELISA).

The virus inoculum, incubation times, and conditions of trypsin treatment are optimized for the cell line studied. The experimental conditions described for either the lymphocytic or neural cell lines yield reproducible results; however, conditions may need to be adjusted, depending on the sensitivity of a given cell line to both viral infection and trypsin treatment.

Representative Results

The initiation of HIV-1 infection in CD4-positive and CD4-negative cells was examined using a sensitive virus internalization assay. After exposure of the cell lines to HIV-1, the cell lysates were analyzed for internalized p24^{gag} antigen as a measure of virus uptake. The internalization of HIV-1 is an energy-dependent process and

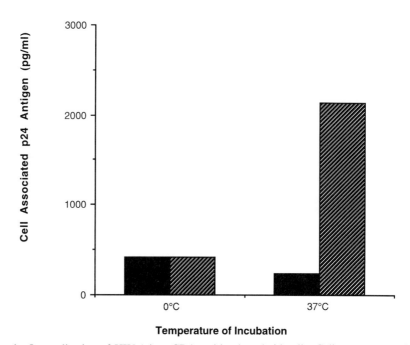

Temperature of Incubation

FIG. 1   Internalization of HIV-1 into CD4-positive lymphoid cells. Cells were exposed to HIV-1 and assayed as described in Methods. The virus uptake assay indicates that virus is taken up in the CEMx174 cell line (striped bars) only at 37°C, not at 0°C. The nonsusceptible B cell line 721.174 (black bars) does not internalize significant amounts of virus at either temperature.

will not occur at temperatures lower than 4°C (1). To control for HIV-1 that may have remained bound to the cell surface, but not internalized, after the extensive wash and trypsin treatment, we exposed the cell lines to virus at 0°C. Figure 1 demonstrates the results of a representative experiment. Significant amounts of intracellular viral antigen were not present in either cell line when the incubation mixture was maintained at 0°C. These results demonstrate the effective removal of noninternalized virus from the cell surface, and the p24^{gag} values at 0°C represent the sensitivity of the internalization assay. The noninfectable CD4-negative 721.174 cell line served as a control for residual inoculum at 37°C, and the level of p24^{gag} antigen detectable in the 721.174 cells was at or below the background values. In contrast, the infectable CEMx174 cells internalized virus well at the permissive temperature, but not at 0°C. To confirm the specific nature of the internalization assay, we used monoclonal antibody Leu3a, which blocks binding of CD4 to the HIV-1 envelope glycoprotein, gp120, thereby inhibiting infection in CD4-positive cells (14, 15). Monoclonal antibody B33.1.1 recognizes determinants on HLA class II protein, and was used as a control antibody. In Fig. 2, the result of a representative experiment is depicted. Internalization of the virus into CEMx174 was significantly inhibited by prior incuba-

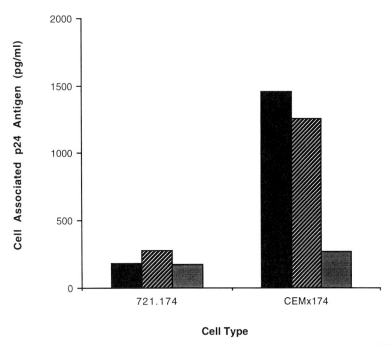

FIG. 2   Internalization of HIV-1 into CD4-positive CEMx174 cells is blocked at 37°C by MAb against the CD4 molecule, Leu3a (gray bars). In comparison, the control MAb B33.1.1 (striped bars) does not have a significant effect. Black bars, phosphate-buffered saline.

tion of cells with Leu3a, whereas prior incubation with B33.1.1 did not inhibit entry of virus. These results are in agreement with published studies on the ability of Leu3a to inhibit interaction of CD4 with the viral envelope protein blocking the entry of HIV-1 into CD4-positive cells (13).

In an attempt to define a cell surface component that may mediate entry of HIV-1 into CD4-negative glioma cells, we screened polyclonal antisera raised against cell surface molecules for their ability to inhibit entry of HIV-1. Figure 3 demonstrates that an antiserum raised against galactosylceramide (Gal-Cer) inhibits entry of HIV-1 into the CD4-negative U373-MG cells at 37°C. This antibody did not inhibit entry of HIV-1 into a CD4-positive HeLa T4 cell line at 37°C. Using the virus internalization assay, we were able to identify a putative receptor for HIV-1 in neurally derived cell lines.

Discussion

Although the infection process of HIV-1 is quite complex, one critical stage is the binding and adsorption of virus onto susceptible cells. We utilized a virus internaliza-

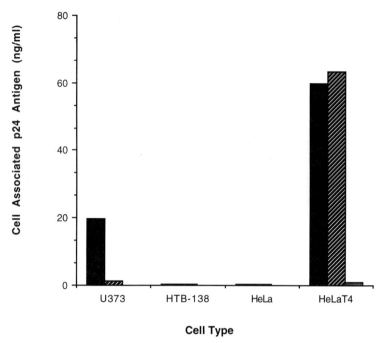

FIG. 3   Polyclonal antiserum against the cell surface glycolipid galactosylceramide inhibits entry of HIV-1 into the CD4-negative, infectable cell line U373-MG at 37°C. The anti-Gal-Cer serum did not inhibit the CD4-dependent entry of HIV-1 in HeLa T4 cells. The noninfectable glioma cell line HTB-138 and the control CD4-negative HeLa cell line did not internalize virus at 37°C. No cell line tested internalized significant amounts of virus at 0°C, indicating p24^{gag} values at 37°C represent internalized virus and not residual inoculum. Black bars, normal rabbit serum (37°C); striped bars, anti-Gal-Cer (37°C); gray bars, anti-Gal-Cer (0°C).

tion assay to study the initiation of virus infection. The results indicate that viral uptake, quantitated as p24^{gag} antigen in cell lysates, is a necessary prerequisite for infection of cells with HIV-1. Although the CD4 molecule is the primary receptor for HIV-1, CD4 does not confer susceptibility to HIV-1 in all human cell lines (19), leading to the suggestion that accessory molecules are involved in the binding of HIV-1 in some human cell lines. In an attempt to elucidate a CD4-independent entry pathway in neurally derived cell lines, we used the internalization assay to study the initial events, binding and internalization, of viral infection. This assay suggested galactosylceramide as a potential receptor for HIV-1 in CD4-negative, yet infectable, cell lines (16). With increased understanding of the specific interactions of HIV-1 with host cells, molecular mechanisms of viral pathogenesis can be better defined.

Acknowledgments

Supported by PHS Grants NS 27405, NS 30606, and NS 31067. J.M.H. is a recipient of a Merck Predoctoral Fellowship. K.S. is a trainee in Virology Training Grant AI 07325.

References

1. D. M. Knipe, *in* "Virology" (B. N. Fields, ed.), p. 293. Raven Press, New York, 1990.
2. A. G. Dalgleish, P. C. L. Beverly, P. R. Clapham, D. H. Crawford, M. F. Greaves, and R. A. Weiss, *Nature (London)* **312,** 763 (1984).
3. P. J. Maddon, A. G. Dalgleish, J. S. McDougal, P. R. Clapham, R. A. Weiss, and R. Axel, *Cell (Cambridge, Mass.)* **47,** 33 (1986).
4. R. C. Collman, B. Godfrey, J. Cutelli, A. Rhodes, N. F. Hassan, R. Sweet, S. D. Douglas, H. Friedman, N. Nathanson, and F. Gonzalez-Scarano, *J. Virol.* **64,** 4468 (1990).
5. G. M. Orloff, S. L. Orloff, M. S. Kennedy, P. J. Maddon, and J. S. McDougal, *J. Immunol.* **146,** 2578 (1991).
6. C. Cheng-Mayer, M. Quirogo, J. W. Tung, D. Dina, and J. A. Levy, *J. Virol.* **64,** 4390 (1990).
7. S. Chiodi, S. Fuierstenberg, M. Godlund, B. Asjo, and E. M. Fenyo, *J. Virol.* **61,** 1244 (1987).
8. J. M. Harouse, C. Kunsch, H. Hartle, M. A. Laughlin, J. A. Hoxie, B. Wigdahl, and F. Gonzalez-Scarano, *J. Virol.* **63,** 2527 (1989).
9. X. L. Ling, T. Moudgil, H. V. Vinters, and D. D. Ho, *J. Virol.* **64,** 1383 (1990).
10. J. M. Harouse, M. A. Laughlin, C. Pletcher, H. M. Friedman, and F. Gonzalez-Scarano, *J. Leuk. Biol.* **49,** 605 (1990).
11. R. D. Salter, D. N. Howell, and P. Creswell, *Immunogenetics* **21,** 235 (1985).
12. R. De Mars, C. C. Chang, S. Shaw, P. J. Reitnauer, and P. M. Sondel, *Hum. Immunol.* **11,** 77 (1984).
13. M. M. Popovic, M. G. Sarngadharan, E. Read, and R. C. Gallo, *Science* **224,** 497 (1984).
14. G. S. Wood, N. L. Warner, and R. A. Wranke, *J. Immunol.* **131,** 1212 (1989).
15. Q. J. Sattentau, A. G. Dalgleish, R. A. Weiss, and P. C. L. Beverly, *Science* **234,** 1120 (1986).
16. J. M. Harouse, S. Bhat, S. Spitalnik, M. Laughlin, K. Stefano, D. Silberberg, and F. Gonzalez-Scarano, *Science* **253,** 320 (1991).
17. M. Raff, R. Minsky, K. Fields, R. Lisak, S. Dorfman, D. H. Silberberg, N. Gregson, S. Leibowitz, and M. Kennedy, *Nature (London)* **276,** 813 (1978).
18. J. A. Benjamins, R. A. Callahan, I. M. Montgomery, D. M. Studzinski, and C. A. Dyer, *J. Neuroimmunol.* **14,** 325 (1987).
19. B. Chesebro, R. Buller, J. Portis, and K. Wehrly, *J. Virol.* **64,** 215 (1990).

[10] Quasispecies Behavior of Human Immunodeficiency Virus Type 1: Sample Analysis of Sequence Data

Carla L. Kuiken, Katja Nieselt-Struwe, Georg F. Weiller, and Jaap Goudsmit

Introduction

Parent–offspring relationships are dealt with in the normal course of evolution. However, with human immunodeficiency virus (HIV), other retroviruses, RNA viruses, and several other organisms this is not so straightforward. The problem is that these organisms multiply and mutate so rapidly that they do not form families in the traditional sense. It has been estimated that the replication mechanism of HIV on average makes one "mistake" (and thus causes one mutation) each time the virus replicates. This means that only in exceptional cases will an exact copy of a viral genome exist. Usually, every replicate will have at least one mutation when compared to the template. Rapid mutation and multiplication rates result in the formation of a so-called quasispecies (1), a swarm of individual virus particles that all differ slightly from each other in their genetic composition.

This quasispecies concept can be applied both to the virus variants that exist in one host, and to the virus variants that exist in a whole population of HIV-infected people. Of course, more variation will exist when we look at virus from different individuals. When the number of mutations is high, the probability increases that the sequences cannot be represented by a family tree. The main cause of this is the occurrence of parallel and back mutations: two mutations at one location, that are counted as only one, or (if the second mutation restores the original nucleotide) are not noticed. There are statistical methods to correct distances for this phenomenon, but they cannot compensate for the accumulation of noise. Because of this noise, the correct branching order of the tree cannot be determined accurately. If this process continues, the sequences can no longer be represented by a tree because the family relationships have become very diffuse, that is, their graphical representation becomes a net.

In HIV, the causative agent of acquired immunodeficiency syndrome (AIDS), an important neutralization epitope we know of is located in the third hypervariable (or V3) region on the external envelope of the virion (2,3). Neutralizing activity against this epitope is highly type specific, meaning that antibody neutralizing one type of HIV does not neutralize other types (4). The antibody recognition is specific, and can in some cases be changed drastically with the mutation of one amino acid (5,6). Many attempts at vaccination are undertaken with vaccines that contain V3-derived pep-

Methods in Molecular Genetics, Volume 4

tides either as the main antigen or as one of the constituents. The V3 region has been studied extensively, both because of its importance as a determinant of various biological properties of the virus and because of its antigenic properties. In many respects, HIV behaves differently from other viruses, even from other RNA-based viruses like the influenza virus and other retroviruses like human T cell-lymphotropic virus type I (HTLV-I), and certainly from other genes. For this reason, many methods designed for the analysis of gene sequences are inapplicable to HIV.

In influenza, another highly variable RNA virus, the combination of detailed study of the epidemiology and extensive sequence analysis has resulted in detailed knowledge of antigenic variation of the virus, which makes the vaccination campaigns of the World Health Organization (WHO) reasonably effective (7). The situation is vastly different in HIV. Apart from the myriad technical problems in vaccination, the extreme and (as yet) untractable variation in the virus (and most importantly in its major antigenic sites) makes design of an effective vaccine a complicated problem. So far, vaccination attempts have been limited to experiments in monkeys, and success has been limited.

We elaborate here on the analysis of a single dataset, consisting of a large number of V3 sequences from different risk groups and different years. All sequences are from virus in recently infected hosts. This is important for two reasons. First, it is known that the virus (at least the V3 region) can change rapidly and drastically in the years after infection. Sampling shortly after infection yields virus that presumably resembles the virus with which the host has been infected, and thus allows us to look at variants that are important in transmission and thus for vaccine design. Second, shortly after infection the diversity of the virus within one person is limited. Later in infection, the sequenced region becomes so diverse that a number of different sequences must be made for each person. We have previously reported elsewhere on the primary analysis of the data (8).

Description of Dataset

Sequences for the dataset described here were obtained from a number of sources. In 1984, an HIV-1 cohort study was set up in Amsterdam among homosexual men (9). Sera were taken from participants every 3 months in 1985, and every 6 months in 1990. Sequences were obtained from 10 sera from people who had seroconverted in 1985 and 10 from 1990. In all, 30 sequences from homosexual men were used in the dataset: 10 from the early 1980s (1980, 1981, 1982), 10 from 1985, and 10 from 1990. The sequences from the early 1980s were obtained from participants in a cohort study to assess the efficacy of an experimental hepatitis B vaccine (10). Retrospectively, participants in the study were asked for permission to test their stored serum samples, taken between 1980 and 1983, for HIV. From these early years, other samples were screened that had been submitted for viral diagnostics for symptoms that were reminiscent of acute HIV infection (lymphadenopathy, influenza-like complaints) or were labeled "gay immune compromised syndrome," which was later identified as acute HIV infection. When no other likely cause of the complaints was found and the sample tested positive for HIV-1 antibody, this was taken as an indication of acute HIV infection, and the sample was used for this study.

Sequences from sera of 32 HIV-1-infected drug users were obtained from another cohort, also in Amsterdam (11). This cohort was started in 1986. The sera from this group were also from seroconverted individuals, but for some of them the period elapsed between the last seronegative and the first seropositive sample is longer, up to 1 year in one case. For most, this period is around 4 months, the standard interval between two samples. The sampling dates of these sequences span the years from 1986 to 1991. Not all participants in this cohort use drugs intravenously, so some of them may have been infected through homosexual or heterosexual contacts.

A third, small set of sequences was taken from a cohort of hemophilia patients (12). Twelve patients were identified as seroconverters between 1984 and 1985 (starting in 1985, blood and blood products were screened for HIV).

In all, the dataset consists of 74 sequences taken shortly after infection from three different risk groups and over a period of 10 years. This provides an opportunity to study the changes in the region that was sequenced, both over time and between the risk groups. The ultimate aim of the study was to try to find patterns that might allow for prediction of future developments in the changes.

The sequencing procedure has been described in detail elsewhere (13).

Data Analysis

Variability

To obtain an initial overview of the dataset, simple measures of variability were calculated. The one used most often for these types of data is the pairwise Hamming distance (14), calculated by scoring the number of different nucleotides between each pair of sequences. If any sequence contains an invalid position, that position is omitted from the analysis in all sequences. In the present dataset, 251 positions remained for the analysis. For purposes of analysis, a distinction is often made between the *V3 region,* which is the whole sequence of 273 nucleotides, and the *V3 loop,* the hydrophobic middle section contained between two cysteines forming a disulfide bridge. The V3 loop presumably forms a separate structure that protrudes from the surface of the viral envelope, and it has been observed to behave somewhat differently from the background sequence (in fact, this was the reason it was first labeled "hypervariable," even though it now appears that the flanking regions are at least as variable, if not more) (15).

In Fig. 1 the variability of the sequences is plotted in a three-dimensional diagram.

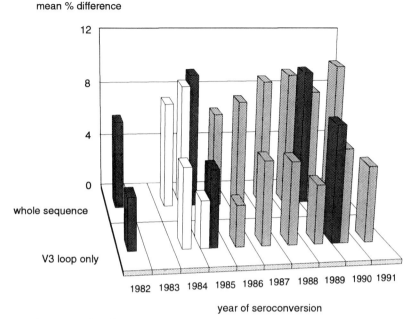

FIG. 1   Mean Hamming distances between the sequences per year and risk group (black bars, homosexuals; gray bars, drug users; white bars, hemophiliacs). In the front row, bars indicate the mean percent difference in only the V3 loop, bordered by the cysteines; the back row shows the distances over the whole 273-base stretch.

The front row shows the variation of the V3 loop (expressed as the Hamming distance divided by the sequence length); the back row shows the variation of the whole V3 region. It can easily be seen that in this dataset the flanking regions are more variable than the loop itself. The variation is split up according to sampling year and risk group. Figure 1 indicates that there is a trend toward greater diversity between the samples from later years in both rows; a statistical significance test on the first and last sample in the two risk groups with a reasonable spread [the drug users (DUs) and the homosexuals] indicates that the difference is significant (DU group 1986 vs 1991, $p < 0.02$; homosexuals, 1982 vs. 1990, $p < 0.001$; Mann–Whitney statistic).

Because the V3 region is known to be highly immunogenic (this is the very feature that makes it of interest for vaccine design) (2,3), it is likely that the effects of selection pressure can be seen in this region. One of the ways in which such selection pressure reveals itself is by a low ratio of silent to nonsilent mutations. If a part of a gene is under pressure to change rapidly, all copies that encode identical amino acids are weeded out. Consequently, all those that survive will show a high number of amino acid-changing mutations. Assuming that the number of silent mutations is independent of any selection pressure from the immune system (as they do not affect the protein and thus have no consequences for the antigenic features), the ratio of silent to nonsilent mutations will be low in this region. To estimate this ratio, we used a simple method proposed by Nei and Gojobori (16). The ratio as calculated over the whole region was 1.25 over the whole dataset. Values around 1 are commonly found for sequences from this region (17,18). For a less immunogenic region of HIV in the *gag* gene, a value of 6.7 was reported (17). A comparison between mouse and rat genes gave silent/nonsilent ratios of around 10 (19).

In some datasets, the results of analyses depend highly on the distance measure used: for instance, some sequences are much more similar when looking at transitions (A/G and T/C differences) than when looking only at transversions. An example in which results of a phylogenetic analysis depended highly on the distance measure (in this case, amino acids vs nucleotide differences) can be found in Louwagie *et al.* (19). To see if any of these effects were present in this dataset, we used the program DIPLOMO (21). This program can be used to calculate and plot a number of different distance matrices pairwise. If the pairwise differences show the same pattern irrespective of which measure is used, the plot shows the dots representing a pair of pairwise differences as a straight line on the diagonal. If the dots do not lie close to the diagonal, this means that one measure results in different distances than the other. A number of different distance measures have been examined (first/second/third codon position changes, silent/nonsilent changes, transition/transversion ratios among others). No systematic differences were found when looking at the sampling years. We did notice a risk group-related difference in the transition/transversion ratio between the two groups, which was higher for the homosexual group. The difference between the means was about one standard deviation, which makes it a "trend" and not a significant difference. The effect remained even when we corrected for biased base composition, and when we examined individual codon positions or silent/nonsilent changes only. The transition/transversion DIPLOMO plot, corrected for biased base composition, is shown in Fig. 2.

Statistical Geometry

Usually, the next step in sequence analysis is to look at family relationships between the sequences. For many datasets, a tree might not be the appropriate topological representation of the dynamic process that gave rise to the data. tRNA sequences from one organism, for example, are better represented by a bundle- or star-like scheme instead of a classic dendrogram (22). Virus isolates from different lineages as well as from one quasispecies often show a netlike evolution process (23,24). Thus for this dataset we need complementary procedures that are able to determine a priori the underlying topology and specifically the treelikeness of the set. The method of statistical geometry is especially suited for this (25,26). It is based on a vertical and

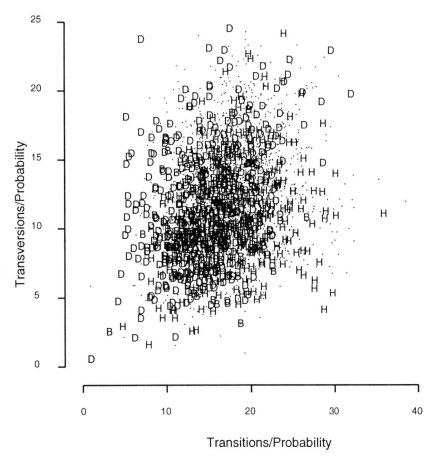

FIG. 2   Plot of transition and transversion distances between all pairs of sequences. Numbers have been corrected for the base composition. Plot symbols: letters are within-group sequence comparisons (B, hemophiliac; H, homosexual; D, drug user); dots represent between-group sequence comparisons.

horizontal analysis of quartets of sequences. It has been shown that to determine whether a dataset is treelike, it is sufficient to analyze all quartet subsets of the dataset (27–29). These quartets are considered as geometric configurations, and numerical parameters describing the geometries are computed. These parameters lead to a graph, which may be used to distinguish between a tree-, bundle-, or netlike geometry of the sequence set.

The general idea of statistical geometry is most easily explained using binary sequences (i.e., sequences with symbols of binary alphabets such as the alphabet of purines and pyrimidines, R/Y). For quartets of binary sequences one may distinguish

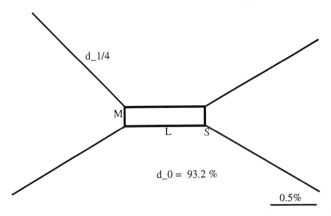

FIG. 3   Representative geometries of the statistical geometry in the binary purine/pyrimidine sequence space of all 72 sequences. All three codon positions were considered. Absolute distances between the sequences are divided into distance segments obtained from their position-wise entry distribution. The parameter d_0 is equal to the (percentage) number of homologous positions of the quartets. Segment d_1 sums up the distance segments comprising the positions, where one sequence differs from the other three. Segment d_2 equals the number of positions with pairwise equal entries. This segment is the sum of the three box dimensions L, M, and S, which are ordered by magnitude. Note that the length of the shortest path between any two nodes in the graph is equal to the Hamming distance between the corresponding sequences. For the graph, the standard length that refers to a relative distance of 0.5% (relative to the total number of positions) is shown.

eight classes of positions in three categories. Category 0 contains all positions with four equal letters, category 1 consists of all positions with one entry differing from the other three (four possibilities). Category 2 comprises all positions having two pairs of equal entries. The last category can be realized in three different ways. Counting the number of positions in each category defines seven distance segments, which can be represented by a graph with four protrusions starting from four mutually diagonal corners of a box with three dimensions (see Fig. 3).

For a set of N sequences, the values of the position classes and/or categories are calculated for each of the possible $\binom{N}{4}$ quartets. The average values of these parameters are then computed, giving rise to the mean quartet geometry of the whole set. The representative graph of the mean geometry, called the statistical geometry of the set, is treelike if one of the three inner box dimensions is large in comparison to the other two. If all three box dimensions are of similar magnitude, that is, the box is almost cubic, then the set of sequences represents a randomized bush. If all distance segments in the graph are of similar magnitude then randomization has proceeded so far that the set is a highly interwoven net.

In a first step of evaluating the statistical geometry of this dataset we took all aligned full-length sequences except for two that contained too many undetermined characters. In all the following analyses we distinguished the three codon positions. We computed the statistical geometry of the sequences in two separate sequence spaces: in the quaternary sequence space over the DNA alphabet and in the binary sequence space over the purine (R)/pyrimidine (Y) alphabet. Note that the geometries of the DNA sequences contain two more distance segments, subsuming all positions with one pair of equal entries and those at which all four entries differ.

The dataset contains some conserved positions: 36.5% in the ACTG alphabet, 67.2% in the RY alphabet. The variable positions are so diverged as to be almost randomized. Figure 3 shows the graph of the statistical geometry of the whole dataset transformed into the binary purine/pyrimidine sequence space. It turned out that for both alphabets (ACTG and RY) there are no expressed differences in the degree of randomization and in the characteristic values for the topologies among the three codon positions. This is remarkable and different in comparison to many other viruses, where often the first and second codon positions of a coding sequence are conserved and the third codon position is much randomized (see, e.g., Refs. 23 and 24). The binary geometries of the RY sequences of this dataset are less randomized, with a slightly higher degree of treelikeness in comparison to the geometries of the DNA sequences. Both graphs have the following features in common: the treelikeness parameter is close to the one of ideal trees, thus a tree reconstruction is appropiate for both sequence sets. On the other hand the trees will have short inner edges, because of the short box dimensions (only approximately 2% for the DNA graph, 1% for the RY geometry of the sequence length). The expected exterior edges of the tree will almost all be equally long, because the variances of the mean values of the distance segments are small and the pairwise distances are all almost equal. Although the geometries show a high degree of conservation, indicated by the number of homologous positions d—0 (see Fig. 3), the degree of nettedness is high. The set represents a homogeneous cluster of a relatively pronounced bundle-like shape with short inner edges, typical for an evolving quasispecies. Omitting the hypervariable 3′ end of the sequences only slightly reduced the bundle-likeness of the graph.

Next we computed the statistical geometries of the two risk groups (drug users and homosexuals) separately. Here again, the differences between the three codon positions are marginal. The graphs of the sequences are similar to the graph of the statistical geometry of the whole dataset, from which we conclude that the differences within the risk groups are not smaller than among all sequences.

Phylogenetic Analyses

The method we used for phylogenetic analysis, the neighbor-joining method, is a fairly new one (30). It does not use an explicit mathematical model, but works under the principle of minimum evolution. It is extremely fast, and it is the only method that can easily deal with large numbers of sequences. The trees that it generates are nearly optimal in most circumstances, and especially with datasets that contain high noise levels (31). This method was used for all phylogenetic analyses discussed in this section.

The neighbor-joining method is based on distance matrices rather than on raw sequences. To some extent, the results of matrix-based phylogenetic methods depend on the choice of the distance measure. Only rarely is a simple counting of the differences used as the analysis unit. The simplest adaption of this so-called "Hamming distance" is a correction for the chance of multiple hits (more than one mutation at one position, which can result in an underestimate of the true difference between two sequences). This correction was first proposed by Jukes and Cantor (32). A second step, for which Kimura (33) provided the mathematical basis, also takes into account the possibility of differential substitution rates for transitions (C/T and A/G) and transversions (all other mutations). When there is no difference in the transition and transversion rates, Kimura's formula yields the same result as the Jukes–Cantor method (34). It has been observed that HIV genomes have a high rate of G/A mutations (35). For this reason, Kimura's formula was used for the estimation of the distances between nucleotide sequences. For amino acids, a simple Hamming distance was used, as well as a recoding method proposed by Smith and Smith (36) that counts only nonconserved changes between amino acids (defined on the basis of their physicochemical properties).

An initial phylogenetic analysis of the whole set of sequences yielded the tree shown in Fig. 4. Several features of this tree are noteworthy. First, as predicted in the statistical geometry analysis, there are no extremely long or short exterior branches. This underlines that all sequences have approximately equal distances from each other. Even sequences in a side group are still separated from each other by sizeable branches, indicating that they do not form separate homogeneous clusters. A second feature (again as predicted) is the fact that the interior branches are short compared to the exterior ones. This was already indicated by the statistical geometry analysis and shows that any groups that may be seen in the tree are weak, in the sense that the "group property" (the variation that separates this group of sequences from the rest) are much less pronounced than the individual sequence variations within the group. Third, although there appears to be a tendency for older sequences (sampled around 1980) to have shorter branches, it is by no means a fixed rule. It looks like there is no strong association between the sampling year (which one might take to be a measure of the age of the virus variant) and the branch length. This may be another indication that in these sequences, there is no straightforward relation between the age of a variant, its descendance, and its distance to other variants. Because selection pressure usually takes place at the amino acid level, in situations in which there is high selective pressure, silent changes are supposed to be a better measure of the real distance between sequences. We calculated the correlation for all sequence pairs between the difference in their sampling years, and number of silent changes. The correlation was 0.42, which is not statistically different from 0, which means that there is no reliable association between age difference of the sequences and their genetic distance.

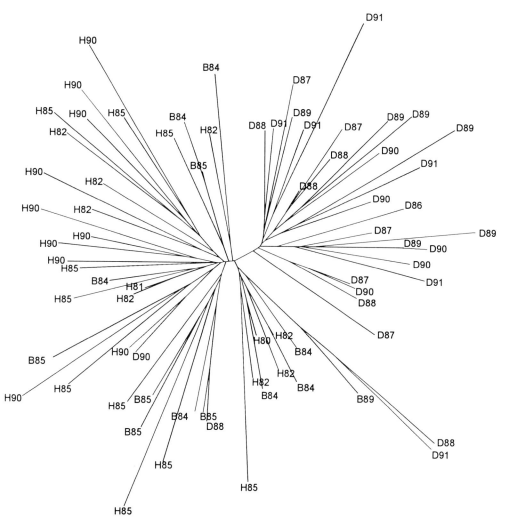

FIG. 4   A dendrogram produced on the basis of all 74 nucleotide sequences. The dendrogram was calculated using the neighbor-joining method and drawn with the DRAWTREE program. Sequence names have been replaced by symbols for the sake of legibility. Letters H, B, and D indicate sequences for homosexuals, hemophiliacs, and drug users, respectively; the numbers are the years of sampling.

In this tree, one obvious and epidemiologically meaningful group division can be discerned, namely the distinction of drug user-derived sequences and the other sequences. A number of drug user sequences end up in the other group, but the separation is clear enough to merit further attention. In this tree, it is the only clustering that can be seen; apart from this, the sequences do not fall into separate groups, but form a swarm that cannot be subdivided. The tree was recalculated omitting the most randomized 3′ end of the sequences. This resulted in a comparable tree, with a more pronounced separation of the two groups and shorter branches within the drug user sequences.

Next, the amino acid translation derived from the (full-length) sequences was analyzed. The main feature of the nucleotide tree, the approximately equal length of the branches, could be seen in this tree (Fig. 4). Again, no clear clusters were present. It is noteworthy that the separation of drug user sequences has completely disappeared in this tree. This indicates that at the amino acid level, the coherence at the nucleotide level in the drug user sequences has almost disappeared. The analysis was repeated using recoded amino acid distances (36). Results from this analysis were comparable to the previous one, in which no clear clusters were formed.

On the basis of this result, we did another analysis to examine the relevance of the amino acid–nucleotide distinction for the tree structure. Two distance matrices were constructed. One contained estimates of the number of silent changes between each pair of sequences, the other estimates of the number of nonsilent changes. Neighbor-joining trees that were calculated on the basis of these two matrices showed the pattern that was expected: in the tree based on silent changes, the drug user group was separated from the rest; in the one based on nonsilent changes, the groups were mixed. It seemed the distinction was present only in positions that are not associated with amino acid changes. The tree based on the silent changes was different from most trees generated in the course of the analysis in that the interior branch separating the drug user group sequences from the rest had a length comparable to the shorter exterior branches, so that the cluster of drug user sequences appeared to be more compact.

A third set consisting of three trees was generated on the basis of the first, second, and third codon positions, respectively. Most of the changes in first codon positions, all of the changes in the second, and only a minority of the changes in the third codon position result in amino acid change. As expected, only the tree based on the third codon position yielded the separation into groups.

Separate analysis of the drug user group and the remaining sequences again did not yield any clusters. This warrants the conclusion that, apart from the drug user/other distinction, no clusters of similar sequences are present in these data. This result was again foreseen on the basis of the statistical geometry analysis.

In conclusion, only one meaningful separation can be seen in the sequences. Furthermore, this separation is caused by differences that are not associated with amino acid differences, and that do not lie within the 5′ domain of the sequences, because it is still present when only the V3 loop and its 3′ flank are used for the analysis.

The next step is to investigate the stability of this separation. The most commonly used method to achieve this is called *bootstrapping*. This is a statistical technique that is also used in many other fields. The principle is to omit a number of randomly selected positions from the dataset, and to duplicate others, so that the total number stays the same. The same phylogenetic analysis is then done on this dataset. This is done many times, each time with a new random selection. The results of these analyses are then summarized in a so-called consensus tree. This tree is constructed so that at each point the majority branching pattern is chosen; in the tree the number of times a particular node was reproduced among the total is shown. If these "bootstrap values" are high enough (a commonly used percentage is 95%), then that particular branch is reproducible, and does not depend much on the particular positions used. This is often taken as an indication that the group in that branch is monophyletic (closely related), but many authors have warned against this interpretation: high numbers mean the tree is stable, but not that it is correct. Lower bootstrap values mean lower reliabilities for the corresponding branches; values below 50% are interpreted to mean that the obtained structure can be attributed mostly to chance.

All groups in the tree had a low bootstrap value. Because tree construction tends to become more difficult (and more subject to error) when many sequences are used, we subdivided the dataset into three subsets that were more or less equivalent in terms of size, group composition, and sampling years. These three subgroups yielded the same risk group distinction in a normal neighbor-joining analysis, but on bootstrapping the consensus tree lost this distinction in one subset, and in the other two again the reproducibility was low.

This merits the conclusion that we cannot have much confidence in our phylogenetic trees. To some extent, this was already known: the dependence of the trees on the similarity measure chosen, and (in most trees) the short interior branches compared to the long exterior ones, had already indicated that only a few nucleotide changes would be needed to move a sequence from one branch to another, and this is exactly what the bootstrap is sensitive to. From this it also follows that the bootstrap is not ideally suited to the problem at hand, which is the reliability of a group distinction that does not depend completely on the exact classification of each single sequence. In "classic" phylogenetic analysis, the aim is to establish the ancestry of sequences, and the ideal test is sensitive to misclassification of a single sequence. In our case, the question is if we can distinguish two groups; whether or not some sequences move from one group to the other is of secondary importance. To determine the reality of the group distinction, we turned to other methods.

Statistical Geometry Revisited

The second feature of the method of statistical geometry allows the testing of an inner edge (and the deduced branching order) in a tree given by an independent cluster analysis. An inner branch forces the set into four subsets, while it might not be clear if the cladistic relationships between the four groups are reliable. In this case the sequences are arranged according to the given subsets. One now computes the sequence space geometries by combining quartets that always contain one member of each subset. The averages of the computed distance segments can then be used to describe the underlying kinship of the four subsets. There is one common node for two subsets, with an edge separating the two nodes, if one of the two mean box dimensions is large in comparison to the other two. If, however, the lengths of the inner edges are all of similar magnitude, then the subsets represent a partly randomized bush, that is, there exists a common node for all four clusters.

From the computed trees we had several models to evaluate. The first and most obvious was to test for the clusters separating the three risk groups: drug users, homosexuals, and hemophiliacs. Because the method requires a partition into four subsets, we evaluated all three combinations of two groups versus two halves of the third. To compute the statistical geometry of these subsets we subdivided the group of the hemophiliacs into two subgroups. All three graphs showed all the typical features of a perfect bundle, with equal box dimensions and almost equal protrusion lengths. Thus we deduce that the four groups all have a common ancestor and the partition in risk groups cannot be deduced from the sequence data. In Fig. 5 we show the graph with the drug user group partitioned into two halves (for all three codon positions). The largest box dimension is the one separating the two groups of drug users from the other groups, but it is only slightly larger than the other two box dimensions, which indicates that the other two possible topologies are almost as feasible. For the DNA alphabet the mean number of positions supporting the separation into the group of drug users and nondrug users is 4.8 against 2.7 and 2.3. For the RY alphabet these numbers reduce to 2.0 against 1.2 and 1.0. We then asked whether this small prefer-

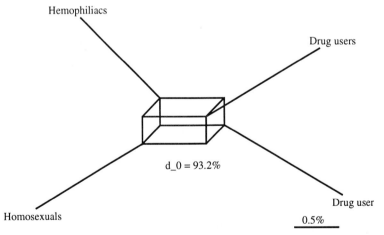

FIG. 5   Statistical geometry of the original dataset divided into two groups of drug users, the group of homosexuals, and the group of hemophiliacs. All positions of the binary RY sequences were considered.

ence for this topology was due to some sequences being constantly misclassified or occurred because of some positions that in all sequences give rise to the other two topologies. Checking this via a positional analysis indicated that there were a number of positions that did not conform to this topology.

Split Decomposition

Split decomposition is a new *nonapproximate* method to represent data and/or its distance matrix by a graph, which may contain exact treelike structures as well as cycles (37). This method associates, to every finite dissimilarity measure defined on a finite set of sequences, a family of weighted splits (i.e., bipartitions). Every dissimilarity measure decomposes into the collection of bipartitions and into a *split-prime* residue d__0, which is not decomposable. The splits suggest how a given dataset may be divided naturally into subfamilies.

The main purpose of the analysis was to use this nonapproximate graph construction method in order to see if the separation of the dataset into the two groups DU and non-DU could be established.

We looked once more at different distance matrices of the dataset. The first matrix resulted from transforming the sequences into the purine (R)/pyrimidine (Y) alphabet, from which we calculated the pairwise Hamming distances. Second, a matrix was determined on the basis of Kimura's correction formula.

It turned out that to compute the split decomposition of the distance matrix the whole set of all sequences was nonrealizable and also unrealistic. Thus we took a representative subset of the original dataset consisting of 26 sequences, which includes 11 DU sequences, 11 sequences of homosexuals, and 4 of hemophiliacs. In Fig. 6, describing the metric space spread out by all 26 sequences, we see a perfect so-called bundle-like graph, with no internal edges and external edges of similar length. This holds for the distance matrix based on Kimura's formula as well as for the Hamming distance matrix based on the RY sequences. In both cases the split-prime residue d__0 is extremely large (40% of the metric space of the DNA sequences and 50% of the metric space of the RY sequences), indicating the high degree of noise (and randomization), which cannot be decomposed via splits. This split-prime residue reveals the amount of nettedness in the geometry. To summarize, the split decomposition could not establish any grouping information within the dataset; neither the separation into risk groups nor the separation into the two groups of drug users and nondrug users was found.

Principal Coordinate Analysis

Principal coordinate analysis is a method that is closely related to principal component analysis (PCA). The purpose of PCA is to reduce a large set of variables to a

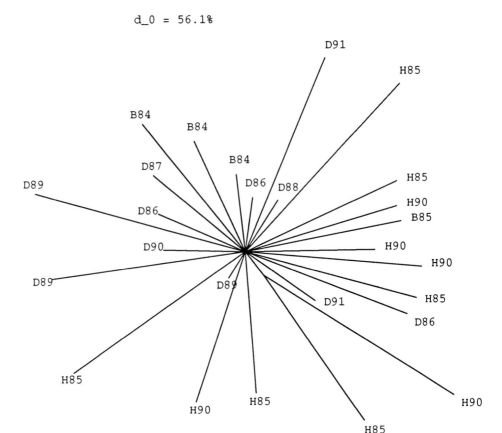

FIG. 6   Graph of the split decomposition of a representative subset of the whole dataset. The distance matrix for the split analysis was computed from Kimura's formula. The parameter d__0 shows the split prime residue, that is, the amount of the distance matrix that is not split decomposable, and which thus represents the amount of noise in the data. Almost the same graph is obtained for the distance matrix obtained by transforming sequences into the binary RY sequence space and then computing the pairwise Hamming distances; the only difference is a slightly larger split-prime residue d__0.

smaller set that is formed of a weighted linear combination of the original set. The technique was developed to be used on covariance matrices between the variables. Principal coordinate analysis (38) was designed to use this same method for distance matrices. The distances must be approximately Euclidean. This is not always the case when comparing sequences, but a transformation on the raw distances ensures that they at least approach Euclidean distances. After this transformation, the principal coordinate analysis determines which combination of variables is most suited to reflect the variation in the data. These combinations are called principal coordinates. When this combination is determined, the scores of all sequences on this combination (this is a total, and thus one number per sequence) are calculated. Of course, these scores will be a poor reflection of the total variation. The remaining variation is determined, and the next combination is found that best reflects this variation. Again the sequence scores are determined, and the search is for a third combination. This process goes on until all variation is enclosed in the combinations, or there are 10 combinations. Each next principal coordinate is less important than the preceding one, because it reflects a smaller portion of the variation in the dataset. Having more than 10 is usually meaningless.

The scores of the sequences on the coordinates are then plotted pairwise: a plot of the scores on coordinate 1 versus coordinate 2, coordinate 1 versus 3, and so on. One hopes for a clear grouping of the objects, based on these component scores; this means that somewhere in the dataset, there is a systematic difference between the groups.

In Fig. 7, the coordinate scores of the sequences in our dataset on the first two axes are shown. The first two coordinates are not strong, meaning they cover only a small percentage of the total variation in the sequences (8 and 5%, respectively). This again indicates that we are dealing with a weak distinction. However, it is striking that the first coordinate extracted (on the horizontal axis) is one that almost perfectly separates the drug user group from the rest, much better than the phylogenetic tree. Four drug user sequences end up in the left part of Fig. 7. All these were also in the "rest" cluster in the first phylogenetic tree. Thus, it appears possible to construct a single score from the positions that can be used to discriminate between drug user sequences and the others, apart from four sequences that are misclassified. The next question, of course, is how this score can be computed.

Finding Discriminating Positions: CART

CART, an acronym for classification and regression trees (39), is a method explicitly designed to find the variables (or, in this case, positions) that can be used to discriminate between two groups that already exist. In our case, we will use it to answer the following question: Given that the two groups are distinct, how can we tell them apart on the basis of these sequences? CART is a method for discriminant analysis on categorical (i.e., nonnumerical) variables. The four nucleotides must be recoded to numbers. CART then uses a simple trial-and-error approach to find the ones that can be used to discriminate between the predefined groups: for each position, the distribution in the two groups is calculated. The position that is most homogeneous within each group and most different between the groups is retained. The groups are reconstructed, as far as possible, on the basis of this position. A rule could be: if position 25 is an "A," then the sequence goes into group 1; if not, then go to the next step. If

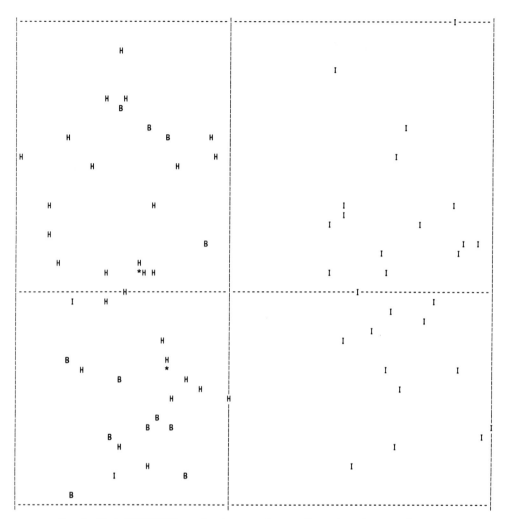

FIG. 7   Plot of PCOORD results. Axes are the two dimensions that were first extracted and explain most variation in the dataset. Plot symbols: H, B, and D, sequences for homosexuals, hemophiliacs, and drug users, respectively.

this rule does not assign each sequence to the right group, the remaining positions are used for further classification.

In our case, the only variable used for grouping was nucleotide 126 of the sequence. This means that adding other positions cannot improve the classification. Position 126 was also found to give the strongest support to the topology with a separate DU group. This nucleotide is in a third position, and changing it never results in an amino acid change, which fits well with the conclusions from the phylogenetic trees. The amino acid that this codon encodes is highly conserved (a glycine in all sequences in the dataset), but its third codon position is taken by a G or A in all the homosexual and hemophiliac sequences, and by a C in all but four of the DU sequences. CART also lists alternative variables that can be used to achieve groupings of similar but slightly lower quality: position 12, another third codon position, was also able to discriminate, but in this case 11 sequences are misclassified; this position is slightly less efficient in discriminating between the groups, but it is highly associated (0.79, on a scale of -1 to 1) with position 126.

There are two positions, therefore, that discriminate between the groups. But do they provide all the grouping information there is? To look at this, the first analysis was a rerun of PCOORD, but using a dataset from which the two positions have been deleted. Looking at the plots, the first two axes extracted in this analysis were similar to the ones from the entire dataset. In addition, the group distinction could still be seen, albeit less clearly. The groups were joined in one swarm in this plot; but still, all the D sequences were on one side of the swarm, and all H sequences on the other. The same four sequences were again misclassified. We therefore conclude that the distinction is not limited to the two position, but is present (diffusely) in the rest of the sequences as well. This can be taken as an indication that the group distinction probably is not a functional one (this would be a more plausible explanation if it were limited to a few biologically important positions) but an evolutionary one, derived from the "founder virus" that established the infection in the DU group.

We have also looked at what happened to the four "misclassified" D sequences in the DIPLOMO analysis, where we found a different transition/transversion ratio between the groups. In this case, these four behaved like intermediate sequences, and could not be assigned unequivocally to the H or D group on the basis of these data. This indicates that the transition/transversion effect must be independent of the group distinction that was found earlier.

Conclusion

We have described the process of analysis of a set of sequences from the human immunodeficiency virus. The main problem with these sequences is the fact that one cannot assume that they will behave: much common-sense knowledge that can be applied in the case of other genes is invalid with HIV sequences. The course of evolution of the sequences is not always clear, because HIV evolves rapidly, so that it is theoretically possible that sequences from later isolates have actually undergone fewer evolutionary events than sequences from earlier isolates. The effect of transmission and of opposing selective pressures (e.g., from the immune system and from replication efficiency factors) may result in intractable variation and an accumulation of mutations that is only weakly related to the chronological age of the virus. This is especially true for the region under scrutiny in this study, the V3 region, because it is antigenic and thus probably highly influenced by immune pressure.

All these factors result in a virus that behaves as a quasispecies. This complicates both the description of the data and their analysis. It cannot be assumed that the sequences behave in a way that can be described by a tree, hence a prior analysis is needed to assess the appropriateness of such a description. In our case, statistical geometry showed that although the dataset had been randomized to a large extent, some insight might still be gained by phylogenetic analysis. The tree then yielded no interpretable clustering except one, associated with risk group. The clustering did not hold up under bootstrap analysis. This emphasizes both the low stability of the trees, and the fact that bootstrap analysis, although well suited for investigating the homogeneity of a certain cluster, cannot be used for investigating the presence of fuzzy clusters in a dataset. Further analysis using principal coordinate analysis did support the two-group distinction in the dataset, and we used CART, a nonnumerical discriminant analysis, to pinpoint the discriminating positions in the sequences.

One of our purposes here, in giving a detailed and process-wise description of the analysis of the data, has been to show that each step in the analysis depends on the results of the previous ones; sometimes, as in the case of the statistical geometry, there is good reason to go back to the results of earlier steps and reinterpret them in the light of new knowledge. The analysis of sequence data is not straightforward, but a trial-and-error process requiring constant reevaluation of all the steps taken, and a large array of analytical tools and programs.

References

1. M. Eigen, *Naturwissenschaften* **58,** 465 (1971).
2. J. Goudsmit, C. Debouck, R. H. Meloen, *et al., Proc. Natl. Acad. Sci. USA* **85,** 4478 (1988).
3. K. Javaherian, A. J. Langlois, C. McDanal, *et al., Proc. Natl. Acad. Sci. USA* **86,** 6768 (1989).
4. T. J. Palker, M. E. Clark, A. J. Langlois, *et al., Proc. Natl. Acad. Sci. USA* **85,** 1932 (1988).
5. T. F. W. Wolfs, G. Zwart, M. Bakker, M. Valk, C. L. Kuiken, and J. Goudsmit, *Virology* **185,** 195 (1991).
6. H. Takahashi, S. Merli, S. D. Putney, *et al., Science* **246,** 118 (1989).
7. B. N. Fields, D. M. Knipe, R. M. Chanock, M. S. Hirsch, J. L. Melnick, T. P. Monath, and B. Roizman, "Virology." Raven Press, New York, 1990.
8. C. L. Kuiken, G. Zwart, E. Baan, R. A. Coutinho, J. A. R. van den Hoek, and J. Goudsmit *Proc. Natl. Acad. Sci. USA* **90,** 9061 (1993).
9. F. de Wolf, J. Goudsmit, D. A. Paul, *et al., Br. Med. J.* **295,** 569 (1987).
10. R. A. Coutinho, N. Lelie, P. Albrecht-van Lent, *et al., Br. Med. J.* **286,** 1305 (1983).
11. J. A. R. van den Hoek, R. A. Coutinho, H. J. A. van Haastrecht, A. W. van Zadelhoff, and J. Goudsmit, *AIDS* **2,** 55 (1988).

12. T. F. W. Wolfs, C. Breederveld, W. J. A. Krone, *et al., Thromb. Haemost.* **59(3),** 396 (1988).
13. G. Zwart, T. F. W. Wolfs, M. Valk, L. van der Hoek, C. L. Kuiken, and J. Goudsmit, *AIDS Res. Hum. Retroviruses* **8,** 1897 (1992).
14. R. W. Hamming, *in* "Coding and Information Theory," 2nd Edition, Prentice Hall, Englewood Cliffs, 1986.
15. B. R. Starcich, B. H. Hahn, G. M. Shaw, *et al., Cell (Cambridge, Mass.)* **45,** 637 (1986).
16. M. Nei and T. Gojobori, *Mol. Biol. Evol.* **3,** 418 (1986).
17. P. Balfe, P. Simmonds, C. A. Ludlam, J. O. Bishop, and A. J. Leigh Brown, *J. Virol.* **64,** 6221 (1980).
18. S. M. Wolinsky, C. M. Wike, B. Korber, *et al., Science* **255,** 1134 (1992).
19. W.-H. Li, *J. Mol. Evol.* **36,** 96 (1993).
20. J. Louwagie, F. E. McCutchan, M. Peeters, *et al., AIDS* **7,** 769 (1993).
21. G. F. Weiller and A. J. Gibbs, "DIPLOMO: Distance Plot Monitor version 1.0." Computer program distributed by the Australian National University, Canberra, 1993.
22. M. Eigen, B. Lindemann, M. Tietze, R. Winkler-Oswatitsch, A. Dress, and A. von Haeseler, *Science* **244,** 673 (1989).
23. J. Dopazo, A. Dress, and A. von Haeseler, *in* "Preprintreihe Sonderforschungsbereich 343, Diskrete Strukturen in der Mathematik," Universiteit Bielefeld, 1990.
24. M. Eigen and K. Nieselt-Struwe, *AIDS* **4,** S85 (1990).
25. M. Eigen, R. Winkler-Oswatitsch, and A. Dress, *Proc. Natl. Acad. Sci. USA* **85,** 5913 (1988).
26. M. Eigen and R. Winkler-Oswatitsch, *in* "Methods in Enzymology," Vol. 183, p. 505. Academic Press, San Diego, 1990.
27. J. M. S. Simoës-Pereira, *J. Comb. Theory* **6,** 303 (1969).
28. K. Zaretskii, *Uspheki Mat. Nauk.* **20,** 90 (1965).
29. P. Buneman, *in* "Mathematics in the Archaeological and Historical Sciences." Proceedings of the Anglo-Romanian Conference" (F. R. Hodson, D. G. Kendall, and P. Tantu, eds.) p. 387. Edinburgh University Press, Edinburgh, 1971.
30. N. Saitou and M. Nei, *Mol. Biol. Evol.* **4,** 406 (1987).
31. A. Rzhetsky and M. Nei, *J. Mol. Evol.* **35,** 367 (1992).
32. T. H. Jukes and C. R. Cantor, *in* "Mammalian Protein Metabolism" (H. N. Munro, ed.). Academic Press, New York, 1969.
33. M. Kimura, *J. Mol. Evol.* **16,** 111 (1980).
34. M. Kimura, *Proc. Natl. Acad. Sci. USA* **78,** 454 (1981).
35. J. P. Vartanian, A. Meyerhans, B. Åsjö, and S. Wain-Hobson, *J. Virol.* **65,** 1779 (1991).
36. R. F. Smith and T. F. Smith, *Proc. Natl. Acad. Sci. USA* **87,** 118 (1990).
37. H. J. Bandelt and A. W. M. Dress, *Adv. Math.* **94,** 47 (1992).
38. D. G. Higgins, *CABIOS* **8,** 15 (1992).
39. L. Breiman, J. H. Friedman, R. A. Olshen, and C. J. Stone, "Classification and Regression Trees." Wadsworth & Brooks, Pacific Grove, California, 1984.

Section II

Retroviruses/RNA Viruses

[11] Gene Trap Retroviruses

Jin Chen, James DeGregori, Geoff Hicks, Michael Roshon,
Christina Scherer, Er-gang Shi, and H. Earl Ruley

Retrovirus Vectors

Retroviruses have been widely used as vehicles for introducing genes into mammalian cells (1). A typical gene transfer strategy using a replication-defective retrovirus is illustrated in Fig. 1. A recombinant provirus (cloned in an *Escherichia coli* plasmid) is transfected into a packaging cell line, which harbors one or more defective helper viruses. The helper virus(es) express all of the proteins required to produce virus particles but lack sequences (designated Ψ) necessary for incorporating RNA into virions. As a result, only transcripts expressed from the transfected plasmid DNA are packaged into infectious virus particles. After infection, the vector RNA is converted into DNA by reverse transcription and integrates into the genome of the target cell. The provirus is defective and, in the absence of a helper virus, cannot be transmitted further.

Recombinant retroviruses vary widely in structure but must contain the following *cis*-acting sequences: (a) Ψ, located in the 5' region extending into *gag*, is required for the incorporation of RNA into virus particles; (b) repeated sequences at each end of the transcript (R) promote strand exchange during reverse transcription; (c) tRNA and polypurine primer sites at the 5' and 3' ends of the genome initiate plus- and minus-strand DNA synthesis, respectively; (d) the ends of U3 and U5 recognized by the viral integrase are required for integration of the proviral DNA (2). The remaining viral sequences can be replaced by foreign genes as long as the total length of the viral transcript does not exceed the packaging limits of the virion (about 9–10 kb). Multiple genes may be inserted, expressed either from the viral long terminal repeat (LTR) or from internal promoters.

The retrovirus host range is determined by the envelope (Env) glycoprotein (2). Murine ecotropic helper viruses produce vectors capable of infecting mouse and rat cells, whereas murine amphotropic helper viruses allow a wide variety of cell types to be infected (including human, monkey, mouse, rat, dog, cat, and chicken). Host range can also be greatly extended by introducing the ecotropic receptor into the target cells (3) or by using pseudotype viruses that contain the VSV G protein (4a).

Because cells are not killed when retroviruses are released, producer clones provide a nearly permanent source of infectious genes. Most established lines of mammalian cells can be infected with amphotropic viruses, although the sensitivity of different cell types to infection varies widely. Titers of 10^6 and 10^7 colony-forming units (cfu)/ml are frequently observed following infection of susceptible cells with vectors expressing a selectable marker, and over 90% of infected cells may stably

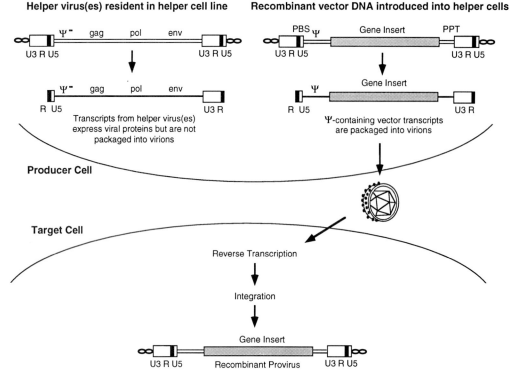

FIG. 1   Retrovirus-mediated gene transfer. A recombinant retrovirus vector (plasmid sequences are not shown for clarity) is transfected into cells expressing one or more packaging defective (Ψ^-) helper viruses. Transcripts from the vector, including an inserted gene (shaded), are packaged into virus particles, using proteins expressed by the helper virus(es). After infection, the viral RNA is reverse transcribed and integrates into the genome of the cell. *gag, pol,* and *env* refer to genes encoding viral proteins; Ψ, *cis*-Acting sequence required to package RNA into virions; PBS and PPT, polymerase-binding site and polypurine tract required to prime plus- and minus-strand cDNA synthesis; U3, R, and U5, regions of the virus long terminal repeat.

express transduced viral genes. Efficient transmission also requires cell proliferation; infectivity drops by three to four orders of magnitude in quiescent cells.

Transmissible helper viruses are sometimes generated due to recombination between the vector and helper viruses, allowing the vector to spread within the infected cell population. To reduce the possibility of such events, producer cell cultures should be frequently reinitiated from cryopreserved stocks and must not be serially passaged for extended periods of time. Moreover, improved packaging lines are now available in which helper functions are provided by two or more noncomplementing viruses (1).

Gene Trap Retroviruses

Retrovirus integration occurs widely throughout the genome and may disrupt cellular genes. To enrich for insertional mutations, several types of retrovirus vectors have been developed that confer selectable phenotypes when the virus integrates into expressed genes. One type contains coding sequences for a selectable marker inserted into the U3 region (Fig. 2). The U3 gene forms a part of the LTRs such that the 5' copy inserts only 30 nucleotides from the flanking cellular DNA. Selection for U3 gene expression gives rise to cell clones in which the proviruses have inserted in or near 5' exons of transcriptionally active genes (4–6). A second gene trap strategy

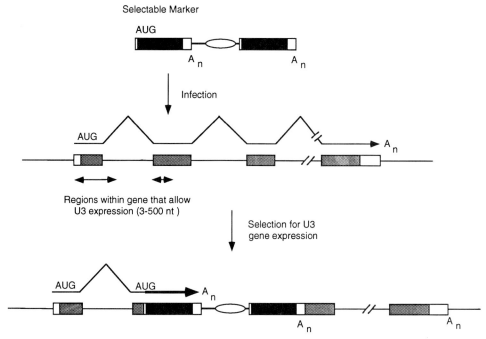

FIG. 2   U3 gene trap vectors. Coding sequences for a selectable marker (shown in black) are inserted into the U3 region. The gene participates in the formation of LTRs such that integration places the leftward (5') copy only 30 nucleotides (nt) from the flanking cellular DNA. A hypothetical cellular gene is shown before and after integration in the second exon. The U3 gene is expressed from fusion transcripts that extend into the virus from the flanking cellular DNA. Transcripts are drawn above expressed exons (rectangles). Coding sequences (shaded areas), initiation codons (AUG), and polyadenylation signals (A_n) are also indicated. The average size of the regions within genes that can activate U3 expression is estimated from the amount of cellular RNA that is appended to U3 transcripts.

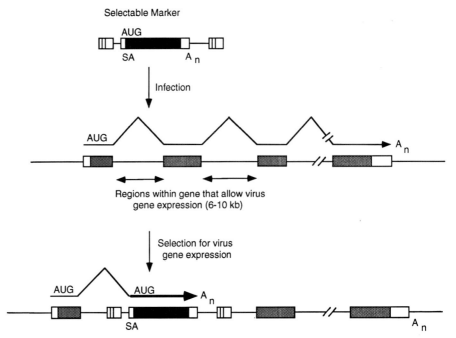

Fɪɢ. 3   Splicing-activated gene traps. Coding sequences for a selectable marker (shown in black) are inserted into a retrovirus vector, just downstream of a splice acceptor sequence (SA). A hypothetical cellular gene is shown before and after integration into the first intron. The viral gene is expressed from cellular transcripts that splice into the virus. Transcripts are drawn above expressed exons (rectangles). Coding sequences (shaded areas), initiation codons (AUG), and polyadenylation signals (A_n) are also indicated. The average size of the regions within genes that can activate virus gene expression is estimated from the proportion of proviruses that express functional fusion genes (see text).

(Fig. 3) relies on splicing to activate the expression of a selectable marker carried by the virus, as can occur when the virus integrates into introns of expressed genes (7,8).

The gene trap vectors were initially referred to as "promoter traps" because the extent to which activating promoters would be associated with cellular genes was not known. In addition, although the enhancer sequences were deleted from the vectors, the possibility existed that virus integration might activate cryptic promoters not normally associated with cellular genes. It is now clear that the transcribed regions that activate viral gene expression have properties of genes transcribed by RNA polymerase II. In particular, they undergo splicing (6,8), are expressed in cells prior to virus integration (6,9), and hybridize to single-copy DNA (6,9). Moreover, several have involved known genes (6,10). Finally, approximately 40% of genes selectively disrupted in embryonic stem cells are linked to obvious recessive phenotypes when introduced into the germline (6,8). Pol I and Pol III promoters do not activate gene traps because the resulting fusion transcripts lack 5′ caps and are not processed or translated efficiently (11,12).

Fusion transcripts generated by gene trap selection typically contain less than 500 nucleotides (nt) of cell-derived RNA (4–8,13,14). Presumably, efficient translation selects for integration events that position the initiation codon of the viral gene near the 5′ end of the resulting fusion transcript (12). In addition, retroviruses may preferentially integrate in 5′ gene regions (15,16). Four genes responsible for activating U3 expression in embryonic stem cells have been characterized (6,10). Regions within genes that allow U3 expression include exons and nearby intron sequences and average less than 500 nt in size, as estimated from the amount of appended cellular RNA (Fig. 2). The small size of these regions explains why U3 genes are expressed by only 1 in 200–2000 proviruses.

The number of genes in the genome that can be disrupted by gene trap selection can be estimated, first, from the fraction of proviruses that express U3 genes and, second, by the frequencies with which single-copy genes are disrupted following gene trap selection (14). In each case, the number of target genes (2×10^4–10^5) is comparable to the total number of expressed genes as estimated by RNA renaturation kinetics (17). The fact that most genes are expressed at relatively low levels (fewer than 10 transcripts per cell) suggests that even weak promoters can activate viral gene expression.

Approximately 10% of splice trap proviruses express functional fusion transcripts (8). This is twice the frequency observed when splice-activated gene traps are introduced by DNA-mediated gene transfer and corresponds to twice the fraction of the genome that is typically transcribed in cells at any one time (17). These discrepancies suggest that retrovirus integration may preferentially involve transcribed regions of the genome. Splice trap proviruses are at least 20 times more likely than U3 gene traps to express functional fusion genes—presumably because the regions within genes that allow viral gene expression (i.e., introns) are much larger. By extrapolation, the average target gene contains 6–10 kb of sequences capable of activating a splice trap provirus (Fig. 3).

Insertional Mutagenesis in Cultured Cells

In principle, retroviruses can be used to identify genes responsible for recessive phenotypes in mammalian cells. The strategy involves (a) isolating large populations of cells in which proviruses have integrated extensively throughout the genome, (b) selecting cell clones for phenotypes that result when gene functions are lost as a result of virus integration, and (c) characterizing specific genes disrupted by the integrated proviruses (18). However, conventional retroviruses are inefficient mutagens as evidenced by the fact that 5×10^6–10^8 integration events have been required to disrupt

TABLE I   Enrichment of Insertional Mutations
Following Gene Trap Selection[a]

| Gene | Integration events to disrupt single-copy gene | | Ref. |
	Without gene trap selection	With gene trap selection	
src	5×10^6	NT	34
hprt	10^8	NT	35
β_2-MG	10^8	NT	36
tk (H1)	8×10^6	2×10^4	14
tk (H4)	4×10^7	1×10^5	14

[a] H1 and H4 are single-copy *tk* genes inserted at different sites in the genome. NT, Not tested; β_2-MG, β_2-microglobulin.

single-copy genes (Table I) (14, 18a–c). Difficulties disrupting specific genes reflect the large size of the genome, potential biases against integration into specific genes, and biological factors (such as phenotypic lag and low cloning efficiencies of cells in selection) that interfere with the recovery of null clones. Consequently, no gene responsible for a recessive phenotype in cultured cells has been identified by insertional mutagenesis.

Gene trap selection enriches for insertional mutations involving expressed genes by two to three orders of magnitude (Table I). This reduces the background of null clones resulting from spontaneous mutations, thus increasing the efficiency of insertional mutagenesis. Because the typical mammalian cell expresses only about 10,000–20,000 genes (17), a library of 10^5 U3-expressing clones is expected to contain proviruses in all readily targeted genes. This number of clones can be isolated even after using a low multiplicity of infection, so that the resulting clones contain only a single provirus. In principle, complete libraries are small enough to permit mutagenesis of diploid genes, because alleles opposite those disrupted by the provirus may be missing as a result of preexisting hemizygosity; lost by gene conversion, nondisjunction, or mutation; or may be transcriptionally inactive. Recovery of clones expressing null phenotypes may be enhanced by increasing selective pressures (19) or by using hypodiploid cells (14).

Several factors influence the probability of disrupting individual genes. First, mutation rates will reflect any preference for or against integration into certain regions of the genome. Second, the size of the region within each gene that allows viral gene expression is greatly affected by the extent to which 5' sequences can suppress translation of a downstream reading frame (12). For example, genes with long, untranslated 5' leader regions should provide larger targets than genes with short leaders.

Highly transcribed genes should also constitute larger targets than weakly expressed genes, because high levels of fusion gene expression can compensate for translational suppression due to appended cellular sequences. Finally, splice trap vectors will preferentially target genes with long 5' introns. It is presently not known whether genes responsible for recessive phenotypes in diploid cells can be identified following gene trap mutagenesis. However, the feasibility of targeting autosomal genes has been demonstrated in Chinese hamster ovary (CHO) cells (14,21a).

Identification of Transcriptionally Regulated Genes

Gene trap retroviruses have employed a variety of selectable markers to disrupt expressed cellular genes (Table II). Those that permit selection both for and against viral gene expression may be used to identify cellular genes that are regulated by hormones or other factors. The strategy (Fig. 4) involves isolating collections of clones selecting either for or against U3 expression. Each type of integration library is then exposed to a hormone or is infected with retroviruses expressing a specific *trans*-acting factor and is placed in selection for the appropriate change in viral gene expression. The resulting clones include those in which the provirus has inserted downstream of transcriptionally regulated promoters.

The minimum size of an integration library sufficient to incorporate every cellular gene into a functional fusion gene (including those genes that are not expressed at the time of integration) can be estimated by dividing the size of the genome (3×10^9 nt) by the average size of the regions within genes that allow viral gene expression. This corresponds to approximately 10^6 and 10^7 clones containing inserted splice trap and U3 trap proviruses, respectively. This number of clones is readily isolated, starting from 10^7 to 10^8 infected cells. Ideally, to eliminate background clones, one would

TABLE II   Genes Used for Retrovirus Gene Trap Selection

Gene	Selection for viral gene expression	Selection against viral gene expression	Ref.
his	L-Histidinol	None	14
hygro	Hygromycin D	None	14
neo	Geneticin (G418)	None	6
lacZ	FACS[a]	FACS	5, 7
B-geo	Geneticin or FACS	FACS	8
tk	HAT[a]	Gancyclovir	14

[a] FACS, Fluorescence-activated cell sorting; HAT, hypoxanthine–aminopterin–thymidine.

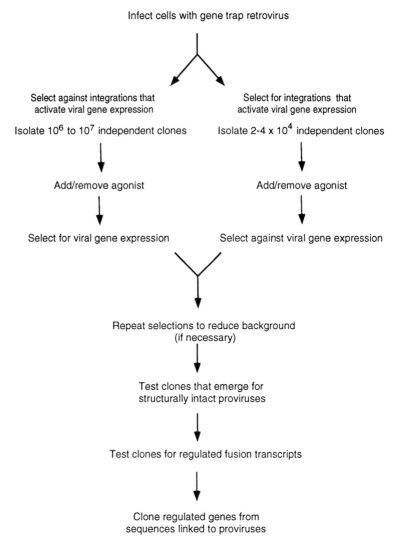

FIG. 4   Strategy for identifying transcriptionally regulated genes. Cells are infected with a gene trap vector and clones are isolated after selecting for or against viral gene expression. Cells from each type of integration library are transferred to medium in which a *trans*-acting factor has been added or deleted and are placed in selection for the appropriate change in U3 gene expression (see Table II for examples). The resulting clones should include those in which the gene trap provirus has inserted downstream of transcriptionally regulated promoters.

want to alternate selection for and against viral gene expression in the presence or absence of the agonist. This may not be possible if, for example, the agonist induces terminal differentiation.

The process has several advantages over cDNA subtraction as a means to isolate differentially regulated genes. First, gene trap selection is sensitive, detecting even weakly expressed genes. This is important because most genes are expressed at levels ($<0.01\%$ of total mRNA) too low for efficient recovery after cDNA subtraction. Second, the process is not especially biased for highly expressed genes. Thus, while highly transcribed genes provided larger targets than weakly expressed genes (see above), the magnitude of the bias is no more than threefold, as estimated from the variation in the size of appended RNA. Indeed, regulated genes have larger untranslated 5′ leaders than constitutively expressed genes, and may also be preferentially targeted. Third, gene traps typically insert near 5′ exons that are often missing from all but full-length cDNA clones. Fourth, differential gene expression is likely to reflect transcriptional control rather than changes in RNA stability or processing. Finally, cells recovered after gene trap selection contain a reporter gene expressed from the natural promoter of the gene and may be used to study factors that regulate transcription.

Genetic Analysis of Mouse Development

The long reproduction cycles and large genomes of mammals generally preclude the types of genetic analysis possible with simpler organisms. Although genes responsible for organismal phenotypes can be isolated, the process is slow, requires relatively detailed physical maps, and is limited to a small number of mutant alleles (20,21). Gene traps provide an alternative means to analyze gene functions in mice. The process involves (a) selectively disrupting genes expressed in embryonic stem (ES) cells, (b) screening clones *in vitro* to identify those in which interesting genes have been disrupted, (c) introducing the disrupted genes into the germline, and (d) assessing gene functions in animals homozygous for the mutations. Large numbers of ES cell clones can be screened for significant mutations, both *in vitro* and *in vivo,* and the affected genes can be cloned from sequences associated with the integrated provirus (24a). The process facilitates a functional analysis of the mouse genome and is expected to provide animal models for human genetic diseases.

An elegant strategy to identify mutations in developmentally regulated genes employs a reporter gene expressing both neomycin phosphotransferase and *E. coli* β-galactosidase activities (8). Mutant clones are isolated by geneticin (G418) selection, and changes in target gene expression associated with cell differentiation are monitored by staining with X-Gal, which is cleaved by β-galactosidase to produce a blue precipitate. Clones expressing developmentally regulated *lacZ* fusion genes have also been identified among ES cells isolated by flow cytometry (13).

Mutant genes can be characterized directly by sequencing DNA adjacent to the integrated provirus. Because U3 gene traps integrate in or near 5′ exons, transcribed regions of the disrupted genes may be isolated rapidly by inverse polymerase chain reaction (PCR) (6). Similarly, segments of genes disrupted by both U3 trap and splice trap vectors can be amplified by a process known as 5′ RACE (rapid amplification of cDNA ends). Computer searches of the sequence databases will then reveal instances in which the provirus has disrupted known genes or transcripts (characterized previously as cDNAs) from which encoded proteins can be deduced. It is presently feasible to isolate a library of ES cell clones of sufficient size (about 20,000) to disrupt most expressed genes. Moreover, by using shuttle vectors, it is possible to isolate and sequence the DNA flanking each provirus in the library. In an initial survey, 12 mutations involving known genes were identified from approximately 120 flanking sequences (6,25a). The flanking sequences will provide functional genetic landmarks, analogous to expressed sequence tags (ESTs), for the emerging genome map, and the stem cell clones will provide a source of mutant alleles to analyze gene functions *in vivo*.

All clones isolated by gene trap selection have thus far contained proviruses inserted into expressed cellular genes. Although transcripts could conceivably splice around proviral sequences, in most cases the proviruses appear to disrupt gene functions. Approximately 40% of fusion genes cause obvious phenotypes (usually embryonic death) when introduced into the germline and bred to a homozygous state (6,8). The proportion is similar to that observed in *Drosophila* and yeast (23–25), and suggests that many mammalian genes are not essential.

Insertional mutations have also been induced (without gene trap selection) in mice following infection of pre- and postimplantation mouse embryos. However, because only 5% of randomly integrated proviruses are associated with obvious phenotypes (21). It is impossible without extensive analysis to determine whether the virus inserted into a nonessential gene or simply failed to disrupt a gene. In contrast, genes disrupted by gene trap selection can be isolated even when they are not linked to obvious phenotypes.

Methods

Preparation of Retrovirus Producer Cells

Helper and virus-producer cells are grown at 37° C in Dulbecco's modified Eagle's medium (D-5648; Sigma, St. Louis, MO) supplemented with 10% (v/v) calf serum (Colorado Serum Co., Denver, CO) in a humidified, 5% CO_2 atmosphere.

Cell lines producing recombinant retroviruses are derived after transfecting helper cells with plasmid DNAs and selecting in the appropriate medium. We routinely use the calcium phosphate coprecipitation procedure to transfer vector DNA into ψ_2 helper cells (4). Individual colonies are trypsinized within cloning cylinders, expanded in mass culture, and analyzed for virus production. For this, 2×10^6 cells from each producer cell clone are seeded onto a 100-mm dish. After 24 hr, the medium is replaced with 2 ml of fresh, incubator-equilibrated medium. After 2 hr, the culture supernatant is filtered through a 0.45-μm pore size nitrocellulose membrane (No. 4184; Acrodisc) and used directly for infection. Virus stocks can be frozen quickly in liquid nitrogen and used later, but titers may drop by a factor of 10.

Target cells (e.g., NIH 3T3 cells) used for titering viruses are seeded 24 hr prior to infection at a density of 10^5/100-mm dish. The medium is then replaced with 1 ml of diluted virus stock in the presence of 8 μg of Polybrene/ml (hexadimethrine bromide; No. 10,768-9, Aldrich, Milwaukee, WI) and the cells are incubated at 37° C for 1 hr with occasional rocking. Fresh medium (9 ml) is added and the cells are incubated overnight prior to growth in selective medium. After 10–14 days, drug-resistant colonies are fixed [10% (v/v) formaldehyde in phosphate-buffered saline (PBS)], stained [0.1% (v/v) crystal violet in equal parts PBS and ethanol], and counted. We routinely titer gene trap retroviruses by selecting for a constitutively expressed gene (if present) and for U3 gene expression. This measures number of transducing viruses and the frequency of gene-trapping events, respectively. The rate of virus production varies widely among individual clones. However, 10–20% of clones are typically high-titer producers ($>5 \times 10^5$ transducing units/ml/10^7 producer cells/2 hr). Therefore, at least 1 high-titer producer line can be established starting from 20 independent clones.

Gene Trap Selection

The fraction of proviruses that expresses U3 genes (representing gene trap events) varies between 1/200 and 1/2000, depending on the vector and target cells used. Target cell lines should be analyzed with regard to levels of drug necessary for selection, sensitivity to virus infection, and the proportion of proviruses that result in gene-trapping events. Drugs should be used at concentrations sufficient to kill most cells within 4–5 days, and monolayer cells should not be allowed to reach confluence before killing starts.

A variety of factors may lower virus titers. Target cells should be actively proliferating and express receptors appropriate for the host range of the virus. Env glycoproteins expressed by some murine cell lines (e.g., L cells) may block infection, in which case the host range of the packaging line may be changed. Polybrene, a polycation used to enhance virus uptake, is toxic to some cells. Some lots of serum, especially fetal bovine serum, also suppress titers. Virus production may also fall as producer cells are passaged in culture.

Infection of target cells is a scaled-up version of the titration protocol. A minimum of 10^7 integration events is required to saturate the genome (3×10^9 bp/genome divided by 300–500 bp, the average size of the regions within genes that activate U3 gene expression) (14). To mutagenize all expressed genes, 10^8 target cells are infected at a multiplicity of infection (MOI) of 0.1 and a library of about 4×10^4 independent U3-expressing colonies is obtained. The library is then expanded as individual pools of cells and may be frozen. Insertional mutations affecting specific genes are isolated by secondary or concomitant selection for phenotypes that result when gene functions are lost.

Rapid Isolation of Sequences Disrupted by U3 Gene Trap Proviruses

Genomic sequences 5' of the U3 gene trap provirus can be isolated by inverse PCR (9). The sequences are generally nonrepetitive and frequently hybridize to transcripts expressed by the disrupted cellular gene (6). In addition, cDNA sequences appended to cell : virus fusion transcripts may be cloned by 5' RACE.

Inverse Polymerase Chain Reaction

Genomic DNAs from His^R and Neo^R cell lines are digested (at 15 μg/ml) overnight with *Hin*fI or *Mse*I, extracted once with phenol–chloroform, once with chloroform, and precipitated with a 1/10 vol of 3 M ammonium acetate and 2.5 vol of ethanol. The precipitated DNA is washed with 70% ethanol, vacuum dried, and resuspended in 50 μl of TE buffer [Tris–ethylenediaminetetraacetic acid (EDTA)], pH 8.0. To circularize restriction enzyme products, 2 μg of digested DNA is ligated overnight at 16°C in a 400-μl reaction containing 8 U of T4 DNA ligase. Ligation reactions are extracted and precipitated as before, but with 1 μg of yeast tRNA added as carrier. [To prevent amplification of fragments from the 3' LTR, the resuspended DNA is digested with *Pvu*II (which cleaves in the provirus just upstream of the 3' LTR); extract and precipitate as before.] DNAs with or without *Pvu*II treatment are resuspended in 20 μl of doubly distilled H$_2$O containing 5 μg of DNase-free RNase A, and incubated at 37°C for 30 min. One microgram of this DNA is used for the PCR in 100-μl reactions containing 10 mM Tris-HCl (pH 8.3), 5 mM KCl, 1.5 mM MgCl$_2$, a 200 μM concentration of each deoxyribonucleoside triphosphate, 1% (v/v) deionized formamide, 2.5 U of Amplitaq (Perkin-Elmer/Cetus, Norwalk, CT), and a 1 μM concentration of each primer. Reactions proceed through 40 cycles of denaturation (95°C for 1.5 min), primer annealing (50°C for 1.5 min), and primer extension (72°C for 3 min). Oligonucleotide primers complementary to *his* or *neo* sequences are used.

> *his:* for *Mse*I circles
> 5' CCAGTCAATCAGGGTATTGA 3'
> 5' GTAAGCTTTAAAACAGAAGTGACAGCGCTACGCG 3'
> *his:* for *Hin*fI circles
> 5' CCAGTCAATCAGGGTATTGA 3'
> 5' GTAAGCTTTGCAGGAAATCTTAACGTCGGCGG 3'
> *neo:* for *Mse*I circles
> 5' CCATCTTGTTCAATCATGCGAAACGATCC 3'
> 5' GTAAGCTTGCTCCCGATTCGCAGCGCATCGCCTT 3'

neo: for *Hin*fI circles

    5' CCATCTTGTTCAATCATGCGAAACGATCC 3'

    5' GTAAGCTTGCGATGCCTGCTTGCCGAATATCATG 3'

Gel-purified PCR products are cleaved with *Nhe*I and *Hin*dIII and ligated to Bluescript KS (−) (Stratagene, La Jolla, CA) plasmids digested with *Xba*I and *Hin*dIII.

5' RACE

cDNA sequences appended to U3Neo or U3His fusion transcripts are cloned by using standard 5' RACE protocols (26,27) with slight modifications.

First-Strand cDNA Synthesis

Fifteen to 20 μg of total cellular RNA (4) is denatured at 70° C for 5 min and annealed to 4 pmol of neoA (5' ATTGTCTGTTGTGCCCAGTCATA 3') or his8 primer (5' GCTGTTCAGGGCTACAGCTGTTCC 3') at 55° C for 10 min in 14 μl of doubly-distilled H_2O. (Sequences of neoA and his8 primers are given in parentheses.) Then 2 μl of 10× PCR buffer [200 mM Tris-HCl (pH 8.4), 500 mM KCl, 25 mM $MgCl_2$, bovine serum albumin (BSA) (1 mg/ml), 1 μl of dNTPs (10 mM each), and 2 μl of 0.1 M dithiothreitol (DTT) are added and incubated (42° C, 2 min), prior to adding 1 μl of SuperScript reverse transcriptase (RNase H⁻; Bethesda Research Laboratories, Gaithersburg, MD), and incubating for 60 min at 42° C. Following first-strand cDNA synthesis, RNA templates are digested with 2 U of RNase H (55° C, 10 min) and the cDNA is purified by using GlassMax cartridges (Bethesda Research Laboratories).

Homopolymeric C Tailing

To one-fifth of the purified first-strand cDNA, add 2 μl of dCTP (2 mM), 1 μl of 10 × PCR buffer, 10 U of terminal transferase, and doubly-distilled H_2O to a final volume of 20 μl, and incubate for 10 min each at 37 and 70° C. Homopolymeric A tailing has worked equally well in our hands.

Amplification of Tailed cDNA

In our experience, amplification of a specific product normally requires two consecutive PCR reactions using nested primers. For the first PCR reaction, 5 μl of tailed cDNA is mixed with 5 μl of 10 × PCR buffer, 1 μl of dNTPs (10 mM), 4 pmol of neoB (5' CGAATAGCCTCTCCACCCAA 3') or his7 primer (5' CCAGTCAAT-CAGGGTATTGA 3'), 0.5 pmol of anchor primer (for C-tail use the Bethesda Research Laboratories anchor Primer; for A-tail use 5' ATTGAATTCTCTAGA-ACGCGTCTCGAGTTTTTTTTTTTTTTTTTTTT 3'), and 1 μl of *Taq* polymerase, in a reaction volume of 50 μl. Reactions are cycled 30 times at 94° C for 45 sec, 57° C for 60 sec, and 72° C for 120 sec. For the second PCR reaction, 5 μl of the first PCR product is mixed with 5 μl of 10 × PCR buffer, 1 μl of dNTPs (10 mM), 20 pmol of neo1 (5′ CCATCTTGTTCAATCATGCGAAACGATCC 3′) or U3 primer (5′ CCTACAGGTGGGGTCTTTC 3′), 2 pmol of adapter primer (for C-tail use the Bethesda Research Laboratories Universal amplification primer; for A-tail use 5′ ATTGAATTCTCTAGAACGCGTCTCGAG 3′), and 1 μl of *Taq* polymerase, in a reaction volume of 50 μl. Reactions are cycled 30 times at 94°C for 45 sec, 57°C for 60 sec, and 72°C for 120 sec. It is important to note that the U3 primer may be used as a nested primer for any U3 fusion transcript. Moreover, it overlays a natural viral reverse transcription stop signal.

Insertional Mutagenesis in Embryonic Stem Cells

Preparation of Mitotically Inactivated Fibroblast Feeder Layers

Isolation of MEFs from Mouse Embryos

We routinely use mouse primary embryonic fibroblasts (MEFs) as feeder layers to maintain ES cell totipotency. All steps are performed in a hood using sterile equipment. Sacrifice a 13- to 14-day gestation mouse by cervical dislocation. Apply a liberal amount of 70% ethanol to the abdomen and cut through the abdominal wall to expose the uterus. Remove the uterine horns, transfer them to a 10-cm bacteriological petri dish, and rinse with sterile PBS. Dissect each embryo away from the uterus and placenta, using a pair of watchmaker forceps, and rinse in fresh PBS. Remove heads and dark colored tissues (e.g., liver) and rinse the remaining embryo in fresh PBS. Transfer the embryos to a fresh dish and mince with two scalpels or a pair of fine scissors. Add 25 ml of trypsin–EDTA (GIBCO, Grand Island, NY) and disrupt tissues further by pipetting up and down through a 5-ml pipette. Transfer the tissue suspension into a 100-ml bottle and incubate at 37°C for 10 to 15 min with vigorous stirring. Add 50 ml of culture medium [Dulbecco's modified Eagle's medium (DMEM), 10% (v/v) fetal bovine serum (FBS)], and allow large clumps and cell debris to settle for 2 min. Viscous material (presumably lysed cells and DNA) in the supernatant does not interfere with the process. Split the supernatant between two 50-ml conical tubes, centrifuge at 1000 rpm for 5 min, resuspend the cells (1 ml for each embryo) in ice-cold freezing medium [DMEM, 50% (v/v) FBS, 12% (v/v) dimethyl sulfoxide (DMSO)], and freeze slowly (-1°C/min).

Expansion of Mouse Embryonic Fibroblasts in Culture

Mouse embryonic fibroblasts have a limited life span in culture and they become senescent after 15 to 20 cell divisions. Mouse embryonic fibroblasts grow relatively quickly when maintained at high cell densities and can be passaged every 2–3 days. Typically, one vial of the above frozen MEF is thawed and plated onto a 15-cm culture dish and expanded to 27 dishes by splitting the cells 1:3 each time the cells become confluent for a total of three passages.

Mitotic Inactivation of Mouse Embryonic Fibroblasts by γ-Irradiation

After expansion, each 15-cm culture dish is washed with PBS and 5 ml of trypsin–EDTA is added. The cells are incubated at 37° C for 2 min, serum is added to a final concentration of 10%, and the cells are centrifuged at 1000 rpm for 5 min. Cell pellets are resuspended in an equal volume of feeder cell medium and an aliquot is counted using a hemacytometer. In a 50-ml conical tube, cells are exposed to 3000-rad γ irradiation, pelleted by centrifugation, and resuspended at 2×10^6 cells/ml in freezing medium. Cells are frozen in 1-ml aliquots, which is sufficient to support the culture of ES cells on one 10-cm dish.

Routine Maintenance of Embryonic Stem Cells

Embryonic stem cells are grown on an irradiated MEF feeder layer prepared as described above. Feeder cells (2×10^6) are seeded onto a 10-cm tissue culture dish precoated with 0.1% (w/v) gelatin. Feeders can be either pre- or coplated with ES cells. If preplating, MEF cells can be thawed directly onto a 10-cm dish, allowed to adhere for 4 hr, and the medium changed to ES medium before adding ES cells. If coplating, the thawed MEFs should be centrifuged and resuspended to remove DMSO, which inhibits the growth of ES cells. Embryonic stem cells are grown in DMEM supplemented with 15% preselected and heat-inactivated FBS, 100 mM nonessential amino acids, 0.1 mM 2-mercaptoethanol, and leukemia inhibitory factor (1000 U/ml) (Esgro; GIBCO). Established ES cells divide rapidly and the medium must be changed every day. Overgrown ES colonies tend to differentiate; therefore the cells should be subcultured at least twice a week. Finally, the medium is routinely changed 2–4 hr prior to passaging or freezing ES cells.

A 10-cm plate is passaged as follows. The cells are washed once with PBS, trypsinized in 2 ml of trypsin–EDTA solution [0.25% (w/v) trypsin, 1 mM EDTA (pH 8.0), in PBS] and incubated at 37° C for 3 to 4 min. The cell suspension is vigorously pipetted up and down to disrupt cell aggregates, incubated at 37° C for another 3 min, and then checked visually under the microscope to ensure that a single-cell suspension is generated. Cell aggregates tend to differentiate and should be avoided. Trypsinization is inhibited by the addition of 5 ml of ES cell medium and the cell suspension is split at a ratio of 1 : 10 on fresh gelatinized plates with feeders.

Selective Disruption of Genes by U3Bgeo Gene Trap Retrovirus

Stocks of the U3Bgeo virus are prepared as described above. Embryonic stem cells are infected at an MOI of 1, as titered on NIH 3T3 cells (the actual titer on ES cells is lower), by adding 1 ml of diluted virus stocks to 10^6 ES cells (plated 12 hr before infection) in the presence of Polybrene (4 μg/ml). The cells are incubated for 1 hr at 37° C with occasional rocking, at which time 9 ml of fresh ES cell medium is added. The following day, cells are placed in ES medium containing active G418 (300 μg/ml) and maintained for 7 days. The medium is changed every day. Culture dishes are marked to indicate the positions of undifferentiated Neo^r colonies, and the colonies are gently washed with PBS. Marked colonies are drawn into a disposable tip of a micropipettor set at 25–50 μl, while visualized at 40 × magnification. We use a Nikon (Southern Micro Instruments, Atlanta, GA) TMS inverted microscope placed in the hood. The cells are placed in a microtiter well containing 100 μl of trypsin–EDTA solution for 5 min, and pipetted up and down to obtain a single-cell suspension. ES medium (100 μl) is added to each well, and the cells are pipetted again and are transferred to a 24-well plate with feeder cells. A multichannel pipettor facilitates processing large numbers of clones. Finally, cells are expanded in mass cultures and cryopreserved until they can be analyzed further.

In Vitro Differentiation of Embryonic Stem Cells and Identification of Regulated Genes

To identify genes that are developmentally regulated, Neo^r ES clones are induced to differentiate in DMEM plus 10% FBS as follows: (a) embryonic stem cells are seeded in high density (6×10^6) on gelatinized plates and grown to confluence in the absence of feeder cells; (b) cell aggregates are dislodged by mild trypsinization, transferred to a bacteriological petri dish, and allowed to grow in suspension for 4 days. The cell aggregates will form simple embryoid bodies consisting of an outer layer of large endoderm cells separated from an inner core of cells by a dark basement membrane. When cultured for longer than 4 days, simple embryoid bodies may differentiate further, forming cystic embryoid bodies that contain an inner layer of ectodermal cells surrounding a fluid-filled cavity; (c) after 6 days, the suspensions containing both simple and cystic embryoid bodies are plated onto gelatinized plates and left undisturbed for 2 days. The embryoid bodies attach and differentiate into a variety of cell types; (d) undifferentiated cells and differentiated embryoid bodies from each Neo^r clone are fixed in PBS–0.2% (v/v) glutaraldehyde–2% (v/v) formaldehyde at 4°C for 15 min, and stained in PBS–X-Gal (1 mg/ml)–2 mM MgCl$_2$–5 mM K$_3$Fe(CN)$_6$–5 mM K$_4$Fe(CN)$_6$ for 4–6 hr at 37°C. Staining for longer periods of time is not recommended in order to reduce the background resulting from endogenous lysosomal β-galactosidase activity. Regulated genes can be identified by comparing the intensity and pattern of X-Gal staining of Neo^r ES clones before and after differentiation *in vitro*.

Construction and Analysis of Germline Chimeras

The general strategy for transmitting genes from ES cells into the germline of mice involves (a) injecting ES cells into preimplantation blastocysts, (b) transferring the blastocysts into the uterus of outbred foster mothers, and (c) breeding the resulting chimeric mice to obtain offspring that have inherited genes from the ES cells (28). The ES cells are derived from a mouse whose coat color differs from that of the strain from which the blastocysts were isolated. Consequently, chimeras and offspring that have inherited genes from ES cells are readily identified on the basis of coat color. Most laboratories use male ES cells, because the efficiency of germ cell chimerism is greater than when female ES cells are used. Moreover, male chimeras generate more offspring than female chimeras. Finally, transfer of male ES cells into a female blastocyst can cause the resulting chimera to develop as a phenotypic male. Consequently, the majority of germ cells are ES cell derived, resulting in a high frequency of germline transmission (29,30).

We routinely use D3 ES cells (31) or a D3 subclone, J1, provided by J. Rossant (Mount Sainai Hospital, Toronto, Ontario, Canada). The cells are derived from the 129 strain and are homozygous for the Agouti coat color marker. Typically, 15–20 ES cells are injected into C57BL6 blastocysts (3.5 days) and implanted into the uterus of pseudopregnant CD1 recipients as described (28). After chimeric males are mated with C57BL6 mice, 3–100% of the offspring from 50% of the male chimeras inherit the Agouti coat color marker. The progeny of female chimeras will occasionally inherit genes from male ES cells. However, the efficiency is too low (about 5% of female chimeras) to justify the expense of testing female chimeras for germline transmission.

Because the ES cells contain a single provirus per diploid genome, 50% of the Agouti offspring will be heterozygous for the U3 fusion gene. However, some transgenes such as *tk* and *his* may not be transmitted efficiently through the germline (6,32). To identify these mice, DNA extracted from tail segments (33) is analyzed by Southern blot hybridization. Mice that carry the provirus are inbred to produce offspring of which 25% are homozygous for the transgene. Homozygous animals are identified by Southern blot analysis, using flanking cellular sequences (isolated by inverse PCR or 5' RACE) as probes. Animals that fail to produce viable homozygous offspring typically harbor recessive mutations that result in embryonic death.

Acknowledgments

We thank Harald von Melchner and Jay Schwartz for providing their modifications of the 5' RACE protocol. The authors are supported by Public Health Service Grants R01HG400684, RO1GM48688, and RO1CA40602 to H.E.R. and by postdoctoral fellowships from the American Cancer Society (J.C.) and the Canadian National Cancer Institute (G.H.).

References

1. A. D. Miller, *Curr. Top. Microbiol. Immunol.* **158,** 1 (1991).
2. J. M. Coffin, *in* "Virology" (B. N. Fields, D. M. Knipe, R. M. Chanock, M. S. Hirsch, J. L. Melnick, T. P. Monath, and B. Roizman, eds.), p. 1437. Raven Press, New York, 1990.
3. L. M. Albritton, L. Tseng, D. Scadden, and J. M. Cunningham, *Cell (Cambridge, Mass.)* **57,** 659 (1989).
4. H. von Melchner and H. E. Ruley, *J. Virol.* **63,** 3227 (1989).
4a. J. C. Burns, T. Friedmann, W. Driever, M. Burrascano, and J. K. Yee, *Proc. Natl. Acad. Sci. U.S.A.* **90,** 8033–8037 (1993).
5. S. Reddy, J. DeGregori, H. von Melchner, and H. E. Ruley, *J. Virol.* **65,** 1507 (1991).
6. H. von Melchner, J. DeGregori, H. Rayburn, C. Friedel, S. Reddy, and H. E. Ruley, *Genes Dev.* **6,** 919 (1992).

7. D. G. Brenner, S. Lin-Chao, and S. N. Cohen, *Proc. Natl. Acad. Sci. USA* **86,** 5517 (1989).
8. G. Friedrich and P. Soriano, *Genes Dev.* **5,** 1513 (1991).
9. H. von Melchner and H. E. Ruley, *Proc. Natl. Acad. Sci. USA* **87,** 3733 (1990).
10. M. Roshon, J. DeGregori, H. von Melchner, and H. E. Ruley, unpublished results (1993).
11. A. S. Banerjee, *Microbiol. Rev.* **44,** 175 (1980).
12. M. Kozak, *J. Biol. Chem.* **266,** 19867 (1991).
13. S. Reddy, H. Rayburn, H. von Melchner, and H. E. Ruley, *Proc. Natl. Acad. Sci. USA* **89,** 6721 (1992).
14. W. Chang, C. Hubbard, C. Friedel, and H. E. Ruley, *Virology* **193,** 737 (1993).
15. S. P. Goff, *Cancer Cells* **2,** 172 (1990).
16. S. B. Sandmeyer, L. J. Hansen, and D. L. Chalker, *Annu. Rev. Genet.* **24,** 491 (1990).
17. B. Lewin, *Cell (Cambridge, Mass.)* **4,** 77 (1975).
18. S. P. Goff, *in* "Methods in Enzymology," Vol. 151, p. 489. Academic Press, Orlando, Florida, 1987.
19. R. M. Mortensen, D. A. Conner, S. Chao, A. A. Geisterfer-Lowrance, and J. G. Seidman, *Mol. Cell. Biol.* **12,** 2391 (1992).
20. A. D. Reith and A. Bernstein, *Genes Dev.* **5,** 1115 (1991).
21. J. Rossant and N. Hopkins, *Genes Dev.* **6,** 1 (1992).
21a. S. C. Hubbard, L. Walls, H. E. Ruley, and E. A. Muchmore, *J. Biol. Chem.,* in press (1994).
22. M. Roshon, J. DeGregori, D. Williams, and H. E. Ruley, unpublished results (1993).
23. M. G. Goebl and T. D. Petes, *Cell (Cambridge, Mass.)* **46,** 983 (1986).
24. L. Cooley, R. Kelly, and A. Spradling, *Science* **239,** 1121 (1988).
24a. J. V. DeGregori, A. Russ, H. von Melchner, H. Rayburn, P. Priyaranjan, N. Jenkins, N. Copeland, and H. E. Ruley, *Genes Dev.* in press (1994).
25. S. G. Oliver, Q. J. M. van der Aart, M. L. Agostoni-Carbone, M. Aigle, L. Alberghina, D. Alexandraki, and G. Antoine, *Nature (London)* **357,** 38 (1992).
25a. G. Hicks, E.-G. Shi, M. Roshon, J. DeGregori, D. Williams, and H. E. Ruley, unpublished results.
26. D. M. Schuster, G. W. Buchman, and A. Rashtchian, *Focus* **14,** 46 (1992).
27. F. M. Ausubel, R. Brent, R. E. Kingston, J. G. Moore, D. D. Moore, J. A. Smith, and K. Struhl, "Current Protocols in Molecular Biology," p. 1. John Wiley & Sons, New York, 1993.
28. E. J. Robertson, "Teratocarcinomas and Embryonic Stem Cells." IRL Press, Oxford, 1987.
29. M. J. Evans, A. Bradley, M. R. Kuehn, and E. J. Robertson, *Cold Spring Harbor Symp. Quant. Biol.* **50,** 685 (1985).
30. M. A. H. Surani, S. C. Barton, and M. L. Norris, *Biol. Reprod.* **36,** 1 (1987).
31. T. C. Doetschman, H. Eistetter, M. Katz, W. Schmidt, and R. Kemler, *J. Embryol. Exp. Morphol.* **87,** 27 (1985).
32. T. M. Wilkie, R. R. Braun, W. J. Ehrman, R. D. Palmiter, and R. E. Hammer, *Genes Dev.* **5,** 38 (1991).
33. P. W. Laird, A. Zijderveld, K. Linders, M. A. Rudnicki, R. Jaenisch, and A. Berns, *Nucleic Acids Res.* **19,** 4293 (1991).
34. H. E. Varmus, N. Quintrell, and S. Ortiy, *Cell (Cambridge, Mass.)* **25,** 23 (1981).
35. W. King, M. D. Patel, L. I. Lobel, S. P. Goff, and C. Nguyen-Huu, *Science* **228,** 554 (1985).
36. W. Frankel, T. A. Potter, N. Rosenberg, J. Lenz, and T. V. Rajan, *Proc. Natl. Acad. Sci. USA* **82,** 6600 (1985).

[12] Regulation of Viral and Cellular Gene Expression in Cells Infected by Animal Viruses, Including Influenza Virus and Human Immunodeficiency Virus Type 1

Glen N. Barber, Michael B. Agy, and Michael G. Katze

Introduction

Virus control of mRNA translation has been increasingly recognized as an important research component of the careful examination of gene regulation. During the past several years we have been utilizing virus–cell systems as models to further our understanding of both viral pathogenicity and the regulation of cellular protein synthesis. Much of our work has focused on understanding the strategies by which influenza virus and human immunodeficiency virus type 1 (HIV-1) regulate protein synthesis during infection. This article focuses on methods designed to understand the translation regulation in cells infected by two distinctly different RNA viruses: influenza virus and HIV-1. We first describe ways in which influenza virus and HIV-1 ensure the selective and efficient translation of their viral mRNA in infected cells. Second, we discuss ways in which influenza virus and HIV-1 may inhibit the double-stranded RNA (dsRNA)-activated protein kinase, PKR, an enzyme thought to play a key role in the establishment of the antiviral state mediated by interferon. Such regulation is necessary because many viral RNAs can activate PKR, an event that can cause a dramatic inhibition of protein synthesis.

Regulation of Gene Expression in Cells Infected by Primate Lentiviruses, HIV-1 and SIV, and by Influenza Virus

Analysis of Regulation of Gene Expression by HIV-1

Background

HIV-1 is a CD4-tropic lentivirus with a complex genomic arrangement containing several regulatory genes in addition to the *gag, pol,* and *env* genes common to all retroviruses (for a review see Steffy and Wong-Staal (1). HIV-1 is also the etiological agent of human acquired immunodeficiency syndrome (AIDS). Despite numerous reports on the function of HIV regulatory gene products and their effect on viral gene expression, few studies had reported on the effects of HIV-1 on host macromolecular synthesis. We therefore decided to analyze cellular and viral gene expression in HIV-1-infected cells, using several human CD4-positive lymphoid T cell lines. We found, in a pattern similar to many other eukaryotic virus infections, a decline in the rates of cellular protein synthesis that was concomitant with an increase in viral protein synthesis during infection (2). By 3 days postinfection, immunofluorescence analyses showed that almost 100% of the cells were infected while immunoprecipitation analysis of infected cell extracts revealed the complete spectrum of viral protein synthesis (Fig. 1). To determine whether the reduction in cellular protein synthesis was due to a decrease in steady state cellular mRNA levels, total RNA was prepared from mock- and HIV-1-infected cells. Northern blot analyses of these RNA preparations revealed that while viral mRNA levels were increasing during infection, cellular mRNA levels were in fact decreasing (interestingly, however, no significant loss of ribosomal RNA was observed). To determine whether the decrease in steady state levels of cellular mRNA was due to decreased transcription rates, nuclear runoff assays with nuclei from mock- and HIV-1-infected cells were performed. We found that the decline in cellular mRNAs was not due to a specific drop in RNA polymerase II transcription rates, but was rather due to degradation of cellular mRNA during HIV-1 infections. This conclusion was further supported by both polysome distribution analysis and *in vitro* translatability of cellular mRNA. Allowing for the decreased levels of cellular mRNA, we found that the polysome distribution in HIV-1-infected cells did not differ from that in the uninfected controls. Further, *in vitro* translation of cellular mRNA revealed that cellular mRNA, although present at lower levels in HIV-1-infected cells than in uninfected controls, were translationally competent, indicating that the observed decline in protein synthesis was due to a decrease in levels of cellular mRNA and not to their translational efficiency (2).

The exact mechanisms responsible for the destabilization of cellular mRNA in HIV-1-infected cells remain unknown. Development of an *in vitro* assay for mRNA degradation, however, may allow us to define the molecular mechanisms accountable for this process. Interestingly, simian immunodeficiency virus (SIV) did not impose a selective block on host macromolecular synthesis like HIV-1 (3). No selective reduction in cellular mRNA stability or protein synthesis was observed in cells infected by SIV_{mne} (both viral and cellular protein synthesis occurred concomitantly). Further, SIV_{mne} infected only cells that were virally transformed (3). On the basis of these results, it is tempting to speculate that the observed differences may be due to the regulatory gene product unique to the HIV-1 and SIV_{cpz} viruses, vpu. vpu has been characterized as a phosphorylated membrane-associated protein that is required for virus maturation and release and not for virus infection (5). In addition, vpu is associated with intracellular trafficking (6) and dissolution of the gp120–CD4 complex through the Golgi and endoplasmic reticulum and the intracellular degradation of CD4 (7). vpu has not, however, been assigned a definitive function. We are therefore examining the effects on cellular gene expression in cells infected by an HIV-1 *vpu*-negative mutant. These studies, as well as experiments in progress with HIV-2

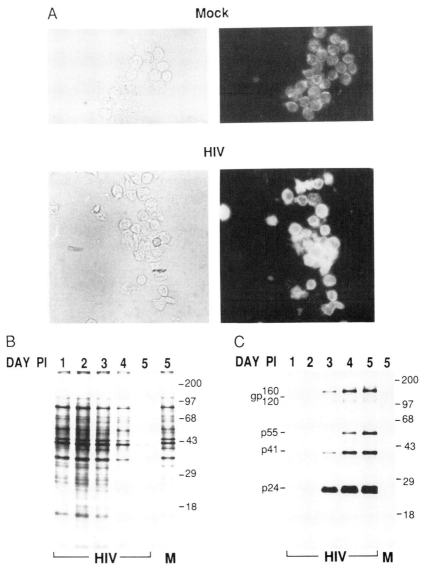

FIG. 1   Immunofluorescence analysis and examination of cellular and viral protein synthesis in HIV-1-infected CEM cells. (A) Photomicrographs of mock-infected cells (left) and cells infected by HIV-1 for 3 days (right). The left-hand members of the photomicrograph pairs are phase-contrast micrographs of the mock- and virus-infected cells; the right-hand members are an immunofluorescent analysis of cells reacted with HIV-specific antiserum and fluorescein-conjugated anti-human IgG. (B) CEM cells were either mock infected (M) or infected with HIV-1 (HIV) and, at the indicated days postinfection (PI), 2.0×10^6 viable cells were removed from the culture and labeled with [^{35}S]methionine (100 μCi/ml) for 2 hr. Cell extracts were prepared and the labeled proteins analyzed by electrophoresis on a 14% polyacrylamide gel. The positions of the molecular weight protein markers ($\times 10^{-3}$) are shown on the right. (C) The labeled extracts described for (B) were subjected to immunoprecipitation analysis with pooled antiserum from HIV-infected patients. The immunoprecipitated proteins were then analyzed by SDS-PAGE. The positions of representative HIV-1-specific proteins are shown on the left and the molecular weight standards on the right.

(which also lacks the *vpu* gene), should provide clues as to the possible involvement of vpu in the degradation of host cell mRNA.

To analyze the regulation of HIV-1, SIV, and cellular gene expression in infected cells, we routinely used the following procedures. Before performing our studies we developed a virus–cell culture system in which 100% of the cells could become nearly synchronously infected. We also had to ensure that our analyses of the infected cells occurred well before extensive cytopathic effects and cell death. The following methodologies allowed us to infect greater than 90% of the target cells by 3 days postinfection as determined by immunofluorescence studies. Moreover, trypan blue exclusion assays and [^{3}H]thymidine uptake analyses revealed that the infected cells remained at least 80% viable through day 5 postinfection.

Methods

HIV-1 and SIV Strains and Cells Used for Infection

For studies with HIV-1, we utilize the viral strain HIV-1$_{LAI}$, formerly LAV or HIV-1$_{BRU}$ (8,9). HIV-1$_{LAI}$ is an uncloned, laboratory or T cell line-adapted strain that can be readily propagated to high titers (2).

For SIV studies, we use the SIV$_{mne}$ strain that was originally isolated from a co-culture of *Macaca nemestrina* lymphoma cells and a human T cell line, HUT-78 (10). A biological clone E11S obtained from a persistently infected HUT-78 cell line is used as a source of SIV$_{mne}$ stock virus (11). Virus produced by these cells accumulates in culture supernatants over the course of 1 week and is collected and analyzed as described below. Titers of approximately 1×10^5 TCID$_{50}$ (50% tissue culture infective dose)/ml are commonly obtained by this method.

It is important that cells utilized for infection with HIV-1 or SIV be prescreened to determine whether complete and synchronous infections can be established. We have examined the susceptibility of a number of human T cell lines to HIV-1$_{LAI}$ and SIV$_{mne}$ infection and have found that whereas all are susceptible for HIV-1 infection, only those T cells known to be virally transformed support SIV replication (3). These specifically include C8166 (12), MT4 (13), AA-2 (14), and CEMX174 (15) for SIV and HIV-1 infection. CEM (16), Jurkat (17), and HUT-78 (18) can be infected by HIV-1 alone. All of these cells are maintained in RPMI-1640 medium containing 10% (v/v) fetal bovine serum (FBS), penicillin–streptomycin, and 10 mM 4-(2-hydroxyethyl)-1-piperazine-ethanesulfonic acid (HEPES) (pH 7.3) at a density of $0.5–1.0 \times 10^6$ cells/ml.

Preparation of HIV-1 and SIV Virus Stocks

Briefly, HIV-1$_{LAI}$ virus stock (approximately $1 \times 10^{6.5}$ TCID$_{50}$/ml in 0.5 ml) is incubated with 5×10^7 CEM cells in 5.0 ml of medium for 0.5 hr at 37° C. Unattached virus is removed by washing and centrifuging with excess phosphate-buffered saline (PBS). Infected cells are then resuspended in 50 ml of RPMI-1640 containing 10% fetal bovine serum and incubated upright in a 75-cm^2 tissue culture flask at 37° C in

5% CO_2. After a 3-day incubation, an additional 50 ml of medium is added to the flask. Infected cells are removed from the culture supernatant (100 ml) by centrifugation on day 4. The virus-containing supernatant is frozen in 1 to 2-ml aliquots at either $-70°C$ or over liquid nitrogen. Following freezing, one sample from each preparation is thawed and diluted for titration (described below).

The titer of a given virus stock may vary by as much as 2 logs, depending on the cell types used for viral growth (lentiviruses are commonly titrated in terms of 50% tissue culture infectious doses [$TCID_{50}$]). For example, titers of HIV-1_{LAI} grown in CEM Cells vary from approximately 1×10^7 $TCID_{50}$/ml in C8166 cells, a human T cell line, to 1×10^5 $TCID_{50}$/ml in fresh human peripheral blood mononuclear cells (PBMCs). Virus preparations are 10-fold serially diluted and added (0.1 ml) to 1×10^6 cells/well in 24-well tissue culture plates. The cells are cultured in 2 ml of medium and monitored visually for evidence of infection [cytopathic effect (CPE), including syncytium formation] and for viral antigen production by antigen capture enzyme-linked immunosorbent assay (ELISA) (see below) during the next 7 days. To compensate for the presence of inoculum virus, which can potentially contribute to high background levels observed through ELISA analysis, samples are taken at time 0 (i.e., immediately following infection and washing) and analyzed with samples taken at 1 week. The verification of infectious virus in the diluted inoculum requires increased viral antigen levels in the culture supernatants after 1 week. The titer is calculated using the Reed and Muench formula (19).

Infection of Cells with HIV-1 or SIV

Two days before infection the cells are split 1 to 2 in fresh medium. We use sufficient virus to achieve a multiplicity of infection (MOI) of 0.5 $TCID_{50}$/cell (2,3). The virus is allowed to attach for 30 min at $37°C$. If the infections are to be monitored by antigen capture or polymerase chain reaction (PCR) (see below), unattached virus must be removed by thoroughly washing the cells in three 100-fold volumes of PBS. Failure to wash away the free virus from the cells results in unacceptably high background antigen and DNA levels at time 0 analysis. Greater than 90% of the cells should become infected by 2–3 days postinfection (pi), as determined by indirect immunofluorescence (see below and Fig 1). Protein and mRNA analyses can be performed on T cell lines infected through 4–5 days, during which time the cultures remain greater than 80% variable as determined by [^{3}H]thymidine labeling of cellular DNA and by trypan blue dye exclusion assays.

Indirect Immunofluorescence Assay for HIV-1 Infection

Suspension cells (T cell lines and primary lymphocytes) are applied to glass well slides to give a uniform cell layer after drying (approximately 1×10^6 cells/slide). The cells are fixed in a 1:1 (v/v) methanol–acetone solution for 10 min at room temperature. Following fixation the slides may be stored at $-20°C$ indefinitely. As is the case for all immunofluorescence assays, proper antiserum dilutions must be determined for each application. We have used pooled sera from HIV-infected humans and SIV-infected macaques for HIV and SIV studies, respectively, as the source for the primary antibody (2,3). The SIV antisera must be preadsorbed on the human T cell line used to prepare the virus for animal inoculation to remove the anti-human cellular protein antibodies. Importantly, antihuman IgG Fab–fluorescein isothiocyanate (FITC) conjugates readily cross-react with monkey antibodies (Fig. 1).

HIV-1 and SIV Antigen Capture Assays

There are several commercially available products that provide a highly sensitive and technically straightforward means to monitor *in vitro* infections by measuring the accumulation of viral proteins in culture supernatants. Both polyclonal and monoclonal antibody-based test kits are available with similar antigen detection sensitivities. Briefly, samples are incubated with immobilized primary antibodies. Unbound proteins are removed and a second antiviral antibody is applied, thus capturing or sandwiching the antigen between the immobilized and secondarily added antibodies. Following a second wash, a third, enzyme-conjucated antibody against the second antibody is applied and color development follows the addition of the substrate of the enzyme. The amount of color produced is directly proportional to the level of accumulated antigen in the sample but does not necessarily correspond to the levels of infectious virus in the sample (i.e., synthesis of incomplete virions would be detected by these assays as well as whole infectious virions).

Immunoprecipitation Analysis

For immunoprecipitation analysis, approximately 2×10^6 viable cells are removed from infected cultures, washed free of medium with PBS, and metabolically labeled with 200 μl of [^{35}S]methionine–cysteine (750 μCi/ml) for 2 hr in medium lacking cold methionine and cysteine. After labeling, the cells are washed free of unincorporated radiolabel in ice-cold PBS. We disrupt the cells in ice-cold lysis buffer containing 1% (v/v) Triton X-100, 10 mM Tris-HCl (pH 7.5), 50 mM KCl, 2 mM MgCl$_2$, 1 mM dithiothreitol (DTT), and 0.2 mM phenylmethylsulfonyl fluoride (PMSF). Labeled cellular and viral proteins are then resolved by electrophoresis on 10 or 14% (w/v) sodium dodecyl sulfate (SDS)-polyacrylamide gels. For immunoprecipitation analysis, cell extracts are diluted in buffer I [20 mM Tris-HCl (pH 7.5), 0.4 mM NaCl, 1 mM DTT, 1 mM ethylenediaminetetraacetic acid (EDTA), 0.2 mM PMSF, aprotinin (100 U/ml), 1% (v/v) Triton X-100, and 20% (v/v) glycerol]. Diluted extracts are reacted with pooled antisera from HIV-1-infected individuals or SIV-infected macaques (1:500 dilution) for 2 hr at 4°C. Protein A–agarose is then added to the lysates and the incubation allowed to continue for an additional 1 hr at 4°C. The precipitates are subsequently washed four times in buffer I and three times in buffer II [20 mM Tris-HCl (pH 7.5), 100 mM KCl, 0.1 mM EDTA, aprotinin (100 U/ml), and 20% (v/v) glycerol]. The washed immunoprecipitates are diluted with an equal volume of 2× protein disruption buffer [1× disruption buffer: 70 mM

Tris-HCl (pH 6.8), 0.5 M 2-mercaptoethanol, 2% (w/v) SDS, 10% (v/v) glycerol] and electrophoresed through a 14% polyacrylamide gel (SDS-PAGE). Gels are fluorographically enhanced, dried, and autoradiographed (Fig. 1).

To assess whether declines in cellular protein synthesis are due to decreases in cellular mRNA levels, Northern blots can be performed according to the following procedure.

RNA Isolation and Northern Blot Analysis

We isolate total RNA from mock- and HIV-1/SIV-infected cells according to the method of Chomczynski and Sacchi, using acidic guanidinium thiocyanate–phenol–chloroform extraction (20). Briefly, cells are lysed in solution D (100 μl/10^6 cells) containing 4 M guanidinium thiocyanate, 25 mM sodium citrate (pH 7.0), 0.5% (v/v) sarcosyl, and 0.1 M 2-mercaptoethanol. Following lysis, 0.1 vol of 2 M sodium acetate, 2 vol of water, and 0.2 vol of saturated phenol and chloroform—isoamyl alcohol (49:1) are added sequentially, vortex mixed, and incubated on ice for 15 min. The chilled mixture is centrifuged at 10,000 g for 20 min at 4°C. The aqueous (top) layer is then removed and an equal volume of 2-propanol added and cooled at −20°C for 1 hr. Precipitated RNA is pelleted by centrifugation as above. RNA pellets are then redissolved in 300 μl of solution D, transferred to a 1.5-ml microcentrifuge tube, and reprecipitated with 300 μl of 2-propanol at −20°C for an additional hour. The RNA is repelleted in a 4°C microfuge at 14,000 rpm for 10 min. The resulting pellet from this precipitation is washed with 75% ethanol, air dried, and dissolved in water. Alternatively, if the RNA is to be used in Northern blot analysis, the pellets are dissolved in 50% RNA denaturation buffer [20 mM morpholinepropanesulfonic acid (MOPS) (pH 7.0), 10 mM EDTA, 50 mM sodium acetate, 6.5% (v/v) formaldehyde, 50% (v/v) formamide]. The quantity and quality of the RNA is determined spectrophotometrically at 260 and 280 nm (21). Total RNA samples are heat denatured (65°C for 5 min) and electrophoresed through a 1% (v/v) agarose gel containing 2.2% (v/v) formaldehyde and transferred to nitrocellulose as described previously (21,22). The locations of the 28S and 18S rRNA are marked on the nitrocellulose blots, and the filters are baked at 80°C for 2 hr, and prehybridized for 2 hr. The RNA-containing filters are then probed with randomly primed ^{32}P-labeled full-length proviral DNA or various cellular DNA clones specific for human β-actin or rat GAPDH (23).

Nuclear Runoff Transcription

To examine whether declines in cellular mRNA levels are due to decreases in rates of RNA polymerase II transcription, nuclei can be isolated from mock- and HIV-1-infected cells and the relative levels of preinitiated transcription measured according to the following protocol. For this assay, we used techniques previously described by Greenberg and Ziff (24) and Groudine et al. (25). Typically, nuclei from 5 × 10^7 mock- and virus-infected cells are isolated. First the cells are washed three times with

PBS. Then the pellet is gently vortexed while 4 ml of Nonidet P-40 (NP-40) lysis buffer [10 mM Tris-HCl (pH 7.4), 10 mM NaCl, 3 mM MgCl$_2$, 0.5% (v/v) NP-40] is slowly added. Extracts are then mixed for a further 10 sec and incubated for 5 min on ice. The nuclei are pelleted at 4°C at a speed of 700 rpm for 10 min. The lysis buffer is then decanted and the nuclei resuspended in 200 μl of ice-cold glycerol storage buffer [50 mM Tris-HCl (pH 8.3), 40% (v/v) glycerol, 5 mM MgCl$_2$, 0.1 mM EDTA]. The condition of the nuclei can be assessed by phase-contrast microscopy. Ideally, nuclei have not clumped together and have a characteristic refractive blue-green appearance. At this point nuclei can be stored in liquid nitrogen until assayed.

Frozen nuclei are thawed at room temperature and 200 μl of nuclear suspension is added to a 15-ml polypropylene centrifuge tube. Immediately, 200 μl of 2 $\times$ reaction buffer is added [10 mM Tris-HCl (pH 8.5), mM MgCl$_2$, 0.3 mM KCl] containing 10 μl each of 100 mM nucleotides (ATP, CTP, and GTP) and 5 μl of 1 M DTT plus 10 μl of [α-^{32}P]UTP (760 Ci/mmol, 10 mCi/ml, i.e., approximately 200 μCi/reaction). Following incubation for 30 min at 30°C, with occasional gentle mixing, 40 μl of RNase-free DNase I (1 mg/ml) in 1 ml of HSB buffer [10 mM Tris-HCl (pH 7.4), 0.5 M NaCl, 50 mM MgCl$_2$, 2 mM CaCl$_2$] is added to the reaction (600 μl/tube). The incubation is then continued for an additional 5 min. Next, 200 μl of SDS–Tris buffer is added [0.5 M Tris-HCl (pH 7.4), 5% (w/v) SDS, 0.125 M EDTA] followed by 10 μl of proteinase K (20 mg/ml). The mixture is incubated for 30 min at 42°C and should be a precipitate-free solution. Samples are extracted with 1 ml of phenol–chloroform–isoamyl alcohol solution, centrifuged at 800 g for 5 min, and the aqueous phase retrieved. To the aqueous phase, 2 ml of H$_2$O is added followed by 3 ml of 10% trichloroacetic acid (TCA) solution containing 60 mM sodium pyrophosphate. Ten microliters of *Escherichia coli* tRNA (10 mg/ml) is also added as carrier RNA to precipitate the runoff transcripts, and the mixture incubated on ice for 30 min. Precipitated RNA is collected by vacuum filtering through 0.45-μm pore size HA Millipore (Bedford, MA) filter disks. The precipitates are then washed three times with 10 ml of 5% TCA containing 30 mM sodium pyrophosphate. The filters are transferred to small glass scintillation vials and incubated with 1.5 ml of DNase I buffer [20 mM HEPES (pH 7.5), 5 mM MgCl$_2$, 1 mM CaCl$_2$] plus 37.5 μl of RNase-free DNase I (1 mg/ml) at 37°C for 30 min. This reaction is stopped with 45 μl of 0.5 M EDTA and 68 μl of 20% SDS. The RNA is eluted from the filters by incubating the solution at 65°C for 10 min. Solutions are retained in 15-ml polypropylene tubes and the elution step is repeated with an additional 1.5 ml of elution buffer and 65°C incubation. (*Note:* This filtration step is a most efficient means to remove unincorporated [^{32}P]UTP from the transcripts.) The second elution is combined with the first, 4.5 μl of proteinase K (20 mg/ml) is added to the 3 ml of elute, and the mixture incubated at 37°C for 30 min. The eluted 3-ml solution containing the runoff transcripts is extracted two times with equal volumes of phenol–chloroform–isoamyl alcohol and chloroform–isoamyl alcohol. Aqueous phases are incubated on ice with 0.75 ml of 1 M NaOH for 10 min prior to the addition of 1.5 ml of 1 M HEPES (free acid). The RNA is precipitated overnight at $-20°C$ after adding 0.53 ml of 3 M sodium acetate and 14.5 ml of ethanol.

Precipitated RNA is then pelleted at 10,000 g for 30 min at 4°C, dried briefly, and redissolved in 0.4 ml of water and 20 μl of 4 M LiCl before being transferred to a 1.5-ml microcentrifuge tube and reprecipitated overnight under 1 ml of ethanol. The RNA is again pelleted, washed in 95% ethanol, dried, and redissolved in 1 ml of TES buffer [10 mM N-tris(hydroxymethyl)methyl-2-aminoethanesulfonic acid (TES) (pH 7.4), 10 mM EDTA, 0.2% (w/v) SDS] by incubating at room temperature for 30 min. The counts per minute (cpm) per milliliter is then determined for each sample. Generally, 1 ml of TES containing equal cpm from each sample (typically around 5×10^5 cpm) is hybridized with 1 ml of TES/NaCl buffer [10 mM TES (pH 7.4), 10 mM EDTA, 0.2% (w/v) SDS, 0.6 M NaCl] to separate filter strips containing target DNA plasmids (see below). After 36 to 48 hr at 65°C, the strips are washed free of unhybridized RNA in repetitive $2 \times$ SSC washes ($1 \times$ SSC is 0.15 M NaCl plus 0.015 M sodium citrate) for 1 hr at 65°C. Blots are dried and exposed to X-ray film. Sufficient target DNA plasmid is linearized, denatured, neutralized, and applied to nitrocellulose filters in a slot blot apparatus to provide 5 μg of target per slot. The filters are baked for 1–2 hr at 80°C and cut into strips such that each strip contains a complete set of targets. Because of the relatively low radioactive signal of each transcript type, it is important to minimize the size of the blot to allow for the highest possible probe concentration during hybridization.

Polysome Analyses

Polysome analyses of uninfected and HIV-1 infected cells are prepared using procedures explained below (see Polysome Analysis, below). Approximately 2×10^8 HIV-1-infected and uninfected cells are recommended as starting material when using this procedure.

Analysis of Regulation of Gene Expression by Influenza Virus

Background

The influenza viruses types A and B are enveloped viruses with negative-stranded RNA genomes consisting of eight segments. In cells infected by influenza virus, host cell protein synthesis is severely diminished (26,27). At the same time, viral mRNAs are efficiently and selectively translated (27) (Fig. 2). In the past several years, the molecular mechanisms underlying this translational control have begun to be elucidated. Newly synthesized cellular mRNAs fail to reach the cytoplasm due to degradation of nuclear cellular transcripts early (1 hr) after infection (28). High levels of functional cellular mRNA remain in the cytoplasm, however, suggesting that influenza virus also must prevent preexisting cellular mRNA from being translated. Following a comparison of polysome-associated specific cellular mRNA in infected and

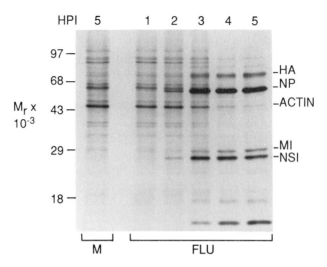

Fig. 2   Shutoff of host cell protein synthesis in cells infected by influenza virus. COS-1 cells were either mock infected (M) or infected with influenza virus (FLU) at a multiplicity of 50 pfu/cell. The cells were metabolically labeled with [^{35}S]methionine for 30 min preceding the indicated hours postinfection (HPI). The labeled proteins were analyzed by SDS-PAGE on a 14% gel. The positions of the representative influenza virus-specific proteins [hemagglutinin (HA), nuclear capsid protein (NP), nonstructural protein (NS1), matrix protein 1 (M1), and actin (a representative cellular protein)] are shown on the right. The positions of molecular weight standards are shown on the left.

uninfected cells, it was observed that the initiation and elongation steps of cellular mRNA translation were inhibited in contrast to influenza viral mRNAs, which escape these blocks (29,30). It remains unclear how influenza viral mRNAs escape the elongation block imposed on viral and cellular mRNA, although evidence is now accumulating that selectivity at the level of initiation may result, at least in part, from the structure of the viral mRNA. This was discovered by constructing a recombinant adenovirus that could express the influenza virus nucleocapsid protein (NP) mRNA in the absence of other influenza viral gene products. Following infection with the recombinant virus, the influenza virus NP mRNA was found on polysomes similar in size to polysomes found for NP mRNA in cells infected by influenza virus alone, or in cells doubly infected by influenza virus and adenovirus (31). Additionally, viral NP constructs transiently transfected into COS-1 cells are not subjected to the host cell protein synthesis shutoff (in comparison to transfected cellular genes) following influenza infection (32). It remains to be seen whether specific primary sequences or secondary or higher order structures mediate the selective translation of influenza virus mRNA. Interestingly, all influenza viral mRNAs do contain a common 12-nucleotide sequence directly downstream from stolen cellular 5′ ends that are known to be important for replication and packaging (26,33). The importance of viral 5-UTR mRNA sequences in the selective translation of influenza mRNA, however, remains to be addressed.

The following protocols have been routinely used for studying the selective translation of influenza viral mRNA in infected cells.

Methods

Influenza Virus and Cells Used for Infection

For our studies we use the WSN strain of influenza virus type A, which is a laboratory-derived strain replication competent in several mammalian cell lines. Madin–Darby bovine kidney (MDBK) cells are used as host cell production of virus (34). MDBK cells are maintained in Dulbecco's modified Eagle's medium (DMEM) containing 10% (v/v) FBS and penicillin/streptomycin.

Preparation of Virus Stocks

Stock virus, diluted 1 : 30,000 (10 ml), is added to confluent layers of MDBK cells in 10 ml of DMEM containing 2% FBS per 150-cm^2 tissue culture flask. The virus attaches for 1 hr at 37°C. After removing unattached virus, 10 ml of complete medium is added per flask and the cells allowed to incubate at 37°C for an additional 40 hr. The medium containing virus is then poured from each flask and centrifuged at 600 rpm for 10 min at 4°C. The supernatant is collected and stored at −70°C.

To determine the titer of the virus stock, confluent layers of Madin–Darby canine kidney (MDCK) cells in 60-mm dishes are washed with PBS and inoculated with 500 μl of diluted virus (10^{-5} to 10^{-8}). Cells are then incubated for 1 hr at 37°C, and the virus inoculent removed (27). Each plate is washed with PBS to remove unattached virus before being overlaid with 5 ml of overlay medium (2 × DMEM complete medium containing an equal amount 1.6% (w/v) Noble agar [DIFCO, Detroit, MI] and a 1 : 200 dilution of trypsin–EDTA solution [GIBCO, Grand Island, NY]). Plates are incubated at 37°C until plaques form (2–3 days).

Analysis of Influenza Virus-Mediated Host Translation Shutoff

Medium is removed from cell monolayers and influenza virus added at an MOI of 50 plaque-forming units (pfu)/cell. We routinely use 35-mm plates containing cells that have been allowed to attach overnight, at a confluency of 80–90%. Cells are then incubated at 4°C for 45 min to allow the virus to attach. The plates are gently rocked every 15 min. After this time, virus is removed and warm medium[DMEM, 2% (v/v) FBS, penicillin/streptomycin, 500 mM HEPES (pH 7.5)] is added to the cells. The plates are incubated at 37°C and protein synthesis analyzed periodically, up to 5 hr postinfection, by briefly labeling the cells with [^{35}S]methionine. By this time host protein synthesis is almost entirely shut off (Fig. 2). Accordingly, at each time point, culture medium is replaced with cold, methionine-free medium containing [^{35}S]methionine (20 μCi/plate). The plate is then allowed to incubate for 30 min at 37°C, and the unincorporated radioactive label removed. Cells are washed twice with Hanks' balanced salt solution and treated with 100 μl of lysis buffer (described above). Extracts are collected and centrifuged for 2 min and the supernatants retrieved. To ensure that equal amounts of protein are analyzed we spot 1 μl of extract from each time point on 3MM Whatman (Clifton, NJ) DEAE paper and TCA precipitate the labeled proteins. Briefly, spotted paper is boiled for 10 min in 10% TCA, rinsed twice in 5% TCA and once with 95% ethanol, and allowed to dry before determining the cpm. Equal cpm or cell equivalents are added to an equivalent volume of 2 × protein disruption buffer and loaded onto 10% SDS-polyacrylamide gels.

Immunoblot and Immunoprecipitation Analysis

Western blot analyses are performed according to Towbin *et al.* (35). Briefly, electrophoresed cellular extracts are transferred to nitrocellulose, blocked with 5% (v/v) milk (dried, non-fat milk, Carnation Company, Los Angeles, CA), and reacted with a mixture of antisera to purified influenza virions. These included antibodies to the polymerase, nucleocapsid, and hemagglutinin proteins (provided by P. Palese and M. Krystal, Mt. Sinai Medical School, New York, NY). Monospecific antibodies to the matrix and two nonstructural proteins can be included in this mix (and are provided by R. Webster, St. Jude's Children's Research Hospital, Memphis, TN). After incubation and washing, ^{125}I-labeled protein A is added to detect the primary antibody. Western blot analysis using the Amersham (Arlington Heights, IL) ECL nonisotopic chemiluminescence system works equally well.

RNA Isolation and Northern Blot Analysis

Following viral infection, cell monolayers are washed with Hanks' balanced salt solution (HBSS) and disrupted in a solution of 140 m*M* NaCl, 1.5 m*M* MgCl$_2$, 10 m*M* Tris-HCl (pH 7.5), 0.5% (v/v) Nonidet P-40, and 5 m*M* vanadyl-ribonucleoside complexes (28). After incubation on ice for 10 min, nuclei and cell debris are removed by centrifugation at 500 *g* for 5 min at 4°C. Cytoplasmic extracts are then diluted with an equal volume of a 2 × proteinase K buffer [0.2 *M* Tris-HCl (pH 7.5), 25 m*M* EDTA, 300 m*M* NaCl, 0.2% (w/v) SDS, proteinase K (500 μg/ml)] and mixtures incubated at 37°C for 30 min. RNA is extracted with phenol–chloroform (1:1) and ethanol precipitated. Cytoplasmic poly(A)$^+$ mRNAs are isolated using oligo(dT) chromatography. For Northern blot analysis, poly(A)$^+$ RNA is electrophoresed on formaldehyde–agarose gels as described previously for analysis of mRNA from cells infected with HIV-1. [^{32}P]UTP-labeled RNA probes, which are antisense to the viral mRNA, are used to detect viral mRNA and not virion RNA (23,36).

Polysome Analysis

Polysomes are prepared from mock- and influenza virus-infected cells 5 hr postinfection as follows (30,37). This procedure, however, works equally well on lentivirus-infected cells. Cells are washed in HBSS containing cyclohexamide (100 μg/ml) to prevent peptide chain elongation and thus polyribosome runoff. The cells are then scraped off the plate and collected in HBSS. After centrifugation at 1000 rpm for five min, cells are collected and lysed with polysome buffer [10 mM Tris (pH 7.5), 10 mM NaCl, 1.5 mM MgCl$_2$, cycloheximide (100 μg/ml), and 5 mM vanadyl-ribonucleoside complexes]. Triton X-100 is added to a final concentration of 0.5% (v/v), and the cells incubated for 3 min on ice followed by addition of 1% (v/v) Tween 40 and 0.5% (v/v) deoxycholate. Cell extracts are next slowly homogenized in an ice-cold stainless steel Dounce homogenizer. The extracts are centrifuged at 10,000 rpm for 5 min and the supernatant layered onto a 10–50% (w/v) linear sucrose gradient containing 10 mM Tris (pH 7.5), 5 mM magnesium acetate, 0.5 M KCl. Polyribosomes are resolved from monosomes and other small structures by centrifugation in a Beckman (Fullerton, CA) Ti-41 rotor at 40,000 rpm for 2 hr at 4°C. Fractions (500 μl) are collected (50 μl is used to determine nucleic acid concentrations) with an 18-gauge needle and added to 400 μl of 2 $\times$ pronase buffer. Fractions are treated with 500 μg of pronase and incubated for 1 hr at 37°C. The RNA is extracted from the pooled fractions by phenol–chloroform extraction, and then precipitated using ethanol containing 0.2 M LiCl. Following centrifugation, the pellets are resuspended in 40 μl of sterile water. For each fraction, cell equivalents of RNA are collected on nitrocellulose filters by vacuum slot blot. Filters are baked and prehybridized as described earlier (32).

Introduction to Interferon-Induced, Double-Stranded RNA-Activated Protein Kinase

Background

Interferon (IFN) is known to mediate its effect on cell growth and differentiation, and on viral replication, through the induction of more than 30 responsive genes (38–40). Following the observation that protein synthesis was inhibited in IFN-treated cell-free systems by addition of dsRNA or viral RNA, the product from one of these inducible genes was isolated from interferon-treated cells with poly(I):poly(C)–Sepharose columns. This enzyme was found to be a serine–threonine protein kinase with a molecular weight of 68,000 in human cells, and is now termed PKR (40,41) (but was previously referred to as P68 or DAI). Subsequent investigations showed that dsRNAs induce PKR to autophosphorylate and, more importantly, to catalyze the phosphorylation of the α subunit of protein synthesis initiation factor 2 (eIF-2) (Fig. 3A). Phosphorylation of the α subunit blocks the eIF-2B-mediated exchange of GDP, in the inactive eIF–GDP complex, with GTP required for catalytic utilization of eIF-2. These events lead to limitations in functional eIF-2, which is an essential component of protein synthesis initiation, and is normally required to bind initiator

A

WHEN PKR IS ACTIVATED

B

PKR PROTEIN KINASE

AMINO ACID # 1 11   77   101   167   274   551

WILD-TYPE	A	B	I -------------------------------- XI
	REGULATORY DOMAINS		CATALYTIC DOMAINS

Fig. 3   Schematic structure and function representations of the interferon-induced, double-stranded RNA-activated protein kinase PKR. (A) Following activation by dsRNA (or poly-anions), PKR autophosphorylates and in turn phosphorylates the subunit of the eukaryotic initiation factor 2 (eIF-2). These reactions, which inhibit protein synthesis, are dependent on divalent cations and ATP. (B) The regulatory regions of PKR reside in the amino-terminus region of the kinase. Similar to other kinases, the 11 catalytic domains are contained in the carboxyl region, starting from amino acid 274.

Met-tRNA (via the ternary complex eIF–2GTP–Met tRNA) to the initiating ribo-somal subunit before mRNA is bound (42,43). Thus, if uncountered, activation of PKR by viral dsRNA could lead to a severe decrease of protein synthetic rates in host cells, which shuts down both cellular metabolism and viral replication. Viral repli-cation, therefore, can occur only if the effects of PKR are neutralized in the infected cell. The majority of work on the human PKR has focused on the possible participa-tion of the kinase during the interferon-induced inhibitory response to virus infection. For example, we have shown that cell lines overexpressing the wild-type PKR (but not a catalytically inactive mutant kinase) could reduce the ability of mouse encepha-lomyocarditis virus to replicate (44). This observation correlated with an increase in eIF-2α phosphorylation levels. Further evidence that PKR is involved in a potential antiviral role arises from observations that many if not most other animal viruses have evolved varied mechanisms to downregulate PKR, presumably to avoid such problems with protein synthesis and regulation (reviewed in Refs. 39, 41, and 45). For several years we have been studying the interaction of PKR with influenza virus, HIV-1, and poliovirus, all of which, we have found, have devised different strategies designed to inhibit PKR function.

Structural Analysis and Cellular Regulatory Roles of PKR

PKR has been found to possess two RNA-binding motifs, approximately 65–68 amino acids long, which are located in the amino-terminus region of the protein (46–51) (Fig. 3B). The consensus motif, similarly represented in a number of other proteins known or thought to bind RNA, is G-X-G-X-S-K-X-X-A-K-X-X-A-A-X-X-A-X-X-X-L (47). It is not yet clear whether both copies of the motifs possess analogous function, although preliminary mutagenic studies suggest that the first is of more importance (49). RNA-binding studies suggest that viral activators such as reovirus dsRNA probably bind to the same site as viral inactivators such as adenovirus VAI RNA (46,52,53). The conserved features of protein kinases, namely the 11 catalytic domains, reside in the catalytic half of the molecule, from amino acid 262 onward (Fig. 3B) (54). Further analysis suggests that PKR belongs to a subfamily of protein serine–threonine kinases that is related to protein tyrosine kinases (54). Several serine–threonine sites in PKR are appropriate candidates for phosphorylation, although they have not yet been identified (54). A single copy of the gene is present on human chromosome 2p21 and on mouse 17E2 (55). PKR also shares extensive homology with three other eIF-2α kinases: HRI, the heme-regulated eIF-2α protein kinase isolated from rabbit reticulocytes (56); GCN2, the eIF-2α protein kinase from yeast (57,58); and TIK kinase, the PKR mouse homolog (59). PKR and TIK share an overall amino acid homology of 62%, whereas the catalytic cores of PKR, HRI, and GCN2 share approximately 40% sequence identity. Although the eIF-2α kinases share extensive homology in certain domains (e.g., six conserved amino acids between catalytic domains V and VI), compared to other serine–threonine kinases, the binding site for eIF-2α recognition has not been identified. Substitution of the invariant lysine residue in catalytic domain II with arginine (K296R) renders the kinase incapable of autophosphorylation and inhibits its ability to phosphorylate the eIF-2α substrate (53). Expression of a functional PKR in several eukaryotic systems proved to be unsuccessful due to the toxicity of this protein (46). In yeast, PKR could functionally substitute for GCN2 and phosphorylate the yeast eIF-2 homolog (sui-2) to inhibit protein synthesis (57,58). To date, functional PKR has been expressed only in cell-free extracts and in bacteria (46,52,53). Apart from experimentation in yeast, several other lines of evidence suggest that PKR may indeed be an evolutionarily important modulator of cellular growth. For example, 3T3 fibroblasts stably overexpressing catalytically inactive mutants proved to be highly tumorigenic when injected into nude mice (60,61). We and others have also confirmed that PKR can regulate protein synthesis at the level of translation *in vivo* (62,63). Inactive kinases probably interfere with the functioning of wild-type PKR in a dominant negative fashion by either competing with activator and/or forming complexes with its counterpart and substrate. That PKR may function through other pathways, such as the regulation of transcription factors, however, has not been ruled out. These data taken together sug-

gest that (a) PKR may be a tumor suppressor protein and play a role in the antiprolif-
erative activity of IFN, and (b) that PKR may be involved in general control of translational regulation (58–60,62). Intense study on the varied roles of PKR is currently
being performed by a number of laboratories. These efforts will no doubt further
unravel the interesting properties of this important enzyme.

Methods

Interferon Treatment of Cells

To increase the levels of PKR, cells can be treated with IFNs, which exert their ac-
tions through species-specific cell surface receptors (38). Human Daudi (64) or 293
(65) cell lines are good sources of PKR and can be treated for 17 hr with 500 units
of human lymphoblastoid IFN (10^8 U/mg; purchased from Hayashibara Biochemistry Laboratories, Okayama, Japan).

Assay of PKR Activity

PKR activity was first measured after purifying extracts from interferon-treated cells
on poly(I):poly(C)–Sepharose columns (66). We assay for PKR activity following
immunoprecipitation of PKR protein from IFN-treated or virus-infected cells by
using a monoclonal antibody developed by A. Hovanessian at the Pasteur Institute
(Paris, France) (66). Alternatively, we have found that polyclonal antibody to PKR
can also be successfully used for this purpose and we are currently characterizing
several newly generated monoclonal antibodies against PKR (46). Briefly, medium
is removed and cells washed once in PBS. Following removal of all PBS, cells are
treated with cold lysis buffer (described above in the HIV-1 section "Immunoprecipitation Analysis"). We use about 200 μl of lysis buffer for approximately $1-2 \times 10^6$
cells. Extracts are prepared and centrifuged at 4°C for 2 min. All the following procedures are performed at 4°C. The supernatant is collected and diluted with 500 μl
of buffer I (described in HIV-1 section "Immunoprecipitation Analysis"). Antibody
(1:500 dilution) is added to the extracts and allowed to incubate with rocking at 4°C.
After 1 hr, 50 μl of protein G–agarose is added to the mixture and allowed to incubate for an additional hour at 4°C. Immunoprecipitated extracts are then briefly centrifuged and the supernatant removed. Pellets are subsequently washed four times in
buffer I and three times with buffer II (described in HIV-1 section "Immunoprecipitation Analysis") prior to being resuspended in 30 μl of buffer III [20 mM Tris-HCl
(pH 7.5), 1 mM DTT, 0.1 mM EDTA, 25 mM KCl, 0.1 mM PMSF, 5% (v/v) glyc-
erol]. We then add 4 μl of buffer IV [10 mM Tris-HCl (pH 7.5), 0.1 mM EDTA, 500
mM KCl, 20 mM MgCl$_2$, 20 mM MnCl$_2$, 12.5 μM ATP, aprotinin (80 μg/ml), BSA
(3mg/ml), 40% (v/v) glycerol] containing 2 μM [^{32}P]ATP (50 Ci/mmol) to each
tube. At this point, various PKR activators can be added. Specifically, dsRNA (0.01

μg/ml), poly(I):poly(C) (Boehringer Mannheim, Indianapolis, IN), or 10 units of the polyanion activator heparin (Sigma, St. Louis, MO) can be added to each sample. Alternatively, immunoprecipitated extracts can be equally divided into two reactions and activator added only to one tube (i.e., reactions observed with and without activator). Reactions are then mixed gently by pipette, and incubated for 15 min at 30° C. Protein loading buffer (50 ςl) containing 100 mM EDTA is used to stop the reactions. Extracts are then boiled for 2 min, centrifuged briefly for 30 sec at 14K at room temperature, and loaded onto 10% SDS-polyacrylamide gels. After electrophoresis, the protein dye front containing free ^{32}P is cut away and the gels carefully dried and exposed to X-ray film at $-70°$ C. On occasion, the effect of PKR activity on its substrate (eIF-2α) is required. We add approximately 0.1 μg/reaction to the ^{32}P-containing mixture, and proceed as above. Purification of eIF-2α has been described previously (67).

Immunoblot and Immunoprecipitation Analysis of PKR

Immunoblot Analysis

Cells are lysed as described above and extracts are denatured in 2× protein disruption buffer. After boiling for 2 min, the proteins are resolved by electrophoresis on 10% SDS-polyacrylamide gels. Following transfer to nitrocellulose paper, blots are blocked as described earlier and incubated with PKR-specific polyclonal antisera or monoclonal antibody diluted 1:500 in PBS containing 10% (v/v) fetal bovine serum and 0.05% (v/v) Tween 20. After a 1-hr incubation at room temperature, blots are washed with PBS containing 0.05% (v/v) Tween 20, and treated with horseradish peroxidase-conjugated anti-mouse antibody. We also find that the ECL chemiluminescence detection kit (Amersham) provides a suitable method for detecting PKR proteins.

Recombinant Expression of PKR and Preparation of PKR-Specific Antisera

The limited availability of PKR for biochemical analysis has led us to attempt to synthesize this enzyme in recombinant systems. We have successfully managed to express functional recombinant PKR in *E. coli,* utilizing the pET series of vectors purchased from Novagen, Madison, WI (52,68) (Fig. 4). However, wild-type PKR has proved to be toxic when expressed in recombinant baculovirus-infected insect cells and in most other eukaryotic expression systems as well (46,52,53). In yeast, for example, PKR has been found to phosphorylate the yeast eIF-2α homolog (sui-2) to inhibit protein synthesis (57,58). For prokaryotic expression, plasmids (Fig. 4) containing wild-type or mutant PKR under control of the T7 promoter are transformed into *E. coli* strain BL21(DE3) pLysS (Novagen) and grown overnight at 37° C in TY medium containing ampicillin (75 μg/ml). Isopropyl-β-D-thiogalactopyranoside (IPTG) (0.5 mM) is used to induce expression, and after 30 min rifampicin (Sigma) (200 μg/ml in methanol) is added. After 2 hr, cells are placed

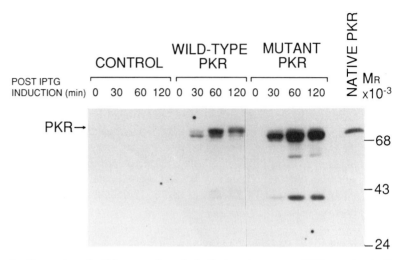

FIG. 4 Expression of wild-type and catalytically inactive mutant PKR proteins in *E. coli.* Bacteria, carrying expression plasmids containing wild-type or mutant kinases (K269R), were grown and induced with IPTG. At various times postinduction, the bacteria were harvested and, following electrophoresis, examined by immunoblot analysis. The blots were incubated with monoclonal antibody specific to human PKR. Molecular weight standards are shown on the right.

on ice and centrifuged. Pellets are treated with lysis buffer (prepared as above) to disrupt the cells. Usually, the mix is diluted in buffer I to reduce the protein concentration. We add $100\mu l$ of DNase I (25 mg/ml) (Boehringer Mannheim) per 500-ml culture, and incubate the lysate on ice to break down the host DNA. Extracts can then be centrifuged at 20,000 rpm. A majority of the recombinant PKR is soluble in *E. coli* with little retained in inclusion bodies. For PKR purification, supernatants can be reacted with cyanogen bromide-activated Sepharose beads (Pharmacia, Piscataway, NJ) coupled to a PKR-specific monoclonal antibody (66). After incubation at 4°C for 2 hr, the beads are washed four times with buffer I and three times with buffer II. The beads are then placed into a disposable gel chromatography column (Bio-Rad, Richmond, CA) and treated with ice-cold 1 *M* KSCN, pH 11. The eluant containing PKR is allowed to drip into a neutralizing buffer [20 m*M* HEPES (pH 7.4), 50 m*M* KCl, 0.1 m*M* EDTA, 1 m*M* DTT, 0.2 m*M* PMSF, aprotinin (10 μg/ml), 10% (v/v) glycerol] and is dialyzed against 2 liters of the same buffer. PKR can be concentrated using a Centriprep 10 device (Centricon, Amicon, Danvers, MA) and analyzed by Coomassie Brilliant Blue-stained 10% SDS-polyacrylamide gels (52). Alternative purification approaches that result in relatively low recovery of functional PKR from IFN-treated cells have been described by Kostura and Mathews (69). To generate antisera specific to PKR, we inject New Zealand White rabbits with approximately 100 μg of recombinant purified kinase mixed with 5 mg of poly(A):poly(U) adjuvant (Sigma). Two similar injections were administered at 1- and 2-month intervals, respectively (46).

Cell-Free Expression of PKR

For the cell-free production of PKR protein (70), *in vitro*-transcribed mRNA can be added to message-dependent reticulocyte lysate translation mixture, supplemented with 50 mM KCl, 0.5 mM magnesium acetate, 150 μM amino acids (minus methionine, unless translation was done in the absence of [^{35}S]methionine), 18 μCi of [^{35}S]methionine, 15 mM phosphocreatine, 2 U of creatine phosphokinase/ml, 10 μM hemin, 262 U of RNasin/ml (53). The reaction mixtures are incubated at 30° C for 1 hr, and terminated by the addition of 2 mM NaOH and 100 μg of RNase A/ml. An equal volume of 2× protein denaturation electrophoresis buffer is added, and the reaction products are then analyzed by SDS-PAGE. Message-dependent reticulocyte lysate is prepared according to the protocol of R. Jagus, described previously (70). Briefly, immature female New Zealand White rabbits are injected subcutaneously with 0.25 ml of a 2.5% phenylhydrazine chloride solution per kilogram body weight on 5 consecutive days. After 2 days, blood is collected, centrifuged at 1000 g to pellet the reticulocytes, which are washed once in complete buffer (0.14 M NaCl, 5 mM KCl, 7.5 mM magnesium acetate, 1 mM glucose). The reticulocytes are washed again in buffer without magnesium acetate or glucose and centrifuged as before. The cells are lysed in 2 vol of ice-cold water for 5 min. Cellular debris is removed by centrifugation and the supernatant is snap frozen in aliquots over liquid nitrogen. Message-dependent lysates are prepared by treating the supernatant with *Staphylococcus aureus* micrococcal nuclease (150 U/ml) in the presence of 1 mM CaCl$_2$ and hemin (5 U/ml). This mixture is incubated at 20° C for 20 min, and the reaction is stopped by adding EGTA to 2 mM. Finally, calf liver tRNA is added to a final concentration of 50 μg/ml.

Regulation of PKR by Selected Viruses

Many viruses have devised strategies to inhibit PKR (39,41,45). We developed the following protocols to study the downregulation of PKR by several RNA viruses.

Regulation of PKR by HIV-1

Background

In our studies of PKR regulation by HIV-1, we found that sequences within the 5′-UTR of HIV mRNA [termed TAR, the Tat responsive region (71)] were responsible for the activation of PKR (72,73). We determined that TAR-containing poly(A)$^+$ (at low concentrations) from HIV-1-infected cells efficiently activated PKR and that TAR-containing RNA synthesized *in vitro* formed a stable complex with the protein kinase. Efficient PKR activation and binding were dependent on the TAR RNA stem structure (72,74,75). In related work we discovered that productive infection by HIV-1 resulted in a significant decrease in PKR levels. Furthermore, PKR steady state levels were reduced in HeLa cell lines stably expressing the HIV-1 Tat regulatory protein as compared to the amount of the kinase in control HeLa cells (72). Thus, the potential translational inhibitory effects of the TAR RNA region mediated by kinase activation may be downregulated by Tat during productive HIV-1 infection.

Methods

PKR Activation Assay

To examine *in vitro* PKR activation by HIV-1-infected and HIV-1-LTR transformed cells, poly(A)$^+$ RNA from these cells is reacted with immunoaffinity-purified PKR in an *in vitro* activation buffer [Tris-HCl (pH 7.5), 75 mM KCl, 2 mM MgCl$_2$, 2 mM MnCl$_2$, 1 μM ATP, 0.8 mM DTT, 1 mM EDTA, BSA (0.3 μg/μl), 10 μCi of [γ-^{32}P]ATP] for 15 min at 30°C. The reaction is stopped with the addition of 20 μl of 3 × electrophoresis disruption buffer, 10 μg of RNase A, and 1 μl of 1 M MgCl$_2$. Following denaturation by boiling, activated PKR is detected by SDS-PAGE on 10% gels.

PKR–TAR Binding Assays

An RNA mobility gel shift assay and PKR-immunoaffinity column assay are used to demonstrate binding of wild-type and mutant TAR transcripts with purified PKR *in vitro* (73). For the gel shift assay, 250–500 ng of immunoaffinity-purified PKR is preincubated in the reaction buffer lacking RNA [10 mM Tris-HCl (pH 7.5), 50 mM NaCl, 1 mM DTT, 1 mM EDTA, RNasin (0.5 U/μl), BSA (0.09 μg/μl), 0.05% (v/v) glycerol] at 30°C for 10 min. Gel-purified ^{32}P-labeled (2000 cpm) TAR containing RNA is added (for a final volume of 20 μl) and incubated for an additional 10 min. The samples are analyzed on a preelectrophoresed 5% polyacrylamide gel containing 5% glycerol in Tris–borate–EDTA buffer at 30-mA constant current for 2.5 hr at 4°C. Following electrophoresis, the gels are dried and autoradiographed. Alternatively, immunoprecipitated affinity-purified PKR from IFN-treated 293 cells is reacted with ^{32}P-labeled TAR RNA (1.5 × 10^4 cpm) for 20 min at 30°C. The complexes are washed in 10 mM Tris-HCl (pH 7.5), 100 mM KCl, 0.1 mM EDTA, aprotinin (100 U/ml), 20% glycerol, and then resuspended in disassociation buffer [50 mM Tris-HCl (pH 7.5), 150 mM NaCl, 0.05% Nonidet P-40, 1% SDS, carrier tRNA]. Following phenol–chloroform extraction and ethanol precipitation, TAR RNA specifically bound to PKR is analyzed by autoradiography following electrophoresis on an 8% polyacrylamide–8 M urea gel.

Regulation of PKR by Influenza Virus

Background

We first found that influenza virus encoded a mechanism to downregulate PKR when cells were coinfected with adenovirus dl331 VAI RNA negative mutant and influenza virus (for reviews see Refs. 26 and 27). Influenza virus was able to reduce the abnormally high levels of PKR autophosphorylation and activity normally detected in mutant adenovirus-infected cells (27,39). We then reported a similar suppression in cells infected by influenza virus alone (39). Utilizing *in vitro* assays for PKR inhibition, we purified, to near homogeneity, a PKR repressor from influenza virus-infected cells (76). The purified product inhibited PKR autophosphorylation as well as phosphorylation of the α subunit of eIF-2 by the kinase (76,77). Interestingly, we found that the purified repressor, which had a molecular weight of approximately 58,000 (thus referred to as P58), was a cellular and not a viral protein (76). This was initially determined because the inhibitor did not react to virus-specific antibodies but, more importantly, has now been purified from uninfected cells (76,77). We also found that P58 was not an ATPase, protease, or phosphatase, nor did it function by degrading or sequestering the dsRNA activator of P68 (77). It remains unclear how influenza virus, which is insensitive to the antiviral effects of IFN unless the host cell possesses the Mx gene (78) [in which case viral transcription but not translation is blocked (78)], induces P58. To investigate the function of P58 further, we have isolated a full-length cDNA clone encoding this PKR inhibitor and plan to express functional recombinant protein to perform further biochemical analysis.

Methods

PKR Inhibition Assays

An *in vitro* PKR inhibition assay was first developed by utilizing a histone phosphorylation assay to measure kinase activity in uninfected and influenza-infected cells (76,77). It has been used successfully to purify the PKR inhibitor we have termed P58. Fractions of influenza-infected cell extracts are mixed with an IFN-treated 293 cell extract (as a source of PKR) for 20 min at 30°C. PKR is then immunoprecipitated as above with PKR monoclonal antibody and then analyzed for function by its ability to phosphorylate exogenously added histones, a suitable *in vitro* substrate and a substitute for eIF-2 (76,77).

Purification of PKR Inhibitor

The PKR inhibitor, P58, was originally isolated from MDBK cells infected with influenza virus (76). This inhibitory activity could not be detected in crude, uninfected cell extracts. However, ammonium sulfate treatment of uninfected cells unmasked this activity and allowed purification of P58 with chromatographic properties identical to those of the P58 obtained from influenza virus-infected cells (76). There is

TABLE I   Purification of p58, Influenza Virus-Activated PKR Inhibitor[a]

Purification step	Volume (ml)	Protein (mg)	Total activity (units)	Specific activity (units/mg)	Purification (-fold)	Yield (%)
Cytoplasmic extract	438.00	2,847.000	83,000	29.2		100.00
100,000 g	430.00	2,193.000	64,500	29.4	1.0	77.70
Ammonium sulfate	30.00	600.000	24,900	41.5	1.4	30.00
Mono Q	5.00	16.000	4,000	250.0	8.6	4.80
Heparin–agarose	3.50	1.820	2,590	1,420.0	48.6	3.10
Mono S	2.80	0.180	1,148	6,370.0	230.5	1.40
Glycerol gradient	1.68	0.013	269	20,690.0	708.0	0.32

[a] Protein was measured by the Micro BCA protein assay, except for the Mono S and glycerol gradient fractions, for which analysis of silver-stained polyacrylamide gels was used to estimate the quantity of protein. One unit of activity is defined as the amount of protein required to cause a 1% inhibition in phosphorylation of exogenously added histones by PKR isolated from IFN-induced 293 cell extracts.

evidence to suggest that P58 is bound to, or regulated by, an "anti-inhibitor," which is "released" during influenza virus infection or by treatment with ammonium sulfate (77). Further characterization of this potential regulator of P58, which we have partially purified, is currently underway. A brief description of our purification procedures to isolate P58 is presented here (Table I). A more detailed description can be found in an article by T.-G. Lee *et al.* (76). MDBK cells are washed twice in Hanks' balanced salt solution and are lysed with disruption buffer (as above). Cytoplasmic extracts are then centrifuged at 100,000 g for 1 hr at 4°C, using a Beckman Ti 60 rotor. The supernatant (S100) is fractionated by ammonium sulfate precipitation into 0–40, 40–60, and 60–80% samples. From our kinase histone assays, we observed that the inhibitory activity was present in the 40–60% fraction. This fraction was then resuspended in buffer III (see above) supplemented with 100 mM KCl and dialyzed twice against the same buffer. Dialyzed samples are applied to a Mono Q HR 10/10 FPLC (fast protein liquid chromatography) anion-exchange column (Pharmacia) and PKR repressor activity is eluted off the column at approximately 280 mM KCl. The peak fractions are added to a heparin–agarose cation-exchange column. Repressor activity is contained in the 300 mM KCl range of fractions, which are then concentrated, using a Centriprep 30 (Centricon), and dialyzed against buffer III containing 25 mM KCl. The dialyzed material is then applied to a fast protein liquid chromatography Mono S column, after which kinase inhibitory activity can be detected in the 250 mM KCl eluate fraction. Active fractions from this column are sedimented on a 10–30% glycerol gradient containing 25 mM KCl and centrifuged at 49,000 rpm for 21 hr at 4°C in a Beckman SW 55 rotor. Fractions are assayed for activity, spot dialyzed, and stored at −70°C. A single protein of 58,000 Da is obtained from this purification. This P58 inhibitor blocks both the autophosphorylation and kinase activity of PKR (76,77).

Regulation of PKR by Poliovirus

Background

We have demonstrated that PKR is both highly autophosphorylated and activated by viral-specific RNA during poliovirus infection (79). In this case we found that PKR levels did not increase but rather dramatically declined in poliovirus-infected cells *in vitro.* Pulse–chase experiments allowed us to confirm that the protein kinase was in fact significantly degraded during virus infection (79). After utilizing an *in vitro* degradation assay we established that although viral gene expression was required for PKR degradation, the major poliovirus proteases 2A and 3C were not directly involved with PKR proteolysis (80). We found that the protease responsible for PKR degradation required divalent cations and was likely composed of both an RNA and protein component, because both trypsin and ribonuclease treatments abrogated the degradation activity (80). Despite the requirements for divalent cations and RNA, we also found that activation of the kinase was not required for proteolysis because a catalytically inactive mutant PKR was also degraded. Mapping of PKR protease-sensitive sites, using *in vitro*-translated and deletion mutants, revealed that sites required for degradation resided in the amino terminus and colocalized to dsRNA-binding domains (Fig. 5) (80). Also, preincubation of cell-free extracts with synthetic dsRNA, poly(I):poly(C), largely prevented PKR proteolysis, providing additional evidence for the critical role played by RNA. Attempts to purify the protease biochemically have proved to be difficult, partially due to its insolubility and partly because of its multicomponent structure (79,80).

Methods

In Vitro Assay for Degradation of PKR by Poliovirus

Poliovirus type 1 is propagated in HeLa cells maintained in Joklik's modified minimal essential medium (MEM) supplemented with 10% fetal bovine serum (FBS) (79). For preparation of virus stocks, suspension HeLa cells are infected at a multiplicity of infection of 5 pfu/cell for 10 hr in MEM containing 2% FBS. Infected cells are pelleted by centrifugation at 1000 g for 5 min at room temperature and resuspended in a minimum volume of MEM with 2% FBS. Virus stock titers are then determined by infecting monolayers of HeLa cells with appropriate virus dilutions. Cells are incubated at 37°C for 1 hr and overlaid with agar as described above for influenza virus. To detect PKR degradation, extracts of infected or uninfected HeLa cells are washed in Hanks' balanced salt solution and disrupted with lysis buffer (as above). As a source of radiolabeled PKR, extracts can be prepared from HeLa cells labeled with [^{35}S]methionine (500 μCi/ml) and treated with IFN, as described above. Labeled extracts are then treated with either mock- or poliovirus-infected cell extracts for 15 min at 30°C. After incubation, extracts are immunoprecipitated with antisera specific for PKR and resolved by SDS-PAGE through a 10% gel. As an alternative source of radiolabeled PKR, *in vitro*-translated PKR can be used equally

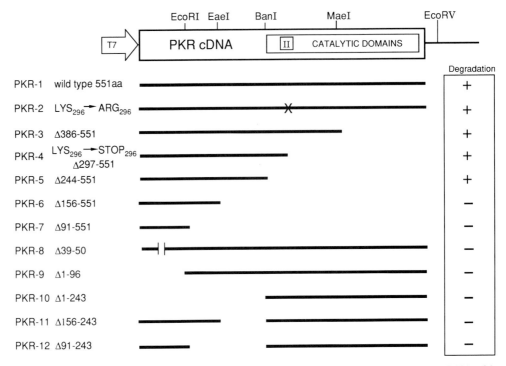

FIG. 5   Mapping PKR protease-sensitive sites in poliovirus-infected cells. PKR cDNA with selected restriction endonuclease sites that were used to produce runoff RNA and truncated PKR proteins is schematically represented at the top. Constructs PKR-1 through PKR-12 were prepared using standard molecular biological techniques as described previously (80). The deleted amino acids (Δ) are indicated on the left. Each construct was translated *in vitro* in the presence of [^{35}S]methionine, mixed with either mock- or poliovirus-infected extracts, and then immunoprecipitated with PKR polyclonal antiserum, which reacts with all of the *in vitro* translated proteins. Degradation is scored on the basis of the relative amounts of immunoprecipitated, truncated PKR proteins recovered from the *in vitro* assay (see text). The degree of degradation was determined by laser densitometric scanning. +, >90% degraded; −, <10% degraded.

well in this assay. This latter *in vitro*-made PKR was used to map the protease-sensitive sites (Fig. 5).

Concluding Remarks

This article has reviewed our methodologies which we routinely use to study virus–cell interactions. For our studies, we have focused on the dissimilar RNA viruses, influenza virus, human immunodeficiency virus, and poliovirus. The use of the described techniques allowed us to investigate viral replication strategies unique to each of these viruses. These analyses have added to our understanding, not only of viral pathogenicity, but also of normal cellular translational regulation. Although the schemes utilized by these viruses to control gene expression in their host cells are being elucidated, determining the precise molecular mechanisms underlying each strategy remains the objective for future studies.

Acknowledgments

We thank past and present members of our laboratory, who performed much of the work described here: Marlene Wambach, Judy Tomita, Tae Gyu Lee, Min Ling Wong, Michele Garfinkle, Tracy Black, Karlissa Foy, and Mary O'Riordan. We also gratefully acknowledge the collaborations of Drs. Naham Sonenberg, Sophie Roy, and Ara Hovenessian. We thank Harry Gerecke and Marjorie Domenowske for assistance in manuscript and figure preparation. This work was supported by NIH Grants AI22646 and RR00166.

References

1. K. Steffy and F. Wong-Staal, *Microbiol. Rev.* **55,** 193 (1991).
2. M. B. Agy, M. Wambach, K. Foy, and M. G. Katze, *Virology* **177,** 251 (1990).
3. M. B. Agy, K. Foy, M. J. Gale, R. E. Benveniste, E. A. Clark, and M. G. Katze, *Virology* **183,** 170 (1991).
4. T. Huet, R. Cheyner, A. Meyerhans, G. Roelants, and S. Wain-Hobson, *Nature (London)* **345,** 356 (1990).
5. T. Klimkait, K. Stebel, M. D. Hoggan, M. A. Martin, and J. M. Orenstein, *J. Virol.* **64,** 621 (1990).
6. R. L. Willey, F. Maldarell, M. A. Martin, and K. Strebel, *J. Virol.* **66,** 226 (1992).
7. R. L. Willey, F. Maldarell, M. A. Martin, and K. Strebel, *J. Virol.* **66,** 7193 (1992).
8. F. Barre-Sinoussi, J. C. Chermann, F. Rey, M. T. Nugeyre, S. Chamaret, J. Gruest, C. Dauguet, C. Axler-Blin, F. Vezinet-Brun, C. Rouzioux, W. Rozenbaum, and L. Montagnier, *Science* **220,** 868 (1983).
9. S. Wain-Hobson, J.-P. Vartanian, M. Henrey, N. Chenciner, R. Cheyner, S. Delassus, L. P. Martins, M. Sala, M.-T. Nugeyre, D. Guetard, D. Klatzmann, J.-C. Gluckman, W. Rozenbaum, F. Barre-Sinoussi, and L. Montagnier, *Science* **252,** 961 (1991).
10. R. E. Benveniste, L. O. Arthur, C.-C. Tsai, R. Sowder, T. D. Copland, L. E. Henderson, and S. Oroszlan, *J. Virol.* **60,** 483 (1986).
11. R. E. Benveniste, R. W. Hill, L. J. Eron, U. M. Casaikl, W. B. Knott, L. E. Henderson, R. C. Sowder, K. Nagashima, and M. A. Gonda, *J. Med. Primatol.* **19,** 351 (1990).
12. S. Z. Salahuddin, P. D. Markham, F. Wong-Staal, G. Franchini, V. S. Kalyanaraman, and R. C. Gallo, *Virology* **129,** 51 (1983).

13. S. Harada, Y. Koyanagi, and N. Yamamoto, *Science* **229,** 563 (1985).
14. S. Chaffe, J. M. Leeds, T. J. Mathews, K. J. Weinhold, M. Skinner, D. P. Bolognesi, and M. S. Hershfield, *J. Exp. Med.* **168,** 605 (1988).
15. R. D. Salter, D. N. Howell, and P. Cresswell, *Immunogenetics* **21,** 235 (1985).
16. G. E. Foley, H. Lazarus, S. Farber, B. G. Uzman, B. A. Boone, and R. E. McCarthy, *Cancer* **18,** 522 (1965).
17. A. Weiss, R. L. Wiskocil, and J. D. Stobo, *J. Immunol.* **133,** 123 (1984).
18. A. F. Gazdar, D. N. Carney, P. A. Bunn, E. K. Russell, E. S. Jaffe, G. P. Schechter, and J. G. Guccion, *Blood* **55,** 409 (1980).
19. L. Reed and H. Meunch, *Am. J. Hyg.* **27,** 493 (1938).
20. P. Chomczynski and N. Sacchi, *Anal. Biochem.* **162,** 156 (1987).
21. J. Sambrook, E. F. Fitsch, and T. Maniatis, "Molecular Cloning," 2nd ed. Cold Spring Harbor Laboratory Press, Cold Spring Harbor, New York, 1989.
22. P. S. Thomas, *Proc. Natl. Acad. Sci. USA* **77,** 5201 (1980).
23. A. P. Feinberg and B. Vogelstein, *Anal. Biochem* **132,** 6 (1983).
24. M. E. Greenberg and E. B. Ziff, *Nature (London)* **311,** 433 (1984).
25. M. Groudine, M. Peretz, and H. Weintraub, *Mol. Cell. Biol.* **1,** 281 (1981).
26. R. M. Krug, F. Alonso-Caplen, I. Julkunen, and M. G. Katze, *in* "The Influenza Viruses" (Krug, ed.), p. 89. Plenum Press, New York, 1989.
27. M. G. Katze and R. M. Krug, *Enzyme* **44,** 265 (1990).
28. M. G. Katze and R. M. Krug, *Mol. Cell. Biol.* **4,** 2198 (1984).
29. M. G. Katze, B. M. Detjen, B. Safer, and R. M. Krug, *Mol. Cell. Biol.* **6,** 1741 (1986).
30. M. G. Katze, D. DeCorato, and R. M. Krug, *J. Virol.* **60,** 1027 (1986).
31. F. V. Alonso-Caplen, M. G. Katze, and R. M. Krug, *J. Virol.* **62,** 1606 (1988).
32. M. S. Garfinkel and M. G. Katze, *J. Biol. Chem.* **267,** 9383 (1992).
33. R. A. Lamb and C. M. Horvath, *Trends Genet.* **7,** 261 (1991).
34. P. R. Etkind and R. M. Krug, *J. Biol. Chem.* **260,** 5493 (1975).
35. H. Towbin, T. Staehlin, and J. Gordon, *Proc. Natl. Acad. Sci. USA* **76,** 4350 (1979).
36. F. M. Ausubel, R. Brent, R. E. Kingston, D. D. Moore, J. G. Seidman, J. A. Smith, and K. Struhl (eds.), "Current Protocols in Molecular Biology." Greene Publishing and Wiley Interscience, New York, 1990.
37. R. Lenk and S. Penman, *Cell (Cambridge, Mass.)* **16,** 289 (1979).
38. G. C. Sen and P. Lengyel, *J. Biol. Chem.* **267,** 5017 (1992).
39. M. G. Katze, *J. Interferon Res.* **12,** 241 (1992).
40. A. G. Hovanessian, *J. Interferon Res.* **11,** 199 (1991).
41. M. G. Katze, *Semin. Virol.* **508,** 1 (1993).
42. B. Safer, *Cell (Cambridge, Mass.)* **33,** 7 (1983).
43. J. W. B. Hershey, *J. Biol. Chem.* **264,** 20823 (1989).
44. E. Meurs, Y. Watanabe, G. N. Barber, M. G. Katze, K. Chong, B. R. G. Williams, and A. G. Hovanessian, *J. Virol.* **66,** 5805 (1992).
45. C. E. Samuel, *Virology* **183,** 1 (1991).
46. G. N. Barber, J. Tomita, M. Garfinkel, A. G. Hovanessian, E. Meurs, and M. G. Katze, *Virology* **191,** 670 (1992).
47. D. St. Johnston, N. H. Brown, J. G. Gall, and M. Jantsch, *Proc. Natl. Acad. Sci. USA* **89,** 10979 (1992).

48. K. L. Chong, K. Schappert, E. Meurs, F. Feng, T. F. Donahue, J. D. Friesen, A. G. Hovanessian, and B. R. G. Williams, *EMBO J.* **11,** 1553 (1992).
49. S. R. Green and M. B. Mathews, *Genes Dev.* **6,** 2478 (1992).
50. S. J. McCormack, D. C. Thomis, and C. E. Samuel, *Virology* **188,** 47 (1992).
51. R. Patel and G. C. Sen, *J. Biol. Chem.* **267,** 7871 (1992).
52. G. Barber, J. Tomita, A. Hovanessian, E. Meurs, and M. G. Katze, *Biochemistry* **30,** 10356 (1991).
53. M. G. Katze, M. Wambach, M.-L. Wong, M. Garfinkel, E. Meurs, K. Chong, B. R. G. Williams, A. G. Hovanessian, and G. N. Barber, *Mol. Cell. Biol.* **11,** 5497 (1991).
54. E. Meurs, K. Chong, J. Galabru, N. Thomas, I. Kerr, B. R. G. Williams, and A. G. Hovanessian, *Cell (Cambridge, Mass.)* **62,** 379 (1990).
55. G. N. Barber, S. Edelhoff, M. G. Katze, and C. M. Disteche, *Genomics* **16,** 765 (1993).
56. J. Chen, J. Pal, M. S. Throop, L. Gehrke, I. Kuo, J. K. Pal, M. Brodsky, and I. London, *Proc. Natl. Acad. Sci. USA* **88,** 7729 (1991).
57. P. L. Icely, P. Gros, J. J. M. Bergeron, A. Devault, D. E. H. Afar, and J. C. Bell, *J. Biol. Chem.* **266,** 16073 (1991).
58. T. E. Dever, J.-J. Chen, G. N. Barber, A. M. Cigan, L. Feng, T. F. Donahue, I. M. London, M. G. Katze, and A. G. Hinnebusch, *Proc. Natl. Acad. Sci. USA* **90,** 4616 (1993).
59. T. E. Dever, L. Feng, R. C. Wek, A. M. Cigan, T. F. Donahue, and A. G. Hinnebusch, *Cell (Cambridge, Mass.)* **68,** 585 (1992).
60. A. E. Koromilas, S. Roy, G. N. Barber, M. G. Katze, and N. Sonenberg, *Science* **257,** 1685 (1992).
61. E. Meurs, J. Galabru, G. N. Barber, M. G. Katze, and A. G. Hovanessian, *Proc. Natl. Acad. Sci. USA* **90,** 232 (1993).
62. G. N. Barber, M. Wambach, M.-L. Wong, T. E. Dever, A. G. Hinnebusch, and M. G. Katze, *Proc. Natl. Acad. Sci. USA* **90,** 4621 (1993).
63. D. C. Thomis and C. E. Samuel, *Proc. Natl. Acad. Sci. USA* **89,** 10837 (1992).
64. K. Klein and G. Klein, *Cancer Res.* **28,** 1300 (1968).
65. F. Graham, J. Smiley, W. Russell, and R. Nairu, *Mol. Cell. Biol.* **36,** 59 (1977).
66. A. G. Laurent, B. Krust, J. Galabru, J. Svab, and A. G. Hovanessian, *Proc. Natl. Acad. Sci. USA* **82,** 4341 (1985).
67. A. Konieczny and B. Safer, *J. Biol. Chem.* **256,** 3402 (1983).
68. F. W. Studier, A. H. Rosenberg, J. J. Dunn, and J. W. Debendorff, *in* "Methods in Enzymology," Vol. 185, p. 60. Academic Press, San Diego, 1990.
69. M. Kostura and M. B. Mathews, *Mol. Cell. Biol.* **9,** 1576 (1989).
70. M. G. Katze, M. Wambach, M.-L. Wong, M. Garfinkel, E. Maurs, K. Chong, B. R. G. Williams, A. G. Hovanessian, and G. N. Barber, *Mol. Cell. Biol.* **11,** 5497 (1991).
71. C. A. Rosen, J. G. Sodroski, W. C. Goh, A. Dayton, J. Lippke, and W. A. Haseltine, *Nature (London)* **319,** 555 (1986).
72. S. Roy, M. G. Katze, N. T. Parkin, I. Edery, A. G. Hovanessian, and N. Sonenberg, *Science* **247,** 1216 (1990).
73. S. Roy, M. B. Agy, A. G. Hovanessian, N. Sonenberg, and M. G. Katze, *J. Virol.* **65,** 632 (1991).
74. I. Edery, R. Petryshyn, and N. Sonenberg, *Cell (Cambridge, Mass.)* **56,** 303 (1989).
75. D. N. SenGupta and R. H. Silverman, *Nucleic Acids Res.* **17,** 969 (1989).

76. T. G. Lee, J. Tomita, A. G. Hovanessian, and M. G. Katze, *Proc. Natl. Acad. Sci. USA* **87,** 6208 (1990).

77. T. G. Lee, J. Tomita, A. G. Hovanessian, and M. G. Katze, *J. Biol. Chem.* **267,** 14238 (1992).

78. J. Povlovic and P. Staehli, *J. Infect. Res.* **11,** 215 (1991).

79. T. Black, B. Safer, A. Hovanessian, and M. G. Katze, *J. Virol.* **63,** 2244 (1989).

80. T. L. Black, G. N. Barber, and M. G. Katze, *J. Virol.* **67,** 791 (1993).

[13] *In Vitro* Biochemical Methods for Investigating RNA–Protein Interactions in Picornaviruses

Holger H. Roehl and Bert L. Semler

Introduction

Picornaviruses are a large family of medically important RNA viruses. They are small, nonenveloped, icosahedral viruses that contain a single strand, positive-sense RNA genome. The picornavirus family consists of the following five genera, and prototypical members of each genus are listed: (a) enteroviruses, which include poliovirus and coxsackievirus, (b) rhinoviruses, which include human and bovine rhinovirus, (c) cardioviruses, which include encephalomyocarditis virus and Theiler's murine encephalomyelitis virus, (d) aphthoviruses, which include foot-and-mouth disease virus, and (e) hepatoviruses, which include hepatitis A virus.

Studies of picornaviruses have yielded many valuable insights into the infectious cycle of RNA viruses. Additionally, picornaviruses are also helpful in investigating macromolecular interactions such as RNA–protein or protein–protein interactions. Several features, discussed below, make picornaviruses interesting and amenable for studying molecular mechanisms of the infectious cycle in general, and for studying macromolecular interactions in particular. Facilitating picornavirus investigations is the fact that they can be propagated in tissue culture cells, which was first shown for poliovirus by Enders *et al.* (1) in 1949. Most picornaviruses can be grown to high titers and can be studied both *in vivo* and *in vitro*.

The genomes of many picornaviruses have been cloned and their cDNAs were shown to be infectious (2, 3). This allows one to perform genetic analyses by transfecting wild-type and mutant full-length cDNAs into tissue culture cells. With the advent of prokaryotic transcription vectors, transfection of *in vitro*-generated full-length positive-strand RNA molecules is now the preferred method for introduction of genetically defined viral genomes into cells (4, 5). Recovery of wild-type, small-plaque, or temperature-sensitive virus and determination of which mutations are lethal have been instrumental in defining the importance of a mutated region for viral functions.

Translation initiation of picornavirus mRNA occurs by internal ribosome entry in the 5′ untranslated region via a cap-independent mechanism. Contrary to nearly all eukaryotic RNAs, the 5′ end of picornavirus RNA does not have a 7-methylguanosine cap structure. The unusually long 5′ noncoding region of picornavirus RNAs (~600–1200 nucleotides) contains a number of computer-predicted stem–loop structures and multiple AUG codons preceding the translational start site.

These features predict a translation initiation mechanism different from the traditional ribosome scanning model. Several groups have provided convincing evidence that ribosomes bind internally to the 5′ noncoding region in poliovirus (6, 7), encephalomyocarditis virus (8–10), foot-and-mouth disease virus (11), and hepatitis A virus (12), and the regions critical for cap-independent translation initiation have been mapped.

Another interesting feature of picornaviruses is the fact that they synthesize one large polyprotein that is proteolytically processed (for a review see Ref. 13). The organization and expression of the poliovirus genome as a prototypical picornavirus are shown in Fig. 1. The poliovirus polyprotein is cleaved cotranslationally by the virally encoded proteinase 2A. Proteinase 2A cleaves the capsid precursor protein P1 from the nascent polypeptide, and the full-length polyprotein is usually not found in cells. Most of the protein processing, however, occurs posttranslationally by the virally encoded proteinase 3C, or its precursor 3CD, generating the structural and nonstructural proteins shown in Fig. 1. The final step in the proteolytic cascade is the maturation cleavage of the capsid precursor protein VP0 to VP4 and VP2 by a mechanism that is still poorly understood.

Most picornaviruses inhibit host cell protein synthesis. It was shown in poliovirus-infected cells that cap-dependent host cell translation initiation is shut down. In-

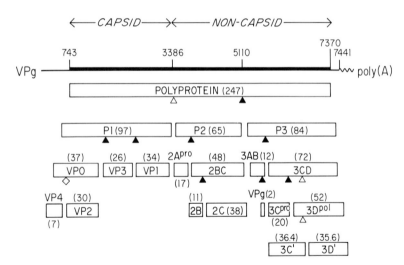

FIG. 1   Simplified map of the poliovirus genome and its translation products. Poliovirus RNA is translated into a large polyprotein that is proteolytically processed. Numbers in parentheses show the molecular mass of viral proteins (in kDa). Open triangles below proteins indicate sites cleaved by the viral protein 2A. Solid triangles below proteins indicate sites cleaved by the viral protein 3C. The open diamond indicates the site of the maturation cleavage of capsid precursor protein VP0. Reprinted with permission from (13a) and the American Society for Microbiology.

hibiting cap-dependent host cell translation enhances cap-independent translation of viral proteins. Shutoff of host cell protein synthesis is thought to occur by proteolytic processing of the p220 subunit of the translation initiation factor eIF-4F (14). It has been suggested that the virally encoded proteinase 2A is indirectly involved in p220 cleavage (15); however, the exact mechanism of p220 cleavage awaits further study.

Picornavirus RNA replication (for a review see Ref. 16) is mediated by a virally encoded RNA-dependent RNA polymerase called 3D polymerase. On entry of virions into a permissive host cell and following primary translation, positive-strand RNA is replicated to generate negative-strand RNA. New infectious positive-strand RNAs are subsequently synthesized by replication of the negative strands. Initiation of negative-strand RNA synthesis is presumably mediated by *cis*-acting sequences localized at the 3' end of positive-strand RNA. In a similar fashion, initiation of positive-strand RNA is thought to be mediated by sequences localized at the 3' end of negative-strand RNA. These sequences, as yet undefined, are thought to be recognized by replication proteins that include the viral RNA polymerase 3D and other viral and/or cellular factors. Even though picornavirus RNA replication has been studied for many years, the mechanism(s) of replication initiation remains unclear.

Besides being medically relevant pathogens, picornaviruses are ideally suited for investigating macromolecular interactions, specifically RNA–protein interactions. Interactions of viral and/or cellular proteins with picornavirus RNA occur during viral translation, viral RNA replication, and packaging of the viral RNA. *In vitro* biochemical investigations of these RNA–protein interactions by methods described here have led to intriguing results that, in many cases, can be subsequently studied by *in vivo* molecular genetic analysis.

RNA Electrophoretic Mobility Shift Assay

The RNA electrophoretic mobility shift assay is a fast and relatively simple assay for studying specific interactions between viral RNAs and virus and/or host cell proteins *in vitro*. The principle underlying RNA electrophoretic mobility shift assays is the observation that the electrophoretic mobility of an RNA molecule with bound protein(s) is retarded in relation to the mobility of a free RNA molecule in a native polyacrylamide gel. This allows one to determine if an RNA molecule forms an RNA–protein complex with crude extracts or proteins from various stages of purification. RNA electrophoretic mobility shift assays were first employed to investigate RNA–protein interactions occurring during the formation of eukaryotic mRNAs. Using this technique, spliceosome formation and the mechanism of RNA splicing to remove intervening sequences have been studied (17). In the study of picornaviruses, RNA electrophoretic mobility shift assays have been instrumental in defining the *cis*-acting sequences in the viral 5' noncoding region required for internal translation initiation and identifying proteins involved in the formation of translation initiation complexes (7, 18–21). RNA electrophoretic mobility shift assays were also used to study the formation of viral replication complexes and the *cis*-acting sequences these complexes interact with during initiation of viral RNA replication (22).

Preparation of Cytoplasmic Extract

To investigate poliovirus translation initiation, cytoplasmic extracts from uninfected HeLa suspension cultures are used. HeLa cell S10 extracts or ribosomal salt wash fractions (which are enriched for translation initiation factors) are prepared using published procedures (23, 24).

To investigate poliovirus replication, cytoplasmic extracts from infected HeLa suspension cultures are used. It has been shown that picornavirus RNA replication is associated with smooth membranes (25). Therefore, this extract is prepared using a Nonidet P-40 (NP-40) lysis protocol (20, 26). The advantage of this procedure is that membrane-associated proteins will not be completely disrupted by the nonionic detergent NP-40, and *in vitro* binding studies using NP-40 lysis extract will mimic *in vivo* conditions more accurately than an S10 extract. Briefly, harvest cells at a suitable time after infection and wash three times in phosphate-buffered saline (PBS). After the last wash, resuspend the cells in an equal volume of lysis buffer [50 mM Tris (pH 8), 5 mM ethylenediaminetetraacetic acid (EDTA), 150 mM NaCl, 0.5% (v/v) NP-40, 0.1 mM phenylmethylsulfonyl fluoride (PMSF)] and incubate for 20 min on ice. Centrifuge the mixture for 10 min at 17,000 g at 4°C. The supernatant, which contains cytoplasmic proteins and membranes, can be further fractionated by performing ammonium sulfate precipitations. Dialyze the extract overnight against a buffer containing 10% (v/v) glycerol and salt concentrations identical to the ones used in RNA electrophoretic mobility shift or ultraviolet (UV) cross-linking assays (see below for buffer composition). Freeze extracts in small aliquots at −70°C until further use.

Preparation of RNA Probes for Binding Studies

1. Prepare the DNA template. Subclone viral cDNA fragments containing sequences of interest into a T7/T3 or SP6 promoter-based transcription vector. Picornaviral RNAs have an elaborate computer-predicted secondary structure. It is important, therefore, to minimize vector sequences between the transcriptional start site and the 5′ end of the viral cDNA insert, which could perturb the secondary structure of the transcript. Often synthetic oligonucleotides are used during cloning to reconstruct the promoter region in such a fashion that there are no vector sequences (or maximally one or two residues necessary for efficient promoter function) between the transcriptional start site and the viral template. Prior to transcription, cleave the

cDNA insert at its 3' end with a suitable restriction endonuclease so that all the transcripts terminate at the same nucleotide. If the restriction endonuclease used generates a 3' staggered end, create a blunt end before the transcription reaction is initiated. This will inhibit foldback of the nascent RNA, which can result in a longer than expected or a heterogeneous population of transcripts. Prior to transcription, extract the template DNA twice with phenol–chloroform, once with chloroform, and precipitate the DNA in ethanol.

2. A typical *in vitro* transcription reaction contains the following RNase-free reagents (final concentrations).

Tris (pH 7.5)	40 mM
MgCl$_2$	6 mM
NaCl	10 mM
Spermidine	2 mM
Dithiothreitol (DTT)	10 mM
RNasin (Promega, Madison, WI)	100 U
GTP, ATP, CTP	0.5 mM each
UTP	25 μM
[α-^{32}P]UTP (specific activity, 3000 Ci/mmol)	50 μCi
RNA polymerase	40 U
Diethyl pyrocarbonate (DEPC)-treated H$_2$O to	100 μl

If nonradiolabeled RNA is to be transcribed, the [^{32}P]UTP is omitted from the reaction mix described above and a 0.5 mM concentration of each cold NTP is used instead. Incubate for 1.5 hr at 37°C.

3. To digest the DNA template, add 30 U of RNase-free DNase (Boehringer Mannheim, Indianapolis, IN) to the transcription mix at the end of the 1.5-hr incubation. Incubate for 15 min at 37°C.

4. Do one phenol–chloroform extraction and transfer the aqueous phase to a fresh tube. Back extract the phenol–chloroform with 100 μl of TE buffer (10 mM Tris, pH 8, 1 mM EDTA) and add to the aqueous phase. Add 100 μl of 7.5 M ammonium acetate and 700 μl of 100% ethanol to precipitate the RNA for about 1 hr at −20°C. Do not precipitate at temperatures lower than −20°C, as unincorporated nucleotides will also precipitate. Spin for 15 min at 16,000 g at 4°C, pour off the supernatant, resuspend the pellet in 200 μl of TE, and repeat the precipitation to ensure removal of unincorporated nucleotides and the DNA template. Resuspend pellet in 10 μl of TE. It is usually advisable to check the size of the synthetic RNA on a polyacrylamide minigel, which can either be ethidium bromide stained or exposed to X-ray film. RNA can be quantitated by determining the number of counts incorporated or, when nonradioactive RNA is transcribed, by determining the OD$_{260}$ in a spectrophotometer. For transcripts between 100 and 700 nucleotides in length, a typical *in vitro* transcription reaction yields about 10–30 μg RNA.

RNA Electrophoretic Mobility Shift Assay

A flow chart of an RNA electrophoretic mobility shift assay is shown in Fig. 2. Here, an assay used to investigate the interaction of cellular and/or viral proteins with picornavirus RNA is described. The concentration of various salts in the binding buffer (see below) or the incubation temperature should be empirically determined, as both will vary depending on the protein(s) studied. All the manipulations described below are performed under RNase-free conditions.

1. In a preincubation step, nonspecific and low-affinity RNA-binding proteins present in the cellular extract are allowed to bind to synthetic poly [r(I-C)] RNA. It is also possible to use tRNA for this purpose. Set up the following mix.

KCl	25 mM
MgCl$_2$	10 mM
NaCl	150 mM
EDTA	0.1 mM
HEPES (pH 7.8)	5 mM
Glycerol	3.8% (v/v)
Heparin	1.5 μg
Poly[r(I-C)]	20 μg
Extract from either poliovirus-infected cells or uninfected cells	50 μg
DEPC-treated H$_2$O to	9 μl

Incubate for 10 min at 30°C.

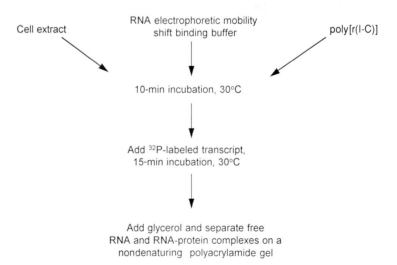

Fig. 2   Flow chart of an RNA electrophoretic mobility shift assay.

2. Add 1 μl of ^{32}P-labeled RNA probe suitably diluted to 20–100 fmol/μl [specific activity, ~1 × 10^4 counts per minute (cpm)/μl]. Incubate for 15 min at 30° C. During this incubation, specific RNA-binding proteins that have a detectable affinity for the RNA probe are allowed to bind.

3. Add glycerol to a final concentration of 10% (v/v) and immediately load the sample on a nondenaturing polyacrylamide (40:1) gel. Also load a standard loading buffer containing bromphenol blue and xylene cyanol in an additional well (without RNA sample) so that migration of the dye can be followed during electrophoresis. For probes up to 200 nucleotides in length, 6% polyacrylamide gels are used, and for probes longer than 200 nucleotides, 4% polyacrylamide gels are used. Gels are cast in 0.5× TBE (Tris–borate–EDTA) buffer, contain 5% (v/v) glycerol, and are electrophoresed in 0.5× TBE buffer. Electrophoresis is performed at 4°C, gels are prerun for ~30 min at a constant 15 V/cm, and electrophoresis separating free and bound RNAs is performed at a constant 5–15 V/cm. Gels are then dried under vacuum and exposed to film.

If the extract used contains one or more protein(s) that can bid the RNA probe, RNA–protein complexes will be separated from unbound RNA during electrophoresis. The RNA–protein complexes can be distinguished from unbound RNA by virtue of their slower migration, that is, they will be closer to the top of the gel than unbound RNA. An advantage of using RNA electrophoretic mobility shift assays is that they are rapid assays to determine whether or not protein(s) in a given extract bind specifically to a given RNA probe. The shortcoming of this assay is that no information can be derived about the molecular weight of the protein(s) or how many proteins are involved in complex formation.

To determine the specificity of complexes formed, it is necessary to perform competition experiments. Different molar excesses (usually between 5- and 50-fold) of nonradioactive specific competitor RNA and nonspecific competitor RNA are added to the preincubation step described under step 1 above. Nonradiolabeled specific competitor RNA will bind significant amounts or all of a specific RNA-binding protein present in the extract during the preincubation. When radiolabeled RNA is subsequently added, considerably less or no radiolabeled RNA will be shifted during electrophoresis. Nonspecific competitor RNA, on the other hand, should have a much lower affinity for the specific RNA-binding protein, and addition of nonspecific competitor RNA should not significantly affect the amount of shifted radiolabeled RNA. RNA electrophoretic mobility shift experiments in the presence of nonradiolabeled competitor RNA allow one to determine whether RNA–protein complexes formed are specific or are fortuitous due to nonspecific RNA–protein interactions.

To ascertain that shifted complexes are indeed due to RNA–protein interactions, cellular extracts can be digested with proteinase K or trypsin. Using such a digested extract subsequently in an RNA electrophoretic mobility shift assay should not yield any shifted RNA complexes. Thus, it can be ruled out that a shifted complex is due to membranes or other RNAs (endogenous to the extract) interacting with the radiolabeled RNA probe. For example, a radiolabeled negative-strand viral RNA probe could hybridize to positive-strand viral RNA present in extract from infected cells. The migratory pattern of this double-stranded RNA (dsRNA) may mislead one to conclude that a shifted RNA–protein complex has been formed. Therefore, it is important to use proteinase-digested extract as a negative control in RNA electrophoretic mobility shift assays.

Super Shift Assays

The "super shift" assay is a modification of the RNA electrophoretic mobility shift assay, which allows one to identify proteins involved in complex formation (19). After incubation of radiolabeled RNA probe with cellular extract, specific antibodies to proteins suspected to be involved in complex formation are added. If the protein that is recognized by the specific antibody is present in the complex, binding of this antibody to the protein will further slow migration of the RNA–protein complex during native polyacrylamide electrophoresis. This effect is called a "super shift."

Super shift assays can be used initially to identify proteins involved in RNA–protein complex formation. Using antibodies directed against poliovirus 3C proteinase and 3D polymerase in a super shift assay, it was shown that both 3C and 3D are involved in putative replication complex formation at the 5′ end of poliovirus RNA (22). If purified proteins are available, their involvement in complex formation can be confirmed using this assay.

Ultraviolet Cross-Linking Assay

The UV cross-linking assay is an additional, more elaborate, assay to study specific interactions between RNA and proteins *in vitro*. Under the conditions of RNA electrophoretic mobility shift assays, radiolabeled RNA is incubated together with cellular extract to allow formation of RNA–protein complexes. After complexes have formed they are exposed to ultraviolet radiation, which permanently cross-links the RNA sequence to proteins that bind directly to the RNA. This occurs by a process that is not entirely understood, whereby amino acid side chains and RNA form covalent bonds via a free radical reaction (27).

In the study of picornavirus RNA–protein interactions, UV cross-linking assays were employed to study internal translation initiation, specifically in defining and identifying host cell proteins involved in viral translation initiation. A cellular 57-kDa protein that appears to be involved in internal translation initiation of encephalomyocarditis virus (28), foot-and-mouth disease virus (29), and poliovirus (30) has been described. Additionally, a 52-kDa protein has been identified that is also thought to play a role in translation initiation of poliovirus protein synthesis (18).

Ultraviolet Cross-Linking Assay

A flow chart of a UV cross-linking assay is shown in Fig. 3. Here, a UV cross-linking assay used to investigate interactions of cellular and/or viral proteins with picornavirus RNA is described. All manipulations described are to be performed under RNase-free conditions. Similar to the RNA electrophoretic mobility shift assay, the composition of the binding buffer and the incubation temperature should be determined empirically, as both will vary with the protein(s) studied. The preparation of RNA probes and cytoplasmic extract to be used in UV cross-linking assays is identical to the preparations described for RNA electrophoretic mobility shift assays. The first two steps, the preincubation and the incubation, are also identical to RNA electrophoretic mobility shift assays.

1. During the preincubation step, nonspecific poly[r(I-C)] RNA (20 μg) and extract (50 μg) from poliovirus infected or uninfected cells are incubated in binding

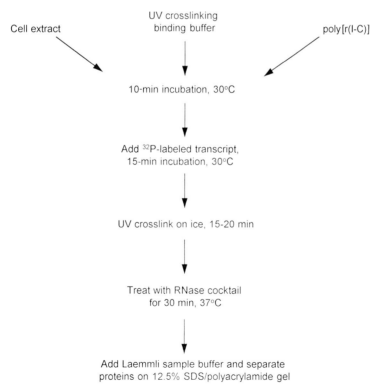

FIG. 3   Flow chart of a UV cross-linking assay.

buffer containing 25 mM KCl, 10 mM MgCl$_2$, 150 mM NaCl, 0.1 mM EDTA, 5 mM HEPES (pH 7.8), and 3.8% (v/v) glycerol. Incubate for 10 min at 37° C. During this incubation, nonspecific and low-affinity RNA-binding proteins bind to the synthetic poly[r(I-C)] RNA.

2. Add 1 μl of ^{32}P-labeled RNA probe suitably diluted to 2–10 pmol/μl (specific activity, 1 $\times$ 10^6 cpm/μl). Incubate for another 15 min at 30° C. During this incubation, RNA-binding proteins that have a detectable affinity for the radiolabeled RNA probe are allowed to bind.

3. Place reaction mixes on ice with open caps and UV cross-link at 254 nm for 10–20 min. This can be done using either a Stratagene (La Jolla, CA) UV-Strata-linker or a hand-held UV lamp. When relatively crude extracts, such as the NP-40 lysis extract, are used, cross-linking times of 15 min have been employed. Using purified poliovirus 3D polymerase for UV cross-linking studies, it was reported that UV irradiation times exceeding 2 min led to detectable degradation of the 3D polymerase protein (31).

4. After UV cross-linking, digest the mixture with an RNase cocktail containing 30 μg of RNase A, 30 U of RNase T1, 3 U of RNase T2, and 7 U of RNase V1. Incubate the reactions for 30 min at 37° C. RNase A cleaves after pyrimidine residues, RNase T1 after guanosine residues, RNase T2 after any residue, and RNase V1 cleaves double-stranded RNA. During this incubation, the free RNA will be digested, and only short stretches of radiolabeled RNA cross-linked to protein(s) will be protected from digestion. Therefore, only the proteins that interacted with the radiolabeled RNA can ultimately be visualized by autoradiography.

5. Using published conditions (32), boil samples in Laemmli sample buffer after RNase digestion and separate on a 12.5% polyacrylamide gel under denaturing conditions. After electrophoresis, dry the gel and subject it to autoradiography.

Proteins that interact with a radiolabeled RNA probe are separated by their molecular weight during electrophoresis and are then visualized by virtue of their labeling with cross-linked [^{32}P]NTP-containing RNA. Using commercially available molecular weight markers or radiolabeled viral protein markers as a reference, the approximate molecular weight of cross-linked proteins can be estimated. This is a major advantage of UV cross-linking assays over RNA electrophoretic mobility shift assays, in which only the formation of an RNA–protein complex, but not the molecular weight of protein(s), can be detected. Ultraviolet cross-linking assays also allow one to determine if several proteins within a complex directly interact with the RNA probe, which is not possible in an RNA electrophoretic mobility shift assay. However, when using UV cross-linking assays, it is not possible to determine whether additional proteins, other than the protein(s) that directly bind to the RNA, are involved in complex formation. For example, a protein that in turn binds to the actual RNA-binding protein could not be detected because no radiolabel would be cross-linked to the protein riding "piggyback." This complex formation would be disrupted when the proteins are boiled and denatured in Laemmli sample buffer.

Competition assays can be performed to determine the specificity of RNA-binding proteins to the radiolabeled RNA probe. Different molar excesses (between 5- and 50-fold) of nonradioactive homologous and heterologous competitor RNA are added to the preincubation reaction described in step 1 above. The differential affinity of an RNA-binding protein for homologous and heterologous competitor RNA allows one to determine whether RNA–protein complexes formed are specific or are fortuitous due to nonspecific RNA–protein interactions.

For investigating the formation of RNA–protein complexes in picornaviruses, it is especially important to include double strand-specific RNase V1 in the RNase cocktail described in step 4 above. Extracts prepared from infected cells contain a significant amount of picornavirus RNA. It is possible that a radiolabeled negative-strand RNA probe hybridizes to the positive-strand RNA endogenous to the extract. This double-stranded RNA is cleaved inefficiently by RNase A, T1, or T2, and it would falsely appear as if an RNA–protein complex had been formed when analyzed by polyacrylamide gel electrophoresis. The same problem can occur if the radiolabeled RNA probe forms secondary structures such as stem–loops, and the double-stranded stems cannot be efficiently digested without inclusion of RNase V1 in the RNase cocktail.

Using cell extracts that have been digested with proteinase K in a UV cross-linking assay is an important negative control experiment to perform. If complex(es) visualized are indeed due to RNA–protein interactions, they should no longer be able to form if proteinase K-digested extract is used. At the end of a UV cross-linking experiment, a polyacrylamide gel can be silver or Coomassie blue stained before drying to ascertain that the proteinase K digest was complete. No proteins should be detectable in the silver-stained gel, and autoradiography of the gel should not yield a signal, as no RNA–protein complexes should have formed.

Immunoprecipitation

After a UV cross-linking assay has been performed, specific antibodies to proteins expected to bind the radiolabeled RNA probe can be used for immunoprecipitation (33). If the protein that is recognized by the specific antibody is indeed a protein that has been cross-linked, it can be immunoprecipitated and visualized by subsequent sodium dodecyl sulfate-polyacrylamide gel electrophoresis (SDS-PAGE). Immunoprecipitations can be used initially to identify proteins binding to the RNA probe. Additionally, if the protein in question has been purified, UV cross-linking assays using the purified protein followed by immunoprecipitation can be used to verify involvement of the protein in RNA binding and complex formation.

An immunoprecipitation protocol employing moderately stringent conditions (34, 35) is described below. This protocol may allow one to detect proteins that are involved in RNA–protein complex formation even if these proteins do not bind directly to RNA, because protein complexes may not be completely disrupted in this protocol.

For example, it is conceivable that a putative poliovirus replication initiation complex contains 3D polymerase but that 3D polymerase itself does not initially interact with the RNA template. By using an anti-3D polymerase antibody in the immunoprecipitation protocol described below, other proteins of this hypothetical replication initiation complex would coimmunoprecipitate together with 3D polymerase. After electrophoresis and autoradiography, only proteins that directly bound to the RNA probe would be visible, but their precipitation with an anti-3D polymerase antibody would be indicative of the 3D polymerase being involved in complex formation. Using antibodies against other poliovirus proteins thought to be involved in, for example, RNA replication initiation, thus enables one to investigate the involvement of viral proteins in complex formation even if these proteins themselves should not bind directly to RNA.

1. After digestion with an RNase cocktail during the UV cross-linking assay, transfer 10 μl (or up to 20 μl, depending on the abundance of the protein in question) to an Eppendorf tube on ice. Add ice-cold extraction buffer to a final volume of 400 μl. Extraction buffer is made up of the following.

Tris (pH 8)	50 mM
EDTA	5 mM
NaCl	150 mM
NP-40	0.5% (v/v)

2. To this mixture, add 20 μl of a 10% suspension of activated *Staphylococcus aureus* cells. Activated *S. aureus* cells are resuspended in 50 mM Tris (pH 7.4), 5 mM EDTA, 150 mM NaCl, 0.5% NP-40 (v/v). Incubate for 10 min on ice. During this 10-min preadsorption, nonspecific *S. aureus*–protein complexes are allowed to form.

3. Spin the mixture for 1 min at 16,000 g at 4° C. This will pellet nonspecific *S. aureus*–protein complexes. Transfer the supernatant to a new tube.

4. Add a suitable titer of antiserum or add a suitable dilution of purified antibody to the tube. Incubate for 1 hr on ice. During this incubation, specific antibodies will bind their cognate proteins if they are present in the reaction mix.

5. Add 40 μl of a 10% suspension of activated *S. aureus* cells. Incubate for 15–20 min on ice. During this incubation, the Fc portion of antibodies will bind to Fc receptor present on the *S. aureus* surface. Thus, UV cross-linked proteins, which were recognized by a specific antibody, can be pelleted by subsequent centrifugation.

6. Centrifuge the mixture for 30 sec at 16,000 g and wash the immunoprecipitates three times in 500 μl of IP wash buffer, which contains the following.

Tris (pH 7.4)	50 mM
NaCl	500 mM
EDTA	5 mM
NP-40	1% (v/v)
Sucrose	5% (w/v)

7. After the last wash, resuspend samples in 50 μl of Laemmli sample buffer. Boil the samples for 3 min and spin for 2 min at 16,000 g. Supernatants (containing putative proteins of interest) are then subjected to electrophoresis on a 12.5% SDS-polyacrylamide gel.

After establishing the specific formation of RNA–protein complexes using a viral RNA probe and extract of interest, it is important to define the *cis*-acting sequences required for complex formation. Binding sequences are typically roughly determined by performing deletion analyses. Once *cis*-acting sequences have been localized to a small fragment of RNA, chemical modification assays, as described previously for studies on spliceosome components (36), can be performed. The rationale of the chemical modification assay is that only RNA sequences that are not bound by proteins or are not involved in secondary structure can be chemically modified. Additional methods to determine binding sites of proteins are RNA footprint analysis (37) and primer extension inhibition footprint analysis (38), the so-called toeprint analysis, which allows one to determine the 3′ border of a protein-binding site. Once a binding site has been delineated, it can be systematically mutated by introducing point mutations to define a binding motif. After identifying proteins and *cis*-acting sequences involved in complex formation *in vitro,* a logical extension of the study is to compare and correlate these findings to *in vivo* events by performing genetic tests. Such *in vitro* biochemical analyses combined with molecular genetic approaches continue to be helpful in defining important RNA–protein interactions as they occur during picornavirus replication, translation, and packaging.

Acknowledgments

We thank Aurelia Haller and Steve Todd for comments on the manuscript. Work on poliovirus RNA–protein interactions in the laboratory of B.L.S. is supported by Public Health Service Grant AI 22693 from the National Institutes of Health. H.H.R. is supported by a postdoctoral fellowship from the American Cancer Society (PF-3756).

References

1. J. F. Enders, T. H. Weller, and F. C. Robbins, *Science* **109,** 85 (1949).
2. V. R. Racaniello and D. Baltimore, *Science* **214,** 916 (1981).
3. B. L. Semler, A. J. Dorner, and E. Wimmer, *Nucleic Acids Res.* **12,** 5123 (1984).
4. S. Mizutani and R. J. Colonno, *J. Virol.* **56,** 628 (1985).
5. S. van der Werf, J. Bradley, E. Wimmer, F. W. Studier, and J. J. Dunn, *Proc. Natl. Acad. Sci. USA* **83,** 2330 (1986).
6. J. Pelletier and N. Sonenberg, *Nature (London)* **334,** 320 (1988).
7. A. A. Haller and B. L. Semler, *J. Virol.* **66,** 5075 (1992).
8. S. K. Jang, H. G. Kräusslich, M. J. H. Nicklin, G. M. Duke, A. C. Palmenberg, and E. Wimmer, *J. Virol.* **62,** 2636 (1988).

9. S. K. Jang, M. V. Davies, R. J. Kaufman, and E. Wimmer, *J. Virol.* **63,** 1651 (1989).

10. A. Kaminski, T. Howell, and R. J. Jackson, *EMBO J.* **9,** 3753 (1990).

11. R. Kühn, N. Luz, and E. Beck, *J. Virol.* **64,** 4625 (1990).

12. M. J. Glass, X. Y. Jia, and D. F. Summers, *Virology* **193,** 842 (1993).

13. M. A. Lawson and B. L. Semler, *Curr. Top. Microbiol. Immunol.* **161,** 49 (1990).

13a. P. G. Dewalt and B. L. Semler, *J. Virol.* **61,** 2162 (1987).

14. D. Etchison, S. C. Milburn, I. Edery, N. Sonenberg, and J. W. B. Hershey, *J. Biol. Chem.* **257,** 14806 (1982).

15. H. G. Kräusslich, M. J. H. Nicklin, H. Toyoda, D. Etchison, and E. Wimmer, *J. Virol.* **61,** 2711 (1987).

16. O. C. Richards and E. Ehrenfeld, *Curr. Top. Microbiol. Immunol.* **161,** 90 (1990).

17. M. M. Konarska and P. A. Sharp, *Cell (Cambridge, Mass.)* **46,** 845 (1986).

18. K. Meerovitch, J. Pelletier, and N. Sonenberg, *Genes Dev.* **3,** 1026 (1989).

19. R. M. del Angel, A. G. Papavassiliou, C. Fernández-Tomás, S. J. Silverstein, and V. R. Raciniello, *Proc. Natl. Acad. Sci. USA* **86,** 8299 (1989).

20. L. Najita and P. Sarnow, *Proc. Natl. Acad. Sci. USA* **87,** 5846 (1990).

21. S. L. Dildine and B. L. Semler, *J. Virol.* **66,** 4364 (1992).

22. R. Andino, G. E. Rieckhof, and D. Baltimore, *Cell (Cambridge, Mass.)* **63,** 369 (1990).

23. B. A. Brown and E. Ehrenfeld, *Virology* **97,** 396 (1979).

24. T. Helentjaris, E. Ehrenfeld, M. L. Brown-Luedi, and J. W. B. Hershey, *J. Biol. Chem.* **254,** 10973 (1979).

25. L. A. Caliguiri and I. Tamm, *Virology* **42,** 112 (1970).

26. P. Sarnow, *Proc. Natl. Acad. Sci. USA* **86,** 5795 (1989).

27. J. Welsh and C. R. Cantor, *Trends Biochem. Sci.* **9,** 505 (1984).

28. S. K. Jang and E. Wimmer, *Genes Dev.* **4,** 1560 (1990).

29. N. Luz and E. Beck, *J. Virol.* **65,** 6486 (1991).

30. T. V. Pestova, C. U. T. Hellen, and E. Wimmer, *J. Virol.* **65,** 6194 (1991).

31. O. C. Richards, P. Yu, K. L. Neufeld, and E. Ehrenfeld, *J. Biol. Chem.* **267,** 17141 (1992).

32. U. K. Laemmli, *Nature (London)* **227,** 680 (1970).

33. M. D. Kattoura, L. L. Clapp, and J. T. Patton, *Virology* **191,** 698 (1992).

34. D. I. H. Linzer and A. J. Levine, *Cell (Cambridge, Mass.)* **17,** 43 (1979).

35. B. L. Semler, C. W. Anderson, R. Hanecak, L. F. Dorner, and E. Wimmer, *Cell (Cambridge, Mass.)* **28,** 405 (1982).

36. D. L. Black and A. L. Pinto, *Mol. Cell. Biol.* **9,** 3350 (1989).

37. J. R. Patton and C. B. Chae, *Proc. Natl. Acad. Sci. USA* **82,** 8414 (1985).

38. D. Hartz, D. S. McPheeters, R. Traut, and L. Gold, *in* "Methods in Enzymology," Vol. 164, p. 419. Academic Press, Orlando, Florida, 1988.

[14] Construction and Characterization of Rotavirus Reassortants

Dayue Chen and Robert F. Ramig

Introduction

Rotaviruses, members of virus family Reoviridae, contain a genome of 11 molecules, or segments, of double-stranded RNA (dsRNA). As for other RNA viruses with segmented genomes, when 2 different rotaviruses infect the same cell, the 11 segments may be assorted to yield 2048 different possible genome segment constellations among the progeny virus. This assortment of genome segments during viral replication is known as *reassortment,* and the progeny viruses with segment constellations different from those of the parental viruses are called reassortants. Among segmented viruses, reassortment is the only known mechanism of genetic recombination. The genetic recombination achieved by reassortment, which does not involve the breakage and reformation of covalent bonds in nucleic acids, is characterized by its high-frequency occurrence. Readers are referred to a review for details on reassortment in rotaviruses and other members of the Reoviridae (1).

Reassortment has played a significant role in our understanding of the rotaviruses (1). Through the analysis of reassortants, phenotypes associated with one of the viral parents can be assigned to a specific genome segment, or combination of segments, from that parent. As an example, reassortants have been used to identify genome segment 4, and its protein product VP4, as the determinant of virus growth in cultured cells (2), cell tropism *in vitro* (3), virulence in the mouse model (4), and as the segment containing the group A temperature-sensitive mutants (5). We describe here the construction and characterization of rotavirus reassortants in a step-by-step manner and discuss considerations to be taken into account when reassortment is used as a tool to study rotavirus genetics.

Considerations in Choosing Parental Viruses

The first step in construction of reassortants is the choice of appropriate parental viruses to allow mapping the phenotype of interest. There are two major considerations. First, the two parental viruses should have contrasting expression of the phenotype of interest, so that phenotype of reassortants can be easily scored. Second, the parental viruses should have distinct migration rates of dsRNA segments, or electropherotype, to allow easy determination of the parental origin of each genome segment in reassortants by polyacrylamide gel electrophoresis (PAGE). The ideal parental viruses have sharply contrasting phenotype and distinct electropherotype to make scoring of both phenotype and parental origin of genome segments as easy as possible.

Mixed Infection for Generation of Reassortants

The prerequisite for the efficient generation of reassortants is mixed infection of each cell with both parental viruses. The percentage of cells mixedly infected with two different parental viruses is given by the expression $(1 - e^{-mA})(1 - e^{-mB}) \times 100$, where mA and mB are the multiplicity of infection (MOI) of parents A and B, respectively. In practice, an MOI of 10 plaque-forming units (PFU) of each parental virus theoretically guarantees that 100% of cells are mixedly infected.

Approximately $5-10 \times 10^4$ MA 104 (rhesus monkey kidney) cells in 1 ml of medium 199 containing 5% (v/v) fetal bovine serum (FBS) are seeded into 1-dram, flat-bottom vials that are loosely capped and incubated at 37°C overnight in a 5% CO_2 incubator. The medium is aspirated from the resulting confluent monolayer and replaced with 1.0 ml of serum-free medium 199, followed by a further 12- to 18-hr incubation at 37°C to allow metabolism of the remaining FBS components that inhibit growth of rotaviruses. The number of cells in representative vials can be determined by trypsinization and counting in a hemacytometer, although the number is generally within the range of $1-2 \times 10^5$ cells/vial. The serum-free monolayers of MA 104 cells are then ready for infection.

Immediately prior to infection the virus inoculum is prepared as follows (assumes 1×10^5 cells/vial). Each parental virus is prepared at a concentration of 2×10^7 pfu/ml by dilution into serum-free medium 199. Virus mixes are then prepared by mixing 0.1 ml each of parents A and B in a small tube. To this tube is then added 0.2 ml of trypsin (10 μg/ml) prepared in serum-free medium 199, and the mixture is incubated for 30 min at 37°C to activate fully the infectivity of the inoculum. The serum-free cell monolayers are then infected by adding 0.2 ml of the mixture of the parental viruses, prepared as above. This protocol results in an MOI of 10 PFU for each parent. The infected monolayer is incubated for 1 hr at 37°C to allow virus adsorption. After adsorption, the inoculum is aspirated from each vial, taking care to avoid cross-contamination, and the cells are fed with serum-free medium 199 containing trypsin (1 μg/ml) (1 ml/vial) and maintained loosely capped in 5% CO_2 at 37°C until complete cytopathic effect (CPE). The infection is terminated by tightly capping each vial and freezing at -20°C.

Isolation and Selection of Reassortant Progeny Viruses

The next step is to isolate individual progeny clones from the yield of the mixed infection. This is done by plaque isolation. For plaque assays, MA 104 cells ($\sim$1 $\times$

10^5 cells/ml) in medium 199 containing 5% FBS are seeded into the wells of six-well plates, 2 ml/well. The plates are incubated in 5% CO_2 at 37°C until the cells form a confluent monolayer. Once confluence is attained, the medium is removed from the wells and replaced with serum-free medium 199 (2 ml/well) and incubated for 12–18 additional hours to allow metabolism of the residual FBS components inhibitory to rotavirus plaque formation.

Immediately before the plaque assay, the frozen yield of the mixed infection is thawed and briefly sonicated to disrupt virus aggregates. Serial 10-fold dilutions of the infected cell lysate are prepared using serum-free medium 199 as diluent. The serum-deprived confluent MA 104 cell monolayers in six-well plates are inoculated (0.1 ml/well) with appropriate dilutions of the mixed infection yield. The plates are incubated for 60 min at 37°C to allow virus adsorption. After adsorption, the monolayers are overlaid with serum-free medium 199 (3 ml/well) containing 1% (w/v) Bacto-agar (Difco, Detroit, MI), DEAE-dextran (25 μg/ml), and pancreatin (0.27 IU/ml) (Oxoid). The plates are then incubated in 5% CO_2 at 37°C for 4–6 days (depending on the time for plaque formation with the viruses used). The plates are then overlaid with serum-free medium 199 (1 ml/well) containing 1% (w/v) agar and neutral red (25 μg/ml), and returned to the incubator for 18–24 hr. Progeny plaques are usually picked 2 days after the second overlay, to minimize contamination with late-forming plaques. Plaques are picked from the dilution of virus yielding well-separated plaques, using a Pasteur pipette to remove a plug of agar and the infected cells. The picked plaque is suspended in 2 ml of serum-free medium 199 containing trypsin (1 μg/ml) and stored at −20°C.

In practice, specific assay conditions are often used to select reassortants with desired genome segment constellations. The most common selective pressure is neutralizing antibody, either polyclonal or monoclonal. The prerequisite for using antibody as selective pressure is that it neutralizes only one of the parental viruses used in the mixed infection. The antibody can specifically and selectively neutralize the progeny virus containing the genome segment(s) encoding the proteins via which the antibodies neutralize. Alternatively, if one or both of the parental viruses are temperature-sensitive mutants. the nonpermissive temperature can be used to select reassortants with desired phenotype.

When antibodies are used to select reassortants with a specific genome segment constellation (genotype), one concern is phenotypic mixing. It has been reported that 40–77% of the progeny virus from cells mixedly infected with two different rotaviruses at high MOI contained phenotypically mixed outer capsids, as determined by plaque reduction neutralization assay using VP7-specific monoclonal antibodies (6). Because neutralization of rotaviruses is determined by the composition of the outer capsid (phenotype) rather than by the genotype, direct interaction of phenotypically mixed progeny virions with neutralizing antibody prior to infection would provide an ineffective genotypic selection. There are two methods to overcome the problem of phenotypic mixing. First, instead of reacting antibody with progeny viruses prior to infection, the neutralizing antibody is added to the agar overlay so that selection occurs during the cell-to-cell spread of plaque formation. Alternatively, progeny virus from the mixed infection is amplified one life cycle at low MOI (0.1 PFU/cell), and then subjected to antibody selection by direct reaction of antibody and virus. Both of these methods serve to rectify phenotype and genotype by virus replication under conditions of low MOI.

Genotypic Characterization of Reassortant Clones

Electrophoretic polymorphism of genome segments from different rotavirus strains is routinely used as the marker to determine the genotype of reassortant viruses (7, 8). The first step of genotypic characterization of reassortant viruses is to amplify the picked plaques to obtain viral dsRNA. Depending on the method of visualization, there are two slightly different ways to amplify viruses. When autoradiography is the chosen method for visualization, genomic dsRNA is metabolically labeled with $H_3{}^{32}PO_4$. Serum-depleted MA 104 cell monolayers in 35-mm plates are infected with half of the picked plaque suspension (1 ml). After a 1-hr incubation at 37°C, 1 ml of serum-free medium 199 containing actinomycin D (10 μg/ml), trypsin (1 μg/ml), and $H_3{}^{32}PO_4$ (100 μCi/ml) is added to each plate. The plates are then incubated at 37°C until CPE is complete. If silver staining is the method for visualization, viruses are amplified as described above except that the 1 ml of serum-free medium 199 added to plates after adsorption contains no radioactive label.

When the amplifying infection is complete, the cells are scraped off the plates, using a sterile rubber policeman, and collected by low-speed centrifugation (1000 rpm) for 10 min at 4°C. The infected cell pellet is retained and lysed in 1 ml of 10 mM Tris buffer (pH 7.4) containing 0.5% (v/v) Nonidet P-40 (NP-40), 150 mM NaCl, and 5 mM MgCl$_2$. The resulting cell lysate is extracted once with an equal volume of H_2O-saturated phenol in the case of visualization by autoradiography, or twice in the case of silver staining. After centrifugation (2500 rpm) for 10 min at 4°C, the aqueous phase is transferred to a clean tube. The viral dsRNAs are adjusted to 250 mM NaCl and precipitated by addition of 2.5 vol of absolute ethanol overnight at −20°C. The precipitated RNAs are collected by centrifugation (3000 rpm) for 1 hr at 4°C. The RNA pellet is retained, dried in a lyophilizer, and dissolved in 50 μl of sample buffer [120 mM Tris-HCl (pH 6.8), 20% (v/v) glycerol, 2% sodium dodecyl sulfate (SDS), 10% (v/v) 2-mercaptoethanol]. The samples are then subjected to electrophoresis in 12% (w/v) polyacrylamide gels prepared according to Laemmli (9). Gels 20 × 23 cm × 0.75 mm thick are run at a constant current of 20 mA/gel for 18 to 20 hr. The dsRNA bands are visualized either by autoradiography or by silver stain. Viral dsRNAs of both parental viruses are always included in the same gel to serve as migrational markers for genotypic scoring of progeny viruses. In general, half of the sample is sufficient for visualization by either method, leaving addi-

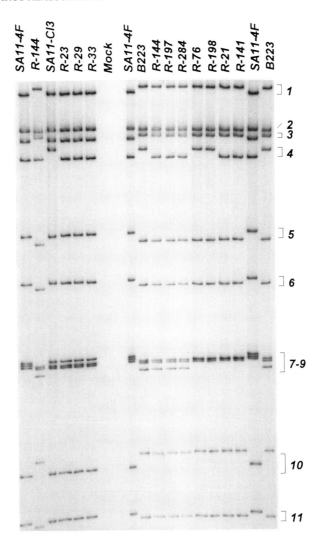

Fig. 1   Genotypic scoring of rotavirus reassortants as visualized by ^{32}P labeling and auto-radiography. Double-stranded RNAs of parental and reassortant viruses were prepared and subjected to PAGE as described. Reassortant viruses are indicated as R-(number). The genome segments are indicated on the right. In the original autoradiograph the distance from segment 1 to segment 11 or R-144 was 23.5 cm. The genotypes of the reassortants deduced from this gel are shown in Table I.

tional sample for confirmatory gels if necessary. Figure 1 shows a typical example of a genotypic scoring gel visualized by ^{32}P labeling; the deduced origin of each genome segment of the progeny viruses is summarized in Table I.

TABLE I   Segregation Analysis of SA11-4F/B223 and Other Reassortants

Viruses	Parental origin of indicated segment[a]											Plaque size (+) protease mean (SD)	Plaque formation (−) protease
	1	2	3	4	5	6	7	8	9	10	11		
Parental													
SA11-4F[b]	F	F	F	F	F	F	F	F	F	F	F	2.8 (0.6)	Yes
B223[b]	B	B	B	B	B	B	B	B	B	B	B	0.8 (0.1)	No
SA11-Cl3[b]	S	S	S	S	S	S	S	S	S	S	S	1.1 (0.2)	No
SA11-4F/B223 reassortants													
R-21[b]	B	B	B	F	B	B	B	B	F	B	B	3.5 (0.6)	Yes
R-141	B	B	B	F	B	B	B	B	F	B	B	3.5 (0.6)	Yes
R-491	F	F	F	B	F	F	F	F	B	F	F	1.0 (0.2)	No
R-144[b]	B	B	B	F	B	B	B	B	B	B	B	0.9 (0.1)	No
R-197[b]	B	B	B	F	B	B	B	B	B	B	B	0.9 (0.1)	No
R-284[b]	B	B	B	F	B	B	B	B	B	B	B	1.0 (0.2)	No
R-004	F	F	F	B	F	F	F	F	F	F	F	1.4 (0.4)	No
R-76[b]	B	B	B	B	B	B	B	B	F	B	B	1.5 (0.6)	No
R-198[b]	B	B	B	B	B	B	B	B	F	B	B	1.5 (0.4)	No
R-179	F	F	F	F	F	F	F	F	B	F	F	1.0 (0.2)	No
R-106	B	B	B	F	B	F	B	B	F	B	B	3.0 (0.4)	Yes
R-497	F	F	F	B	F	B	F	F	B	F	F	1.2 (0.1)	No
R-600	B	B	B	B	B	F	B	B	B	B	B	1.1 (0.2)	No
R-51	F	F	F	F	F	B	F	F	F	F	F	2.8 (0.6)	Yes
R-202	F	F	F	F	F	B	F	F	F	F	F	3.2 (0.6)	Yes
Other reassortants[c]													
R-23[b]	S	S	S	F	S	S	S	S	S	S	S	3.2 (0.5)	Yes
R-29[b]	S	S	S	F	S	S	S	S	S	S	S	3.4 (1.0)	Yes
R-33[b]	S	S	S	F	S	S	S	S	S	S	S	3.3 (1.0)	Yes
R-924	F	F	F	S	F	F	F	F	F	F	F	1.3 (0.3)	No

[a] Parental origin of genome segments: F, SA11-4F; B, B223.

[b] Genotypic scoring of clone shown in Fig. 1.

[c] Other reassortants were derived from a mixed infection of SA11-4F reassortant R-144 and SA11-Cl3 (R-23, -29, and -33), or R-004 and SA11-Cl3 (R-924).

In some cases, not all 11 genome segments between two parental viruses can be separated by SDS-PAGE. In these cases, DNA–RNA or RNA–RNA hybridization can be used to determine the complete genotype of the progeny viruses or to determine parental origin of the segment(s) not scoreable by PAGE. When partial or full-length cDNA of the genome segments is available, ^{32}P-labeled cDNA can be used as probe in a dot hybridization assay to determine the parental origin of this particular genome segment of the reassortant viruses. The stringency of the hybridization assay should be predetermined using the two parental viruses to ensure the probe hybridizes only with the parental virus from which the cDNA was derived. When cDNA clones are not available, ^{32}P-labeled mRNA generated by *in vitro* transcription can be used as probe to determine the parental origin of genomic segments of reassortants. Purified single-shelled particles at a concentration of 100 μg/ml are generally used to generate ^{32}P-labeled mRNA in a standard protocol (10). For detailed information on using mRNA as the probe in hybridization assay, the reader is referred to Flores *et al.* (11).

After initial screening, two additional rounds of plaque purification are performed to ensure genetic homogeneity of the progeny reassortants with genotypes of interest. Standard plaque assays are performed using serial 10-fold dilutions of the original picked plaque suspension, as described above. Two to five plaques are picked for each reassortant, and each plaque is suspended in 2 ml of serum-free medium 199. These second round (2°) plaques are subjected to a third round of plaque purification, as above. Two plaques are picked for each 2° plaque, resulting in a total of 4–10 triply plaque-purified subclones (3°) of each original reassortant of interest. The genotypes of the 3° plaques are confirmed by SDS-PAGE and/or hybridization as described above. The additional rounds of plaque purification are performed to ensure the genetic homogeneity of the resulting 3° plaques. This is necessary to prevent the use of mixed populations of virus in the subsequent experiments.

The triply plaque-purified clones of the reassortants of interest must then be amplified to produce high-titer stocks for subsequent experiments. To make first-passage virus stock, 1 ml of 3° plaque suspension is activated by addition of 5.0 μl of trypsin (1 mg/ml) (5 μg/ml final concentration) and incubated for 30 min at 37°C to activate the virus. A confluent, serum-free MA 104 cell monolayer in a 150-cm^2 flask is then infected with the trypsin-activated, plaque-purified virus diluted to a total volume of 5.0 ml with medium 199. Following a 1-hr incubation at 37°C, the inoculum is removed and the monolayer is fed with serum-free medium 199 containing trypsin (1.0 μg/ml) (25 ml/flask). The infected flasks are incubated at 37°C until complete CPE, subjected to three cycles of freezing and thawing, and aliquoted in vials for storage at -20°C. The first-passage virus is titered in a standard plaque assay, and then used to infect MA 104 cell monolayers at an MOI of 0.1–0.5 PFU/cell to make a second-passage virus stock. The remaining first-passage virus stock is retained and frozen at -20°C to provide material for production of additional second-passage virus stocks. Second-passage viruses are used to characterize the phenotypes of the reassortant viruses. It is generally recommended that the genotype at every passage be verified to ensure the virus population remains genotypically homogeneous. Figure 2 schematically illustrates the steps from mixed infection to acquisition of characterized reassortants.

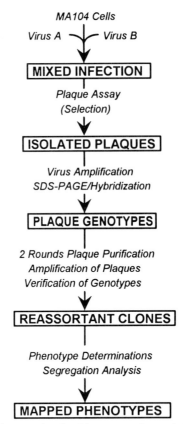

FIG. 2   Flow diagram of the steps involved in construction and characterization of rotavirus reassortants.

Phenotypic Characterization of Reassortant Clones

Phenotypic characterization of genotypically characterized reassortant viruses is the last step in the segregation analysis that allows a phenotype(s) to be assigned (mapped) to a specific genome segment(s) and its encoded protein(s). Depending on the phenotype(s) under study, scoring phenotypes of reassortant viruses can be carried out either *in vitro* by observing phenotypic expression in infected cells, or *in vivo* by observing phenotypic expression in infected animals. In either case, the two parental viruses are always included in the assay as controls. The following example illustrates a typical segregation analysis in which two biological phenotypes (plaque size and ability to plaque without protease) of the SA11-4F variant are successfully assigned to specific genome segment(s).

SA11-4F forms extremely large plaques in the presence of protease, and forms small, clear plaques in the absence of protease. In contrast, virus strains B223 and SA11-CI3 form small plaques in the presence of protease and fail to form plaques in the absence of protease. The contrast in these phenotypes (Table I) makes SA11-4F and B223 good candidate viruses with which to investigate the genotypic determinants of plaque size and ability to plaque without protease. Furthermore, SA11-4F and B223 differ in the mobility of all 11 genome segments (Fig. 1), so that parental origin of each genome segment in reassortants can be determined by simple PAGE.

With the aim of performing segregation analysis to identify the SA11-4F genome segment(s) responsible for large plaque size and formation of plaques in the absence of protease, a battery of SA11-4F/B223 reassortants was constructed and characterized using the methods outlined above. Table I shows the genotypes and phenotypes expressed by the resulting reassortants. Segregation analysis revealed that formation of large plaques in the presence of protease and formation of small, clear plaques in the absence of protease, like that characteristic of the SA11-4F parent, required the presence of both genome segments 4 and 9 from the SA11-4F parent (reassortants R-21, R-141, R-51, and R-202). Neither segment 4 (reassortants R-1144, R-284, R-179, and R-497) nor segment 9 (R-004, R-76, and R-198) was sufficient to confer the phenotypes. Because VP4, the product of segment 4, is known to be the protease-sensitive protein of the virus, this result was unexpected. To examine a possible role of genetic context on expression of these phenotypes, additional reassortants were made between SA11-4F/B233 reassortants as the first parent and SA11-CI3 as the second parent (Table I, other reassortants). In these reassortants, the plaque size phenotype and plaque formation in the absence of protease phenotype segregated with segment 4 alone.

Both sets of reassortants shown in Table I gave clear results to the genome segments that encoded the phenotypes under examination. The surprising result was that the two different sets of reassortants gave different answers for the determinants of the phenotypes of interest; in one case two SA11-4F segments were required, in the other case a single SA11-4F segment was sufficient to transfer the phenotype. The striking take-home lesson from this series of experiment was that the recipient genetic background, onto which donor genome segments are moved by reassortment, can affect phenotypic expression of the donor segment(s) (12). This result was subsequently confirmed, and interactions between structural proteins of heterologous parental origin were shown to be the basis of unexpected phenotypic expression in some reassortants (13).

General Considerations

There are several general considerations that should be taken into account when reassortants are used to map phenotypes of dsRNA viruses. These considerations apply not only to rotaviruses, but also to other segmented genome viruses.

Independent Isolates of Reassortant Segment Constellations

RNA viruses have high spontaneous mutation rates. Although the precise rate of spontaneous mutation in dsRNA viruses has not been systematically investigated, the high frequency of reversion of simian rotavirus *ts* (temperature-sensitive) mutants suggests that it is high. Spontaneous mutation to *ts* phenotype has been observed in rotavirus reassortants derived from two non-*ts* parental viruses. However, in this case, other independently derived reassortants of identical genome segment constellation were not *ts* (D. Chen and R. F. Ramig, unpublished results, 1993). In other studies, the unusual phenotype of a reassortant was confirmed by examining independent isolates (12, 13). This example illustrates the possibility that spontaneous mutation can affect the results of segregation analysis, and emphasizes the importance of constructing and examining the phenotype of several reassortants with identical genome segment constellation. This is particularly true in the case of the reassortant constellation(s) critical to assignment of a phenotype to a segment(s). To isolate truly independent, and not sister reassortants, each duplicate reassortant should be isolated in an independent process beginning with independent mixed infections of the parental viruses.

Reassortants with Reciprocal Genome Segment Constellations

If possible, reassortants with reciprocal genome segment constellations should be isolated and analyzed in phenotypic mapping studies. Results from phenotypic mapping studies that lack reciprocal reassortant constellations should be interpreted with caution, because segments not segregated among the reassortants could mask determinants of, or modifiers of, the phenotype under study. For example, in studies to map the human rotavirus protein responsible for inducing neutralization antibodies (14, 15), none of the reassortants contained segment 4 (VP4) from the human parent and genome segment 9 (VP7), which was segregated, was identified as the determinant of neutralization. Subsequent studies, using reciprocal reassortant constellations, revealed that VP4 was also a determinant of neutralization (16, 17). Thus, ideally, reciprocal reassortant constellations should be examined, and if this is not possible the data should be interpreted cautiously.

Effects of Recipient Genetic Background on Phenotypic Expression

Most phenotypic mapping studies use reassortants derived from a single pair of parental virus strains. However, it has been demonstrated that a phenotype can segregate with different reassortant genome segment constellations when different pairs of parental viruses are used to generate the reassortants (12, 13). This is also shown in Table I, where the phenotypes segregate with segments 4 and 9 in SA11-4F/B223 reassortants, but with segment 4 alone in SA11-4F/SA11-CI3 reassortants. These results suggested that the genetic background of the recipient virus was able to affect phenotypic expression of the donor segments moved onto that background (14). It was proposed that, in some parental virus pairs, the recipient genetic background affected phenotypic expression through altered protein–protein interactions among proteins of heterologous origin. Evidence for protein–protein interactions being at least one of the causes of genetic background effects has been presented (13).

Host Cell Effects on Reassortant Formation and Selection

It has been reported that reassortants with different constellations of genome segments were isolated when the yield of a single mixed infection was plated on different monkey kidney cell lines (18). Although many of the constellations were isolated from assays on both cell lines, many other constellations were isolated from only one of the cell lines. Thus, if the desired reassortant constellations of genome segments are not obtained using the protocols given here, alternative cell lines should be considered.

Virus Strain Effects on Reassortant Formation and Selection

Rotaviruses have been divided into genogroups based on the degree of cross-hybridization between strains (19). Reassortment between virus strains assigned to the same genogroup appears to occur at higher frequency than between the more distantly related viruses from different genogroups (1). Thus, if the parental viruses for which reassortants are desired belong to different genogroups, selection of reassortants should be possible but may be a tedious process.

General Applicability of Protocols

The protocols presented here are generally applicable to members of other genera within the Reoviridae, among which the general properties of reassortment are similar to those of rotaviruses (1). Indeed, reassortment has also been a major tool in the examination of viruses belonging to the genera Orthoreovirus, Orbivirus, and Coltivirus (20–24). The major adjustments that must be made to the protocols given here involve the use of different host cell lines, and the considerations related to growth of those cells and plaque assays using those cells as indicator monolayers.

References

1. R. F. Ramig and R. L. Ward, *Adv. Virus Res.* **39,** 163 (1991).
2. H. B. Greenberg, A. R. Kalica, R. W. Wyatt, R. W. Jones, A. Z. Kapikian, and R. M. Chanock, *Proc. Natl. Acad. Sci. USA* **78,** 420 (1981).
3. R. F. Ramig and K. L. Galle, *J. Virol.* **64,** 1044 (1990).
4. P. A. Offit, G. Blavat, H. B. Greenberg, and H. F. Clark, *J. Virol.* **57,** 46 (1986).
5. J. L. Gombold and R. F. Ramig, *Virology* **161,** 463 (1987).
6. R. L. Ward, D. R. Knowlton, and H. B. Greenberg, *J. Virol.* **62,** 4385 (1988).
7. A. R. Kalica, M. M. Sereno, R. G. Wyatt, C. A. Mebus, D. H. Van Kirk, R. M. Chanock, and A. Z. Kapikian, *Virology* **74,** 86 (1976).
8. R. F. Ramig, R. K. Cross, and B. N. Fields, *J. Virol.* **22,** 726 (1977).
9. U. K. Laemmli, *Nature (London)* **227,** 680 (1970).
10. J. Cohen, J. Laporte, A Charpilienne, and R. Scherrer, *Arch. Virol.* **60,** 177 (1979).
11. J. Flores, H. B. Greenberg, J. Myslinske, A. R. Kalica, R. G. Wyatt, A. Z. Kapikian, and R. M. Chanock, *Virology* **121,** 288 (1982).
12. D. Chen, J. W. Burns, M. K. Estes, and R. F. Ramig, *Proc. Natl. Acad. Sci. USA* **86,** 3743 (1989).
13. D. Chen, M. K. Estes, and R. F. Ramig, *J. Virol.* **66,** 432 (1992).
14. A. R. Kalica, H. B. Greenberg, R. G. Wyatt, J. Flores, M. M. Sereno, A. Z. Kapikian, and R. M. Chanock, *Virology* **112,** 385 (1981).
15. H. B. Greenberg, J. Flores, A. R. Kalica, R. G. Wyatt, and R. Jones, *J. Gen. Virol.* **64,** 313 (1983).
16. Y. Hoshino, M. M. Sereno, K. Midthun, J. Flores, A. Z. Kapikian, and R. M. Chanock, *Proc. Natl. Acad. Sci. USA* **82,** 8701 (1985).
17. P. A. Offit and G. Blavat, *J. Virol.* **57,** 376 (1986).
18. A. Graham, G. Kudesia, A. M. Allen, and U. Desselberger, *J. Gen. Virol.* **68,** 115 (1987).
19. O. Nakagomi, T. Nakagomi, K. Akatani, and N. Ikegami, *Mol. Cell. Probes* **3,** 251 (1989).
20. L. A. Schiff and B. N. Fields, *in* "Virology" (B. N. Fields, D. M. Knipe, R. F. Chanock, M. S. Hirsch, J. L. Melnick, T. P. Monath, and B. Rolzman, eds.), p. 1275. Raven Press, New York, 1990.
21. D. L. Knudson and T. P. Monath, *in* "Virology" (B. N. Fields, D. M. Knipe, R. F. Chanock, M. S. Hirsch, J. L. Melnick, T. P. Monath, and B. Rolzman, eds.), p. 1405. Raven Press, New York, 1990.
22. H. A. Gonzalez, "The genetic relatedness of Orbiviruses: RNA–RNA Blot Hybridization and *in Vitro* Reassortment," Ph.D. Thesis, Yale University, New Haven, Connecticut, 1987.
23. R. F. Ramig, C. Garrison, D. Chen, and D. Bell-Robinson, *J. Gen. Virol.* **70,** 2595 (1989).
24. A. El-Hussein, R. F. Ramig, F. R. Holbrook, and B. J. Beaty, *J. Gen. Virol.* **70,** 3355 (1989).

[15] Dynamics of Viral mRNA Translation: Identification of Ribosome Pause Sites by Primer Extension Inhibition

Charles E. Samuel and James P. Doohan

Introduction

Ribosome activity can vary significantly along the length of a messenger mRNA. As a consequence, different mRNAs are often translated with widely different efficiencies (1,2). The molecular bases of translational control include mechanisms that involve a differential interaction of ribosomes and associated factors of the host translational machinery with mRNA. One approach for analyzing the dynamic interaction of ribosomes with mRNAs during the course of translation involves the use of a sensitive ribosome protection assay to monitor ribosome movement (3,4). The results of such analyses reveal that the distribution of translating ribosomes on a given mRNA indeed is not necessarily uniform (3–7).

The concept of primer extension inhibition is the basis of the ribosome protection assay described here. This assay has the capacity to detect ribosome pause sites, that is, those regions on mRNA where ribosomes spend a disproportionately longer period of time during steady state translation. Regions of a mRNA involved in slow steps of translation are protected from nuclease degradation at a greater frequency than other regions. Thus, slow steps of translation can be identified by mapping the positions of the ribosome-protected fragments, using a primer extension inhibition assay. Slow steps of translation that have been identified by this assay include initiation and termination (3,4,7), as well as elongation arrests mediated by overlapping open reading frames (ORFs) of polycistronic mRNAs (4,7), translational frameshifting from one ORF to another (6), and signal recognition particle interactions during targeting of nascent secretory proteins to the endoplasmic reticulum (3,5).

We describe herein procedures for analyzing the dynamics of ribosome interaction with mRNA during the translation process, both *in vitro* in cell-free protein synthesizing systems (3–7) and *in vivo* in virus-infected or vector-transfected cells (7). The procedures are illustrated with two reovirus mRNAs, the polycistronic s1 mRNA and the monocistronic s4 mRNA. The distribution of ribosomes on these mRNAs was determined using the primer extension inhibition assay and cDNA clones of the reovirus S1 and S4 genes.

Strategy

The strategy (4,7) for mapping the distribution of translating ribosomes on an mRNA such as the reovirus s1 or s4 mRNAs is summarized in Fig. 1. Briefly, the assay involves the addition of cycloheximide to intact cells or cell-free protein synthesizing systems to inhibit elongation and subsequently "freeze" translating ribosomes on mRNA. A mixture of ribonucleases is then used to trim away those fragments of the mRNA not protected by ribosomes. Ribosomes typically protect approximately 30 nucleotides (3,8). The ribosome-protected mRNA fragments (RPFs) are then purified and annealed to a single-stranded antisense cDNA copy of the mRNA under investigation. The annealing is performed under conditions such that a maximum of one RPF is annealed per single-stranded DNA (ssDNA) template. The position of each RPF is then determined by extension with T7 DNA polymerase of a 5′-labeled oligonucleotide primer annealed to the cDNA template. Because T7 polymerase will not unwind the heteroduplex formed between the mRNA fragment and ssDNA template (4,9,10), the extension product synthesis terminates at the 5′ end of the RPF. The primer extension products, which thus represent the map positions of the RPFs, are size fractionated electrophoretically on 7 M urea-polyacrylamid gels that include appropriate sequence ladder standards.

If the rate of ribosome movement along an mRNA during translation is constant, then all possible RPFs will be equally represented and the pattern of the primer extension products will appear as a ladder of uniformly intense bands (Fig. 1). However, if the rate of ribosome movement along the mRNA during translation is not constant, then ribosomes will protect those mRNA regions associated with slow translational steps to a greater degree than the other mRNA regions. These over-represented RPFs are characterized within the pattern of primer extension products by bands of enhanced intensity (Fig. 1). These enhanced bands correspond to the positions of the trailing (5′) edge of the ribosomes on the mRNA. Regions of ribosome pausing are therefore observed as intense bands relative to the background band pattern.

Materials and Methods

Reagents

All chemicals used in the following procedures are of reagent grade. Solutions are prepared with deionized, glass-distilled water. The pH of buffers is measured at 25° C.

Tris-HCl (pH 7.9)	1.0 M
Tris-HCl (pH 7.5)	1.0 M

PROTEIN SYNTHESIZING SYSTEM

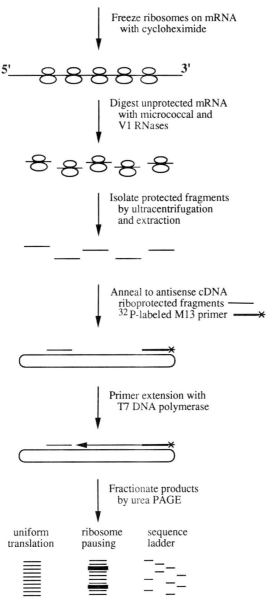

FIG. 1   Schematic summary of the strategy used for the identification of ribosome pause sites by primer extension inhibition.

HEPES (pH 7.2)	0.5 *M*
NaCl	5.0 *M*
KCl	1.0 *M*
MgCl$_2$	1.0 *M*
CaCl$_2$	0.05 *M*
NaOAc	3.0 *M*
KOAc	2.0 *M*
Mg(OAc)$_2$	0.1 *M*
EDTA	0.5 *M*
Dithiothreitol (DTT; Sigma, St. Louis, MO)	0.1 *M*
ATP, GTP, UTP, CTP (Sigma or P-L Biochemicals, Milwaukee, WI)	0.1 *M*
L-[^{35}S]Methionine (Du Pont-New England Nuclear, Boston, MA)	1200 Ci/mnol
[α-^{35}S]dATP (Du Pont-New England Nuclear)	500 Ci/mnol
[α-^{32}P]UTP (ICN, Costa Mesa, CA)	650 Ci/mnol
[γ-^{32}P]ATP (ICN)	7000 Ci/mnol

Rabbit reticulocyte lysate (micrococcal nuclease treated) (Promega, Madison, WI)

Wheat germ cell-free extracts [prepared as described by Levin and Samuel (11)]

T7 DNA polymerase (United States Biochemical Corp., Cleveland, OH)

Sequenase version 2.0 (United States Biochemical Corp.)

SP6 RNA polymerase (New England BioLabs, Beverly, MA)

T7 RNA polymerase (United States Biochemical Corp.)

T4 DNA ligase (Bethesda Research Laboratories, Gaithersburg, MD)

Restriction endonucleases (New England BioLabs)

Micrococcal nuclease (Boehringer Mannheim, Indianapolis, IN)

Snake venom ribonuclease V$_1$ (Pharmacia, Piscataway, NJ)

Ribonuclease A (Sigma)

RNase-free DNase (RQ1) (Promega)

Proteinase K (Bethesda Research Laboratories)

RNasin (Promega)

Transfer RNA from *Escherichia coli* (Sigma)

Cycloheximide (Sigma)

Guanosine 5'-[β,τ-methylene]triphosphate (GMPPCP) (Fluka, Ronkonkoma, NY)

7-Methyl-G(5')ppp(5')G (New England BioLabs)

7-Methyl-GDP (Sigma)

Fetal bovine serum (FBS) (HyClone, Logan, UT)

Dulbecco's modified Eagle's medium (DMEM; GIBCO, Grand Island, NY)

Eagle's minimal essential medium (EMEM; GIBCO)

PANSORBIN *Staphylococcus aureus* cells (Calbiochem Corp., La Jolla, CA)

Vector Constructions

The vector constructions containing the wild-type and mutant versions of the reovirus S1 and S4 genes (serotype 1, Lang strain) used to illustrate the procedures described here have been described (4,7,12–14).

Cell Maintenance, Infection, and Transfection

Infections are carried out with mouse L fibroblast cells in monolayer culture maintained at 37°C in EMEM containing 5% (v/v) FBS and actinomycin D (0.2 μg/ml). The Lang strain of reovirus serotype 1 is used at a multiplicity of infection of 10 (15). Transfections are performed with monkey COS-1 cells (12,13) maintained at 37°C in DMEM containing 5% FBS. The reovirus pSVS4 expression vector constructions (7, 16) are introduced into COS-1 cells in monolayer culture by the DEAE-dextran–chloroquine phosphate transfection method using DNA (5 μg/ml (17).

Preparation of RNA Transcripts in Vitro

Reovirus mRNA transcripts are prepared as *in vitro* transcription products using either T7 or SP6 RNA polymerase and pGEM4 vector constructions containing reovirus cDNA inserts (4,14,18). The standard reaction mixture (100 μl) contains 40 mM Tris-HCl (pH 7.9), 10 mM NaCl, 6 mM MgCl$_2$, 10 mM dithiothreitol, 2 mM spermidine, bovine serum albumin (BSA) (100 μg/ml), RNasin (0.5 unit/μl), 0.5 mM CTP, 0.5 mM ATP, 0.5 mM ^{7}mGpppG and 0.1 mM GTP, [α-^{32}P]UTP (500 μCi/ml) and 0.1 mM UTP, 2 μg of linearized DNA template, and 20 units of either SP6 or T7 RNA polymerase. The reaction mixture is incubated for 60 min at 37°C (T7 RNA polymerase) or 40°C (SP6 RNA polymerase). RNase-free DNase (5 units) is then added, and the reaction mixture incubated for an additional 15 min at 37°C. RNA transcripts are isolated by extensive extraction with phenol–chloroform, followed by ethanol precipitation. The amount of RNA synthesized is quantitated by trichloroacetic acid (TCA) precipitation on Whatman (Clifton, NJ) GF/C filter disks; the radioactivity is measured with a liquid scintillation system [Beckman (Fullerton, CA) LS230]. The integrity of the RNA transcripts is then ascertained by agarose–formaldehyde gel electrophoresis followed by autoradiography, and/or by *in vitro* translation.

Translation in Vitro

Translation *in vitro* of reovirus mRNA catalyzed by rabbit reticulocyte lysates is performed according to the manufacturer recommendations (Promega). Conditions for protein synthesis *in vitro* with wheat germ extracts are as described by Levin and Samuel (11). Reaction mixtures (25 μl) containing 100 ng of reovirus mRNA synthesized *in vitro* are incubated for 30 min in the presence of [^{35}S]methionine (400 μCi/ml), either at 30°C for the reticulocyte system or 25°C for the wheat system. RNase A is then added to a final concentration of 0.4 μg/ml and the incubation continued for an additional 15 min at 37°C. Samples are then diluted fivefold with radioimmunoprecipitation assay (RIPA) buffer [0.65% (v/v) Nonidet P-40 (NP-40), 10 mM Tris-HCl (pH 7.5), 140 mM NaCl, 1.5 mM MgCl$_2$] and immunoprecipitated. Immunoprecipitations are carried out using rabbit polyclonal antiserum prepared against purified reovirions (for σ1 and σ3 immunoprecipitations) or against a TrpE– σ1NS fusion protein (for σ1NS immunoprecipitations). Immune complexes are precipitated with PANSORBIN *S. aureus* cells and the ^{35}S-labeled proteins analyzed by sodium dodecyl sulfate-polyacrylamide gel electrophoresis (SDS-PAGE) and autoradiography (13,15). Quantitation of protein yield is determined by scanning the autoradiograms with an LKB (Bromna, Sweden) Ultrascan XL laser densitometer.

Measurement of Protein Synthesis in Vivo

The analysis of reovirus protein synthesis *in vivo* is carried out at 20 hr postinfection or 48 hr posttransfection (7,12,13). Briefly, cells are starved for 30 min in methionine-free EMEM and then labeled with [^{35}S]methionine (50 μCi/ml) for either 30 min (infections) or 1 hr (transfections). Cells are then harvested in RIPA buffer. Immunoprecipitation, SDS-PAGE analysis, and quantitation of ^{35}S-labeled reovirus σ1 and σ3 proteins are performed as described above for the *in vitro* translation procedure.

Preparation of Ribosome-Protected mRNA Fragments from Cell-Free Protein-Synthesizing Systems

Ribosome-protected fragments (RPFs) of mRNA are prepared from cell-free protein synthesizing systems essentially as described by Wolin and Walter (3). However, because of the high degree of secondary structure that can exist within mRNAs such as the reovirus s1 mRNA (14), a double-stranded RNA-specific nuclease such as V$_1$ is used in addition to micrococcal nuclease in order to completely digest mRNA not protected by ribosomes (4). To summarize briefly the standard RPF preparative procedure, ^{32}P-labeled mRNA (specific activity about 10^7 cpm/μg; final concentration of 4 μg/ml) is translated in a rabbit reticulocyte lysate as descried above. For controls, parallel translations are carried out (a) in the absence of exogenously added reovirus mRNA, and (b) with added reovirus mRNA and in the presence of a saturating concentraton of a protein synthesis inhibitor such as 1 mM cycloheximide, 10 mM ^{7}mGDP, or 5 mM GMPPCP added prior to the reovirus mRNA (pretreated).

Following 25 min of incubation at 30°C, translation reaction mixtures are placed on ice and cycloheximide is added to a final concentration of 1 mM. Samples are then incubated for 30 min at 30°C with a mixture of micrococcal (15 units/μl) and V$_1$ (0.02 unit/μl) nucleases in the presence of 3 mM CaCl$_2$ and 3.5 mM Mg(OAc)$_2$. These are saturating concentrations of the nucleases as measured with the s1 mRNA; complete digestion of some mRNAs can be achieved with lower concentrations of micrococcal nuclease alone. Micrococcal nuclease is then inactivated by addition of 60 μl of buffer T (3) [20 mM HEPES (pH 7.2); 150 mM KOAc; 10 mM Mg(OAc)$_2$; 5 mM EGTA; and 2 mM dithiothreitol]. The samples are then layered onto a 60-μl cushion of 0.25 M sucrose in buffer T, present in a 1.5-ml Eppendorf tube. Ribosomes are then pelleted by centrifugation at 55,000 rpm for 1 hr (Beckman TLA-100.3 fixed-angle rotor and Beckman TL-100 bench-top ultracentrifuge). Following centrifugation, the upper 120 μl of the supernatant fraction is removed. One hundred microliters of a proteinase K solution [proteinase K (200 μ/ml), 50 mM Tris-HCl (pH 7.5); 5 mM ethylenediaminetetraacetic acid (EDTA); 50 mM NaCl; and 0.5% SDS] is added to the lower 40 μl, which includes the ribosomal pellet. This mixture is then incubated for 30 min at 37°C. The samples are then extracted first with an equal volume of phenol–chloroform–isoamyl alcohol (24:24:1m, v/v) followed by chloroform–isoamyl alcohol. The RNA is then precipitated with ethanol in the presence of 20 μg of *E. coli* tRNA and 0.3 M NaOAc. Following centrifugation at 55,000 rpm for 1 hr (Beckman TL100), the pellet containing the RPFs of reovirus mRNA is suspended in H$_2$O (10 μl) and stored at −70°C.

Preparation of Ribosome-Protected mRNA Fragments from Virus-Infected or Vector-Transfected Cells in Culture

For the preparation of ribosome-protected fragments from virus-infected or vector-transfected cells, extracts are prepared from cultured cells essentially as described by Walden *et al.* (19). Briefly, monolayer cultures (60-mm dishes) are harvested 20 hr postinfection or 48 hr posttransfecton. The culture maintenance medium is removed, the cells are rapidly cooled by the addition of 0.8 ml of ice-cold polysome buffer to the culture dish [200 mM sucrose, 20 mM Tris-HCl (pH 7.4), 50 mM KCl, 5 mM Mg (OAc)$_2$, and cycloheximide (100 μg/ml)] and the dish is then placed on an ice–salt bath (−5°C). Cells are harvested by scraping and collected by centrifugation at 800 g for 5 min at 4°C. Cell pellets are washed once by suspending in 0.8 ml of ice-cold polysome buffer and collecting by centrifugation as before. The washed pellet of cells is suspended in 0.2 ml of polysome buffer. Cell suspensions derived from four parallel culture dishes are typically pooled. Nonidet P-40 is then added to a final concentration of 0.65% (v/v). Cells are then disrupted by homogenation with a tightly fitting Dounce homogenizer (Wheaten, Millville, NJ). The supernatant solution obtained following centrifugation of crude extracts at 5000 g for 5 min at 4°C is then used either directly as the source of polysomes or stored at −70°C in 0.1-ml aliquots.

To isolate ribosome-protected fragments, reaction mixtures (120 μl) containing 90 μl of crude extract are incubated at 30°C for 30 min with a mixture of micrococcal nuclease (15 units/μl) and V$_1$ nuclease (0.2 unit/μl) in the presence of 3 mM CaCl$_2$ and 100 mM KOAc. Control samples are then treated with 20 mM EDTA (37°C for 15 min) in order to dissociate ribosomes from mRNA. The micrococcal nuclease is then inactivated by the addition of 80 μl of buffer T. Ribosomes are isolated and the ribosome-protected mRNA fragments recovered by proteinase K digestion, phenol–chloroform–isoamyl alcohol extraction, and ethanol precipitation as described above for the isolation of RPFs from cell-free protein synthesizing systems.

Mapping mRNA Positions of Ribosome-Protected Fragments

The mapping of RPFs of reovirus mRNA is performed by a modification of the procedure of Wolin and Walter (3) as described by Doohan and Samuel (4). Among the most significant modifications is the use of T7 DNA polymerase instead of T4 DNA polymerase in the primer extension mapping reaction. T7 DNA polymerase, which does not require the accessory proteins 44/62 and 45, yields results comparable to those obtained with T4 DNA polymerase, which does require the accessory proteins (3,4).

To summarize the modified procedure (4), RPFs and a 5'-labeled M13 oligonucleotide sequencing primer (typically the -20 primer) are first annealed to an appropriate M13 antisense single-stranded DNA template. When the mRNA under investigation is large relative to the number of bases that can be accurately resolved on a standard sequencing gel [such as the reovirus s1 and s4 mRNAs, which are 1463 and 1196 nucleotides (nt), respectively], then M13 templates that possess inserts corresponding to the 5', internal, and 3' regions of the mRNA are used as necessary to map the RPFs accurately.

A typical annealing reaction mixture contains 20 ng of M13 construct as the template, 0.1 ng of ^{32}P-labeled primer (M13 -20 primer), 1 to 3 μl of RPFs, 88 mM KPO$_4$ (pH 7.5), 6.7 mM MgCl$_2$, and H$_2$O to a 9-μl final volume. The mixture is then heated to 65°C for 5 min and allowed to cool slowly to <37°C over the course of about 1 hr, annealing conditions giving optimal results being empirically determined. Equivalent results also are obtained with reovirus mRNA-derived RPFs when the annealing is carried out at 55°C for 1 hr. These latter conditions minimize the likelihood of complications due to potential secondary structure formation within either the RPFs or the ssDNA template that might otherwise adversely affect their annealing with each other.

After annealing, 10 μl of extension mix (88 mM KPO$_4$, 6.7 mM MgCl$_2$, 10 mM 2-mercaptoethanol, and 0.3 mM dNTPs) is added to the reaction mixture. Following the addition of 1 μl of T7 DNA polymerase (final concentration, 60 units/ml), the reaction mixture is incubated for 15 min at 37°C. The mixture is then diluted to

100 μl with TE buffer [10 mM Tris-HCl (pH 7.5); 1 mM EDTA] and the primer extension products extracted with an equal volume of phenol–chloroform–isoamyl alcohol (24:24:1, v/v). The resultant aqueous phase is extracted with chloroform–isoamyl alcohol (24:1, v/v) prior to ethanol precipitation. The precipitate is washed with 70% ethanol, dried, and suspended in gel loading buffer [95% (v/v) formamide, 10 mM EDTA, 0.1% (v/v) bromophenol blue, 0.1% (v/v) xylene cyanol]. Samples are heated to 65° C for 4 min and then fractionated on an 8% (w/v) polyacrylamide gel containing 7 M urea. Dideoxy sequencing reactions (20) are carried out in parallel with the single-stranded M13 DNA template, using Sequenase version 2.0 and [α-^{35}S]ATP in order to generate marker standards.

Results and Discussion

The procedures described herein for analysis of the dynamics of mRNA translation have been used by our laboratory to identify ribosome pause sites during translation of two viral mRNAs, the polycistronic s1 mRNA and the monocistronic s4 mRNA of reovirus (4,7).

Ribosome Pausing on Reovirus mRNAs in Cell-Free Protein-Synthesizing Systems

Ribosome pausing analyses have been carried out with both wild-type and site-directed mutant reovirus mRNAs. Such mRNAs, prepared as *in vitro* transcripts from cloned cDNAs, are routinely assessed first for their structural integrity prior to utilization in cell-free protein synthesizing systems (4,7). This is done in two ways: (a) by size analysis of the RNAs on agarose–formaldehyde gels under denaturing conditions, and (b) by verification of their ability to program the synthesis in cell-free protein synthesizing systems of the expected protein products (σ1, σ1NS, and σ3). A single RNA species of the predicted size should be detected on the denaturing gels for each of the transcripts (4). The 1463-nt polycistronic s1(wt) mRNA directs the synthesis of two proteins, the 51-kDa minor capsid protein σ1 and the 14-kDa nonstructural protein σ1NS (Fig. 2, lane 3). The 1196-nt monocistronic s4(wt) mRNA directs the synthesis of a single product, the 41-kDa major capsid protein σ3 (Fig. 2, lane 8). Several s1 mRNA mutants have been constructed that affect utilization of either the 5'-proximal ORF (ORF1 encoding σ1) or the internal overlapping ORF (ORF2 encoding σ1NS) (Fig. 2, lanes 4–6 and 10). These mutants represent important reagents for characterization of the pausing assay.

Production of Ribosome-Protected Fragments from s1 mRNA in Vitro
Ribosome-protected fragments prepared from translation mixtures containing uniformly ^{32}P-labeled s1(wt) mRNA are about 25 to 30 nt in length (Fig. 3). Such RPF,

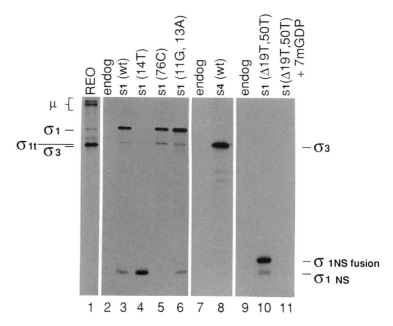

FIG. 2   Autoradiogram of reovirus σ-class polypeptides synthesized *in vitro* in the rabbit reticulocyte system programmed with pGEM vector-derived reovirus s1 and s4 transcripts. REO, Marker reovirion polypeptides immunoprecipitated from an extract of virus-infected mouse L cells; endog, no reovirus mRNA added to the cell-free system; S1(wt) and S4(wt), full-length wild-type s1 and s4 mRNAs, respectively; S1(14T), S1(76C), S1(11G,13A), and S1(Δ19T, 50T), site-directed mutant S1 mRNAs. S1(14T) specifies s1 mRNA in which the σ1 AUG14 initiation codon is changed to a noninitiation UUG codon. S1(76C) specifies s1 mRNA in which the σ1NS AUG75 initiation codon is eliminated by mutation to a noninitiation ACG codon. S1(11G,13A) specifies s1 mRNA in which the nucleotides flanking the σ1 AUG14 initiation codon are changed from a weak to a strong context. S1(Δ19T, 50T) specifies s1 mRNA in which the σ1 ORF shifts to the σ1NS ORF at nt 19. Translations were carried out in the absence (lanes 2–10) or presence (lane 11) of 10 m*M* 7-methyl-GDP (⁷mGDP), a cap analog inhibitor of translation. [From J. P. Doohan and C. E. Samuel, *Virology* **186,** 409 (1992). Reprinted with permission.]

can be generated with micrococcal nuclease alone from some mRNAs (3–5). However, for other mRNAs, micrococcal nuclease alone is not sufficient to completely digest away those regons of mRNA not protected by ribosomes. For example, in order to completely digest unprotected regions of mRNAs, such as the reovirus s1 mRNA that possesses significant secondary structure, a double-stranded RNA-selective RNase such as V_1 is required in addition to the micrococcal nuclease (4).

Ribosome-protected fragments prepared from cycloheximide (CHX) pretreated cell-free protein-synthesizing systems are expected to correspond to functional initiation sites. In the presence of CHX, ribosomes can scan mRNA and initiate transla-

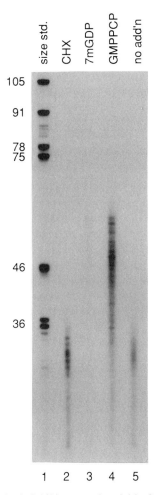

FIG. 3   Effect of protein synthesis inhibitors on the yield of s1(wt) mRNA fragments protected by ribosomes (RPFs) from digestion by micrococcal and V_1 nucleases. The standard reticulocyte translation system containing wild-type s1 [^{32}P]mRNA was incubated for 25 min in the absence or presence of the indicated protein synthesis inhibitor; RPFs were then isolated and fractionated on an 8% polyacrylamide gel containing 7 M urea. The translation reaction mixture contained the following: lane 2, 1 mM cycloheximide (CHX); lane 3, 10 mM ^{7}mGDP; lane 4, 5 mM GMPPCP; lane 5, no added inhibitor. Lane 1, ^{32}P-labeled size standards generated from pGEM4 plasmid DNA digested with Sau3A. [From J. P. Doohan and C. E. Samuel, *Virology* **186,** 409 (1992). Reprinted with permission.]

tion, but elongation is blocked because peptidyltransferase is inhibited (2, 21). By contrast, RPFs prepared from systems in the absence of translation inhibitors are expected to correspond to all randomly protected sites on the mRNA, and thus reflect the steady state distribution of translating ribosomes. The sizes of the RPFs generated both in the presence (Fig. 3, lane 2) and the absence (Fig. 3, lane 5) of CHX are similar, and about 25 to 30 nt in length. These RPFs represent protection of the s1(wt) mRNA by 80S ribosomes.

Ribosome-protected fragments are not detected in the presence of the ^{7}mGDP cap analog because ribosomes cannot access the s1 message (Fig. 3, lane 3). This result indicates that the production of mRNA fragments is indeed dependent on the protection of the message by ribosomes from nuclease degradation. By contrast, RPFs are detected in the presence of the nonhydrolyzable GTP analog, GMPPCP. GMPPCP interferes with translation by inhibition of the association of 40S and 60S subunits to form 80S ribosomes. 40S subunits can scan mRNA and form preinitiation complexes at AUG initiation sites in the presence of GMPPCP, but they cannot initiate translation (2,21,22). Thus, RPFs prepared from GMPPCP-pretreated protein synthesis systems correspond to functional initiation sites protected by 40S ribosomal subunits. Ribosome-protected fragments prepared in the presence of GMPPCP are 35 to 55 nt in length (Fig. 3, lane 4), somewhat larger than the 25- to 30-nt RPF obtained in the absence of inhibitor or in the presence of CHX (Fig. 3, lanes 2 and 5). The inhibitor-dependent difference in RPF size from s1(wt) mRNA (Fig. 3) is consistent with the conclusion that 40S subunits protect larger sized fragments of mRNA than do 80S ribosomes (22).

Mapping of Ribosome-Protected Fragments Prepared
from in Vitro Protein-Synthesizing Systems

Five major regions of ribosome pausing are detected on the reovirus polycistronic s1(wt) mRNA (4,7). They correspond to the pausing of ribosomes during the initiation and termination of translation within the two ORFs, and also to the stacking of 80S ribosomes (4). This is exemplified in Fig. 4 by the intense primer extension signal in the region designated I, which corresponds to σ1 ORF1 initiation at AUG14, and the intense signal at the region designated III, which corresponds to

---

FIG. 4   Ribosome pausing during initiation of translation of wild-type and mutant reovirus mRNAs in the presence of cycloheximide. (A) Ribosome-protected fragments (RPFs) derived from s4(wt), s1(wt), and s1(mutant) mRNAs were mapped by primer extension on the antisense ssDNA template mp8S1c5-4. Ribosome-protected fragments were prepared from reticulocyte translation mixtures that contained 1 m*M* cycloheximide and were programmed with the reovirus mRNAs described in the caption to Fig. 2. A dideoxy sequencing ladder, generated with the mp8S1c5-4 template, is included as a standard (lanes 8–11). The major regions of ribosome pausing are denoted by I, II, III, and IV. (B) Ribosome pausing pattern observed for RPFs derived from s1(wt) mRNA, but prepared from rabbit reticulocyte translation mixtures that contained protein synthesis inhibitors (lane 2, CHX; lane 3, ^{7}mGDP) as described in the caption to Fig. 3. Ribosome-protected fragments were mapped by primer extension on the S1 antisense ssDNA template mp8S1c5-4. [From J. P. Doohan and C. E. Samuel, *Virology* **186,** 409 (1992). Reprinted with permission.]

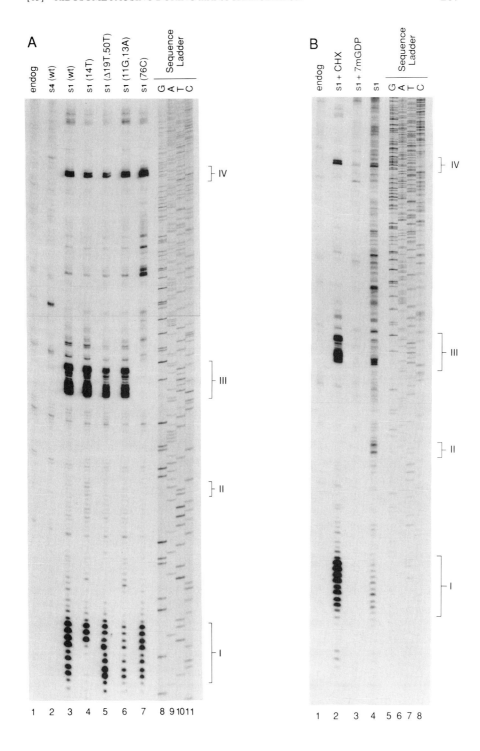

σ1NS ORF2 initiation at AUG75. Results obtained with the rabbit reticulocyte and wheat germ cell-free protein synthesizing systems are comparable (4).

The extension product signal detected at region I (Fig. 4A) is located at nt −2 to +3, where the 5′-terminal nucleotide of the s1 mRNA is defined as +1. (The s1 mRNA has a relatively short 5′ UTR; the AUG codon for σ1 initiation is at nt +14). Because RPFs are approximately 25 to 30 nt in length, and if one assumes that ribosomes are centered over the codon that they are translating, then the region I signal is consistent with an 80S ribosome paused directly over the AUG14 initiation codon utilized for σ1 protein synthesis (Fig. 4A, lane 3). Likewise, the intense signal observed at region III (Fig. 4A) occurs approximately 15 nt on the 5′ side of the AUG75 codon utilized for σ1NS ORF2 initiation.

From the analysis of ribosome pausing patterns obtained from CHX-pretreated systems programmed with site-directed mutant s1 mRNAs, it is possible to identify unequivocally the signals that correspond to ribosomes paused during the initiation of translation (Fig. 4A). For example, both the lower portion of the region I pause signal (Fig. 4A, lane 4) and the synthesis of σ1 protein (Fig. 2, lane 4) are eliminated in the S1(14T) mutant, that is, when the ORF1 AUG14 that initiates σ1 synthesis is converted to a UUG14 noninitiator codon. The S1(76C) mutant possesses a noninitiator ACG75 codon in place of AUG75, the site of σ1NS protein synthesis initiation in ORF2. The S1(76C) mutation eliminates both the region III pause (Fig. 4A, lane 7) and the synthesis of σ1NS protein (Fig. 2, lane 5). It should be noted that the S1(76C) mutation does not affect either the region I pause or the synthesis of σ1, and conversely the S1(14T) mutation does not affect either the region III pause signal or the synthesis of σ1NS (Fig. 2 and 4A).

The intensity of the primer extension inhibition signals caused by pausing at AUG initiation codons (e.g., regions I and III) is significantly increased when RPFs are isolated from cell-free translation systems incubated in the presence of cycloheximide (Fig. 4B, lane 2) as compared to steady state translation conditions in the absence of inhibitor (Fig. 4B, lane 4). Because RPFs generated in the presence of CHX represent protection of functional initiation sites confirmed by mutagenic analysis (Fig. 4A), these results provide confirmation that the region I and III signals also observed in the absence of CHX indeed correspond to initiation events, and suggest that the initiation events are slow events in the translational process.

Ribosome Pausing on Reovirus s1 and s4 mRNAs in Virus-Infected and Vector-Transfected Cells

The primer extension inhibition mapping results obtained with RPFs prepared from reovirus-infected mouse L cells *in vivo* are comparable to those obtained with RPFs prepared from rabbit reticulocyte lysates (*in vitro*) programmed with either s1 or s4 reovirus mRNAs (Fig. 5). The major region of ribosome pausing detected on the s4 mRNA during steady state translation, both *in vivo* and *in vitro,* corresponds to the site of translation initiation at AUG33 (Fig. 5, lanes 8 and 10). Both the size and the relative intensity of the ribosome pause signal attributed to σ3 initiation at AUG33 (denoted by the arrow on the right side of the gel in Fig. 5) are identical between the *in vivo* and the *in vitro* samples. Furthermore, the pattern of ribosome pausing detected *in vivo* on reovirus s4 mRNA is comparable between virus-infected and pSVS4 vector-transfected cells (7).

Analysis of the steady state distribution of translating ribosomes on the s1 mRNA likewise reveals that the pattern of ribosome pausing obtained with *in vivo*-derived RPFs from reovirus-infected cells (Fig. 5, lane 7) is essentially identical to that obtained with *in vitro*-derived RPFs from s1 mRNA-programmed reticulocyte lysates (Fig. 5, lane 9). This is exemplified by the pausing during steady state translation observed at the s1 mRNA regions designated I and III, both *in vivo* and *in vitro,* which correspond to the ORF1 σ1 AUG14 (region I) and ORF2 σ1NS AUG75 (region III), as denoted by the arrows on the left side of the gel (Fig. 5).

The region IV pause on s1 mRNA (Fig. 6, lanes 3, 4, and 7) corresponds to the initiation of translation of a newly identified protein synthesized from ORF1, the σ1t protein that initiates at AUG230 (4). Significant ribosome pausing on s1 mRNA is also observed at UAG432 of s1 mRNA in virus-infected cells (7). UAG432 is the site of termination of translation of the σ1NS protein. The σ1, σ1NS, and σ3 initiation (Fig. 5) and termination (not shown) pauses are not detected with RPFs isolated from mock-infected cells (Fig. 5, lanes 5 and 6) or from reticulocyte lysates in the absence of exogenously added s1 or s4 mRNAs (Fig. 4B, lane 1; Fig. 6, lane 6).

A high degree of ribosome protection of reovirus s1 mRNA is observed, both *in vivo* and *in vitro,* between nucleotide positions $+75$ to $+432$ of the s1 mRNA (Fig. 5, lanes 7 and 9). This is the region of s1 mRNA where the σ1 and σ1NS open reading frames overlap. Ribosomes are far more uniformly distributed on the monocistronic s4 mRNA than on the polycistronic s1 mRNA in virus-infected cells, in excellent agreement with the results for *in vitro* RPFs prepared from reticulocyte lysates programmed with s1 or s4 mRNA (Fig. 5).

Controls for Primer Extension Inhibition Obtained with Ribosome-Protected Fragments

Several controls can be performed in order to strengthen the significance of the T7 DNA polymerase-catalyzed primer extension results obtained with ssDNA templates and RPFs derived from virus-infected cells or cell-free protein synthesizing systems. These controls are necessary to ascertain the specificity of the mapping technique.

First, it is important to carry out the primer extension analysis using the antisense

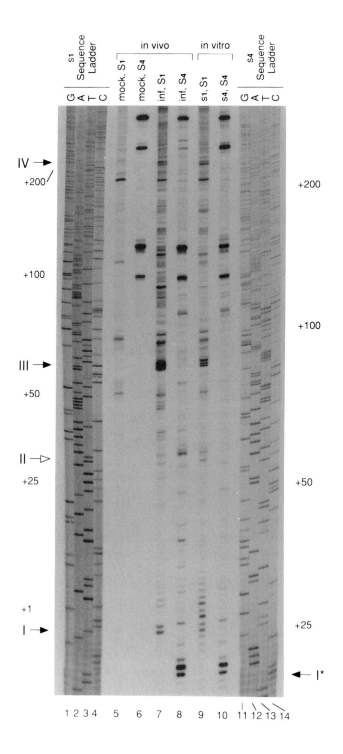

template of the mRNA under study (e.g., S1 template) with "endogenous" RPFs isolated from protein synthesizing systems not programmed with exogenously added mRNA [e.g., s1 mRNA (Fig. 4A, lane 1; Fig. 6, lane 6)] or prepared from parallel cultures of uninfected cells (mock) as compared to virus-infected cells (Fig. 5, lanes 5 and 6). The extension products obtained with endogenous and mock RPF preparations should be reduced or entirely eliminated (Fig. 5, lane 5 vs lane 7 for S1, and lane 6 vs lane 8 for S4). However, as shown by the endogenous (mock) analyses, the T7 DNA polymerase does stop occasionally at positions that are independent of RPFs derived from exogenously added reovirus mRNA or from reovirus-infected cells. These signals likely arise from the protection of endogenous RNAs from nuclease degradation, RNAs that generate fragments sharing limited sequence homology with the reovirus genes and that copurify with the reovirus-specific RPFs.

Second, the effect of EDTA treatment should be examined during the preparation of RPFs to verify that the derived reovirus-specific mRNA fragments are indeed generated by virtue of their protection by paused ribosomes from nuclease degradation. Treatment of cytoplasmic extracts with EDTA dissociates ribosomes from mRNA. EDTA treatment prior to isolation of ribosomes by ultracentrifugation greatly reduces the yield of RPFs. Thus, the mRNA-specific RPF-dependent primer extension products are greatly reduced as shown for s1 mRNA (Fig. 6), both for the RPFs derived from reovirus-infected L cells (lane 2 vs lanes 1 and 3) and from s1 mRNA-programmed reticulocyte lysates (lane 5 vs lanes 4 and 6).

Third, it is important to carry out the primer extension analysis with RPFs isolated from protein synthesizing systems programmed with exogenously added mRNA, but also treated with an inhibitor of translation such as ^{7}mGDP, which prevents both the cap-dependent synthesis of protein (Fig. 2, lane 11) and the cap-dependent protection of mRNA by ribosomes (Fig. 4B, lane 3). As an additional negative control, no signal above the endogenous background should be observed when RPFs prepared from a

---

FIG. 5 Comparison of ribosome pausing during translation of reovirus wild-type mRNA *in vivo* and *in vitro*. Ribosome-protected fragments were mapped by primer extension inhibition on the S1 and S4 antisense ssDNA templates mp8S1c5-4 and mp9S4. *In vivo*-derived RPFs were prepared from mouse L cells, either reovirus infected (lanes 7 and 8) or mock infected (lanes 5 and 6). *In vitro*-derived RPFs were prepared from the reticulocyte translation system programmed with either s1 (lane 9) or s4 (lane 10) reovirus mRNAs. Dideoxy sequencing ladders, generated with either mp8S1c5-4 or mp9S4 and the −20 primer, are included as standards (lanes 1–4 and 11–14). Nucleotide sequence numbers of the s1 (left) and s4 (right) mRNAs are shown, with the first nucleotide of the 5′ UTR designated as +1. Closed arrows indicate positions of pausing that are associated with initiation of translation. The open arrow indicates a site of ribosome stacking (4). [From J. P. Doohan and C. E. Samuel, *J. Biol. Chem.* **268,** 18313 (1993). Reprinted with permission.]

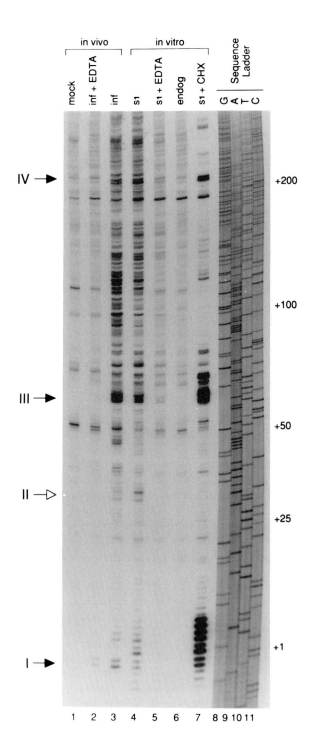

heterologous mRNA (e.g., s4 mRNA) are mapped with the antisense template (e.g., S1) used in the extension reaction (Fig. 4A, lane 2).

Finally, it is important with the primer extension inhibition mapping assay to ascertain that the ssDNA template is present at vast molar excess over the RFPs in the reaction mixture. This ensures that only a single RPF is annealed to each ssDNA template. As shown in Fig. 7 for RPFs derived from reovirus-infected L cells and mapped on the S1 ssDNA template, the yield of specific primer extension product is dependent on fragment concentration. The RPF-specific primer extension inhibition products that correspond to the initiation of $\sigma 1$ (region I) and $\sigma 1NS$ (region III) translation clearly increase in a concentration-dependent manner, but the nonspecific extension products do not increase (Fig. 7).

Concluding Comments

By use of a sensitive ribosome protection assay based on the inhibition of primer extension by RPFs, regions on mRNAs where ribosomes spend a disproportionately longer period of time during translation, that is, the slow or pause sites, can be identified (3–7). The distribution of translating ribosomes on mRNAs at steady state is not uniform. Major positions of ribosome pausing observed on mRNAs represent translation initiation and termination sites. In addition, elongation arrests are observed at positions of overlapping reading frames, frameshifting, and signal particle recognition. The ribosome pausing assay also provides a sensitive technique for identification of candidate translational initiation sites hitherto unrecognized within mRNAs derived from cDNA clones (4,23) and for the study of initiation factor-mediated mRNA–ribosome interactions (24).

Analysis of the dynamics of ribosome–mRNA interaction during translation *in vivo* in whole cells may have potential advantages over analyses performed using *in vitro* protein synthesizing systems. The localized microenvironments and organized subcellular structures characteristic of a whole cell, parameters that cannot be reproduced well in *in vitro* protein synthesizing reaction mixtures, may play crucial roles

---

FIG. 6   Effect of EDTA treatment on yield of ribosome-protected fragments of reovirus wild-type s1 mRNA *in vivo* and *in vitro*. The RPFs were mapped by primer extension inhibition on the S1 antisense ssDNA template mp8S1c5-4. *In vivo*-derived RPFs were prepared from mouse L cells, either reovirus infected (lanes 2 and 3) or mock infected (lane 1). *In vitro*-derived RPFs were prepared from reticulocyte translation system programmed with the following: lanes 4 and 5, s1 mRNA; lane 6, no exogenously added mRNA; lane 7, s1 mRNA plus 1 mM cycloheximide. The RPFs mapped in lanes 2 and 5 were derived from EDTA-treated samples. The numbers and arrows are as described in the caption to Fig. 5. [From J. P. Doohan and C. E. Samuel, *J. Biol. Chem.* **268**, 18313 (1993). Reprinted with permission.]

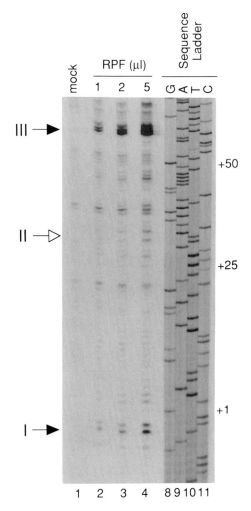

FIG. 7   Effect of concentration of ribosome-protected fragments on the yield of s1 mRNA-specific primer extension products. One, 2, or 5 μl of RPF preparation isolated from reovirus-infected L cells (lanes 2–4) or 3 μl of RPF preparation isolated from mock-infected L cells (lane 1) was mapped on the S1 antisense ssDNA template mp8S1c5-4. The numbers and arrows are as described in the caption to Fig. 5. [From J. P. Doohan and C. E. Samuel, *J. Biol. Chem.* **268**, 18313 (1993). Reprinted with permission.]

in translational control. Cells in culture can be infected with viruses and they can be treated with regulatory agents such as interferons or growth factors, which mediate their effect through cell surface receptors. The primer extension inhibition assay provides an approach to examine the effect of perturbing the physiological state of the cell by virus infection or cytokine treatment on the dynamics of ribosome interaction with specific mRNAs.

Acknowledgments

This work was supported in part by Research Grants AI-12520 and AI-20611 from the National Institute of Allergy and Infectious Diseases, U.S. Public Health Service.

References

1. J. W. B. Hershey, *Annu. Rev. Biochem.* **60,** 717 (1991).
2. M. Kozak, *J. Biol. Chem.* **266,** 19867 (1991).
3. S. L. Wolin and P. Walter, *EMBO J.* **7,** 3559 (1988).
4. J. P. Doohan and C. E. Samuel, *Virology* **186,** 409 (1992).
5. S. L. Wolin and P. Walter, *J. Cell Biol.* **109,** 2617 (1989).
6. C. Tu, T.-H. Tzeng, and J. A. Bruenn, *Proc. Natl. Acad. Sci. USA* **89,** 8636 (1992).
7. J. P. Doohan and C. E. Samuel, *J. Biol. Chem.* **268,** 18313 (1993).
8. J. A. Steitz, *Nature (London)* **224,** 957 (1969).
9. R. L. Lechner and C. C. Richardson, *J. Biol. Chem.* **258,** 11185 (1983).
10. R. L. Lechner, M. J. Engler, and C. C. Richardson, *J. Biol. Chem.* **258,** 11174 (1983).
11. K. H. Levin and C. E. Samuel, *Virology* **106,** 1 (1980).
12. J. A. Atwater, S. M. Munemitsu, and C. E. Samuel, *Virology* **159,** 350 (1987).
13. B. A. Belli and C. E. Samuel, *Virology* **193,** 16 (1993).
14. J. R. Bischoff and C. E. Samuel, *Virology* **172,** 106 (1989).
15. S. M. Munemitsu and C. E. Samuel, *Virology* **136,** 133 (1984).
16. S. M. Munemitsu and C. E. Samuel, *Virology* **163,** 643 (1988).
17. Luthman and G. Magnusson, *Nucleic Acids Res.* **11,** 1295 (1983).
18. D. A. Melton, P. A. Krieg, M. R. Rebagliati, T. Maniatis, K. Zinn, and M. R. Green, *Nucleic Acids Res.* **12,** 7035 (1984).
19. W. E. Walden, T. Godefroy-Colburn, and R. E. Thach, *J. Biol. Chem.* **256,** 11739 (1981).
20. F. Sanger, S. Nicklen, and A. R. Coulson, *Proc. Natl. Acad. Sci. USA* **74,** 5463 (1977).
21. D. Vazquez, *Mol. Biol. Biochem. Biophys.* **30,** 51 (1979).
22. M. Kozak and A. J. Shatkin, *J. Biol. Chem.* **252,** 6895 (1977).
23. D. C. Thomis, J. P. Doohan, and C. E. Samuel, *Virology* **188,** 33 (1992).
24. D. D. Anthony and W. C. Merrick, *J. Biol. Chem.* **267,** 1554 (1992).

[16] Analysis of RNA Replication in Plant Viruses

A. L. N. Rao, R. Duggal, F. C. Lahser, and T. C. Hall

Introduction

The majority of viruses infecting plants, animals, and humans contain single-stranded RNA genomes of positive polarity. An important prerequisite for developing methods of virus disease control is to understand how these organisms replicate and proliferate in their host organisms. The advent of recombinant DNA technology has provided many new approaches for such studies and it is now relatively simple to manipulate and analyze RNA genomes to determine the functions of their gene products in the viral life cycle. Two major advances simplified the maneuverability of single-stranded, (+)-sense RNA genomes: (a) conversion of single-stranded RNAs, (ssRNAs) into double-stranded DNA by reverse transcription, and (b) the development of vectors with bacteriophage promoters that allow the synthesis *in vitro* of RNA transcripts from cloned cDNA.

Brome mosaic virus (BMV) ranks among the better-studied RNA viruses of eukaryotic organisms (1). It is a single-stranded (+)-sense RNA virus pathogenic to many monocotyledonous plant species. The 8.2-kb genome is distributed among three single-stranded RNA species; this division makes it especially convenient for analyzing the functions of individual RNAs and their gene products in viral replication. Brome mosaic virus RNA-1 [3234 nucleotides (nt)] and RNA-2 (2865 nt) are monocistronic and encode nonstructural proteins that are integral to viral replication. RNA-3 (2117 nt) is dicistronic. The gene proximal to the 5' end encodes a nonstructural protein that is thought to mediate virus movement between plant cells. The viral capsid protein encoded by the translationally silent downstream coding region present on genomic RNA-3 is expressed from a subgenomic RNA (RNA-4) that is produced during replication of the virus *in vivo*. The 3' end of all four RNAs contain a highly conserved, multifunctional sequence motif that is essential for initiation of (−)-strand synthesis and also displays several tRNA-associated functions (2–4). Additional information on the molecular biology of BMV is provided in reviews by Marsh *et al.* (5) and Ahlquist (6).

The nonstructural regions of BMV exhibit extensive homology with the nonstructural proteins encoded by Sindbis and members of the alphavirus superfamily (7). Consequently, BMV has proved to be an excellent system for studying replication mechanisms common to many (+)-stranded RNA viruses and, although the methods described here were developed for studies on BMV, we are confident that they are broadly applicable to other viruses.

Methods in Molecular Genetics, Volume 4

Production *in Vitro* of Full-Length Capped RNA Transcripts of Brome Mosaic Virus

Following the development of recombinant DNA techniques, several full-length cDNA clones of plant RNA viruses were constructed. Because cDNA clones are not directly infectious, placement of the viral sequence downstream of a promoter of an RNA polymerase is necessary for *in vitro* synthesis of capped infectious RNA transcripts. This approach is particularly important for viruses such as BMV, because physical separation of individual genomic RNAs is impractical or laborious. Using an *Escherichia coli* promoter, full-length cDNA clones permitting transcription of each of the three BMV genomic RNAs were constructed by Ahlquist and Janda (8). Because RNA polymerases of bacteriophages such as SP6, T7, or T3 are stable and at least 10 times more active *in vitro* than *E. coli* RNA polymerase, Dreher *et al.* (9) placed full-length cDNA clones of BMV RNAs under the control of a T7 promoter. Wild-type and mutant RNA transcripts synthesized *in vitro* for mapping *cis*- and *trans*-acting regions that are essential for replication of BMV have been used successfully (9–14). We describe here a simple procedure to synthesize large quantities of capped, full-length infectious transcripts.

Procedure

1. Linearize the recombinant plasmid template with a restriction enzyme that cleaves 3′ to the cloned insert, usually in the distal portion of the polylinker. Following complete digestion, the plasmid is extracted once with a pheno–chloroform mixture (1:1) and the DNA precipitated with ethanol. Wash the pellet with 70% ethanol, dry in a Speed-Vac (Savant model SVC 100H; Savant, Hicksville, NY), and resuspend at 1 μg/μl in TE buffer [10 mM Tris and 1 mM ethylenediaminetetraacetic acid (EDTA), pH 8].

2. In a sterile 1.5-ml Eppendorf tube, mix the following:

Stock	Volume added	Final concentration
Template DNA (1 μg/μl)	10 μg	10 μg/100 μl
1 M Tris-HCl, pH 8.3	4 μl	40 mM
0.1 M Dithiothreitol (DTT)	5 μl	5 mM
0.2 M MgCl$_2$	7.5 μl	15 mM
20 mM ATP	5 μl	1 mM
20 mM CTP	5 μl	1 mM
20 mM UTP	5 μl	1 mM
1 mM GTP	7.5 μl	75 μM

(continued)

Stock	Volume added	Final concentration
Bovine serum albumin (BSA) (5 mg/ml)	10 μl	500 μg/100 μl
m^7 GpppG (5 mM; New England BioLabs, Beverly, MA)	10 μl	500 μM
T7 RNA polymerase (85 U/μl; Pharmacia, Piscataway, NJ)	1 μl	85 U/100 μl
RNAguard (40 U/μl; Pharmacia)	2 μl	80 U/100 μl
Water	28 μl	—

3. Incubate the reaction mixture at 37° C for 90 min.
4. Stop the reaction by adding 2 μl of 0.25 M EDTA.
5. Extract twice with an equal volume of phenol–chloroform mixture.
6. Add 1/10 vol of 3 M NaOAc, pH 5, and 2.5 vol of cold ethanol.
7. Leave at -80° C for 30 min.
8. Pellet the RNA–DNA mixture at 4° C in an Eppendorf centrifuge at 10,000 rpm for 20 min. Wash the pellet once with 70% ethanol and dry in a Speed-Vac. If the RNA transcripts are to be used to infect whole plants, dissolve the RNA in TE buffer. However, for transfection of protoplasts, the DNA template must be removed, because RNA transcripts containing DNA are poorly taken up by protoplasts.

Removal of DNA Template

Although the DNA template can be removed by using RNase-free DNase, it is our experience that such RNA transcripts are poorly infectious. A better alternative is to free the RNA from the DNA template by precipitation with lithium chloride. The procedure described below yields RNA having excellent infectivity.

1. After ethanol precipitation (from step 8 above), completely dissolve the RNA–DNA pellet in 0.1 M NaCl, 10 mM Tris-HCl (pH 7.5), and 1 mM EDTA.
2. Add lithium chloride to a final concentration of 2.7 M.
3. Thoroughly vortex, then incubate the mixture on ice (4 °C) for at least 16 hr.
4. Centrifuge for 10,000 rpm in an Eppendorf centrifuge for 20–30 min at 4 °C.
5. Wash the pellet once with 70% ethanol, dry in a Speed-Vac and suspend in TE buffer at a volume of 1 μl/μg of starting template.
6. Check the integrity of RNA transcripts by agarose gel electrophoresis.

Infectivity Assays

The replication and infectivity of most plant viruses can be analyzed by using three assay systems. These are (a) transfection of plant protoplasts, and inoculation of (b) systemic and (c) local lesion hosts. These systems have been widely used in molecular plant virology to understand the role of various gene products encoded by the viral genomes. The following section describes a detailed methodology in utilizing these assay systems, and also narrates some of the major advantages and inherent disadvantages that are associated with each assay system.

Protoplast System for Replication Studies

The discovery by Takabe and Otsuki (15) that single plant cells can be isolated by enzymatic treatment and infected with either purified virus or viral RNA has greatly facilitated investigation of plant virus replication strategies. Merits of the protoplast system include the establishment of synchronous infections, the addition of known quantities of inoculum to a defined number of cells, uniform sampling, as well as facile isolation and characterization of viral products. Northern blots of extracted RNAs followed by hybridization with strand-specific RNA probes (9–12) permit sensitive, quantitative analysis of synthesis and relative temporal accumulation of individual viral (+)- and (−)-sense RNAs. The only disadvantage of this system is that viral movement functions cannot be analyzed. The following section (adapted from Loesch-Fries and Hall, Ref. 16) describes isolation and transfection of barley mesophyll protoplasts and various methods we commonly use to analyze progeny (+)- and (−)-sense RNAs.

Isolation and Transfection of Barley Protoplasts with Polyethylene Glycol

Solutions and Reagents

Mannitol (0.55 M), pH 5.9: Weigh 100.2 g of mannitol and dissolve in 800 ml of double-distilled water. Adjust to pH 5.9 with 0.1 M KOH; make to 1 liter. Sterilize by autoclaving

Sucrose (0.55 M): Weigh 18.82 g of sucrose and dissolve in 80 ml of sterile distilled water. Adjust volume to 100 ml and sterilize by filtration

Polyethylene glycol (PEG; M_r 1540), 40%: Weigh 40 g of PEG-1540 and add 50 ml of sterile distilled water, 10 ml of 5% morpholineethanesulfonic acid (MES) buffer (pH 5.9), and 1 ml of 0.3 M $CaCl_2$. Briefly warm the mixture in a microwave oven and stir the mixture at room temperature until all the PEG is completely dissolved. The pH of this will be about 5.8; if necessary, adjust the pH with 0.1 M KOH. Finally, adjust the volume to 100 ml with sterile water. Filter the PEG solution initially through a 0.45-μm pore size filter and then through a 0.2-μm pore size filter. Store at room temperature. *Note:* Because PEG solutions turn acidic with age, it is not advisable to use solutions stored for more than 3 weeks at room temperature

Cellulase: Obtain from Calbiochem (La Jolla, CA) as Cellulysin

Macerozyme: Obtain from Yakult Honsha Mfg. Co., Ltd. (Tokyo, Japan)

BSA (fraction V): Obtain from Sigma (St. Louis, MO)

Protoplast culture medium

Solution A (1000×): Weigh 0.00249 g of $CuSO_4$, 0.0166 g of Kl, and 24.648 g of $MgSO_4$, and dissolve in 100 ml of 0.55 M mannitol, pH 5.9. The solution should be about pH 5.8. Sterilize by autoclaving

Solution B (1000X): Dissolve 2.7218 g of KH_2PO_4 and 10.111 g of KNO_3 in 100 ml of sterile distilled water and adjust the pH to 6.5 with 1 N KOH. Sterilize by autoclaving

Solution C (100X): Weight 14.702 g of $CaCl_2$ and dissolve in 100 ml of 0.55 M mannitol, pH 5.9. The pH of this solution should be about 6.3. Sterilize by autoclaving

Gentamicin: Prepare a 10-mg/ml stock solution in sterile water. Keep at 4 °C

Cephaloridine: Prepare a 30-mg/ml stock solution in sterile water. Keep at −20 °C

Preparation and Inoculation of Barley Protoplasts

1. Prepare the enzyme solution (per 0.5 g of leaf tissue): in a sterile beaker, weigh 0.25 g of cellulase, 12.5 mg of BSA, and 12.5 mg of macerozyme. Add 12.5 ml of 0.55 M mannitol, pH 5.9, and stir at room temperature. Adjust the pH of the solution to 5.9 with 0.1 M KOH and sterilize by filtration through a 0.2-μm pore size filter.

2. Harvest 0.5 g of 5 to 6-day-old barley plants. Avoid any dry or yellow plants. With a sharp razor, slice the leaves lengthwise and then crosswise until they are in 1-mm^2 sections. Incubate them in the enzyme solution for 3–3.5 hr at 28 °C.

3. After incubation at 28 °C, decant the enzyme solution containing the protoplasts into a beaker. Gentle swirling will facilitate release of the protoplast. Filter through gauze (300–350 μm) into another beaker. Transfer the solution to a sterile, 50-ml polypropylene tube. Underlay with 10 ml of 0.55 M sucrose. Use 12.5 ml of protoplast solution per tube.

4. Centrifuge in a Beckman (Fullerton, CA) J6B centrifuge at 400 rpm for 7 min at 20° C.

5. Carefully remove the protoplasts from the interface of mannitol and sucrose. They will constitute a dark green band at the interface. Transfer them into another 50-ml tube containing 10–15 ml of 0.55 M mannitol, pH 5.9. Mix gently and centrifuge in the Beckman J6B centrifuge at 600 rpm for 4 min at 20° C. Remove most of the supernatant and gently resuspend the cells in the remaining liquid. Add 10 ml of 0.55 M mannitol, pH 5.9, and repeat the wash as described above.

6. Resuspend the protoplasts carefully in a known volume of 0.55 M mannitol, pH 5.9 (e.g., 2 ml) and determine the number of viable protoplasts with a hemacytometer, as follows.

7. Stain the protoplasts with fluorescein diacetate by combining 1–2 drops of protoplast suspension with 1–2 μl of fluorescein diacetate (5 mg/ml in acetone). View after about 2 min with an ultraviolet (UV) microscope. (Turn on the power supply to about 5 V.) Count the bright fluorescent cells (Fig. 1B). Determine the number of cells per cubic centimeter (10 cells/nm^2 = 10^5 cells/cm^3). As there is now no enzyme present, the protoplast walls begin to reform, so speed and accuracy are essential. Determine the dilution necessary to attain 10^5 cells/ml. Place the appropriate amount of protoplasts in each tube (15-ml capacity polypropylene tubes) and add 1 ml of mannitol.

8. Centrifuge in the Beckman centrifuge J6B at 600 rpm for 3 min at 20° C.

9. Remove as much of the supernatant as possible and add either virus or RNA, keeping the volume quite small (5–10 μl).

FIG. 1   (A) Barley protoplasts photographed with transmitted light and (B) UV light after staining with fluorescein diacetate. (C) Local lesions induced on the inoculated leaves of *Chenopodium hybridum* by BMV. Barley plants healthy (D) and infected with BMV (E).

10. Gently shake the pellet to resuspend the protoplasts in the RNA and then add 150 μl of 40% PEG.

11. Gently shake the protoplasts for 10 sec. Add 2 drops of 0.55 M mannitol, pH 5.9. Gently mix.

12. Continue to add mannitol dropwise over the next 5–10 min until the volume has reached 1.5 ml. Incubate on ice for 15 min.

13. Pellet the protoplasts at 600 rpm for 3 min at 20° C.

14. Wash once with 1 ml of 0.55 M mannitol, pH 5.9.

15. Resuspend 1 × 10^5 protoplasts in 1 ml of culture medium containing 0.55 M mannitol (pH 5.9), a 1× concentration each of solutions A, B, and C, 10 μg of gentamicin, and 0.3 mg of cephaloridine.

16. Place the transfected protoplasts in a culture plate (24-well plate from Corning, Corning, NY) and keep under a fluorescent lamp covered with muslin cloth for 20–24 hr.

Extraction of Viral RNA from Protoplasts

1. After incubation for 20–24 hr, carefully collect the transfected protoplasts by centrifugation at 600 rpm for 3 min at 20° C.

2. Discard the supernatant and simultaneously add 250 μl of solution containing 100 mM glycine, 10 mM EDTA, 100 mM NaCl (pH 9.5), 2% SDS, bentonite (2.5 mg/ml), and 250 μl of phenol–chloroform; vortex.

3. Centrifuge at 4 °C in an Eppendorf centrifuge for 10 min.

4. Collect the supernatant and repeat the phenol–chloroform extraction.

5. Adjust the sodium concentration to 0.3 M with 3 M NaOAc, pH 5, and add 2.5 vol of cold ethanol. Mix the contents by vortexing and keep at − 80 °C for 15–30 min.

6. Pellet the RNA at 4 °C in an Eppendorf centrifuge for 20 min. Wash the pellet once with 70% ethanol. Dry and dissolve the RNA in 20 μl of sterile water.

Northern Analysis of Progeny Viral RNAs

Agarose Gel Electrophoresis of Viral RNA after Denaturation with Glyoxal

Northern hybridization (Fig. 2) is usually performed by immobilizing the denatured RNAs on either nitrocellulose or nylon membranes. A number of methods have been developed to denature RNAs prior to transferring RNAs to membranes. The most commonly used denaturants include formaldehyde (17), methylmercuric hydroxide (18), and glyoxal and dimethyl sulfoxide (DMSO) (19). Although complete denaturation of RNA is attainable with formaldehyde and methyl mercury, the extreme volatility and toxicity associated with these reagents requires carrying out all the opera-

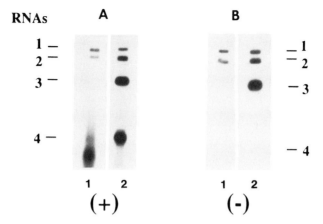

FIG. 2   Autoradiograph of BMV progeny RNAs. Northern blot analysis of progeny (+)- and (−)-sense RNA samples isolated from barley protoplasts transfected with wild-type RNAs 1 + 2 (lanes 1) and RNAs 1 + 2 + 3 (lanes 2). RNAs extracted from approximately 2.5×10^5 barley protoplasts were electrophoresed in 1% agarose gels after denaturation with glyoxal, and transferred to nylon membranes. The blots were then hybridized with ^{32}P-labeled RNA probes of (−)-sense (A) and (+)-sense (B) RNAs representing the homologous 3′ region present on each of the four BMV RNAs (4, 9). The positions of the four wild-type BMV RNAs are shown on the left (A) and right (B) sides.

tions in a chemical hood. These complications are, however, circumvented by using glyoxal and DMSO. We routinely denature RNAs isolated from various sources by using glyoxal and DMSO, and the method described below is essentially adapted from McMaster and Carmichael (19).

Buffers and Reagents

Phosphate buffer (0.2 *M*), pH 7: Mix 32.4 ml of 0.2 *M* NaH_2PO_4 and 67.6 ml of 0.2 *M* Na_2HPO_4. Sterilize by autoclaving and store at room temperature
Dimethyl sulfoxide (DMSO)
Glyoxal: Obtain a deionized, 40% solution (6 *M*) of glyoxal from Clontech Laboratories (Palo Alto, CA)
Loading buffer: 50% (v/v) glycerol, 10 m*M* phosphate buffer (pH 7), and 0.4% (w/v) bromophenol blue. Sterilize by autoclaving and store at room temperature

Note: To minimize the activity of RNases, it is important to keep electrophoresis gel boxes exclusively for RNA work. We observed that gel boxes and accessories can easily be freed from contaminating nucleases by treating them with a solution containing 0.2% (w/v) SDS and 0.2% (w/v) EDTA for at least 16 hr.

Procedure

1. In a sterile Eppendorf tube mix the following:

RNA (in sterile water)	3.7 μl
DMSO	8.0 μl
Phosphate buffer, pH 7 (0.1 M)	1.6 μl
Glyoxal (6 M)	2.7 μl

2. Vortex the reaction mixture and centrifuge briefly.

3. Incubate the tubes at 50°C for 1 hr.

4. While RNA samples are denaturing, prepare 1% agarose gels in 10 mM phosphate buffer, pH 7. It is not necessary to add glyoxal to the gels. Gels are run in 10 mM phosphate buffer, pH 7.

5. Cool the samples and add 1.6 μl of loading dye and load the samples into the wells of the gel.

6. Electrophorese the samples initially for 30 min at 3 V/cm.

7. Increase the current to 7 V/cm. To prevent the formation of pH gradient, it is essential to circulate the buffer during electrophoresis. It is our experience that this can be circumvented by using a Bio-Rad (Richmond, CA) DNA Sub-Cell apparatus; constant stirring of the buffer with stir bars is sufficient.

8. Stop electrophoresis when bromophenol blue has migrated approximately 80% of the length.

9. Transfer the fractionated RNA samples to nitrocellulose or nylon membranes by electroblotting. No other treatment of the gel is required.

Electroblotting of Fractionated RNAs to Nytran

After fractionation on agarose gels, RNAs frequently are immobilized on nitrocellulose or nylon membranes by a variety of methods. Some of the commonly used methods include capillary elution, vacuum blotting, and electroblotting. In our hands electroblotting yields excellent results and virtually 100% of the RNAs can be transferred in less than 4 hr. The method described utilizes Nytran (a nylon membrane from Schleicher & Schuell, Keene, NH) filters and the Hoefer (San Francisco, CA) scientific electroblotting apparatus.

1. Prepare 5 liters of electroblotting buffer [50 ml of 100× electroblotting buffer, containing 0.8 M Tris, 0.4 M sodium acetate, 20 mM EDTA (pH 7.3) (adjust pH with glacial acetic acid), and 3.8 ml 0.5 M EDTA] and cool to 4°C.

2. Cut one piece of Nytran and three pieces of Whatman (Maidstone, England) 3 mm paper to the exact size of the gel. Care must be taken not to touch the Nytran with greasy fingers; always wear gloves when handling filters.

3. Saturate the blotting paper in transfer buffer and Nytran in sterile water.

4. Prior to transfer, soak the Nytran filter in $6 \times$ SSC ($1 \times$ SSC contains 0.15 M NaCl and 0.015 M sodium-citrate).

5. The order of the transfer assembly starting from the cathode is as follows: (a) one side of the gel cassette, (b) two sheets of blotting paper, (c) the gel, (d) the Nytran filter, (e) one sheet of blotting paper, (f) a sponge, completely wet in transfer buffer, and (g) the second side of the gel cassette. To remove any trapped air bubbles, gently roll across each layer as it is laid down.

6. The assembled cassette is placed in the blotting apparatus, Nytran filter toward the anode. Introduction of air bubbles can be avoided by keeping the assembled cassette horizontal until placing into the transfer apparatus.

7. Conduct the transfer at 10 V for 30 min, followed by 30–40 V for an additional 2–3 hr, or at 15 V overnight. It is essential to cool the buffer during electroblotting by circulating $-4\,°C$ water–methanol (or engine antifreeze) through a prechilled cooling coil. Alternatively, the entire blotting can be done conveniently in a cold room. The buffer in the apparatus must be stirred during the entire transfer process.

8. After the transfer, remove the Nytran filter and bake it for 1 hr at 80 °C.

9. Thomas (18) found that removal of glyoxal adduct from the filters is likely to enhance the sensitivity of detection. This can be conveniently done by boiling the filters for 10 min in 10 mM Tris-HCl, pH 8.0.

10. Proceed with prehybridization and hybridization, using either ^{32}P or biotin-labeled riboprobes.

Preparation of Radiolabeled and Nonradiolabeled RNA Probes

During replication, ($+$)-sense RNAs accumulate at elevated levels, permitting rapid detection by radiolabeled probes (Fig. 2A). Because ($-$)-sense RNAs are less abundant, their detection is more difficult and is often neglected. However, there are several germane reasons to examine ($-$)-strand production. The detection of ($-$)-strands indicates that the observed ($+$)-strand RNAs are the result of *de novo* synthesis (Fig. 2B). For some viruses the generation *in vivo* of ($+$)-sense subgenomic RNAs is indicative of viral replication. However, to distinguish input inoculum from replicated progeny, especially in those viruses that do not produce subgenomic RNAs, detection of ($-$)-sense RNAs is an obligatory control. Further, quantitation of replication levels can be obscured by the presence of input inocula, especially in the case of mutants that replicate inefficiently.

Several plasmids are commercially available that contain highly specific promoter sequences for transcription *in vitro* by DNA-dependent RNA polymerases. Examples include SP6, T7, and T3 promoters. The promoter is adjacent to a polylinker cloning site. If a fragment cloned 3' to this promoter is in the polylinker, the RNA polymerase will transcribe the cloned sequence in an *in vitro* reaction using the purified recombinant plasmid as a template. When labeled ribo-NTPs are included in the transcription reaction, RNA probes can be synthesized. These RNA probes, which can be made in large quantity, can be used in Northern and Southern hybridizations. Moreover, RNA–RNA hybrids are more stable than RNA–DNA hybrids. An additional advantage of the RNA probes is their single-stranded nature; they cannot reassociate, like double-stranded, nick-translated DNA probes. The only disadvantage of RNA probes is their susceptibility to degradation by ubiquitous contaminating RNases. Although the sensitivity of biotin-labeled probes is similar to that of ^{32}P-labeled probes (20), the shorter time required to detect viral bands and prolonged storage without losing activity enhances the attractiveness of biotin-labeled riboprobes. Here we describe reliable procedures to synthesize large quantities of strand-specific riboprobes labeled with either ^{32}P or biotin and their application in the detection and quantitation of progeny viral RNAs (Fig. 2).

Procedure

1. In a sterile Eppendorf tube mix the following at room temperature:

Transcription buffer (5×) [100 mM Tris-HCl (pH 8.3), 250 mM KCl, 25 mM MgCl$_2$, BSA (1 mg/ml)]	4 μl
DTT (0.1 M)	2 μl
RNAguard (40 U/μl)	1 μl
ATP, CTP, GTP (10 mM each)	4 μl
UTP (100 μM)	2.4 μl
[^{32}P]UTP (~3000 Ci/mmol) or 10 mM biotin-11-UTP	4 μl
DNA template (1 μg/μl)	1 μl
T7 or T3 RNA polymerase (25 U/μl)	1 μl
Water	0.6 μl

2. Mix the contents of tube by vortexing and briefly centrifuge.
3. Incubate the reaction mixture at 37 °C for 60 min.
4. Terminate the reaction by adding 20 μl of TE buffer.
5. Extract once with an equal volume of phenol–chloroform. This step should be omitted if riboprobes are to be synthesized in the presence of biotin-11-UTP.
6. Collect the supernatant, add 1 μl of carrier yeast RNA (1 μg/μl), 1/2 vol of 7.5 M ammonium acetate, and 2 vol of cold ethanol. Incubate at -80 °C for 15–30 min.
7. Centrifuge at 4 °C for 20 min.
8. Wash the pellet with 70% ethanol, dry, and suspend in 50 μl of TE buffer.
9. Determine the number of counts per microliter, using the Cerenkov method.

Note: RNA probes labeled with biotin can be quantitated by incorporating [³H]UTP into the transcription mixture.

Hybridization with Riboprobes

Prehybridization

1. Prepare prehybridization buffer as below:

Deionized formamide	10 ml
Sodium dodecyl sulfate (SDS) (10%, w/v)	2 ml
Denhardt's (50×)	2 ml
Denatured salmon sperm DNA (10 mg/ml)	300 μl
NaCl (5 *M*)	4 ml
Double-distilled H₂O (sterile)	1.7 ml

2. Gently slip Nytran membrane containing the target RNA into a heat-sealable bag.
3. Add prehybridization buffer, approximately 0.2 ml for each square centimeter of Nytran membrane.
4. Remove the air bubbles by squeezing the bag and seal the open end of the bag with a heat sealer.
5. Incubate the bag containing the Nytran membrane for 5–16 hr at 42°C.
6. Replace the prehybridization buffer with hybridization buffer containing either ³²P- or biotin-labeled RNA probe.

   Hybridization buffer

Deionized formamide	10 ml
SDS (10%)	2 ml
Denhardt's (50×)	2 ml
Denatured salmon sperm DNA (10 mg/ml)	300 μl
NaCl (5 *M*)	4 ml
Yeast RNA (50 mg/ml)	0.2 mg/ml
³²P-Labeled riboprobe (1 × 10⁶ cpm/μl)	20 μl
or biotin-labeled riboprobe	150 ng/ml
Double-distilled H₂O (sterile)	to 20 ml

7. Hybridize the Nytran membranes for 16 hr at 42 °C.
8. Following hybridization, perform the following posthybridization washes:
   a. Two times, 10 min each time, in 2× SSC at room temperature.

  b. Two times, 30 min each time, in $0.1 \times$ SSC containing 0.1% SDS at 65 °C.

  c. Two times, 5 min each time, in $0.1 \times$ SSC at room temperature.

9. Dry the filters and expose to X-ray film.

Colorimetric Detection of Biotinylated RNA Probes

Buffer A: 0.1 M Tris-HCl (pH 9.5), 1 M sodium chloride

TSMT buffer: 0.1 M Tris-HCl (pH 7.5), 1 M NaCl, 2 mM MgCl$_2$, 0.05% (v/v) Triton X-100

Buffer B: 0.1 M Tris-HCl (pH 9.5), 0.1 M NaCl, 50 mM MgCl$_2$

10. Following posthybridization washes as described above (step 8), wash the filters for 10 min each time at room temperature in buffers A and TSMT.

11. The filters are then blocked in 3% BSA (fraction V; Sigma) prepared in TSMT buffer for at least 1 hr at 65 °C. *Note:* We have tried dry milk powder as an alternative to BSA. It is our experience that filters blocked with milk powder develop an undesirable background.

12. Discard the blocking solution and add 10 ml/100 cm^2 of a 1-μg/ml solution of streptavidin–alkaline phosphatase diluted in buffer A. Roll gently for 10 min at room temperature.

13. Wash the filters two times, 15 min each time, in a large volume (20 times the amount in step 12) of buffer A.

14. Then wash once for 10 min in buffer B.

15. Place the filters in a hybridization bag and add substrate solution containing 33 μl of BCIP (5-bromo-4-chloro-3-indolyl phosphate) (0.33 mg/ml) and 25 μl of NBT (nitroblue tetrazolium) (0.16 mg/ml) in 7.5 ml of buffer B. Incubate the filters in the dark for required period of time.

16. Stop the color development by adding TE buffer, pH 8.

Local Lesion Host Assay for Infectivity

A majority of plant viruses often induce local chlorotic or necrotic lesions on primary leaves at the site of inoculation (21). Local lesion assay is quantitative and the number of lesions produced is directly proportional to relative infectivity of the applied inoculum. Because each lesion originates from a single infection event, the assay is analogous to plaque induced by bacteriophages. Brome mosaic virus induces local necrotic lesions in *Chenopodium hybridum* (Fig. 1C) approximately 3–4 days after inoculation. Because BMV replicates and accumulates to high levels within the discrete lesions (22), the approach provides facile isolation and characterization of progeny RNA and other viral gene products. We have successfully utilized *C. hybridum* for analyzing the replicative competence of several BMV mutant RNAs, and the results (23, 24) exemplify the value of *C. hybridum* for studying RNA recombination and other sequence modifications that occur during the replication and infection process. The following section provides a detailed description of reliable methods for isolation and characterization of progeny RNA from single lesions induced in *C. hybridum.*

Preparation of Chenopodium hybridum Seeds for Sowing

This procedure is according to B. M. J. Verduin (Department of Virology, Agricultural University, Wageningen, the Netherlands).

1. Place a small quantity of *C. hybridum* seed in a 100-ml beaker and place the beaker in an ice bucket.
2. Pour 4 ml of concentrated sulfuric acid on the seed and stir gently with a glass rod for 4 min.
3. Pour the mixture containing the seed and sulfuric acid into a 1-liter beaker containing water and ice.
4. Collect the seed on cheesecloth assembled in a funnel.
5. Rinse the seed thoroughly with running tap water for at least 2 hr.
6. Dry the seed on blotting paper and store at 4 °C.
7. Sow the seed on a soil mixture and lightly cover with either sand or soil mix.
8. Moisten the soil and cover with Saran wrap.
9. When the plants touch the Saran wrap, transfer individual seedlings into bigger pots.

Inoculation of Chenopodium hybridum with Viral RNAs

1. Prior to inoculation, keep *C. hybridum* plants (at the two-leaf stage) in the dark overnight.
2. Adjust the viral RNA concentration to 150 μg/ml in a buffer containing 50 mM Tris-HCl (pH 7), 250 mM NaCl, 5 mM EDTA, and bentonite (3 mg/ml).
3. Lightly sprinkle carborundum (320 grit) on *C. hybridum* leaves to be inoculated and rub gently with a protected forefinger.
4. Dispense 10 μl of the inoculum with a sterile pipette tip onto the leaf and rub gently.
5. Hold the inoculated plants 4–7 days in an environmental chamber or greenhouse for lesion development.

Isolation of Total RNA from Local Lesions

1. Using a sterile cork borer, excise distinct and discrete single lesions and place in a sterile Eppendorf tube or a glass tissue grinder.
2. Pour liquid nitrogen over the leaf disk and homogenize thoroughly with a sterile spatula.
3. Add approximately 100 μl of RNA extraction buffer containing 100 mM gly-

cine, 10 mM EDTA, 100 mM NaCl (pH 9.5), and bentonite (2.5 mg/ml). Grind the tissue until no more leaf pieces are left.

4. Add 100 μl of phenol–chloroform mixture and vortex for 5 min.
5. Centrifuge for 10 min at 10,000 rpm at 4 °C.
6. Collect the supernatant and repeat the phenol–chloroform extraction.
7. Adjust the total sodium concentration to 0.3 M with 3 M sodium acetate, pH 5, and add 2 vol of ice-cold ethanol. Thoroughly vortex and leave the mixture at − 80 °C for 15–30 min.
8. Centrifuge for 20 min at 10,000 rpm at 4 °C.
9. Wash the pellet with cold 70% ethanol.
10. Dry the RNA pellet in a Speed-Vac and suspend in 20 μl of sterile distilled water.
11. Store the RNA at − 80 °C.

RNA can be used either for Northern analysis (see above) or for constructing cDNA libraries (see below).

Reverse Transcription and Polymerase Chain Reaction

The methodological approach for constructing cDNA libraries corresponding to viral RNA components present in the total RNA isolated from single lesions involves the synthesis of first-strand cDNA by reverse transcription (RT) followed by the amplification of DNA by polymerase chain reaction (PCR). The procedure described below is relatively simple and an entire experiment can be completed in a single Eppendorf tube.

Procedure

1. To a sterile, 0.5-ml Eppendorf tube add the following:

Total RNA isolated from single lesion	5 μl
3' Primer (2 OD units) complementary to the 3' end of the viral RNA	2 μl
Sterile distilled water	5 μl

2. Heat the sample at 70 °C for 10 min and quickly chill on ice.
3. After a brief centrifugation, add the following:

Buffer (5×) [250 mM Tris-HCl (pH 8.3), 375 mM KCl, 15 mM MgCl$_2$]	4 μl
DTT (0.1 M)	2 μl
Mixed dNTP stock (10 mM each of dATP, dCTP, dGTP, and dTTP)	1 μl
RNase inhibitor (40 U/μl)	1 μl

4. Mix the components by gently vortexing and centrifuge briefly.

5. Place the tube at 45 °C for 2 min to equilibrate the temperature.
6. Add 1 μl (200 units) of Superscript RNase H$^-$ reverse transcriptase (GIBCO-BRL, Gaithersburg, MD).
7. Incubate at 45 °C for 1 hr.
8. Inactivate the reverse transcriptase by boiling the tube at 95 °C for 10 min.
9. Quickly chill on ice and briefly centrifuge to collect condensed reaction products.
10. Add the following:

PCR buffer [100 mM Tris-HCl (pH 8), 500 mM KCl, 15 mM MgCl$_2$, 0.1% (w/v) gelatin, 2 mM concentration of each dNTP, and 1% Triton X-100]	8 μl
3′ Primer (2 OD units)	3 μl
5′ Primer (2 OD units)	3 μl
Ampli Taq DNA polymerase (0.5 U/μl; Perkin-Elmer, Norwalk, CT)	1 μl
Sterile water	65 μl

11. Mix the components and centrifuge briefly. Overlay with 100 μl of mineral oil.
12. Place the reaction tube in a thermocycler and subject to 40 PCR cycles. Each cycle consists of a denaturation at 90 °C for 30 sec, annealing at 50 °C for 1 min, and elongation at 70 °C for 2 min.
13. Analyze a 10-μl sample by agarose gel electrophoresis.

These PCR products can be cloned into suitable vectors for sequencing (e.g., T3/T7 *lacZ;* Pharmacia) after treating the products with Klenow fragment and polynucleotide kinase.

Infectivity Assays Using Systemic Host

Many plant viruses can be introduced into a host organism by simple mechanical inoculation. Following the establishment of infection at the site of inoculation, depending on the host species, viruses can either spread to uninoculated upper leaves and produce disease symptoms (systemic hosts) or be confined to the inoculated leaves and develop local lesions (local lesion host). The most common systemic symptom induced by a virus is the development of light and dark green areas, giving a mosaic effect in infected leaves. The nature and the pattern of symptoms vary widely for different host–virus combinations. For detailed symptomatology induced by various plant virus groups, the reader is advised to refer to Matthews (21). Barley (*Hordeum vulgare* cv. Dickson) is a commonly used laboratory host for investigating the systemic movement of BMV. Characteristic systemic symptoms induced by BMV are shown in Fig. 1E.

Inoculation of Barley Plants

1. Prior to inoculation, approximately 6-day-old barley plants are kept in the dark for at least 12 hr. Shading plants in the dark has proved to enhance the susceptibility to viral infection (21).
2. Lightly dust the plants with carborundum (320 grit; Fisher Scientific, Pittsburgh, PA) and create injury by gentle rubbing.
3. Dispense 10 μl of either purified virus (1 mg/ml) or RNA (150 μg/ml) onto each leaf, and gently rub with a protected forefinger.
4. Keep the inoculated plants for observation at desired light (16 hr of constant light) and temperature (23 °C) conditions.
5. Record symptoms 7 to 10 days after inoculation.

Purification of Virus from Infected Leaves

The purification procedure described here is adapted from Lane (22).

> Extraction buffer: 0.5 M sodium acetate, 0.08 M magnesium acetate, pH 4.5 (adjusted with acetic acid)
> Virus suspension buffer: 0.05 M sodium acetate, 0.008 M magnesium acetate, pH 4.5

1. Approximately 5 g of barley leaves showing characteristic mosaic symptoms is thoroughly ground in a mortar and pestle, using 20 ml of extraction buffer with 1% (v/v) β-metacaptoethanol. Addition of acid-washed sand facilitates grinding.
2. Strain the mixture through a double layer of cheesecloth.
3. Add an equal volume of chloroform and stir at 4 °C for 30 min.
4. Centrifuge the mixture for 15 min at 10,000 rpm at 4 °C.
5. Collect the supernatant and layer over a 10% (w/v) sucrose cushion prepared in virus suspension buffer.
6. Centrifuge at 4° C for 60–90 min at 50,000 rpm in a Beckman ultracentrifuge.
7. Discard the supernatant and suspend the virus pellet in 0.5 to 1 ml of virus suspension buffer.
8. Determine the virus concentration by spectrophotometer.

The virus can be used to infect protoplasts and plants or RNA can be isolated using the SDS–phenol–chloroform method (16).

Preparation and Assay of *in Vitro* Replicase Fraction from Virus-Infected Leaves

The use of an *in vitro* replication system for BMV has permitted the detailed examination of eukaryotic RNA virus promoter sequences (25–27), and provides an additional method for characterizing *cis*-acting RNA sequence elements, with results that are generally in agreement with those from protoplast studies (9). The ability to isolate a catalytically active fraction will also be crucial in the future identification of host proteins that, in association with viral factors, make up the replicase complex. For example, a subunit of the eukaryotic initiation factor 3 (eIF-3) has been identified in association with highly purified BMV replicase fractions (28). A general characteristic of the active extracts obtained from infected plants so far [reviewed by Hall *et al.* (29) and Quadt and Jaspars (30)] is that they are all membrane associated. These extracts catalyze the production of strands complementary to the input viral RNA and also (+)-sense subgenomic RNA synthesis (31). Further purification of the replicase fraction, yielding (+)-sense genomic RNAs, has been claimed by Hayes and Buck (32) using cucumber mosaic virus RNA as template.

Preparation of Viral Replicase

The following procedure is derived from Hardy *et al.* (33), Bujarski *et al.* (34), Miller and Hall (35), and Dreher and Hall (2).

1. Five to 6 days after inoculation of barley seedlings (6 days old) with BMV, symptomless secondary leaves are harvested for use. Leaves can be stored at $-80\,°C$ or used immediately for the preparation.

2. Using a prechilled (at $-80\,°C$) mortar and pestle, grind the leaves with powdered, sterile glass (Pasteur pipettes are good) and buffer A, using 15 ml of buffer per 2 g of leaves. This mixture will take on a green, doughlike texture, which will liquefy with continued grinding.

> Buffer A: 50 mM Tris-HCl (pH 7.4), 10 mM KCl, 1 mM EDTA, 10 mM Mg(OAc)$_2$, 15% (v/v) glycerol, 10 mM DTT (add fresh)

3. Centrifuge the mixture at 3000 rpm for 10 min in a Beckman JA 20 rotor at 4 °C to remove large cell fragments and glass. Collect the supernatant and centrifuge at 60,000 rpm for 20 min in a Beckman Ti 60 rotor at 4 °C.

4. Suspend the pellet in buffer A containing 1% (w/v) dodecyl-β-D-maltoside (12-M), a nonionic detergent, using 1 ml of buffer per 1 g of starting leaf material. Stir in a small beaker for 90 min at 4 °C to resuspend the pellet material completely.

5. Centrifuge the mixture at 40,000 rpm for 30 min in a Beckman Ti 40 rotor at 4 °C.

6. Suspend the pellet in buffer A containing 0.1% (w/v) 12-M, using 1 ml per 2 g of starting leaf material.

7. Layer 2-ml aliquots of the resuspension onto 40% sucrose gradients made up

in buffer A containing 0.1% (w/v) 12-M; open-top tubes capable of holding 30 ml are preferable. Centrifuge at 16,000 rpm for 2 hr in a Beckman SW-27 rotor at 4 °C.

8. Suspend the pellet in buffer B, using 1 ml per 2 g of starting leaf material. This preparation, divided into 200-μl aliquots, is stable for months at -80 °C, although experience has shown that template specificity is slightly lower after 1 month (2).

> Buffer B: 50 mM Tris-HCl (pH 8.0), 0.5 mM Mg(OAc)$_2$, 0.1% (w/v) 12-M (add fresh), 10 mM DTT (add fresh)

Assay of Replicase

Assay Conditions

Replicase activity was determined using an [α-^{32}P]UMP incorporation assay described by Dreher *et al.* (25).

1. Treat an aliquot of replicase fraction, thawed on ice, with 2.5 μl of 50 mM Ca(OAc)$_2$, and 12.5 μl of micrococcal nuclease (10 U/μl) to remove any endogenous viral RNA templates from the extract. Incubate for 30 min at 30 °C. Micrococcal nuclease activity is stopped by the addition of 6.9 μl of 200 mM EGTA.

2. Aliquot the assay components into separate tubes. Dry the contents to a pellet, using a Speed-Vac concentrator (Savant Instruments), and resuspend in a 5-μl total volume containing the desired amount of RNA (virion derived or transcribed *in vitro*) and RNase-free water.

Assay Components

Each tube contains a final concentration of the following: 2 μM UTP, 500 μM CTP, GTP, and ATP, actinomycin D (80 μg/ml), and 10 mM Mg(OAc)$_2$, with 30 μCi of [α-^{32}P]UTP. The use of actinomycin D controls against radiolabel incorporation due to DNA-dependent RNA polymerases.

3. Add 20 μl of micrococcal nuclease-treated replicase extract to each tube, mix, and incubate for 30 min at 30 °C. The reaction is stopped by the addition of 12.5 μl of 200 mM EDTA, followed by phenol–chloroform extraction and ethanol precipitation with NH$_4$OAc. Total precipitable products can be visualized by standard electrophoretic techniques and autoradiography. We can obtain clear and distinct bands using 4% nondenaturing polyacrylamide gels run in 0.8 $\times$ Tris-borate-EDTA (1 $\times$ = 89 mM Tris, 89 mM boric acid, 2 mM EDTA); time of separation depends on the size of the expected products (Fig. 3) (30). Kinetic analyses can be performed by measuring trichloroacetic acid-precipitable products and scintillation counting.

4. To test for the double-stranded nature of these products, we utilize a mild RNase treatment [RNase A (200 μg/ml) and RNase T$_1$ (200 U/ml)] in 30 mM so-

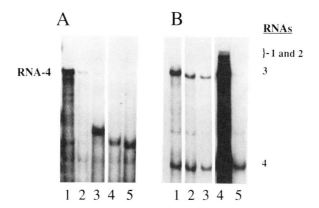

FIG. 3   Autoradiographs of *in vitro* replicase products. The *in vivo* replication characteristics of the mutant RNA-3 derivatives shown are described in Ref. 36. (A) Double-stranded (i.e., RNase resistant in high salt) radiolabeled RNA products made *in vitro* using as input templates equimolar amounts of *in vitro* transcripts corresponding to the 3′-terminal 380 nt of (+)-sense RNA-3. Input RNAs used for (A) include BMV RNA-4 (lane 1), no RNA (lane 2), and 3′-terminal transcripts of wild-type RNA-3 (lane 3), deletion mutant ΔGHIJ RNA-3 (lane 4), and ΔGHIK RNA-3 (lane 5). (B) *In vitro* replication products made using full-length RNA-3 transcripts as templates. Input RNAs used include full-length transcripts of wild-type RNA-3 (lane 1), ΔGHIJ RNA-3 (lane 2), ΔGHIK RNA-3 (lane 3), as well as purified virion RNA (lane 4) and no input RNA (lane 5). In (A) and (B) the amounts of template remaining after the micrococcal nuclease treatments are evident [compare to (A), lane 2, and (B), lane 5].

dium citrate (pH 7.2) and 300 m*M* NaCl (2× SSC), incubated for 30 min at 30 °C. Undigested double-stranded products can be analyzed by the methods listed in step 3 above.

Acknowledgments

Many of these procedures were developed with funding from the NIH (Grant AI22354) and NSF (Grant DMB-8921023) and the Texas A & M Office of University Research Services.

References

1. P. Kaesberg, *in* "Molecular Biology of the Positive Strand RNA Viruses." Academic Press, New York, 1987.
2. T. W. Dreher and T. C. Hall, *J. Mol. Biol.* **201,** 31 (1988).
3. T. W. Dreher and T. C. Hall, *J. Mol. Biol.* **201,** 41 (1988).

4. A. L. N. Rao, T. W. Dreher, L. E. Marsh, and T. C. Hall, *Proc. Natl. Acad. Sci. USA* **86,** 5335 (1989).

5. L. E. Marsh, G. P. Pogue, C. C. Huntley, and T. C. Hall, *Oxford Surv. Plant Mol. Cell. Biol.* **7,** 297 (1991).

6. P. Ahlquist, *Curr. Opin. Genet. Dev.* **2,** 71 (1992).

7. R. Goldbach, R. Le Gall, and J. Wellink, *Semin. Virol.* **2,** 19 (1991).

8. P. Ahlquist and M. Janda, *Mol. Cell. Biol.* **4,** 2876 (1984).

9. T. W. Dreher, A. L. N. Rao, and T. C. Hall, *J. Mol. Biol.* **206,** 425 (1989).

10. A. L. N. Rao and T. C. Hall, *J. Virol.* **64,** 2437 (1990).

11. A. L. N. Rao and T. C. Hall, *Virology* **180,** 16 (1991).

12. R. Duggal, A. L. N. Rao, and T. C. Hall, *Virology* **187,** 262 (1992).

13. R. French, M. Janda, and P. Ahlquist, *Science* **231,** 1294 (1986).

14. P. Traynor and P. Ahlquist, *J. Virol.* **64,** 69 (1990).

15. I. Takabe and Y. Otsuki, *Proc. Natl. Acad. Sci. USA* **64,** 843 (1969).

16. L. S. Loesch-Fries and T. C. Hall, *J. Gen. Virol.* **47,** 323 (1980).

17. N. Rave, R. Crkvenjakov, and H. Boedtker, *Nucleic Acids Res.* **6,** 3559 (1979).

18. P. S. Thomas, *in* "Methods in Enzymology," Vol. 100, p. 255. Academic Press, New York, 1983.

19. G. K. McMaster and G. G. Carmichael, *Proc. Natl. Acad. Sci. USA* **74,** 4835 (1977).

20. A. L. N. Rao, C. C. Huntley, L. E. Marsh, and T. C. Hall, *J. Virol. Methods* **30,** 239 (1990).

21. R. E. F. Matthews, "Plant Virology." Academic Press, New York, 1991.

22. L. C. Lane, *Adv. Virus Res.* **19,** 151 (1974).

23. A. L. N. Rao, B. P. Sullivan, and T. C. Hall, *J. Gen. Virol.* **71,** 1403 (1990).

24. A. L. N. Rao and T. C. Hall, *J. Virol.* **67,** 969 (1993).

25. T. W. Dreher, J. J. Bujarski, and T. C. Hall, *Nature (London)* **311,** 171 (1984).

26. J. J. Bujarski, T. W. Dreher, and T. C. Hall, *Proc. Natl. Acad. Sci. USA* **82,** 5636 (1985).

27. L. E. Marsh, T. W. Dreher, and T. C. Hall, *Nucleic Acids Res.* **16,** 981 (1988).

28. R. Quadt, C. C. Kao, K. S. Browning, R. P. Hershberger, and P. Ahlquist, *Proc. Natl. Acad. Sci. USA* **90,** 1498 (1993).

29. T. C. Hall, W. A. Miller, and J. J. Bujarski, *Adv. Plant Pathol.* **1,** 179 (1982).

30. R. Quadt and E. M. J. Jaspars, *Virology* **178,** 189 (1990).

31. W. A. Miller, T. W. Dreher, and T. C. Hall, *Nature (London)* **313,** 68 (1985).

32. R. J. Hayes and K. W. Buck, *Cell (Cambridge, Mass.)* **63,** 363 (1990).

33. S. F. Hardy, T. L. German, L. S. Loesch-Fries, and T. C. Hall, *Proc. Natl. Acad. Sci. USA* **76,** 4956 (1979).

34. J. J. Bujarski, S. F. Hardy, W. A. Miller, and T. C. Hall, *Virology* **119,** 465 (1982).

35. W. A. Miller and T. C. Hall, *Virology* **125,** 236 (1983).

36. F. C. Lahser, L. E. Marsh, and T. C. Hall, *J. Virol.* **67,** 3295 (1993).

[17] Applications of Ribonuclease Protection Assay in Plant Virology

Peter Palukaitis, Marilyn J. Roossinck, and Fernando García-Arenal

Introduction

The ribonuclease protection assay (RPA) has been used as a means of identifying both the presence of a particular RNA (1, 2) as well as variation in the RNA sequence of a population (3, 4). As a technique, it provides more information than nucleic acid blot hybridization, but not as much information as nucleic acid sequence determination; however, the RPA is much less labor intensive than cloning and sequencing of nucleic acid samples. Thus, information on sequence variation in RNA populations can be gathered much more rapidly using an RPA. Specific applications of the RPA in plant virology include (a) detecting the expression of viral sequences in transgenic plants, (b) quantitation of viral sequences present in nucleic acid extracts, (c) detecting and mapping variation in strains or isolates of a virus, and (d) detecting and mapping heterogeneity in RNA populations; for example, in studies involving virus evolution and virus epidemiology. Here we describe the basic RPA method, and examine some of the parameters that need to be considered when using the RPA for the above applications.

Methods

It is necessary to have at least one cDNA clone of the viral RNA of interest. In addition, it is preferable, but not essential, to know the nucleotide sequence of the cDNA clone. The cDNA insert should be in a plasmid containing dual RNA transcription promoters (T3, T7, SP6) flanking the polylinker region. Numerous such plasmids are available from various suppliers. It is also preferable to have several cDNA clones if sequence variation within a population is a factor in, or is the aim of, the analysis. For most analyses, it is also a good idea to have several subclones (250–500 bp) of the original cDNA. This is an optimal size range for the analysis of the sequence variation, and also facilitates the mapping of sequence heterogeneity in longer RNA molecules.

The basic procedure involves four steps: (a) synthesis of the probe from a linearized plasmid, (b) annealing of the probe to the target RNA, (c) incubation of the RNA with ribonuclease, and (d) detection of protected RNA fragments by gel electrophoresis.

Methods in Molecular Genetics, Volume 4
Copyright © 1994 by Academic Press, Inc. All rights of reproduction in any form reserved.

Synthesis of Probe

1. Combine the following:

DNA template (1 μg)	1 μl
Transcription buffer (5 $\times$): 0.4 M HEPES–KOH (pH 7.5), 60 mM MgCl$_2$, 10 mM spermidine, 0.2 M dithiothreitol (DTT)	2 μl
NTP minus UTP (30 mM ATP, 30 mM CTP, 30 mM GTP)	1 μl
UTP (0.75 mM)	1 μl
[α-^{32}P] UTP (10 μCi) or [α-^{35}S]UTP (30 μCi)	1 μl
RNase inhibitor (5–10 units)	1 μl
RNA polymerase (20–30 units) (SP6, T3, or T7)	1 μl
Inorganic pyrophosphatase (1 unit)	1 μl
Water	1 μl
	10 μl

2. Incubate at 37°C for 2 hr.
3. Add 0.5 μl (2.5 units) of RNase-free DNase I.
4. Incubate at 37°C for 10 min.
5. Add 5 μl 0.1 M ethylenediaminetetraacetic acid (EDTA), 35 μl of water, and 50 μl of phenol–chloroform (1 : 1, v/v).
6. Vortex for 1 min to emulsify and centrifuge to separate the phases.
7. Remove the aqueous phase and back extract the organic phase with 50 μl of 0.1 mM EDTA.
8. Combine the aqueous phases and add 100 μl of 5 M ammonium acetate and 500 μl of ethanol.
9. Store overnight at -20°C.
10. Centrifuge the sample to pellet the RNA.
11. Wash the pellet with 70% ethanol.
12. Resuspend the dried pellet in 20 μl of 0.1 mM EDTA.

In addition to the radiolabeled probe, a positive control RNA exactly complementary to the probe needs to be prepared to standardize the assay. This control RNA is prepared from the same cDNA clone (the plasmid linearized at the other end of the insert) using the RNA polymerase promoter on the other side of the cDNA insert. In this case, the only changes to the protocol are the exclusion of the radioisotope and the use of UTP at the same concentration as the other three nucleotides.

The above procedure incorporates some of the modifications made to the *in vitro* transcription reaction described in the RiboMAX (Promega, Madison, WI) transcription system (5), that is, the use of HEPES as the buffer and inorganic pyrophosphatase (Cat. No. I-1891; Sigma, St. Louis, MO), both of which increase the yield of transcripts (5).

^{32}P-Labeled probes should be used within several days after their preparation, and certainly not beyond 1 week. ^{35}S-Labeled probes can be used for at least 1 month.

For some applications, it may be necessary first to purify the probe by polyacryl-amide gel electrophoresis and elution. However, in most cases this step is unnecessary.

Annealing of Probe

Annealing buffer

Formamide (deionized)	850 μl
NaCl (5 M)	85 μl
EDTA (0.5 M)	2 μl
PIPES–KOH, pH 6.5 (0.5 M)	64 μl
	1 ml

1. Combine the following:

[^{32}P]RNA (10,000 cpm) or [^{35}S]RNA (100,000 cpm)	1 μl
Viral RNA (200 ng of purified viral RNA or 2 μg of total plant RNA)	1 μl
Annealing buffer	30 μl

2. Incubate at 85° C for 2 min, then at 50–55° C for 1–3 hr.
3. Chill on ice or place in the freezer until ready for the assay.

For the positive control RNA, use 50 ng of unlabeled anticomplementary sense transcript. For the negative control RNA, use either 200 ng of yeast (or some other unrelated) RNA, or 2 μg of total RNA from a healthy (or nontransformed) plant.

The annealing time will vary, depending on both the length of the probe and the concentration of the target RNA. Generally, shorter probes [less than 400 nucleotides (nt)] can be efficiently annealed in 2 hr, whereas long probes may require more time; however, longer hybridization times may increase the appearance of background bands. For the detection of viral RNA transcripts in transgenic plants, overnight in-cubations generally are required.

Ribonuclease Digestion of Annealed RNAs

RNase digestion buffer

Water	916 μl
NaCl (5 M)	60 μl

Tris-HCl, pH 7.5 (1 M)	10 μl
EDTA (0.5 M)	10 μl
RNase A (10 mg/ml)	2 μl
RNase T1 (2.5 units/μl)	2 μl
	1 ml

1. Add 300 μl of RNase digestion buffer to each sample of annealed RNA (32 μl).
2. Incubate at one of several temperature/time combinations (see below; e.g., 34° C for 30 or 60 min).
3. Add 20 μl of 10% (w/v) sodium dodecyl sulfate (SDS) and 10 μl of protease E (10 mg/ml).
4. Incubate at 37° C for 15 min.
5. Add 5–10 μg of carrier (yeast or transfer) RNA, and extract with phenol–chloroform (1:1, v/v).
6. To the aqueous phase, add 40 μl of 3 M sodium acetate and 1 ml of ethanol.
7. Incubate at −20° C for several hours to overnight, or at −80° C for 15 min.
8. Recover the RNA by centrifugation and resuspend the sample in 10–25 μl of either 0.1 mM EDTA or gel loading buffer (the composition of which depends on the type of gel to be used for the analysis).

The system needs to be optimized for the salt concentration during the digestion, as well as for the temperature of incubation and the time of digestion. The optimal procedure is the one with the lowest salt concentration and the longest incubation time that still results in complete protection of the probe. This will ensure the most complete cleavage of the probe in the presence of a heterologous target RNA. These parameters will be considered below.

[35]S-Labeled probes require slightly longer incubation times for complete cleavage (e.g., 45 min with [^{35}S]RNA vs 30 min with [^{32}P]RNA).

The ribonucleases used can also be varied. The specificity of the nucleases can affect the result. In addition, the digestion conditions can affect the cleavage profile. These effects will also be considered below.

Detection of Protected RNA Fragments by Gel Electrophoresis

For RNAs in which the fully protected duplexes are expected to be greater than 1 kb, it is preferable to analyze the denatured duplexes on agarose gels (e.g., see Ref. 6). For duplexes less than 300 bp, denaturing polyacrylamide gel electrophoresis should be used for the analysis of the fragments (e.g., see Refs. 4 and 6). For RNAs between 300 bp and 1 kb, the gel matrix that should be used for the analysis may depend on the number and size of fragments to be expected; however, generally polyacrylamide gel electrophoresis is the preferred system.

Agarose Gel Electrophoresis

Gels of 1.5 to 2% (w/v) agarose in 20 mM MOPS–acetate, 5 mM sodium acetate, 1 mM EDTA, pH 7 (with or without 6% formaldehyde in the gel) are used to analyze large RNase-protected fragments. Prior to electrophoresis, the samples are incubated at 70°C for 5 min in a solution containing 1 × electrophoresis buffer, 50% (v/v) formamide, and 6% (v/v) formaldehyde to denature the duplexes. After electrophoresis, the gel is dried onto Whatman (Clifton, NJ) 3MM paper and autoradiographed.

Polyacrylamide Gel Electrophoresis

Samples are dissolved in 99% (v/v) formamide, 0.1 mM EDTA, 0.025% (w/v) bromphenol blue, and 0.025% (w/v) xylene cyanol blue. The samples are incubated at 85°C for 2 min, cooled on ice, and loaded onto a 5–8% (w/v) polyacrylamide gel in 89 mM Tris–borate, 1 mM EDTA, pH 8.3, containing 7 or 8 M urea. After electrophoresis, the gel is dried and autoradiographed.

We have used polyacrylamide gels of varying dimensions, from 9.5 × 0.7 × 0.05 cm [BioRad (Richmond, CA) Protean II minigels] to 20 × 40 × 0.04 cm (sequencing gels). As a guide, 500 cpm (minigels) or 5000 cpm (14 × 17 × 0.08 cm gels) of ^{32}P-labeled RNAs is loaded per lane. The samples are quantified by Cerenkov radiation while still dry, and are then dissolved in the appropriate volume to facilitate gel electrophoresis.

More [^{32}P]RNA can be loaded and the time for autoradiography can be reduced. Thin gels can also be autoradiographed without drying. Using ^{35}S-labeled probes, the resolution of the bands in the gel is superior to ^{32}P; however, at least 10-fold more radioisotope must be used per assay, and the gels must be fixed to remove urea before drying. Detection of ^{35}S-labeled bands can be enhanced by fluorography [e.g., impregnating the gel with sodium salicylate (7)].

Although we have never tried this method with a nonradioactive probe, others (8) have done the RPA using digoxygenin-rUTP instead of [α-^{32}P]UTP, and the protected fragments were transferred to a nylon membrane and visualized using a chemiluminescence detection system. Apparently, the RNase digestion step was similar to that used for ^{32}P-labeled probes.

Optimization of RNase Digestion Conditions

Ribonuclease

The original protocols for the RPA used a combination of RNase A plus RNase T1 (1–4, 6). Nuclease P1 can be substituted for RNase A when the probes are rich in AU sequences (9). Nuclease P1 must also be removed with phenol–chloroform. In an RPA kit (Cat. No. 1410) from Ambion (Austin, TX), guanidinium isothiocyanate and ethanol are used to inactivate RNases A plus T1 and to precipitate the RNA, thus eliminating the phenol extraction step. RNase ONE (Promega) can also be used to obviate the need for protease digestion and phenol extraction, because RNase ONE can be inactivated by the addition of SDS (10). However, in our hands the addition of SDS to 0.1% (w/v) (the recommended amount) to stop the digestion resulted in complete degradation of the probe after denaturation, whereas addition of SDS to 0.6% (w/v) resulted in complete inactivation of RNase ONE. It is also worth remembering that RNase ONE is inhibited by 0.2 M NaCl, and thus the supplier recommends using 0.2 M sodium acetate for the digestion at temperatures of 20–30°C (10).

RNase T1 apparently contributes little to detecting mismatches in the duplexes, but rather contributes to the digestion of unhybridized probe and helps to reduce the background (1). We have always used RNase T1 from Calbiochem (La Jolla, CA), because this enzyme digests very thoroughly. RNase T1 from other sources is now generally calibrated in units defined by the generation of a sequencing ladder, that is, a partial digestion, whereas the above source uses the original definition of complete digestion of the substrate.

Concentrations of RNase A used for digestion vary approximately eight-fold, from 8 μg/ml (11) to 60 μg/ml (1). Our experience is that at any of these concentrations RNase A is not a limiting factor in the reaction. A standard concentration of RNase A (20 μg/ml) (Cat. No. R5125, 87 Kunitz units/mg; Sigma) and RNase T1 (5 units/ml) (Cat. No. 556785; Calbiochem) has consistently worked well in our hands for a variety of systems. However, the digestion activity on mismatched duplexes may vary with the source of the enzyme. For example, RNase A from Pharmacia (Piscataway, NJ) (27-0316-01, 80 Kunitz units/mg) at the above concentration resulted in the complete degradation of the annealed probe under the same digestion conditions.

Salt, Temperature, and Time

Recommended concentrations of salt (either sodium chloride or sodium acetate) for the digestion step vary from 0.1 to 0.3 M. The salt concentrations may be varied, depending on the GC content of the RNA, the specific RNase(s) to be used, or the specific temperature of the digestion. Digestion at 15°C in 0.1 M NaCl has been reported to optimize the recognition of small mismatches (12). The results obtained using this temperature depend on the salt concentration and the digestion time. Thus, in 0.1 M NaCl, the optimal time for digestion of heteroduplexes is 30 min; at 60 min the fully protected, positive control duplexes show signs of degradation. However, in 0.3 M NaCl, the digestion of heteroduplexes reaches completion only after overnight incubation. In the latter case, the RPA pattern is the same for digestions at 15°C overnight as for digestions at 30°C for 2 hr.

For optimal digestion, the salt concentration and temperature of choice may be determined empirically. A general outline of the several variables that must be considered when optimizing the assay is given in Table I. It may not be necessary to test all of the parameters for every system; for example, for an RNA with a high GC

TABLE I   Optimization of Ribonuclease Protection Assay[a]

Salt concentration[b] (M)	Temperature[c] (°C)	Hybridization time[d] (hr)	Digestion time[e] (hr)
0.1	15	2, 4, 16	0.5, 1
0.2	15	2, 4, 16	0.5, 1
0.2	30	2, 4, 16	0.5, 1
0.3	30	2, 4, 16	0.5, 1
0.3	37	2, 4, 16	0.5, 1

[a] Based on data using the ca. 340-nt satellite RNA of cucumber mosaic virus (CMV).
[b] Salt concentration for NaCl.
[c] Temperature of digestion.
[d] Three standard hybridization times.
[e] Two alternative digestion times.

content, one could eliminate the higher salt concentrations from the test assay, and the more stringent digestion conditions should be attempted first. These variables are more critical for studies requiring the maximization of detection of single base pair mismatches. As a convenient starting point, the digestions can be done at 15°C in 0.3 M NaCl, and incubating for either 1 hr or overnight, depending on the sequences to be compared. Complex patterns that may be difficult to compare can arise if the probe is greater than 1 kb, and/or the sequence variation between the target RNA and the probe is in the range 1–5% (e.g., see Refs. 13 and 14). In comparing the pattern obtained, differences between digestion at 15°C for 1 hr and 30°C for 2 hr are mainly due to less complete cleavage of the same mismatches under the former conditions, although a small number of mismatches cleaved at 30°C are not cleaved at 15°C.

Variation in Ribonuclease Protection Assay Profile: Cleavage Efficiency

The variation in the RPA pattern with digestion conditions is because not all existing mismatches are cleaved, and not all those detected are cleaved to completion. The efficiency of cleavage of a mismatched base pair depends on its nature and on the sequence context in which it appears. From data in the literature (15, 16) and our unpublished results, an estimation can be made of the efficiency of cleavage at single base pair mismatches for digestion at 30–34°C for 1–2 hr (Table II). In spite of the high error rates to be expected from small sample sizes that do not allow a precise determination to be made of the cleavage efficiencies for some types of mismatches, clear trends are perceived: about 60% of all mismatches are cleaved; those mismatches involving C:C, G:A, or point insertions/deletions are recognized in almost 100% of the cases; G:U pairs are rarely recognized; and C:A mismatches are cleaved

TABLE II   Cleavage of Mismatches by RNase A in RNA–RNA Hybrids[a]

Mismatch	Number of mismatches cleaved				
	I[b]	II[c]	III[d]	Total	%
C:A	9/13	2/10	33/38	44/61	72.13
C:C	2/2	2/2	3/3	7/7	100
C:U	5/5	2/2	2/4	9/11	81.8
U:U	5/6	—	6/9	11/15	73.3
A:A	4/5	0/1	7/8	11/14	78.6
G:A	4/4	—	9/9	15/15	100
G:G	0/1	—	4/4	4/5	80
G:U	2/14	0/2	5/51	7/67	10.4
Δ:N	—	1/1	25/26	26/27	96.3
Double mismatch	—	—	15/17	15/17	88.2
Triple mismatch	—	—	12/12	12/12	100
Total:	21/40	7/18	101/152[e]	129/210	61.4

[a] Number of mismatches cleaved by the enzyme per total number present. Any detectable cleavage (i.e., fragment detection, regardless of relative intensity) is considered positive.
[b] Data from analysis of mutations at codons 12, 13, and 61 of the human c-ki-ras gene, and from different strains of influenza virus (15).
[c] Data from variants of satellite of tobacco mosaic virus (elaborated by Kurath *et al.* [16]).
[d] Data from variants of CMV satellite RNA (M. Aranda, A. Fraile, and F. García-Arenal, unpublished observations, 1993).
[e] Only single mismatches considered.

in about 70% of the cases, whereas the cleavage efficiency of the remainder of the mismatch types varies between 70 and 80%. Double or triple mismatches are almost always cleaved.

The above data were obtained using RNases A plus T1, which recognize unpaired C plus U, or G, respectively, but not A. No such data are available for RNase ONE, which can recognize all four nucleotides (10).

Specific Applications

Viral Population Heterogeneity and Strain Variation

An example of the use of RPA both in detecting strain variation and population heterogeneity is illustrated in Fig. 1. The probe used for this assay is derived from a cDNA clone of cucumber mosaic virus (CMV) strain Fny, RNA 2 (17). The target RNAs include a field isolate, CMV241, that showed a range of symptom responses varying from mild mosaic to severe necrosis on *Cucurbita pepo*. This strain was passaged twice through sugar beet (*Beta vulgaris*), a local lesion host for CMV, and isolated lesions were assessed on *C. pepo* for symptom type (17). Ribonuclease pro-

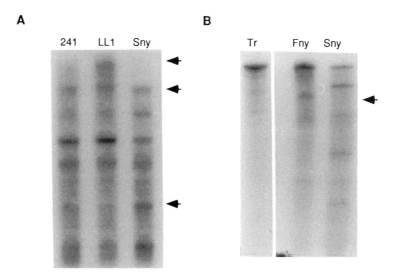

FIG. 1   Polyacrylamide gel electrophoretic analysis of heterogeneity in closely related strains of cucumber mosaic virus (CMV). Each lane represents the protected fragments for the indicated strain. (A) RNA 2 probe, derived from a full-length cDNA clone of Fny-CMV RNA 2. Arrows indicate bands found in the parental strain CMV241, which are more prominent in the different subisolates. (B) Partial RNA 1 probe derived from a cDNA clone of Fny-CMV RNA 1. The arrow indicates a band in the Fny-CMV lane, which is probably due to heterogeneity in the viral RNA population. Tr is the (+)-sense transcript RNA.

tection assays of CMV241 and two of the local lesion isolates, LL1-CMV (a necrotic variant) and Sny-CMV (a mild mosaic variant), showed that at least two variants of RNA 2 existed in the original CMV241 population, and these could be separated by local lesion passage (Fig. 1A). However, it could not be determined whether the large number of additional bands were due to either further heterogeneity in the CMV241 RNA 2 population, or differences between CMV241 and the Fny-CMV RNA 2 clone. To obtain a more accurate estimation of the extent of population heterogeneity, it is preferable to have a cDNA clone from the strain being examined (here, CMV241).

An important control in these assays is the (+)-sense transcript derived from the same cDNA clone as the probe. The RPA pattern of this positive control RNA helps to distinguish bands that occasionally arise from cleavage at AU-rich complementary regions vs cleavage at actual points of heterogeneity. For example, Fig. 1B shows an RPA using an RNA transcript from a partial cDNA clone of Fny-CMV RNA 1 as the probe, and the (+)-sense transcript of the same RNA as the positive control. This control shows a predominant band corresponding to full-length protection of the probe, with only minor bands of faster mobility. When Fny-CMV RNA was used as the target RNA, most of the probe was fully protected; however, a few minor bands appeared that were not observed in the control RNA, indicating some heterogeneity in the Fny-CMV RNA population. The Sny-CMV target RNA shows several additional bands in RNA 1, again indicating either differences vis-à-vis Fny-CMV RNA 1, or differences due to population heterogeneity in Sny-CMV RNA 1. When two virus strains give characteristic banding patterns such as this, the RPA is useful in distinguishing strains in a mixed infection (17).

Quantitation of RNA Levels

A quantitative RPA can be established using densitometric measurements of the intensities of individual fragments or completely protected probes, when the target RNA is in the range of 50–500 ng. For smaller amounts (0.5–50 ng), quantitation can be done by using 200,000 cpm of high specific activity (2.5×10^6 cpm/μg) probes in each assay; higher RNA amounts (5–500 ng) should be quantified with 50,000 cpm of low specific activity probe (1×10^5 cpm/μg) (18).

Quantitation of Sequence Variation

It is beyond the scope of this technical review to describe the theoretical basis for using the RPA quantitatively in determining genetic and phylogenetic relationships. However, the following comments are relevant to the quantitative use of the RPA.

1. Ribonuclease protection assay data can be used to estimate nucleotide substitution values between variant RNAs. In an analysis of variation in CMV satellite RNA, an estimation of the extent of nucleotide substitution (d_{ij}) between two RPA types (i and j) in an RNA population was established, and correlated with the $\hat{d}_{ij}$ estimated from sequence data (19). $\hat{d}_{ij} = (m_i + m_j - 2m_{ij})/N$, where m_i and m_j are the number of fragments in patterns i and j, m_{ij} is the number of fragments common to patterns i and j, and N is the length, in nucleotides, of the probe. Thus, for example, with G-sat RNAs (Fig. 2, lanes 3) a total of 22 fragments (of lengths 88, 90, 96, 98, 99, 101, 102, 154, 156, 162, 171, 174, 180, 183, 187, 188, 201, 209, 268, 300, 315, and 339 nt) were obtained, and for K8-sat RNA (Fig. 2, lanes 5) a total of 13 fragments (of lengths 73, 74, 75, 109, 110, 111, 118, 119, 187, 190, 201, 220, and 235 nt) were seen. Only two fragments (187 and 210 nt) were common to the two RPA patterns. Thus, the estimated d_{ij} for K8- and G-sat RNAs will be $\hat{d}_{ij} = (13 + 22 - 4)/339 = 0.0914$. To minimize the possibility that two different fragments are considered as one, the electrophoretic resolution of the RPA was increased by electrophoresis on two different gels for each RNA, in which the xylene cyanol blue dye migrated 20 and 40 cm from the origin. This approach was applied to natural populations of CMV satellite RNAs from field epidemics, as shown in Fig. 2. Once values of nucleotide substitution per site were estimated, it was possible to calculate parameters describing the population structure, such as nucleotide substitution per site between any two types, within or between subpopulations (intra- and interpopu-

A   **B**

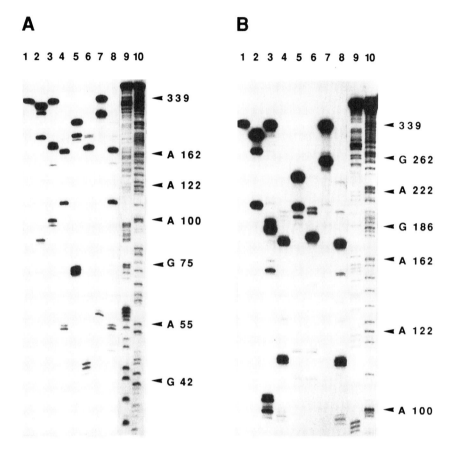

Fig. 2   Autoradiograms of RPA of a cRNA probe complementary to B2-sat RNA by CMV-satellite RNA variants B2 (lane 1), B3 (lane 2), G (lane 3), IX (lane 4), K8 (lane 5), WL1 (lane 6), WL2 (lane 7), and R (lane 8). Lanes 9 and 10 are molecular size markers from sequencing reactions of 5′ end-labeled B2-sat RNA transcripts done with RNase T1 (lane 9) or RNase φM (lane 10). From each RPA two electrophoreses were done, in which xylene cyanol blue migrated 20 cm (A) or 40 cm (B) from the origin. The sequences of satellite RNAs in lanes 1–8 have been published.

lation diversities), and to obtain a more precise view of the population structure and evolution of CMV satellite RNA in the field (20).

2. A major drawback of the RPA for the quantitative study of relationships among RNA variants, that is, the estimation of genetic distances, and of the genetic structure of their populations, arises from the observation that many complex patterns are largely due to partial digestion at mismatches. Thus, from the number of detected

bands it is not easy to determine the number of cleaved sites, which in turn means that a simple relationship linking RPA data (for two RNA variants) with genetic distance (nucleotide substitutions/site) has not been developed. In the analysis of the population structure of the satellite of tobacco mosaic virus, Kurath *et al.* (16) were able to identify the positions of cleaved mismatches in the satellite of tobacco mosaic virus sequence, and thus, to determine the number of cleaved sites from the number of fragments. Also, by sequencing a substantial subset of the RPA datasets, they concluded that 50% of the existing mismatches (i.e., the mutations respective to the probe) were cleaved.

3. Nonlinearity between the estimation of genetic distances based on the RPA and on nucleotide sequence analysis may occur, depending on a number of factors (21). This is primarily due to the difficulty in relating the number of observed fragments to the number of cleaved mismatches, because of an unknown number of partial cleavages. It is possible to overcome this problem by using 3′ end-labeled probes instead of uniformly labeled probes. The use of end-labeled probes and the analysis of samples incubated for different digestion times [as initially proposed for the detection of restriction endonuclease sites (22)] in our hands allow for the unequivocal determination of cleavage sites, and thus provide a ready determination of genetic distances between any two sequences. This approach makes the RPA analysis somewhat more complicated, lessening one of the main advantages of this technique. Also, mismatches occurring near the labeled end of the probe are either poorly detected or not detected at all. Moreover, the labeled end is lost with increasing digestion time before a constant pattern (plateau) is reached, probably due to "breathing" at the ends. Thus, the probe should not be coterminal with the RNAs to be analyzed.

Mapping Sites of Nucleotide Sequence Variation

As detailed in several studies, the mapping process involves making partial transcripts from the same cDNA clones and also transcripts derived from cDNA subclones. By matching bands on the gels common to different length probes, the order of the fragments can be deduced (6). This process is also in part subject to the same pitfalls enumerated above for quantitating sequence variation, that is, the problem of comigrating bands and incomplete digestion. Nevertheless, it is still possible to create maps of sequence heterogeneity between strains or isolates of the same virus (6).

Concluding Remarks

The application of the RPA to various facets of diagnostic, molecular, epidemiological and evolutionary virology allows for the rapid analysis of large numbers of samples for microsequence heterogeneity. Although this technique does not provide as much information as nucleotide sequence variation, the RPA can be used as an initial screen of large population samples for variation, to determine which samples should be subjected to cloning and nucleotide sequence determination. Whereas direct nucleotide sequencing of an RNA determines the sequence of the dominant member of a population, often masking the presence of variants present at levels of less than 20% of the population, the sequencing of cDNA clones allows these variants to be detected and identified. However, the proportion of cDNA clones that reflect the sequence of a variant in a population does not necessarily correspond to the actual proportion of this variant in the population. Here, the RPA can first be used to provide an initial estimation of the extent of variation in the population and then, using RNA transcripts derived from a variant cDNA clone, the RPA can be used to determine the concentration of that variant in a population. The limit of detection of a particular variant is in the range of 1–5% of the population, the sensitivity depending on the background observed with the positive control.

Although the RPA does not provide an estimation of rates of mutation, it can be used to estimate the rate of selection of particular variants in a population, and to examine the effects of host genotype (23) or environmental factors on the rate of selection of variants. As with any technique, the RPA has its limitations and these should be borne in mind, especially when using the RPA quantitatively for more sophisticated analyses in taxonomy and phylogeny. As more data become available comparing the results from an RPA with a nucleotide sequence analysis, the limitations will become more certain, and the conclusions drawn from an RPA will become more definitive.

Acknowledgments

Unpublished work described herein was supported in part by Grant AGR-90-0152 Comisión Interministerial de Ciencia y Tecnología, Spain, and USDA Grant No. 91-37303-6426.

References

1. E. Winter, F. Yamamoto, C. Almoguera, and M. Perucho, *Proc. Natl. Acad. Sci. USA* **82**, 7575 (1985).
2. G. Kurath and P. Palukaitis, *Mol. Plant-Microbe Interact.* **2**, 91 (1989).
3. C. López-Galíndez, J. A. López, J. A. Melero, L. de la Fuente, C. Martinez, J. Ortín, and M. Perucho, *Proc. Natl. Acad. Sci. USA* **85**, 3522 (1988).
4. G. Kurath and P. Palukaitis, *Virology* **173**, 231 (1989).
5. G. S. Beckler, *Promega Notes* **39**, 12 (1992).
6. J. Owen and P. Palukaitis, *Virology* **166**, 495 (1988).
7. J. P. Chamberlain, *Anal. Biochem.* **98**, 132 (1979).
8. I. Wundrack and S. Dooley, *Electrophoresis* **13**, 637 (1992).
9. G. Brewer and J. Ross, *in* "Methods in Enzymology," Vol. 181B, p. 202. Academic Press, San Diego, 1990.

10. G. Brewer, E. Murray, and M. Staeben, *Promega Notes* **38,** 1 (1992).
11. D. K. Grange, G. S. Gotterman, M. B. Lewis, and J. C. Marini, *Nucleic Acids Res.* **18,** 4227 (1990).
12. K. Kirkegaard, *Promega Notes* **12,** 3 (1988).
13. J. Cristina, J. A. López, C. Albó, B. García-Barreno, J. García, J. A. Melero, and A. Portela, *Virology* **174,** 126 (1990).
14. C. López-Galíndez, J. M. Rojas, R. Nájera, C. C. Richman, and C. Perucho, *Proc. Natl. Acad. Sci. USA* **88,** 4280 (1991).
15. M. Perucho, *Strat. Mol. Biol.* **2,** 37 (1989).
16. G. Kurath, M. E. C. Rey, and J. A. Dodds, *Virology* **189,** 233 (1992).
17. M. J. Roossinck and P. Palukaitis, *Mol. Plant-Microbe Interact.* **3,** 188 (1990).
18. E. Moriones, I. Díaz, E. Rodríguez-Cerezo, A. Fraile, and F. García-Arenal, *Virology* **186,** 475 (1992).
19. T. H. Jukes and C. R. Cantor, *in* "Mammalian Protein Metabolism" (H. N. Munro, ed.), p. 21. Academic Press, New York, 1969.
20. M. A. Aranda, A. Fraile, and F. García-Arenal, *J. Virol.* **67,** 5896 (1993).
21. J. Dopazo, F. Sobrino, and C. López-Galíndez, *J. Virol. Methods* **45,** 73 (1993).
22. H. O. Smith and M. L. Birnstiel, *Nucleic Acids Res.* **3,** 2387 (1976).
23. G. Kurath and P. Palukaitis, *Virology* **176,** 8 (1990).

Section III

DNA Viruses

[18] Baculoviruses and Baculovirus Expression Systems: Use in Expression of Proteins

David H. L. Bishop

Introduction

The use of expression vectors for the synthesis of proteins has expanded rapidly over the past few years. There is now an abundant choice available to the researcher and to the manufacturer. Initially, bacteria were employed as vectors due to the ease of their cultivation and the knowledge that existed concerning their transcription promoters and translation mechanisms. Also, they offered the opportunity for the induction of high-level expression of proteins from genes encoded in bacterial plasmids. Subsequently, other systems were investigated and established in order to address a variety of needs that the bacterial systems could not accommodate (e.g., protein folding, glycosylation). Among these other systems is the use of insect-specific baculoviruses and their hosts, insect cells and larvae. In the baculovirus system there have been some significant advances, including developments not addressed in earlier reviews and manuals that describe the use of these expression vectors. One such development concerns the ability to recover recombinants at frequencies that approach 100%. Another is the use of multiple gene expression vectors that allow the synthesis of four or more foreign proteins from the same virus. In addition to a general background description of baculoviruses, introductions of specific topics, and descriptions of relevant procedures, this article summarizes what is known about the properties of the proteins made in the baculovirus system. The descriptions are limited to those that pertain to *Autographa californica* nuclear polyhedrosis virus (AcNPV). The reader is referred to the manuals for additional background information (references, etc.) and basic molecular biology procedures, reagents, and equipment that are required (1–3).

What Are Baculoviruses?

Baculoviruses are arthropod-specific viruses that infect particular species of insects, or crustacea (4). Some baculoviruses are pathogenic for their hosts. As a consequence they have been employed for over 100 years as biological control agents (e.g., insecticides), thereby mimicking their natural role in the environment as regulators of arthropod populations. Individual baculoviruses may exhibit a highly restricted host range in nature (e.g., infecting only one or a few insect species). For example, the nuclear polyhedrosis virus (NPV) of *Euproctes chrysorrhoea,* the brown tail moth, infects only that one lepidopteran species. It does not infect larvae of other moth species, even alternative members of the same lepidopteran family. Nor does it infect other members of the Class Insecta, let alone noninsect invertebrates, or microbes, or plants, or vertebrates such as humans. Other baculoviruses may infect a small number of insect species within, for example, the Order Lepidoptera. However, none of these viruses is known to infect a large number of such insects, or members of different orders of the Class Insecta (Lepidoptera and Hymenoptera, or Coleoptera, or Neuroptera, etc.), nor any other animal, plant, or microbial species. As a laboratory or manufacturing system, therefore, baculoviruses are among the safest systems with which to work, requiring only good microbiological practices (2, 3).

In the natural course of events with the lepidopteran-specific baculoviruses (4), larvae of susceptible species become infected when they eat virus-contaminated materials (e.g., foliage). Following infection, viruses are released from dead larvae and dispersed by a variety of procedures, thereby contaminating the food of subsequent generations of the host. Even for those baculoviruses that may infect several types of insect, some of these species can be productively infected only at very high virus doses (e.g., $>10^5$ viruses per second instar caterpillar), whereas other insect species are infected at low doses (e.g., 10–100 viruses per second instar caterpillar). Consequently, in nature, the effect of these viruses on insect populations is also relatively specific (4).

The family Baculoviridae is divided into two main groups (5), the polyhedrosis viruses (suggested genus name, *Polyhedrovirus*), and the granulosis viruses (suggested genus name, *Granulovirus*). Both produce virions that are occluded by a large crystalline matrix of protein (polyhedrin, granulin). In addition, there are some nonoccluded viruses assigned to the family (5). For the occluded viruses, the particles are either called polyhedra or granules. The multisided polyhedra vary in size and shape and can be >1000 nm in diameter, occluding >100 virions. Granules generally have an ovoid shape (about 600 nm long, 350 nm wide) and generally occlude single virus particles (4, 5).

For both viruses, the virions within the occlusion bodies are enveloped by a lipid membrane that is formed during virus morphogenesis within the infected cell nucleus (4). This is also the site where the numerous polyhedra and granules are produced (>50 per cell, depending on the virus and the cell type). Because there are other, completely different viruses that make polyhedra in the cytoplasm of infected cells (cytoplasmic polyhedrosis viruses, *Cypovirus,* Reoviridae) (5), baculoviruses that produce polyhedra are called nuclear polyhedrosis viruses (NPVs).

At least for the NPVs, the fully formed occlusion body is covered by a calyx that is largely composed of carbohydrate but also possesses a virus-coded phosphoprotein species. In addition, both NPVs and GVs produce enveloped forms of their virions that bud from the surface of an infected cell (4, 5). The synthesis of budded viruses is initiated earlier in the infection course than the synthesis of occluded particles.

Budded virions have a virus-coded glycoprotein that is not associated with the occluded virions, although the latter have at least one other glycoprotein. As indicated above, the purpose of the occluded form of a baculovirus is to protect the virus in the environment (i.e., between hosts). The purpose of the budded form is to spread an infection within a host. Occluded baculoviruses may remain infectious for many years in soil and other locations, depending on the circumstances (4). By contrast, budded viruses rapidly degrade in the environment.

Infection Course of Baculoviruses

The genome of a baculovirus consists of a double-stranded, circular DNA (2–5). The size of this DNA depends on the virus. For AcNPV, the size is about 134 kilobase pairs (kbp). It codes for >100 gene products, only one of which, IE-1, has been shown to contain an intron (6). Although genes are located on one or other of the DNA strands, it appears that few overlap. Purified virus DNA is infectious per se (7, 8).

Briefly, the infection course of a virus such as AcNPV involves the attachment of the virion to cell receptors, uncoating of the nucleocapsid from the virus envelope, and its transport to the cell nucleus (4). How these processes occur and how the virus DNA is introduced into the nucleus are not known. For insect-specific viruses derived from polyhedra, or granules, the process must be preceded by digestion of the occlusion body in the alkaline environment of the insect midgut (pH >9), followed by the release of the virion(s) to infect certain midgut cells (4). Polyhedra are not involved in the spread of infection between cells and organs within infected insects (where this occurs).

Once in the nucleus of an infected cell the virus DNA serves as a template for the synthesis of mRNAs representing a subset of the baculovirus genes. These genes have been described as immediate early (IE) genes (9–12). As mentioned above, one of the IE mRNA species is spliced (6), in contrast to most, if not all, of the other mRNA species that are made later in the infection. Because IE mRNA synthesis is catalyzed by an existing cellular enzyme, it is insensitive to the presence of protein synthesis inhibitors. Certain of the IE proteins *trans*-activate some 1000-fold a second set of mRNA species (9–12). These are the so-called delayed early (DE) genes. The promoters of both types of early genes have upstream TATA boxes and generally resemble those of chromosomal genes (13). Their mRNAs are initiated near a conserved CAGT motif. The increase in the DE gene synthesis is sensitive to the presence of protein synthesis inhibitors.

Following the synthesis of the early gene products, the late-phase products are made (14). Virus DNA replication is initiated and involves a virus-coded DNA polymerase (15). The structural proteins required for nucleocapsid assembly (in the cell nucleus) and virus morphogenesis at the cell surface are produced. To form budded virions, nucleocapsids are transported from the nucleus to the plasma membrane (16), where they acquire an envelope and a virus-coded glycoprotein species (gp67 for AcNPV).

Finally, during the very late phase both polyhedrin and another very late protein, p10, are made (4). Importantly, the syntheses of these two proteins are high, accounting for about 50% of the total protein in an infected cell (or larva). Both proteins are involved in the occlusion process whereby nucleocapsids in the cell nucleus are enveloped, occluded, and acquire a calyx with associated proteins (4). When eventually the organism dies, occluded virus particles are liberated (4). For different baculoviruses and other hosts, the course of infection may vary. Generally, however, in permissive hosts baculovirus infections are highly prolific, for example, yielding up to 10^9 polyhedra per larva (4).

At the present time the processes involved in the transcriptional control of the various baculovirus genes are understood only in outline. It is clear that the IE gene products are transcribed by cellular enzymes. This is demonstrated by the fact that a productive infection can be initiated by transfection of cells with purified virus DNA. The observation that early mRNA synthesis is sensitive to α-amanitin suggests an involvement of the RNA polymerase II enzyme of the host. As noted above, the sequences that regulate the transcription of the early mRNAs appear to be similar to those of many chromosomal genes of eukaryotes (upstream TATA boxes, etc.). Apart from the observation that the IE gene products *trans*-activate DE genes, little is known about the later processes of virus mRNA transcription. Interestingly, the synthesis of viral mRNA becomes insensitive to α-amanitin as the infection proceeds (17). Also, transcription of the late and very late genes involves signals and initiation sites (generally within a consensus TAAG sequence) that differ from those utilized by the early genes (2, 3).

In summary, there are four phases of transcription and translation that can be defined during a baculovirus infection (see Fig. 1). These are the IE and DE phases followed by the late and very late phases. Although this operational separation of gene expression is a useful concept, the phases are not entirely distinct, and some proteins are made in more than one phase. Often this involves the use of different transcriptional promoters during the different phases (2, 3).

Cell Culture

A number of insect cell lines support baculovirus infections. For AcNPV these include *Spodoptera frugiperda* cells [IPLB-SF-9 available from ATCC (Rockville, MD) CRL1711, and IPLB-SF-21). Other lines that support AcNPV replication have been derived from *Trichoplusia ni* (available commercially) and *Mamestra brassicae* (2, 3). Claims have been made that higher protein yields are obtained with the latter two cell lines. However, such claims must be considered against actual experience

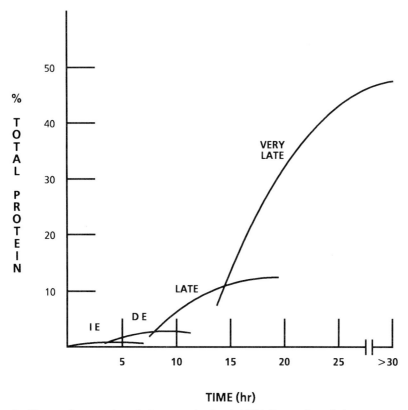

FIG. 1   Temporal expression of virus proteins by AcNPV. Expression of virus genes representing the immediate early (IE), delayed early (DE), late, and very late genes is shown as the percentage of total cellular protein that they represent.

and the general utility of these alternative cells (adaptability to fermentation, use in plaque assays, stability, ease of handling, etc.). We and others recommend the use of the SF-21 (or SF-9) cells as the first choice (2). If yields of a foreign protein are unsatisfactory, then other cell lines may be investigated; however, we have not yet identified better lines.

SF cell lines can be used for plaque assays as well as for monolayer and stirred cultures in Grace's, TC100, or TNM-FH media supplemented with 5–10% (v/v) fetal calf serum (FCS), or (after adaptation) the serum-free IPL-41, SF900-II, or Ex-Cell 401 media, as well as others that are available commercially (2, 3). These are highly enriched media buffered to pH 6.2. The SF cells can be grown at temperatures <29° C and stored frozen at −70° C. For storage, cells are resuspended at concentrations of 1–2 × 10⁶ in TC100 containing 10% FCS and 10% (v/v) dimethyl sulfoxide

(DMSO) (culture grade), or in SF900-II, cooled to 4°C and transferred to liquid nitrogen, preferably using a procedure that involves a gradual reduction in temperature. Cells are recovered by fast-thawing and diluting fivefold in fresh TC100 containing 10% FCS (or SF900-II). For monolayer cultures, the TC100 medium is replaced with fresh medium after 2 hr and the culture monitored and passaged when the cells are confluent. For suspension cultures in SF900-II, or other serum-free media, the cells are recovered after 2 hr and resuspended in fresh SF900-II medium (etc.). Antibiotics [penicillin (500 units/ml) and streptomycin (500 μg/ml)] may be included in the medium.

Cells are passaged following removal of spent medium, by resuspension in fresh medium and dispensing into new containers (T-flasks, Roux bottles, round-bottom flasks with magnetic stirrers, etc.) following a 10 to 20-fold dilution. It is important that the viability of cells is maintained at levels that are >98% when they are employed for virus infections. Because cells may be stored for weeks at room temperature, prior to use they should be passaged and brought to a high viability as demonstrated by trypan blue exclusion.

Plaque Assays

Budded viruses spread a baculovirus infection from cell to cell. In some insect cell cultures (e.g., SF-21 and SF-9) they produce plaques of infected cells under a semisolid overlay, plaques that can be identified visually, or by a simple staining procedure. For plaque assays, monolayers of SF-21 or SF-9 cells (but not those adapted to suspension culture in serum-free media) are infected with a suitable dilution of virus (in TC100) when the cells are just confluent and 1 hr later overlaid with TC100 medium containing 5% FCS, antibiotics, and 1% (w/v) SeaKem [or 1.5% (w/v) SeaPlaque; FMC, Philadelphia, PA] agarose. Some investigators add a small amount (1 ml per 35-mm dish) of medium to the surface of the overlay after it has gelled. Plates are incubated at 28°C in a humid atmosphere for 3 days and inspected visually for the presence of plaques and whether polyhedra can be identified within the infected cells. Cells may be stained with 0.025% (v/v) neutral red in phosphate-buffered saline (PBS). Plaques may be recovered using a Pasteur pipette and resuspended in TC100 medium for replaquing, or to prepare virus stocks. The latter usually involves two or three consecutive passages before high-titered stocks are obtained (see below).

Virus Infections (Monolayers, Stirred, and Fermenter Cultures), Seed, and Working Stocks of Virus

For preparing virus stocks, monolayers of cells are infected when they are just confluent, using low multiplicities of virus [<0.01 plaque-forming units (pfu) per cell]. For protein expression they are infected at high multiplicities (5 pfu/cell) in order to ensure that all the cells are infected at the same time. For stirred cultures in TC100, or other serum-supplemented medium, and although the cells will grow to titers of about 2×10^6/ml, they are usually infected when the cells reach a density of about 8×10^5/ml. Using serum-free medium in stirred cultures of medium-adapted SF cells, the cells will grow to about 7×10^6/ml and they can be infected when they reach a density of about 2×10^6/ml.

In fermenters using serum-free media (e.g., SF900-II, Ex-Cell 401), cells will grow to titers of 4×10^7/ml. They are infected when they reach a density of $<1 \times 10^7$/ml. For fermentation, in addition to high viability, three points are critical if high cell titers are to be obtained. First, the cells must be previously adapted to growth in the medium that is to be used. This is achieved by a weaning process (i.e., passaging cells in mixtures of the new and the previously used medium) followed by further passage and maintenance in the new medium. Second, the by-products of cell growth must be removed either using spin-filter or hollow fiber filtration technologies. With SF900-II, a 0.2-μm hollow fiber, and a 1.2-liter volume culture of adapted cells, we usually replace medium at a rate of 2% volume exchange per hour up to titers of 7×10^6 cells/ml. This is then increased to 4% per hour until just before infection (see below). Third, the dissolved oxygen level must be kept high. On the basis of that water-saturated air contains 20% dissolved oxygen, the oxygen level in the fermenter is maintained at 15% dissolved oxygen, using an oxygen-fed sparger (i.e., 75% of the air value). With the available serum-free media, fermenters employing stirred reactors are satisfactory provided these three points are addressed. It should be noted, however, that the growth rates of SF cells vary according to the culture conditions and the medium that is employed. In stirred cultures with TC100, the doubling time is about 24 hr. In fermenters with adapted cells and SF900-II medium, the doubling time is about 38 hr.

Infection of monolayer cultures usually involves the application of a virus stock (or a dilution thereof in TC100 medium) to the culture 1 hr prior to addition of the maintenance medium. Virus stocks are prepared by harvesting the supernatant fluids at 3–4 days postinfection. When infections are initiated at low multiplicities, virus stocks from stirred cultures are usually of the order of 1×10^8 pfu/ml. After clarification, viruses are stored at 4°C until required, or frozen in aliquots. It is advisable to prepare seed stocks of virus from which working stocks are produced for routine use (preparation of virus DNA, protein expression, etc.). The seed stocks should come from plaque-purified virus that has been passaged no more than three or four times in order to avoid problems with defective interfering particles. We also recommend undertaking sterility assays on the seed stocks to avoid subsequent contamination problems.

At high multiplicities of infection of monolayer or stirred cultures, cell growth is rapidly arrested and the cells are harvested at an appropriate time (<4 days postinfection). This time is determined empirically and depends on the growth rate of the virus (for the preparation of stocks), or the expressed protein yields and the integrity of these products (evidence of breakdown, etc.). Using polyhedrin or p10 promoters, expression is not detected until about 15 hr postinfection. However, depending on the virus, the maximum yields of product may occur anytime from 30 to 72 hr postinfection using these promoters. Some investigators prefer to infect at high cell concentrations (e.g., 10^7 cells/ml) by recovering cells and suspending them in fresh medium prior to the addition of virus. This allows spent medium and cell by-products to be removed. After virus adsorption the cells are then diluted 10-fold in fresh medium for the rest of the infection course.

Infection of cells in fermenters is more complicated due to the high cell densities involved. On the basis of our experience to date, we recommend that cells be perfused at an increased rate immediately prior to infection (e.g., 16% volume exchange per hour for 4 hr), the virus stock then added to give a multiplicity of 5 pfu/cell (if possible) while perfusion is stopped (for 1 hr). Perfusion is then restarted, at 16%/hr for 4 hr, and then maintained at a rate of 4%/hr until harvest. At all times the culture is stirred and the dissolved oxygen is maintained at 15% (75% of the air value). In other fermenters (media, cells, etc.) other conditions may be more appropriate.

Purification of Virus and Infectious Virus DNA

Cells are infected in stirred cultures (or in fermenters) and the supernatant fluids clarified by low-speed centrifugation. These are then centrifuged at 100,000 g for 1 hr at 4°C and the virus pellet resuspended in TE buffer [10 mM Tris-HCl, 1 mM ethylenediaminetetraacetic acid (EDTA) (pH 8.0)]. If required, virus may be further purified by centrifugation at 100,000 g over a step sucrose gradient [5 ml of 50% (w/v) sucrose, 5 ml of 10% sucrose, each in TE buffer] and recovered from the interface prior to dilution fivefold in TE buffer and pelleting. Following suspension in 2 ml of TE buffer, DNA is extracted in the presence of TE and 4% N-laurylsacrosine, incubated at 60°C for 30 min, and aliquots of 0.5 ml layered on top of CsCl gradients composed of 5 ml of 50% (w/w) CsCl containing ethidium bromide (25 μg/ml). After centrifugation at 150,000 g for 18 hr at 20°C, the two DNA bands (supercoiled and relaxed DNA) are collected. Ethidium bromide is removed by four successive butanol extractions involving equivolume mixtures and phase mixing by inversion. The upper butanol phases are discarded. The aqueous phase is then dialyzed against two successive 500-ml volumes of TE buffer (4°C) to remove the CsCl and the DNA concentration determined spectrophotometrically. Finally, the virus DNA is stored at 4°C (i.e., not frozen or precipitated) in order to maintain its infectivity.

Single Gene Replacement Expression Vectors Using p10 or Polyhedrin Promoters

The baculovirus genes that are active in the very late phase, in particular those that are involved in the occlusion process (e.g., polyhedrin, p10), are redundant in relation

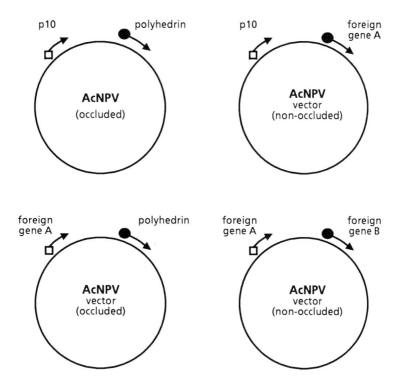

FIG. 2   Schematics of the gene organization of AcNPV (upper left) and expression vectors
with foreign genes (*A* and *B*) substituted into the polyhedrin (upper right), p10 (lower left), or
polyhedrin and p10 sites (lower right). The promoters (polyhedrin, filled circle; p10, open box)
and directions of transcription (arrows) are shown. Viruses lacking polyhedrin make only nonoccluded (i.e., budded) viruses. Viruses with polyhedrin make occluded and nonoccluded
viruses.

to the formation of budded virions. Therefore, and without inhibiting the latter, expression vectors can be made by replacing one of these very late genes by a foreign
gene, provided that the appropriate transcription control elements are retained.

   Shown in Fig. 2 is a schematic of the circular genome of AcNPV with the positions
indicated for the resident p10 and polyhedrin genes and their promoter elements (not
to scale). Not shown in this schematic are the transcription terminators and polyadenylation signals that are located downstream of the two genes, nor the other virus
genes that are located in between p10 and polyhedrin. Also shown in Fig. 2 are schematics of the arrangements of expression vectors in which the p10 and/or the polyhedrin coding sequences are replaced by foreign genes. Replacement of the polyhedrin gene results in the formation of nonoccluded progeny. Although the p10 gene is
involved in an aspect of the occlusion process, when it is replaced occluded progeny
are still made. However, these appear to be less stable than those made in the presence

of p10. The p10 protein forms large fibrous masses in the cytoplasm and nuclei of infected cells. The functions of these structures are not known.

Using a microscope, polyhedra can be observed late in infection in insect cells. Therefore, the procedures that were initially adopted for obtaining baculovirus expression vectors involved the identification and isolation of polyhedrin-negative recombinants (18, 19). These were produced by replacement of the amino-terminal portion of the polyhedrin coding region with a foreign gene. The latter was inserted either out of frame with the residual carboxy-terminal portion of the polyhedrin protein, or separated from that sequence by a translation termination codon in order to prevent the formation of chimeric products. Subsequently, replacement vectors were developed in which all of the polyhedrin coding region was removed. Depending on the gene, and provided the promoter is left intact, such vectors may express foreign proteins to high levels, indicating that the DNA sequences within the polyhedrin coding region are not components of the promoter. For some foreign genes, evidence has been presented that chimeras of the amino terminus of polyhedrin with downstream foreign protein sequences can lead to higher expression levels. Most probably this is simply due to a more stable product rather than to an involvement of the retained DNA sequences in the promoter.

Instead of the laborious search that is required to identify polyhedrin-negative progeny (which occur at frequencies of about 1–2%), other procedures have been developed to obtain and detect recombinants. These include the use of reporter genes (see later), as well as foreign gene and gene product assays (2, 3). Generally these methods have now been superseded by procedures that yield recombinants at frequencies that approach 100% (see "BacPAK6").

Using p10 or polyhedrin replacement vectors, and depending on the foreign gene that is expressed, it is possible to obtain high-level expression of a foreign protein. In our experience some 70% of the foreign genes that are expressed using polyhedrin or p10 promoters give products that exceed 20% of the total cell protein, as evidenced by analysis of stained gel patterns. On a molar basis several of these proteins are made at levels that correspond to those of the p10 or polyhedrin gene products made from wild-type virus, that is, representing some 50% of the total protein recovered from an infected cell.

Nonreplacement Expression Vectors Using Duplicated p10 or Polyhedrin Promoters

As an alternative to using replacement vectors, a duplicated baculovirus promoter can be employed to direct the expression of a foreign gene (20). To achieve this, a cassette of a promoter and associated foreign gene is suitably positioned in the virus genome (i.e., not located within an essential gene of the virus) to allow expression of the foreign sequences in a viable virus. The cassette may be introduced into an intergenic region, or within a nonessential coding or noncoding sequence. It is usual to include in the cassette downstream sequences that provide polyadenylation sites and signals in order to reduce the possibility that antisense RNA is made that could inhibit expression of genes encoded on downstream, complementary sequences. Polyadenylation sites may be provided from copies of baculovirus sequences, or from other eukaryotic sources [e.g., simian virus 40 (SV40), hemoglobin genes].

An example of an expression vector with a duplicated p10 promoter is shown in Fig. 3, with the introduced sequence positioned upstream and in the orientation opposite to the resident polyhedrin gene and its promoter (21). This expression vector makes occluded as well as budded viruses. In the case shown in Fig. 3 the orientation of the introduced promoter is probably not critical. However, if identical sequences are placed in juxtaposition (e.g., a duplicated polyhedrin promoter next to, and in the same orientation as, the resident polyhedrin promoter), and if the intervening sequences do not include essential genes, recombination between the aligned homologous sequences could result in the deletion of the foreign gene. Although this appears to be an infrequent event, the preferred option is to position the copy of the promoter (and other duplicated host elements) in an orientation opposite to the resident sequences and ideally in another region of the virus genome, so that there are essential virus sequences in between the copies. As mentioned previously, downstream polyadenylation sites and transcription termination signals are usually provided. One

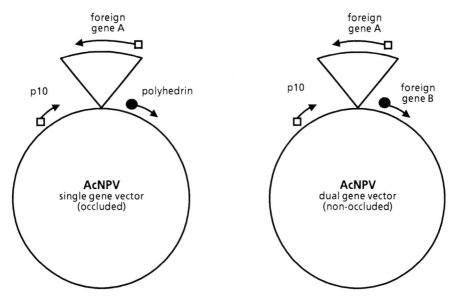

FIG. 3   Schematics of expression vectors with duplicated p10 promoters located upstream of the polyhedrin site. For details see Fig. 2 and text.

place that has been employed to insert additional promoters in the AcNPV genome is the *Eco*RV site in the *Eco*RI "I" fragment of that virus (20). This site is immediately upstream of the polyhedrin promoter and within the promoter element of a nonessential gene.

Multiple Gene Expression Vectors Using p10 and Polyhedrin Promoters

Dual gene vectors can be made by employing both the polyhedrin and p10 promoters to express foreign genes. Examples are given in Figs. 2 and 3. In Fig. 2 one of the schematics shows both the p10 and polyhedrin genes replaced by foreign genes (22). In this case the genes may be the same foreign sequences or different genes. In Fig. 3 one foreign gene is shown under the control of the resident polyhedrin promoter and the other is shown under the control of a copy of the p10 promoter. In this vector the p10 locus is not used. In the case in which the foreign gene is the *Escherichia coli lacZ* it can act as a reporter to detect dual gene recombinants (i.e., *lacZ* and another foreign gene). Such recombinants are selected when virus plaques are stained. They give blue plaques when the β-galactosidase substrate (e.g., X-Gal) is added to the overlay of the plaque assay.

Further examples of multiple expression vectors are given in Figs. 4 and 5. In Fig. 4, triple gene expression vectors are shown in which foreign genes (A, B, and

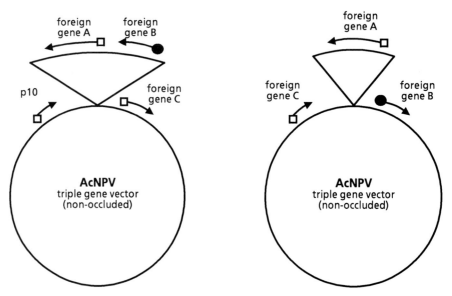

FIG. 4   Schematics of expression vectors that express three foreign genes (A, B, and C). For details see Fig. 2 and text.

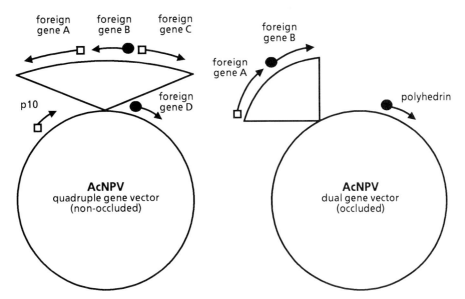

FIG. 5   Schematics of expression vectors that express four foreign genes at the polyhedrin locus (A–D; upper panel), or two foreign genes at the p10 locus (A and B; lower panel). For details see Fig. 2 and text.

C) are placed under the control of either the resident promoters (p10, polyhedrin), or copies of these promoters placed at the *Eco*RV site that is upstream of the polyhedrin promoter (23). For each of these genes the necessary polyadenylation sequences are provided either from baculovirus genes or from other eukaryotic sources. In one of the examples given in Fig. 5, a dual vector is shown with two foreign genes located at the p10 site. In this case the polyhedrin gene is left intact so that occluded viruses are made. In the other example a quadruple vector is shown with four foreign genes at the polyhedrin locus (23). In this case the p10 site is left intact.

For all of the examples provided in Figs. 2–5, expression vectors have been constructed and demonstrated to produce the expected products. Although it has yet to be reported, there appears to be no reason why six gene vectors could not be made with the available transfer vectors (see below) and based on combinations of the arrangements shown in Fig. 5, that is, using both the p10 and polyhedrin loci.

For simplicity, it is usual to introduce several foreign genes at one locus (p10 or polyhedrin), using the available multiple gene transfer vectors (see Table 1). Otherwise, to prepare a multiple gene expression vector, a recombinant with foreign gene(s) at one site (e.g., the polyhedrin locus) must be isolated and subsequently used to insert foreign gene(s) at another site (i.e., p10). Provided the opportunities for deletion of genes by recombination are minimized, the use of available multiple gene transfer vectors for insertion of up to four foreign genes is easier.

Expression Vectors Involving Other Promoters

A disadvantage of using the p10 and the polyhedrin promoters for foreign gene expression is the fact that these are very late promoters that are most active during the
terminal phases of an infection. During this period the nucleus increases in size relative to the cytoplasm and many of the subcellular organelles show evidence of degradation. Some foreign proteins, such as glycoproteins, require extensive processing
during their maturation, involving transport and modification through the endoplasmic reticulum and the Golgi apparatus. For these proteins, posttranslational processing can be affected as the host cell approaches lysis; and even before that time,
modifications of highly expressed foreign genes are often limited by the capacity of
the cell. This is of particular concern for highly expressed and glycosylated proteins.

Alternative vectors have been made that allow foreign proteins to be synthesized
earlier in the infection cycle (e.g., during the DE and/or late phases). The expression
level is usually lower than that obtained using either the p10 or polyhedrin promoters.
However, depending on the protein, the completeness of protein processing may be
better.

Although there are nonessential baculovirus genes expressed during the DE and
late phases, so far the vectors that have been constructed have employed duplicated
copies of the promoters of essential genes (Fig. 6). Examples include the promoters
of the basic protein (24), the gp67 protein (25), and the 39K protein (26). The basic
protein is a late-phase, arginine-rich protein that is associated with the virus nucleocapsid. It is an important structural element of the baculovirus particle. The gp67
protein is a late-phase glycoprotein that is present on the surface of budded virions.
The function of the 39K protein is not known. Interestingly, the 39K protein is synthesized during both the DE and late phases, employing separate promoters for each
phase. The locations of these three proteins in the AcNPV genome are indicated in
Fig. 6 (not to scale). Duplication of the respective promoters and insertion of copies
at the polyhedrin locus, and in lieu of the polyhedrin promoter (Fig. 6), have provided
expression vectors that can be used to express proteins from the late period of an
infection (basic and gp64), or from the DE and late periods (39K).

So far it has proved possible to insert only foreign genes into baculoviruses that
are amenable to growth and plaque assay in tissue culture. This limits the viruses that
can be used as vectors to a handful of species. To introduce foreign genes into the
large baculovirus genome (134 kbp in the case of AcNPV), a gene is initially inserted
into a plasmid transfer vector (e.g., pAcYM1; see Fig. 7). This transfer vector (19)
contains a modified portion of the AcNPV genome (i.e., the 7.3-kbp *Eco*RI I fragment) positioned at an appropriate site in a bacterial plasmid (e.g., pUC8). The plasmid provides a selection system and the requirements for replication and amplification in a suitable host (e.g., *E. coli*). The modified transfer vector is used to prepare
a recombinant transfer vector so that foreign sequences contained therein can be introduced into the parent baculovirus.

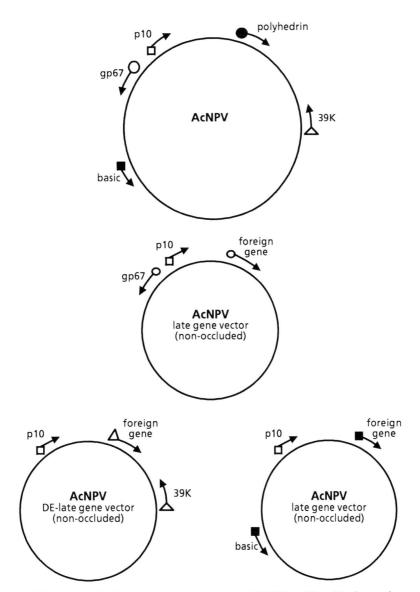

FIG. 6   Schematics of the locations of the polyhedrin, p10, 39K, gp67, and basic protein genes of AcNPV (upper panel) and single-gene expression vectors that utilize duplicated copies of certain promoters to express the foreign gene. For details see Fig. 2 and text.

To prepare a transfer vector, the AcNPV fragment is initially modified. For replacement gene expression vectors that were initially prepared on the basis of the polyhedrin promoter, this modification entailed the removal of some (e.g., pAc373)

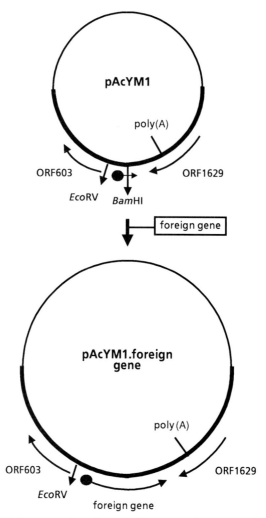

FIG. 7   Schematic of the preparation of a recombinant transfer vector at the polyhedrin locus. *Top:* Arrangement of the pAcYM1 transfer vector with the upstream ORF603 and downstream ORF1629 relative to the retained polyhedrin promoter and a unique *Bam*HI insertion site. The positions of the polyhedrin poly(A) site and a unique *Eco*RV site are indicated. *Bottom:* Foreign gene inserted into the transfer vector so that it is under the control of the polyhedrin promoter. For details see Fig. 2 and text.

or all (e.g., pAcYM1) of the polyhedrin coding sequences, and the retention of part (pAc373) or all (pAcYM1) of the polyhedrin promoter. A unique restriction enzyme site (e.g., *Bam*HI) was provided to insert a foreign gene. The restriction site was positioned immediately following the promoter (e.g., the polyhedrin promoter; see Fig. 7). The downstream polyadenylation site and signals were retained.

The complete polyhedrin promoter includes about 20 nucleotides upstream of the transcription initiation site of the gene and about 50 nucleotides downstream. As noted previously, there is no evidence to indicate that promoters of the polyhedrin or p10 genes involve the coding sequences of the respective proteins, although this cannot be assumed for other genes. For pAcYM1, the sequence upstream of the transcription initiation site represents the normal virus sequence and the downstream sequence extends to the A of the polyhedrin ATG translation initiation site (19). For pAc373, part of the promoter is missing and this can affect protein yields, depending on the foreign gene (18). Following the *Bam*HI site, the rest of the AcNPV sequence in pAcYM1 represents the natural sequences downstream of the polyhedrin coding region (19).

For vectors involving additional promoters, a cassette is constructed in the transfer vector consisting of a duplicated sequence encompassing the promoter, a unique downstream restriction site for foreign gene insertion, and a polyadenylation site and signal. The cassette may be linked with other cassettes for multiple gene expression (23). Usually these cassettes are in an orientation opposite to similar sequences in order to reduce the opportunity for gene deletion by recombination of genes in the derived expression vector (see above). Foreign genes are introduced into the cassettes in the transfer vector by using the appropriate restriction enzyme sites. The arrangements may involve the replacement of the resident gene and its promoter (see Figs. 4–6), or the introduction of additional genes and promoters in addition to the resident polyhedrin gene (see Fig. 3) by insertion of the cassette(s) at the *Eco*RV site (see the previous section). The promoters of other genes have not been well characterized. Therefore, to be sure that all the sequences that may form part of the promoter are incorporated, generally between 350 and 150 nucleotides upstream of the gene transcription start site and all of the sequence between that site and the translation initiation codon are included in the promoter construct. Probably much shorter upstream sequences could be employed; however, this issue has not been investigated in detail. A list of some of the transfer vectors that can be used to insert one or more foreign genes into AcNPV is provided in Table I (27). Several of these transfer vectors are available from commercial sources. Restriction maps of some of them are published elsewhere (2, 3).

Preparation of Recombinant Transfer Vectors

Transfer vector DNA species are prepared and purified by standard procedures (1). They are digested with the appropriate enzyme(s), dephosphorylated, and ligated with the foreign DNA sequence (1). Competent *E. coli* cells are transformed by the products, using electroporation or $CaCl_2$ precipitation (1) and bacterial colonies

TABLE I   Baculovirus Transfer Vectors to Insert Foreign Genes into AcNPV[a]

Transfer vectors	Comments
	Single-gene transfer vectors
Polyhedrin promoter at polyhedrin site; very late expression; polyhedrin negative	
pAc373	9.8 kbp; incomplete promoter; polyhedrin coding residues −7 to +172 deleted; BamHI insertion site
pAcYM1[b]	9.2 kbp; complete promoter; polyhedrin coding residues +2 to +751 deleted; BamHI insertion site
pAcRP23	9.8 kbp; complete promoter; polyhedrin coding residues +1 to +172 deleted; BamHI insertion site
pAcRP25	9.2 kbp; complete promoter; polyhedrin coding residues +3 to +751 deleted; BamHI insertion site
pEV55	6.2 kbp; complete promoter; polyhedrin coding residues +1 to +629 deleted; contains multicloning BglII, XhoI, EcoRI, XbaI, ClaI, Asp718 sites to insert genes (similarly pEVmXIV with modified promoter)
pAcCL29	7.8 kbp; complete promoter; polyhedrin coding residues +2 to +751 deleted; BamHI insertion site; single-strand DNA capability
pVL941	9.8 kbp; complete promoter; polyhedrin coding residues +36 to +629 deleted; ATG mutated to ATT; SspI, BspMII, BamHI, or SmaI insertion sites (pVL945 is similar)
pPAK1	5.5 kbp; complete promoter; polyhedrin coding residues +2 to +751 deleted; BamHI insertion site; single-strand DNA capability; reduced size plasmid
pJV NheI	13.6 kbp; complete promoter; polyhedrin coding residues +51 to +172 deleted; ATG mutated to ATT; NheI insertion site; also with lacZ under p10 promoter with SV40 terminator; single-strand DNA capability (similarly 13.9-kbp pJVP10 and 10.2-kbp pP10; other vectors use ETL [10.2-kbp pETL, 14.0-kbp pJVETL], or HSP70 promoter for lacZ and BamHI insertion site [pAcDZ1])
p10 promoter at polyhedrin locus; very late expression; polyhedrin positive	
pAcUW21[b]	9.3 kbp; complete polyhedrin promoter and gene present; upstream: complete p10 promoter present to residue +1 then BglII insertion site; single-strand DNA capability (also pAcATM4 and pAcATM5 with, in lieu of p10 promoter, the GP64 promoter with or without the GP64 signal sequence; also 5.7-kbp pSyn VI + wtp with "synthetic" promoter and polylinker insertion site upstream of polyhedrin gene and promoter, likewise 5.8-kbp pSynXVI VI+ with expression under upstream tandem "synthetic" and XIV promoters, again with polylinker insertion site)
Basic protein promoter at polyhedrin site; late expression; polyhedrin negative	
pAcMP1	10.1 kbp; polyhedrin promoter and coding sequence +1 to +172 are deleted, replaced by basic protein promoter to its residue +1, BamHI insertion site
pAcMLF1	7.8 kbp; same as pAcCL29 but with basic promoter in lieu of polyhedrin promoter

Properties	Construction
39K protein promoter at polyhedrin site; delayed early and late expression; polyhedrin negative pAcJP1	11.1 kbp; polyhedrin promoter and coding sequence $+1$ to $+172$ are deleted, replaced by 39K protein promoter to its residue -1, *Bam*HI insertion site
GP64 promoter at polyhedrin site; late expression; polyhedrin negative pAcATM3	7.9 kbp; polyhedrin promoter and coding sequence $+2$ to $+751$ are deleted, replaced by GP64 promoter to its residue -1, *Bam*HI insertion site; single-strand capability
GP64 promoter and signal at polyhedrin site; late expression; polyhedrin negative pAcATM2	8.0 kbp; polyhedrin promoter and coding sequence $+2$ to $+751$ are deleted, replaced by GP64 promoter to its residue $+114$, *Bam*HI insertion site; single-strand capability
Polyhedrin promoter and GP64 signal at polyhedrin site; late expression; polyhedrin negative pAcATM1	7.9 kbp; complete polyhedrin promoter, polyhedrin coding sequence $+2$ to $+751$ replaced with GP64 signal sequence ($+1$ to $+114$ residues), *Bam*HI insertion site; single-strand capability
Tandem basic and polyhedrin promoter at polyhedrin site; late and very late expression; polyhedrin negative pAcMLF2	8.7 kbp; basic promoter placed in tandem with resident polyhedrin promoter, polyhedrin coding sequence $+2$ to $+751$ deleted, *Bam*HI insertion site; single-strand capability
Tandem basic and p10 promoter at polyhedrin site; late and very late expression; polyhedrin positive pAcMLF8	9.4 kbp; complete polyhedrin promoter and gene present; upstream: basic promoter placed in tandem with p10 promoter, *Bgl*II insertion site; single-strand capability
Tandem capsid and polyhedrin promoters at polyhedrin site; late and very late expression; polyhedrin negative pC\pS1	Size (?); capsid protein promoter placed in tandem with polyhedrin promoter at polyhedrin site
p10 promoter at p10 site; very late expression; polyhedrin positive pAcUW1*	4.5 kbp; complete p10 promoter present, p10 coding residues $+2$ to $+151$ deleted, *Bgl*II insertion site
pAcAS3	9.8 kbp; complete p10 promoter present, p10 coding residues from $+2$ deleted, *Bam*HI insertion site; *lacZ* gene under control of HSP70 promoter for selection

continues

TABLE I (continued)

Transfer vectors	Comments
Multigene transfer vectors	
Polyhedrin promoters at polyhedrin locus; very late expression; polyhedrin negative	
pAcVC3	About 12 kbp; complete polyhedrin promoter and associated termination sequence placed at upstream *Eco*RV site and in opposite orientation to resident polyhedrin promoter; both polyhedrin coding sequences +2 to +751 deleted; *Bgl*II and *Bam*HI insertion sites (also 5.2-kbp pSynXIV wVIminus [and others] with dual expression under polyhedrin and fused "synthetic" and pXIV promoters, each with polylinker insertion sites)
Polyhedrin and p10 promoters at polyhedrin locus; very late expression; polyhedrin negative	
pAcUW3	About 10 kbp; complete polyhedrin promoter, polyhedrin coding region +3 to +751 deleted, *Bam*HI insertion site; complete p10 promoter, p10 coding region from +2 deleted, *Bgl*II insertion site, SV40 terminator
pAcUW51[b]	8.5 kbp; similar to pAcUW3 but with single-strand DNA capability (also pAcUW31)
Polyhedrin and basic-p10 tandem promoters at polyhedrin locus; late and very late expression; polyhedrin negative	
pAcML F7	8.7 kbp; complete polyhedrin promoter, polyhedrin coding residues 2 to +751 deleted, *Bam*HI insertion site; upstream: tandem basic-p10 promoter, *Bgl*II insertion site, SV40 terminator; single-strand DNA
Polyhedrin and p10 promoters at p10 locus; very late expression; polyhedrin positive	
pAcUW41[b]	10.0 kbp; complete p10 promoter, p10 coding sequences from +2 to +151 replaced with *Bgl*II insertion site and SV40 terminator; downstream: complete polyhedrin promoter (to polyhedrin +2), *Bam*HI insertion site followed by rest of p10 region and terminator; single-strand DNA capability

[a] From Ref. (27).
[b] Available from PharMingen (San Diego, CA).

grown on selective agar (depending on the vectors employed). The colonies that are recovered from the ligated samples are compared to controls of nonligated products and a number recovered and assayed for the presence and orientations of the inserted sequences. Usually DNA recovered from transformed bacterial colonies is initially screened by Southern hybridization, using the foreign gene sequence as a probe (1). Plasmid DNA is recovered from positive colonies and analyzed by restriction enzyme digestion. It is advisable to sequence the ends of the inserted gene to confirm their orientation, and so on.

Introduction of Foreign Genes into Baculoviruses

To obtain an expression vector, the recombinant transfer vector is transfected into insect cells in the presence of virus DNA. The objective of the cotransfection is to produce recombinants by natural, presumably double cross-over, procedures. Infectious wild-type DNA was employed originally (18, 19). Using this DNA, recombinants represent about 1–2% of the total progeny; the rest are wild-type viruses. Progeny are plaqued in permissive insect cells and recombinant plaques identified.

BacPAK6 (BaculoGold)

The earlier, pre-1990, methods for obtaining and selecting recombinants have now been superseded by the use of linearized virus DNA (27, 28). Linearized viral DNA has a greatly reduced infectivity by comparison to circular viral DNA. By contrast, the use of linearized DNA in cotransfections does not appear to reduce significantly the total number of recombinants that are formed. Because the background of wild-type virus progeny is reduced, the percentage of recombinants is increased about 20-fold (to about 30%). Although AcNPV has a large genome, it lacks sites that are recognized by the restriction enzyme *Bsu*361. When a site for this enzyme is engineered into the virus genome, the DNA can then be linearized (27, and Table II).

The *E. coli lacZ* gene contains a natural *Bsu*361 site. As recorded in Table II, replacement expression vectors have been prepared with *lacZ* at the polyhedrin locus (under the control of the polyhedrin promoter), or at the p10 locus (under the control of the p10 promoter). *lacZ* expression can be identified using a suitable substrate as described above (X-Gal, i.e., through the production of blue plaques). The use in cotransfections of (recombinant) virus DNA linearized at the *lacZ Bsu*361 site allows recombinants that replace the *lacZ* sequence to be identified by their white plaque phenotypes (i.e., nonblue plaques).

A further refinement to the procedures available for the recovery of recombinants is the use of a recombinant AcNPV named BacPAK6 (29) (BaculoGold PharMingen,

TABLE II   DNA Viruses Used with Recombinant Transfer Vectors to Select Recombinant
Expression Vectors[a]

Viruses	Comments
AcNPV C6[b]	Clone 6 of wild-type AcNPV; use with any AcNPV transfer vector
AcRP6-SC	Polyhedrin-negative AcNPV clone 6 with *Bsu*361 restriction site between polyhedrin promoter and residual polyhedrin gene coding sequences; linearize DNA with *Bsu*361; use with any single (e.g., pAcYM1[b]), or multiple-gene transfer vector (e.g., pAcUW21[b]) at polyhedrin site or locus (Table I); recombinants obtained at ca. 30% of total progeny; useful for selection of polyhedrin-positive expression vectors (i.e., use with transfer vectors containing polyhedrin gene, e.g., pAcUW21[b], pAcMLF8)
AcRP23.*lacZ*[b]	Polyhedrin-negative AcNPV clone 6 with *lacZ* gene under control of polyhedrin promoter[b] (gives blue plaques on staining with X-Gal); linearize DNA with *Bsu*361; use with any single (e.g., pAcYM1[b]) or multiple-gene transfer vector (e.g., pAcUW21[b]) at polyhedrin site or locus; recombinants (white) obtained at ca. 30% of total progeny, >90% of white plaques
AcUW1.*lacZ*[b]	Polyhedrin-positive AcNPV clone 6 with *lacZ* gene under control of p10 promoter (gives blue plaques on staining with X-Gal); linearize DNA with *Bsu*361; use with any single (e.g., pAcUW1[b]) or multiple-gene transfer vector (e.g., pAcUW41[b]) at p10 site or locus; recombinants (white) obtained at ca. 30% of total progeny, >90% of white plaques
BacPAK6[b]	Polyhedrin-negative AcNPV clone 6 with *lacZ* gene under control of polyhedrin promoter (gives blue plaques on staining with X-Gal); has two extra *Bsu*361 sites in adjacent upstream sequence and an essential downstream gene; linearize DNA with *Bsu*361; use with any single (e.g., pAcYM1[b]) or multiple-gene transfer vector (e.g., pAcUW21[b]) at polyhedrin site or locus; recombinants (white) obtained at ca. 100% of total progeny

[a] From Ref. (27).
[b] Available from PharMingen (San Diego, CA).

San Diego, CA). This virus is shown in schematic form in Fig. 8. It has additional *Bsu*361 sites that have been engineered upstream and downstream of the *lacZ* gene that are under the control of the polyhedrin promoter. One *Bsu*361 site is in the upstream ORF303, the other is within the coding region of the downstream ORF1629. The latter is an essential gene. Although ORF603 is not essential, the provision of an extra *Bsu*361 site in this gene is considered to be an insurance for linearizing the virus DNA and an aid to the removal of viral and *lacZ* sequences. Digestion of BacPAK6 by *Bsu*361, if taken to completion, removes part of an essential gene as well as all of the *lacZ* gene and part of ORF603. The number of progeny viruses representing the parent species is significantly reduced using BacPAK6. In cotransfections with recombinant transfer vectors, recombinants are obtained in which the percentage of white plaques approaches 100% (generally >95%).

In summary, the best procedure to obtain recombinant expression vectors is to perform cotransfections using linearized BacPAK6 DNA and recombinant transfer vectors based on the *Eco*RI I fragment. Cotransfections are generally undertaken in the presence of lipofectin and progeny viruses recovered from the supernatant fluids some 2–3 days later. These progeny are then plaqued in permissive insect cells (see

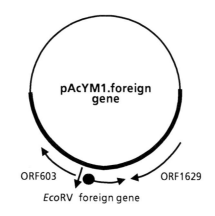

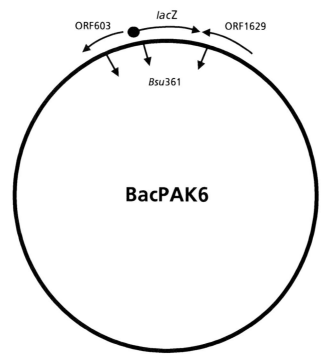

FIG. 8   Schematic of the use of a recombinant transfer vector (see Fig. 7) to insert a gene into BacPAK6 after linearization of that virus DNA with *Bsu*361. For details see text.

below), stained with X-Gal, and several white plaques recovered. Selected viruses are then replaqued a further two or three times, and used to prepare virus stocks for the expression of the foreign gene(s).

Preparation of Linearized Virus DNA

Infectious virus DNA (1 μg) from BacPAK6, AcRP6.SC, or AcRP23.*lacZ,* or AcUW1.*lacZ,* is suspended in 50 μl of *Bsu*361 digestion buffer [10 mM Tris-HCl, 0.1 M NaCl, 10 mM MgCl$_2$, 10 mM 2-mercaptoethanol, bovine serum albumin (BSA) (0.1 mg/ml), pH 7.5] and incubated with 5 units of enzyme (New England BioLabs, Beverly, MA), at 37°C for 6 hr. The reaction is terminated by heating at 70°C for 15 min and the extent of digestion checked by agarose gel electrophoresis. Linearized DNA is stored at 4°C until required. Alternatively, linearized DNA representing BacPAK6, or AcRP23.*lacZ,* or AcUW1.*lacZ* may be purchased from commercial sources (e.g., PharMingen).

Cotransfections of Insect Cells

SF-21 or SF-9 cells are seeded into 35-mm petri dishes in 2 ml of TC100 containing 5% FCS and incubated for 2 hr at 28°C. For lipofectin transfections (30) it is important that serum be removed, also that glass or polystyrene containers be used for the lipofectin solutions (not polypropylene). The growth medium is removed from the petri dishes and the cells washed twice with TC100 minus serum. Finally, 1 ml of TC100 medium lacking serum is added. Meanwhile, purified recombinant transfer vector DNA (0.5 μg) and the corresponding (see Tables I and II) linearized viral DNA (0.1 μg) are suspended in 50 μl of sterile deionized water and mixed with an equivolume of a 10-fold dilution of lipofectin (GIBCO, Grand Island, NY), prepared in sterile water. The mixture is kept at room temperature for 15 min, then added dropwise to the cells. After overnight incubation, 1 ml of TC100 containing 10% FCS is added and the incubation continued for a further 48–60 hr at 28°C. Progeny viruses in the supernatant fluids are screened by plaque assays as described above and at dilutions of 10^0 to 10^3. Where appropriate (see Table II) plaques are stained in the presence of X-Gal to identify parental viruses that contain the *lacZ* gene.

Insect Cell Infections and the Expression of Foreign Protein(s)

The yields of foreign protein that are obtained from an expression vector vary depending on the gene that is expressed, the time of expression, and the promoter that is employed, and so on. These issues are discussed below in greater detail. Even for the same vector the expression level can be influenced by the conditions of growth (monolayers or stirred cultures, fermentation conditions, media, etc.). Expression can also be affected by the other foreign or viral proteins that are made. For this reason we prefer not to include extraneous genes (e.g., *lacZ*) in expression vectors.

To verify the presence of the expected recombinant protein, we recommend that infected cells be analyzed at an early stage (e.g., during the preparation of stock viruses) by immunofluorescence, using a suitable antibody preparation [polyclonal, or monospecific, or a monoclonal antibody (MAb)]. Uninfected and wild-type virus-infected cells should serve as controls. Both fixed and unfixed cells should be employed to provide evidence of the location(s) of the reactive antigens. If the antibody preparations exhibit cross-reactions with the controls, then it is advisable to pretreat the antibody with extracts of insect cells, or those prepared from wild-type virus infections, until the cross-reacting components are removed (by centrifugation) and so that a specific antibody preparation is obtained. This is rarely a problem with MAbs.

It is beyond the scope of this article to describe the many procedures that may be investigated for the recovery of recombinant proteins from infected insect cells. There are, however, some particular conditions that need to be considered, including the provision of protease inhibitors in the initial extraction process (see below). Generally, however, purification can usually be effected without too much difficulty if the product is soluble within the cell cytoplasm or nucleus, or secreted to the supernatant fluids and particularly if there is an affinity procedure available for purification (e.g., an MAb, or a tag sequence). Differential ammonium sulfate precipitation and ion-exchange column chromatography on cell extracts are usually the first methods of choice in purification schemes. These allow some measure of purification from other cellular components. Tags may be designed into the product to aid purification (e.g., at the amino terminus or the carboxy terminus) and involve either a six-membered histidine tag (for purification via a nickel chelate matrix), or an immunogenic peptide (for which an MAb is available and, preferably, a corresponding peptide sequence that will elute the bound protein), or glutathione S-transferase (GST, which can be purified using glutathione bound to agarose beads). As required, factor X, or thrombin, cleavage sites may be included to remove the tag. It should be borne in mind, however, that chimeric products may not be processed in the same manner as the native species. Also, folding and other processing events may be altered, particularly for tags at the amino terminus. Therefore, a rigorous comparison should be undertaken of the final product to compare it to the native species.

Particulate structures resembling natural species, such as virus-like particles, are often purified by simple precipitation and centrifugation procedures (31).

Apart from soluble forms of glycoproteins, such as those from which the transmembrane anchor domains have been removed, glycoproteins are among the more difficult proteins to purify because they are located in membranes. Also, care must be taken not to disrupt the three-dimensional integrity of proteins during the purification procedures. For this reason, purification by procedures that involve denaturation (6 M urea, gel electrophoresis under denaturing conditions, etc.) should be methods of last resort unless denatured proteins are adequate for the research purposes, or unless a suitable refolding process is available. Generally, refolding of proteins is problematic and must be addressed on a case-by-case basis.

Extraction of Proteins from Infected Insect Cells

As noted above, the time that cells are harvested is determined empirically and is influenced by the amount of protein made and its integrity. Infected cells are harvested by low-speed centrifugation (2500 g for 5 min at 4° C), washed carefully in ice-cold isotonic saline (0.15 M NaCl, 10 mM Tris-HCl) to remove serum, and resuspended at a suitable concentration in ice-cold isotonic saline containing 1% (v/v) Triton X-100 [or Nonidet P-40 (NP40), or Triton N-101]. To the lysate protease inhibitors are added, for example, a one-hundredth volume of 100 mM PMSF (dissolved in 100% ethanol and kept at $-20°$ C), or a one-hundredth volume of a mixture (in 100% ethanol) of aprotinin (1.0 mg/ml), benzamidine hydrochloride (1.6 mg/ml), leupeptin (1.0 mg/ml), pepstatin A (1.0 mg/ml), and phenanthroline (1.0 mg/ml). These inhibitors should be added only if preliminary experiments indicate that protease-induced degradation of the protein product is a problem. After about 30 min at 4° C the cellular debris is removed and the supernatant and pelleted materials are analyzed for the presence and distribution of the protein(s).

For cells grown and infected in serum-free media, other protocols may be employed. For example, in the absence of serum it may not be necessary to wash the cells prior to lysis. Rather, the suspension may be adjusted with the appropriate detergent (and other additives) and processed directly. This has the advantage that any products that are in the supernatant due to cell lysis prior to the selected time of harvest will not be discarded. Depending on the time of harvest, 50% of the cells may have lysed and thus their products released to the supernatant fluids. It is important that the integrity of the product be monitored in selecting the appropriate procedure of purification as well as the time of harvest, as mentioned above.

Factors That Affect Expression of Proteins

The transcription level of a foreign gene depends on the promoter that is used. Although only a few studies have been undertaken to date, transcription of a foreign gene appears to be similar to that of the natural gene transcribed by the promoter that is employed. This may be affected when multiple copies of similar promoters are employed due to competition for enzyme components and precursors. The rate-limiting steps in transcription of viral or foreign genes (availability of enzyme, secondary structure of the transcribed sequences, transcription termination, etc.) are not known. It is probably worthwhile to remove residual polyhedrin and p10 genes (i.e., their unused promoters and coding sequences) from expression vectors in order to reduce the opportunity for competition for precursors in transcription (and translation). This is an area for future research.

The proteins made in insect cells are subjected to the processes of translation and posttranslational activities and modification that characterize these cells (2, 3). Generally, the first AUG in the transcribed mRNA is used to initiate protein synthesis. In some cases downstream AUG codons of genes in the same or overlapping reading frames are used (2, 3). However, the efficiency of translation at a second site may differ from that obtained when the same genes are translated in mammalian cells. The sequences flanking the AUG codons also influence the efficiency of translation, as described by Kozak for other eukaryotic systems (32). Furthermore, short upstream ORFs can reduce expression from ORFs that are located downstream. In view of these characteristics, it is preferable to remove unnecessary sequences and to provide a sequence context that resembles the natural baculovirus genes (polyhedrin, p10).

Ribosomal frameshifting between overlapping reading frames has been documented for certain genes expressed in insect cells (see Ref. 33). Although too few genes have been analyzed, there may be some codon usage differences between AcNPV genes and genes expressed in mammalian cells. For example, of 43 known AcNPV genes, only 7 and 9% of the arginine codons are CGG or AGG, respectively, only 8% of the glycine codons are GGG, and 8, 8, and 9% of the leucine codons are CTC, CTT, or CTA, respectively. Whether these are by chance or represent the tRNAs that are available for use by the virus is not known. However, for foreign genes that are expressed poorly, codon usage may be a reason for reduced translation efficiencies.

Depending on the protein, posttranslational processing of expressed proteins occurs (see Ref. 2 for a comprehensive review of this subject). Such processing may involve removal of the amino-terminal methionine and modification of the end sequence (acetylation, etc.), formation of disulfide bonds and other protein-folding activities, phosphorylation, acylation, polyisoprenylation, amidation, protein transfer to the nucleus, or to the lumen of the endoplasmic reticulum, removal of signal sequences, transfer to compartments of the Golgi apparatus and the plasma membrane, protein secretion, or deposition in the cytoplasm. Proteins may undergo proteolytic cleavage in addition to signal sequence removal. Both O- and N-glycosylation of proteins have been documented. For proteins with a large number of glycan side chains, during the very late phase glycosylation is often incomplete. This can be exacerbated for glycoproteins that are highly expressed. To reduce this problem, promoters operative during the delayed early or late phase may be employed, as described above.

Analyses of the glycans added to proteins indicate that they can differ by comparison to the glycans added to proteins in vertebrate cells. Both high- and low-mannose side chains are made but not the complex forms that characterize some glycans made in higher organisms. It appears that where high- or low-mannose side chains are present in the mammalian species, they are present at similar positions in the insect-derived products. Where complex forms are present, in the insect systems low-mannose glycans occur. This is most probably due to the lack of the requisite enzymes in insect cells for the addition of carbohydrates to form complex glycans. Different insect cell lines may differ in the form of glycosylation that occurs. The question of whether the genes coding for the mammalian enzymes involved in the processing of low-mannose to complex structures may be introduced into the insect cells or the baculovirus expression system has yet to be addressed. In view of the differences in glycan structures it is recommended that the antigenicity and immunogenicity of expressed products be investigated and compared with proteins made in their natural hosts.

Some mammalian cells have proteases that are involved in the proteolytic processing of proteins. These too may be absent in the insect cells, so that only the unmodified species is made. Examples include the protease involved in influenza HA modification to HA1 and HA2, and the enzyme involved in HIV gp160 processing to gp120 and gp41. Again, it may be possible to introduce the requisite genes into the insect cell expression system to effect the required processing.

Apart from the above issues, the posttranslational modifications that characterize the processing of proteins in higher organisms also generally occur in insect cells.

References

1. F. M. Ausubel, R. Brent, R. E. Kingston, D. D. Moore, J. G. Seidman, J. A. Smith, and K. Struhl (eds.), "Current Protocols in Molecular Biology," Greene Publishing and John Wiley & Sons, New York, 1993.
2. D. R. O'Reilly, L. K. Miller, and V. A. Luckow, "Baculovirus Expression Vectors, a Laboratory Manual." W. H. Freeman, New York, 1992.
3. L. A. King and R. D. Possee, "The Baculovirus Expression System, a Laboratory Guide." Chapman & Hall, London, 1992.
4. R. Granados and B. Frederici (eds.), "The Biology of Baculoviruses," CRC Press, Boca Raton, Florida, 1986.
5. R. I. B. Francki, C. M. Fauquet, D. L. Knudson, and F. Brown (eds.), "Classification and Nomenclature of Viruses. Fifth Report of the International Committee on Taxonomy of Viruses," Archives of Virology Supplementum 2. Springer-Verlag, Wien and New York, 1991.
6. G. E. Chisholm and D. J. Henner, *J. Virol.* **62,** 3193 (1988).
7. J. P. Burand, M. D. Summers, and G. E. Smith, *Virology* **101,** 286 (1980).
8. E. B. Carstens, S. T. Tjia, and W. Doefler, *Virology* **101,** 386 (1980).
9. L. A. Guarino and M. D. Summers, *J. Virol.* **57,** 563 (1986).
10. L. A. Guarino and M. D. Summers, *J. Virol.* **61,** 2091 (1987).
11. L. A. Guarino and M. D. Summers, *J. Virol.* **62,** 463 (1988).
12. A. M. Crawford and L. K. Miller, *J. Virol.* **62,** 2773 (1988).

13. L. A. Guarino and M. Smith, *J. Virol.* **66,** 3733 (1992).
14. N. E. Huh and R. F. Weaver, *J. Gen. Virol.* **71,** 2195 (1990).
15. L. K. Miller, J. E. Jewell, and D. Browne, *J. Virol.* **40,** 305 (1981).
16. D. Kelly, *J. Gen. Virol.* **52,** 209 (1981).
17. M. A. Grula, P. L. Buller, and R. F. Weaver, *J. Virol.* **38,** 916 (1981).
18. G. E. Smith, M. J. Fraser, and M. D. Summers, *Mol. Cell. Biol.* **3,** 2156 (1983).
19. Y. Matsuura, R. D. Possee, H. A. Overton, and D. H. L. Bishop, *J. Gen. Virol.* **68,** 1233 (1987).
20. V. C. Emery and D. H. L. Bishop, *Protein Eng.* **1,** 359 (1987).
21. U. Weyer, S. Knight, and R. D. Possee, *J. Gen. Virol.* **71,** 1525 (1990).
22. U. Weyer and R. D. Possee, *J. Gen. Virol.* **72,** 2967 (1991).
23. A. S. Belyaev and P. Roy, *Nucleic Acids Res.* **21,** 1219 (1993).
24. M. S. Hill-Perkins and R. D. Possee, *J. Gen. Virol.* **71,** 971 (1990).
25. A. M. Merryweather and R. D. Possee, Manuscript in preparation.
26. J. Pullen and R. D. Possee, Manuscript in preparation.
27. D. H. L. Bishop, *Semin. Virol.* **3,** 253 (1992).
28. P. A. Kitts, M. D. Ayres, and R. D. Possee, *Nucleic Acids Res.* **18,** 5667 (1990).
29. P. A. Kitts and R. D. Possee, *Bio/technology,* **14,** 810 (1993).
30. P. Felgner, T. Gadek, M. Holm, R. Roman, H. Chan, M. Wenz, J. Northrop, G. Ringold, and M. Danielson, *Proc. Natl. Acad. Sci. USA* **84,** 7413 (1987).
31. T. J. French, J. J. A. Marshall, and P. Roy, *J. Virol.* **64,** 5695 (1990).
32. M. Kozak, *J. Mol. Biol.* **196,** 947 (1987).
33. S. Morikawa and D. H. L. Bishop, *Virology* **186,** 389 (1992).

[19] Viral Oncoprotein Interactions with Cellular Proteins: Simian Virus 40 Large T-Antigen and p53 as Models

Janet S. Butel and Michelle A. Ozbun

Introduction

Natural selection has provided certain viruses with the ability to abrogate negative regulators of cell growth that normally control proliferation and presumably contribute to the differentiated state of cells. These virally encoded functions induce cell DNA synthesis and maximize virus production in infected cells. In the event of an infection that is nonpermissive for the complete viral life cycle, the viral growth-promoting proteins may become constitutively expressed, resulting in uncontrolled cellular proliferation and, ultimately, transformation of cells. Small DNA tumor viruses and their growth-promoting/transforming gene products have been studied extensively. The transforming proteins of these viruses appear to have evolved a common mechanism of antagonizing the growth-suppressing activities of normal cells (Table I), namely an ability to form complexes with two cellular tumor suppressor proteins, the retinoblastoma susceptibility gene product (pRB) and p53 (for review see Ref. 1). However, pRB and p53 are not the only cellular proteins with which these viral transforming proteins can interact, so these oncoproteins can be used as tools to define other cellular proteins and interactions that are important for the normal control of cell growth or other cellular processes. We focus here on the interaction of simian virus 40 (SV40) large T-antigen with the p53 tumor suppressor protein as a model for the definition and characterization of direct interactions between viral proteins and cellular proteins. These experiments can be undertaken using either virally infected cells or cells in which the viral protein of interest is expressed transiently or stably. Similar approaches can be applied to systems in which nonviral proteins are being expressed. The reader is directed to reviews for detailed discussions of the properties of SV40 large T-antigen and p53 (2–5). Publications from our laboratory that illustrate the described methods are cited here. Those reports may be consulted for original references.

Indirect Immunofluorescence and Immunohistochemistry

One of the first steps in characterizing cellular products with which viral proteins interact is to determine the subcellular localization of the viral gene products. Cells

Methods in Molecular Genetics, Volume 4

TABLE I   Transforming Gene Products of Small DNA Tumor Viruses[a]

Viruses	Oncoproteins	Tumor suppressor gene association
SV40	Large T-antigen	pRB
		p53
Adenovirus types 5, 12	E1A (289 aa[b]; 243 aa)	pRB
	E1B (55 kDa; 19 kDa)	p53
Human papillomavirus types 16, 18	E7	pRb
	E6	p53

[a] Reviewed by A. J. Levine and J. Momand, *Biochim. Biophys. Acta* **1032**, 119 (1990).

[b] aa, Amino acids; kDa, kilodaltons.

must be fixed and permeabilized to detect intracellular proteins, whereas unfixed cells can be used to detect surface-exposed proteins (Fig. 1) (6, 6a). Indirect immuno-fluorescence requires either a polyclonal or a monoclonal antiserum specific for the viral protein and a secondary antibody, which will react with the primary antiserum, conjugated to a fluorochrome such as fluorescein isothiocyanate or rhodamine. Many of these reagents are commercially available.

The *p53* tumor suppressor gene product is typically difficult to detect using im-munochemical methods on normal cells (5, 7); however, when p53 is complexed with a viral oncoprotein, such as SV40 large T-antigen, the protein is stabilized and readily detectable in cells (5, 8–11). Overexpression of p53 in cells not expressing T-antigen is generally interpreted to indicate that the protein is mutated (5, 7, 12, 13).

It is important to note that immunofluorescent and immunohistochemical ap-proaches per se do not reveal data about sizes and modification of proteins or if the target antigen is interacting with other proteins.

Materials

Indirect Immunofluorescence: Fixed Cells

Sodium carbonate (2%, w/v)

Ethanol and acetone (95%, v/v)

Elvanol: Add 80 g of Elvanol polyvinyl alcohol powder, grade 51-05 (Du Pont, Wilmington, DE), to 300 ml of Elvanol buffer and stir overnight at room temperature. Add 160 ml of glycerol and stir overnight at room temperature. Centrifuge at 10,000 rpm for 15 min at room temperature; collect supernatant and aliquot

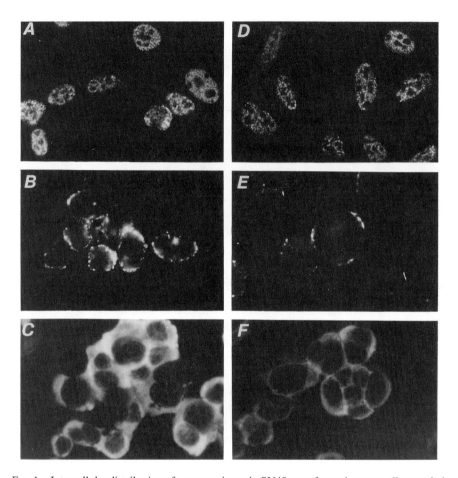

FIG. 1   Intracellular distribution of tumor antigens in SV40-transformed mouse cells revealed by immunofluorescence. (A)–(C) SV40 large T-antigen; (D)–(F) cellular protein p53. Cells (A, B, D, E) were transformed by wild-type SV40 or (C, F) by the cT mutant defective for nuclear transport of T-antigen; cells (A, C, D, F) were grown on coverslips and fixed in acetone before staining or (B, E) were viable and tested in suspension. Note the normal distribution of T-antigen in the nucleus (A) and plasma membrane (B) of transformed cells and the inability of the cytoplasmic T-antigen to be transported to the nucleus in the cT-transformed cells (C). The distribution of p53 (D–F) parallels that of T-antigen (A–C). Magnification: ×365 (A–D, F); ×815 (E). [From J. S. Butel, *Cancer Surveys* **5**, 343 (1986). Reprinted with permission of Oxford University Press.]

Elvanol buffer, pH 7.2: 150 mM NaCl, 10 mM KH_2PO_4, 100 mM Na_2HPO_4; adjust pH with NaOH

Tris-buffered saline (TBS): 2 mM Tris (pH 7.4), 0.4 mM Na_2HPO_4, 6 mM dextrose, 0.7 mM $CaCl_2$, 0.5 mM $MgCl_2$, 5 mM KCl, 140 mM NaCl

Antisera: Primary (both preimmune and immune) and fluorochrome-conjugated secondary

Indirect Immunofluorescence: Unfixed Cells

Glycerin [in phosphate-buffered saline (PBS)] (90%, v/v)

TBS or PBS [10 mM Na$_2$HPO$_4$ (pH 7.2), 145 mM NaCl] with 0.5% (w/v) NaN$_3$

Dulbecco's minimal essential medium (DMEM) containing 10% heat-inactivated fetal bovine serum (FBS)

Antisera: Heat-inactivated primary antiserum and fluorochrome-conjugated secondary antiserum

Immunohistochemistry (p53)

Methanol, acetone, and xylene (100%)

Diaminobenzidine: 1.5 mg in 2.5 ml of TBS plus 25 μl of 1% (v/v) H$_2$O$_2$. *Caution:* Handle this carcinogen with care and dispose of properly

Osmium tetroxide (0.1%, v/v)

Methyl green counterstain (0.1%, v/v)

Permount mounting medium

TBS

Rabbit polyclonal antiserum directed against murine p53 and normal rabbit serum control

Biotinylated goat anti-rabbit antiserum

Avidin–biotin complex reagents (commercially available)

Notes

1A. Special coverslips must be used, usually denoted as "No. 1 thickness" in supply catalogs. Standard coverslips used in histology are not acceptable. We prefer to use round coverslips, 15 mm in diameter.

2A. It is desirable to have cells at ~75–90% confluency at the time of fixation. It is difficult to obtain interpretable reactions when the cells are overly confluent.

3A. It is important to keep track of which side of the coverslips the cells are on. If necessary, the cell side can be determined by making a small scratch in the monolayer with a sharp instrument.

4A. Other useful fixation methods include ethanol at −70°C, fresh 2% (v/v) paraformaldehyde/0.2% (v/v) glutaraldehyde at 4°C, and 0.3% (v/v) glutaraldehyde at room temperature. The most appropriate fixation conditions must be determined empirically for each antigen.

5A. Storage: Fixed cells on coverslips can be stored at 4°C under desiccated conditions until they can be stained. The length of safe storage depends on the particular antigen. Stained coverslips can be stored for several weeks at 4°C, protected from light, but the reactions tend to fade. The stained, unfixed cells suspended in the glycerin solution can usually be stored at 4 or $-20°$C overnight without marked weakening of the reaction.

6A. It is possible to carry out simultaneous dual stains on a sample if primary antibodies are raised in different species and the secondary antibodies, labeled with different fluorochromes, react specifically with only one of the primary reagents. The different staining reactions are visualized by adjusting the filters on a fluorescence microscope.

7A. It is important that the unfixed cells be in a single-cell suspension. If the cells clump together, it is almost impossible to interpret the surface reactions.

Methods

Preparation of Coverslips

1. Place the coverslips (see note 1A) into a 2-liter beaker containing 100 ml of 2% sodium carbonate; boil for 15 min. (*Caution:* Solution boils violently.) Rinse the coverslips well: 10 times in distilled H_2O, 3 times in 95% ethanol, and again 10 times in distilled H_2O.

2. Place the coverslips into a 2-liter beaker containing 500 ml of distilled H_2O, and boil for 15 min. Spread the coverslips on a laboratory tissue-covered rack to dry. Occasionally shake the coverslips apart as they dry; otherwise, they will be difficult to separate.

3. Place the dry coverslips into 60-mm glass petri dishes and sterilize.

Preparation of Cells

1. Dispense growth medium into 35-mm tissue culture dishes. We routinely use six-well plates (2 ml of medium per well).

2. Aseptically transfer two sterile coverslips to each well; push to the bottom of the well, using sterile forceps. The forceps may be flamed with ethanol to maintain sterility during these maneuvers.

3. Seed the cells, usually 10^6 cells/well (see note 2A); proceed with described experimental protocol with respect to incubation times, infection or transfection procedures, and so on.

Indirect Immunofluorescence: Fixed Cells

1. At termination of the experiment, remove the coverslips from the petri dishes and load them into porcelain carriers (see note 3A). Coverslips can be handled individually with forceps if carriers are not available. Wash three times with TBS by soaking the carriers (2 min each wash).

2. Fix the cells (see notes 4A and 5A). First, air dry the coverslips; then incubate them in acetone for $2-10$ min at room temperature. Rehydrate the coverslips by soaking in TBS for 3 min.

3. Place the coverslips cell side up on a Parafilm strip. Add 100 μl of primary antiserum diluted appropriately in TBS; incubate for 30–60 min at room temperature (see note 6A). The antiserum should not run off the edges of the coverslips. Wash three times by soaking in TBS (2 min each wash).

4. Place the coverslips cell side up on a dry Parafilm strip. Add 100 μl of conjugated secondary antiserum; spread evenly over the coverslips and incubate for 30–60 min at room temperature, protected from light. The antiserum should not run off the edges of the coverslips. Wash three times by soaking in TBS for 2 min each wash, and then once in distilled H_2O for 2 min.

5. Air dry the stained cells, protected from light. We routinely lean the coverslips against a folded paper towel to speed drying.

6. Mount the coverslips cell side down, on a slide, using a drop of Elvanol for each. Protect from light. Observe with a fluorescence microscope with appropriate filter adjustments for the fluorochrome used.

Indirect Immunofluorescence: Unfixed Cells

1. Wash the dispersed cells with PBS or TBS containing 0.5% NaN_3 and distribute in aliquots of $1–2 \times 10^6$ cells into 15-ml conical plastic centrifuge tubes (see note 7A). Handle the cells gently throughout the procedure to minimize cell loss.

2. Resuspend the cells in 5–200 μl of heat-inactivated primary antiserum diluted (1:5 to 1:20) in Dulbecco's modified Eagle's medium (DMEM) containing 10% (v/v) heat-inactivated FBS and allow to react for 30–90 min at 0°C in an ice bath. Either place the tubes on a shaker or gently shake by hand every few minutes. Wash the cells three times with PBS or TBS, using 10–15 ml each wash. Pellet the cells at 600 g for 3 min.

3. Resuspend the cells in 50–200 μl of the appropriate fluorochrome-conjugated anti-globulin antiserum diluted in DMEM plus 10% FBS and incubate for 30–60 min at 0°C as above. Wash the cells three times with PBS or TBS as above. To avoid dilution of reagents, drain the tubes well after pelleting the cells and wipe the inside of each tube with a tissue.

4. Resuspend the cells in a minimal volume of 90% glycerin. Place a drop of this suspension on a microscope slide and place a coverslip on top of the liquid (see note 5A). Seal the edges of the coverslip with clear fingernail polish (optional). View the slides immediately, using a fluorescence microscope.

Immunohistochemistry

1. Fix the coverslips in methanol (see note 4A) overnight at 4°C. Wash in acetone for 90 sec at 4°C.

2. Process as follows, rinsing well with TBS between incubations: 45 min in primary antibody, 30 min in secondary antiserum, 30 min in avidin–biotin complex, 6 min in diaminobenzidine, 1 min in osmium tetroxide, 30 sec in methyl green counterstain.

3. Dehydrate the coverslips as follows: rinse three times in distilled H_2O; rinse twice in 95% ethanol; rinse three times in 100% ethanol. Allow to sit in xylene for 5–10 min.

4. Mount the coverslips cell side down on slides, using a drop of Permount for each; observe with a bright-field microscope.

Surface Labeling

Proteins exposed on the cell surface can be specifically labeled using a lactoperoxidase-catalyzed surface iodination protocol (14, 15). These proteins can be further analyzed by immunoprecipitation and sodium dodecyl sulfate-polyacrylamide gel electrophoresis (SDS-PAGE) as described in the section Metabolic Labeling below.

Materials

Lactoperoxidase [1 mg/ml in Dulbecco's phosphate-buffered saline (DPBS)]: Prepare fresh, and protect from light

H_2O_2 stock (30%): For use, dilute 1 μl in 10 ml of DPBS (see note 1 below)

Carrier-free Na^{125}I (>500 mCi/ml; 1 mCi/1.4 ml DPBS for each plate to be labeled)

TBS: At room temperature and 4°C

DPBS (10×): Dilute solution A, pH 7.5 (80 mM Na$_2$HPO$_4$, 15 mM KH$_2$PO$_4$, 27 mM KCl, 1.37 M NaCl) and solution B (9 mM CaCl$_2$, 5 mM MgCl$_2$) 1 : 5 and mix together in equal volumes

Notes

1B. H_2O_2 deteriorates on storage. Older stocks (even in unopened bottles) may yield variable results. Prepare fresh solutions of 1× DPBS, lactoperoxidase (1 mg/ml), and ^{125}I-labeled DPBS.

2B. *Caution:* Free iodine is extremely volatile. Steps 2–5 below must be performed in a chemical fume hood.

Methods

1. Grow cell monolayers in 100-mm plastic petri dishes. For transformed cells, label at ~70–90% confluency. For infected cells, ~24 hr postinfection is often optimal, but timing will vary with experimental conditions. Wash the monolayers twice

in TBS and once in DPBS. Drain excess fluid to the edge of the petri dishes and pipette off, leaving minimal residual fluid on the monolayers.

2. Add 1.4 ml of ^{125}I-labeled DPBS to each plate, tilt the dishes, and add 28 μl of lactoperoxidase to each.

3. Add 28-μl aliquots of H_2O_2 to each dish four times at 2-min intervals, rocking the plates every 30 sec to mix. Time the addition of H_2O_2 carefully.

4. At 8 min, pipette off the labeling medium and discard into a radioactive waste receptacle. Wash the monolayers once with a small volume of TBS (e.g., 1 ml).

5. Place the dishes on ice, wash three times with cold TBS, and drain well. The wash fluids must be disposed of as radioactive waste.

6. Extract the cells and process for immunoprecipitation (see below). At this point, bound iodine in cell extracts poses no volatility-related health risk to the laboratory worker.

Protein Labeling, Extraction, and Immunoprecipitation

Radioisotopic labeling of intracellular proteins can yield a diverse array of information, depending on the labeling procedure, length of labeling time, and type of radioisotope used. The steady state level of a protein can be determined by metabolically labeling the cells (6, 7, 10, 13, 16). The kinetics of processing and the protein turnover rate ("half-life") can be estimated by pulse radiolabeling followed by a time-intervaled "chase" of the isotope with unlabeled cell medium (7, 8, 13, 17, 18). [^{35}S]Methionine or Tran^{35}S-label (cysteine and methionine; ICN Biomedicals, Inc., Costa Mesa, CA) is the most commonly employed isotope for such experiments. Alternatively, ^{32}P$_i$ or ^{3}H-labeled sugars can be utilized to investigate posttranslational modifications of proteins (10, 19, 20).

When studying protein–protein interactions, the extraction conditions must be stringent enough to solubilize the proteins and yet not so harsh as to disrupt the protein complexes of interest. The solubility of the protein can be dependent on its subcellular localization. An investigation of various extraction conditions may not only optimize recovery of the protein(s) of interest, but may also indicate that different cellular proteins are complexed to the viral gene product, depending on the subpopulation of molecules being recovered (6, 10, 16). Three examples of detergent extraction buffers [Nonidet P-40 (NP-40), radioimmunoprecipitation assay (RIPA), and Empigen] are included here (19, 20) as well as a butanol extraction buffer (10).

Immunoprecipitation is commonly used to verify protein–protein interactions. If two or more proteins form complexes, in theory antiserum directed against one of the proteins should coprecipitate the other proteins in the complex. Such is the case for T-antigen and p53 (Fig. 2). When employing monoclonal antisera, one must keep in mind that epitopes may be masked or blocked in multiprotein complexes.

It is beyond the scope of this article to discuss *p53* with respect to cancer etiology,

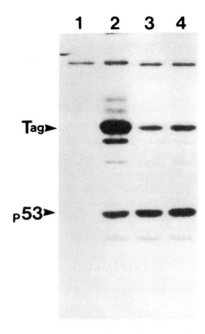

FIG. 2   Comparison of proteins immunoprecipitated from NP-40 detergent extracts of mouse cells, using antisera against T-antigen or p53. Proteins in clarified extracts from ^{35}S-labeled mouse cells growing in suspension were precipitated with control antibody (lane 1), T-antigen antibody PAb430 (lane 2), p53 antibody 200.47 (lane 3), or p53 antibody PAb421 (lane 4). Note that antibody against either member of a complex coprecipitates both components.

but it is noteworthy that *p53* is mutated in a variety of human and animal cancers (for a review see Ref. 21). Many mutant forms of p53 typically have an extended half-life (from ~15 min to ≥2 hr) and have sustained a conformational change resulting in an antigenically distinct protein (reviewed in Ref. 5). In the murine system, mutant forms of p53 frequently lack the epitope recognized by monoclonal antibody PAb246 (22) and have acquired the ability to react with monoclonal antibody PAb240 (which recognizes both mouse and human mutant p53) (23). Conversely, wild-type murine p53 reacts with PAb246 but not with PAb240 (7, 13, 23). Similarly, PAb1620 reacts only with wild-type human p53 (24).

Materials

NP-40 [10% (v/v) in PBS]
1-Butanol [2.5% (v/v) in PBS]
Leupeptin (2 mM in TBS; store at $-20°$C)

Tran[35]S-label (^{35}S; >1000 Ci/mmol), or ^{32}P$_i$ ($\sim$400–800 mCi/ml), or ^{3}H-labeled sugars (e.g., [^{3}H]glucosamine; > 20 Ci/mmol)

TBS and PBS

Empigen BB extraction buffer: 50 mM Tris-HCl (pH 9.0), 5 mM MgCl$_2$, 25 mM KCl, 1 mM [ethylene bis(oxyethylenenitrilo)]tetraacetic acid (EGTA), 1% (v/v) Empigen BB (N-dodecyl-N,N-dimethylglycine; Albright and Wilson, Ltd., Whitehaven, Cumbria, England)

NP-40 extraction buffer: 50 mM Tris-HCl (pH 8.0), 100 mM NaCl, 1% (v/v) NP-40, 1% (v/v) aprotinin, 0.2 mM leupeptin

RIPA extraction buffer: 50 mM Tris-HCl (pH 8.0), 0.15 M NaCl, 5 mM ethylenediaminetetraacetic acid (EDTA), 0.5% (v/v) sodium deoxycholate (DOC), 1% (v/v) NP-40, 0.1% (w/v) SDS, 1% (v/v) aprotinin, 0.2 mM leupeptin

Wash buffer: 50 mM Tris-HCl (pH 8.0), 100 mM NaCl, 1% (v/v) DOC, 1% (v/v) NP-40, 0.1% (w/v) SDS, 1% (v/v) aprotinin

RIPA wash buffer: 50 mM Tris-HCl (pH 8.0), 150 mM NaCl, 5 mM EDTA, 0.5% (v/v) DOC, 1% (v/v) NP-40, 0.1% (w/v) SDS, 1% (v/v) aprotinin

Disruption buffer: 62.5 mM Tris (pH 6.8), 5% (v/v) glycerol, 2% (w/v) SDS, 2% (v/v) 2-mercaptoethanol, 0.005% (v/v) bromophenol blue

MEM (2%): MEM containing 2% (v/v) dialyzed fetal calf serum. The formulation for MEM can be found in commercial catalogs

Methionine-free 2% MEM, or phosphate-free 2% MEM, or glucosamine-free 2% MEM

Antibodies for immunoprecipitation, including nonimmune control sera

Formalin-fixed *Staphylococcus aureus* strain Cowan I (SACI): See note 3C (25)

Notes

1C. Add leupeptin to buffers just before use. Other protease inhibitors may be used in place of, or in addition to, aprotinin. We have found leupeptin to be useful to minimize proteolysis of T-antigen during cell extractions (16).

2C. For SV40-transformed cell lines, monolayer cells should be growing at $\sim$80% confluency; suspension cells should be labeled 2 days after passaging and at a concentration of $\sim$4 $\times$ 10^6 cells/ml. For SV40-infected cells, 24 hr postinfection is often optimal, although timing will depend on the cells and multiplicity of infection used.

3C. Other reagents used to capture the antigen–antibody complexes are commercially available. Examples include protein A (or G)–Sepharose and secondary antibodies linked to magnetic beads. These products are generally more expensive but give less background.

Methods

Metabolic Labeling

1. Wash the cells three times with warm TBS; drain thoroughly and pipette off the excess fluid.

2. Add radiolabel to the cells, using small volumes of medium (1 ml of methionine-free MEM plus 200 μCi of ^{35}S, or 1 ml of glucosamine-free MEM plus 100 μCi of [^{3}H]glucosamine for a 75-mm^2 flask or 4×10^6 suspension cells).

3. Incubate for desired time (e.g., 3 hr) at 37°C; rock the flasks every 15 min to keep the monolayer cells moist. For longer incubation times, larger volumes of media should be used to maintain cell viability. Extract the cells (see Extraction, below).

Pulse–Chase Labeling

1. Wash the cells three times with warm TBS; decant the excess.

2. Starve the cells for 2–3 hr at 37°C, with medium (10 ml for a 75-mm^2 flask or 4×10^6 suspension cells) lacking the precursor to be used as radiolabel (e.g., methionine free). This will facilitate better incorporation of radiolabel into the protein to be analyzed.

3. Radiolabel the cells; for example, 200 μCi of ^{35}S in 1 ml of methionine-free medium per flask, 30 min, 37°C.

4. At the end of the labeling period, harvest (extract as described below) one set of samples.

5. For the remaining samples: wash three times with warm TBS, and add 10 ml of medium lacking radiolabel per 75-mm^2 flask.

6. Extract replicate samples at various time points (e.g., 30 min or hourly intervals).

7. Process all samples by immunoprecipitation (see below).

Extraction

Wash the cells (both monolayer and suspension) three times with cold TBS; drain thoroughly and pipette off the excess fluid.

Detergent Extraction

1. Add detergent extraction buffer (1 ml/75-mm^2 flask or 4×10^6 suspension cells) containing 0.2 mM leupeptin. Rock the flasks to distribute fluid evenly over the cell monolayer; resuspend the pelleted suspension cells well. Incubate for 30 min on ice.

2. Drain the fluid containing cells and cell debris to a corner of each flask. Be careful that cells are not inadvertently left in a flask. Transfer samples to a microcentrifuge tube, and clarify the extract by pelleting the cell debris by centrifugation (10–15 min at 4°C).

3. Decant the supernatant to a clean tube. The supernatant can usually be frozen for later analysis without damage to protein of interest. The pellet of nonsolubilized debris is discarded at this point, unless further analysis is planned.

Butanol Extraction

1. Add 1 ml of 2.5% butanol/PBS to the washed cells (see Extraction, above), and rock the flasks to distribute fluid evenly over the monolayers or resuspend the pelleted suspension cells gently. Incubate the cells for 10–15 min at 37° C, on a horizontal shaker at moderate speed. (Observe how cells detach/resuspend, i.e., as single cells or clumps, as clumps do not extract well. Monolayer cells detach from flasks and become suspended; these conditions do not lyse cells but are cytotoxic.)

2. Immerse the cell suspension in ice for 5 min and dilute butanol to a final concentration of 1% (v/v), using PBS. Pellet the cells and cellular debris in a table-top centrifuge (2000 rpm, 5 min, room temperature).

3. Place the sample tube on ice; remove the supernatant and aliquot into microcentrifuge tubes. Discard the pellet.

4. Clarify the supernatant in a microcentrifuge (10 min) to remove any remaining cell debris. Decant the clarified extract into a clean microcentrifuge tube and add 10% NP-40 to a final concentration of 1% (v/v).

Immunoprecipitation

1. From this point on, keep the samples on ice or work in a cold room.

2. Preclear the extract by adding 50–100 μl of SACI (see note 3C) to 1 ml of clarified extract and incubate for 30 min. This is done to reduce background by removing cellular proteins that nonspecifically bind to SACI.

3. Pellet the SACI in a microcentrifuge for 30–60 sec; transfer the clarified supernatant to a clean tube.

4. Immunoprecipitate by adding the appropriate amount (predetermined) of antiserum (usually 5 to 15 μl) to clarified extract; incubate from 1 hr to overnight, keeping the sample mixed well by using a mechanical shaker or by manually inverting intermittently. For SV40 T-antigen and p53, the incubation time can vary widely without adversely affecting protein recovery.

5. Add SACI ($\sim$10$\times$ volume of antiserum used); incubate for 30 min with constant agitation.

6. Pellet the immunosorbent containing antigen–antibody complexes in a microcentrifuge for 30 sec; discard the supernatant. Wash the pellet three times with 1 ml of wash buffer for NP-40 extracts or with 1 ml of RIPA wash for butanol extracts; vigorously mix using a vortex mixer for 30 sec, pellet, and discard the supernatant. Remove excess fluid from the final SACI pellet.

7. Add 35–40 μl of disruption buffer, and mix using a vortex mixer. Heat the samples in a 100° C oven for 10 min (or boil for 3 min); pellet by centrifugation for 10 min.

8. Carefully save the supernatant in a fresh tube or analyze immediately by SDS-PAGE (15, 26). If refrozen, repeat the heating and centrifugation steps (see step 7) before SDS-PAGE.

In Vitro Transcription and Translation

In vitro synthesis of proteins is frequently used to facilitate characterization of a protein. Typically, the cDNA of the protein of interest is cloned downstream of a bacterial promoter (e.g., SP6, T7, or T3) and the appropriate RNA polymerase is added along with nucleotides to produce mRNA transcripts. The transcripts are then translated in a cell-free system (e.g., rabbit reticulocyte lysate or wheat germ extract) to facilitate synthesis of the protein. Reticulocyte lysate is typically more efficient at translating longer mRNA species and is generally employed when cotranslational processing of proteins is investigated using microsomal membranes. There are numerous commercially available kits that can be used for both *in vitro* transcription and translation. Consequently, we do not include the details of the transcription/translation procedures per se. It is important to remember that proteins synthesized *in vitro* may not be fully modified, and thus may lack important functional epitopes. We have employed this technique in conjunction with immunoprecipitation, Western transfer, and immunoblotting to investigate protein modifications (20) and protein–protein interactions (S. Ames and J. Butel, unpublished data, 1993) by mixing two or more *in vitro* transcribed/translated proteins.

Materials

NH₄OAc (7.5 *M*) and 100% ethanol

Phenol–chloroform–isoamyl alcohol (25:24:1, v/v) and chloroform–isoamyl alcohol (24:1, v/v)

Amino acid mixture (minus methionine) (1 m*M*)

RQ1 DNase (RNase free)

Tran³⁵S-label (³⁵S; 1200 Ci/mmol)

Purified recombinant plasmid containing appropriate promoters and cDNA of interest

Translation mixture (rabbit reticulocyte lysate or wheat germ extract) containing a 20 *μM* concentration of the methionine-free amino acid mixture

Notes

1D. The integrity of the transcribed RNA can be analyzed by electrophoresis on an agarose gel.

2D. Both the RNAs and the proteins can be labeled *in vitro*. Here we describe labeling of the protein.

3D. It is important to use RNase-free reagents and laboratory ware and to handle all of the reagents with gloved hands to prevent degradation of the RNA transcripts.

4D. Certain cotranslational and initial posttranslational modifications can be investigated by adding microsomal vesicles, typically canine pancreatic microsomal membranes. These protein modifications include signal peptide cleavage, membrane insertion, translocation, and core glycosylation.

Methods

1. Linearize the plasmid and transcribe the mRNA of interest.

2. Treat the reaction with RNase-free DNase for 15 min at 37°C, to remove the plasmid DNA template from the nascent RNA transcript.

3. Extract once with phenol–chloroform and once with chloroform; then remove the unincorporated nucleotides and concentrate the RNA by precipitating with 0.5 vol of NH_4OAc followed by 2 vol of ethanol. Wash the pellet with 1 ml of 75% ethanol. Dry the pellet and resuspend in 20 μl of RNase-free distilled H_2O.

4. Follow manufacturer directions for adding the RNA substrate to the translation mixture containing 800 μCi of ^{35}S/ml and a 20 μM concentration of the methionine-free amino acid mixture.

5. The modification(s) of the protein may be assessed by SDS-PAGE and observing a shift in molecular weight.

6. To determine protein–protein interactions, mix varying amounts of two or more *in vitro*-translated proteins (normally only one of the proteins is labeled). Immunoprecipitate and analyze by SDS-PAGE as described above or perform a Western transfer and immunoblot (see the next section).

Western Transfer and Immunoblotting

Transfer of proteins to a nylon or nitrocellulose membrane following SDS-PAGE separation and subsequent probing with specific antisera can be helpful in protein identification. This technique is useful in the demonstration of coimmunoprecipitating proteins. For example, p53 and tubulin were coprecipitated with an anti-T-antigen antibody and subsequently verified by probing the blot with an anti-p53 or an anti-tubulin antibody followed by ^{125}I-labeled protein A (10). This approach has the advantage that the cells or cell extracts need not be radiolabeled prior to immunoprecipitation. This is helpful when tissues are used or if a protein of interest does not radiolabel well. One limitation is that the use of denaturing PAGE to separate the proteins requires the blotting antiserum to be reactive against denatured protein epitopes.

Several systems are available for detection of the specific interaction of the blotting antiserum and the antigen of interest. We routinely use, and describe here, iodinated protein A. Also useful are nonradioisotopic commercial kits that employ avidin/horseradish peroxidase or alkaline phosphatase detection. For each system, the optimized conditions for maximal sensitivity and minimal background must be empirically determined.

Materials

Iodination of Protein A (27)

Protein A solution: 1 mg/ml in phosphate buffer; commercially available
Glycine: 0.2 M in phosphate buffer
Phosphate buffer: 50 mM Na$_2$HPO$_4$, pH 8.0
Veronal-buffered saline: 0.25 mM CaCl$_2$, 0.8 mM MgCl$_2$, 145 mM NaCl; add 0.1% (w/v) gelatin
Bolton–Hunter reagent
Sephadex G-25M column

Western Transfer/Immunoblotting

Ponceau S stain: 0.5% (v/v) Ponceau S, 1% (v/v) acetic acid
Transfer buffer, 4°C: 25 mM Tris, 192 mM glycine, 20% (v/v) methanol
Blocking buffer: TBS containing 0.1% (v/v) NP-40, 0.25% (w/v) gelatin
^{125}I-Labeled protein A (see below)
Nitrocellulose membrane or nylon membrane suitable for protein transfer (0.45-μm pore size)

Notes

1E. Optimal transfer conditions may require empirical determination. The conditions given work well for the transfer of T-antigen and p53 (10, 28).

2E. Prior to blocking, the gel lanes can be visualized by staining with Ponceau S stain for 10–15 min, followed by destaining in distilled H$_2$O for 5 min. This is helpful if the membrane is to be cut into strips for incubation with multiple antisera.

Methods

Iodination of Protein A

1. Dry 0.5 mCi of Bolton–Hunter reagent to a 1.5-ml microcentrifuge tube, using a gentle air stream.

2. Add 100 μl of protein A solution (100 μg of protein) and incubate for 20 min at room temperature.

3. Quench the reaction by adding 500 μl of 0.2 M glycine in phosphate buffer. Incubate for 5 min at room temperature.

4. Separate the unincorporated Bolton–Hunter reagent from the labeled protein A on a Sephadex G-25M column equilibrated in Veronal-buffered saline plus gelatin. Collect fractions containing 30 drops each.

5. Pool the peak fractions as determined by scintillation counting.

6. Determine the specific activity of the pool, using a γ counter.

Western Transfer/Immunoblotting

1. Resolve proteins by SDS-PAGE (26, 29).

2. Fill a Western transfer unit with transfer buffer and cool to 4°C.

3. Place the gel in contact with the nitrocellulose membrane, carefully removing air bubbles. Position the gel and nitrocellulose into the Western unit and transfer for 18–24 hr, with circulating coolant. Initially transfer at 0.2 A for 2–3 hr, changing to 0.5–0.75 A for overnight.

4. Remove the membrane and place into blocking buffer, using sufficient volume to float the membrane (15–20 ml); incubate for 1 hr at 37°C, with agitation (see note 2E). Decant the blocking buffer and rinse the membrane with fresh blocking buffer.

5. Dilute the antibody appropriately in blocking buffer and add to the membrane (10–15 ml); incubate for 2 hr at 37°C, with gentle agitation. Decant the antibody solution; rinse the membrane three times with blocking buffer.

6. Add ^{125}I-labeled protein A (0.5 μCi/gel lane in 10 ml of blocking buffer); incubate for 1 hr at 37°C, with gentle agitation. Rinse the membrane three times with blocking buffer, then twice with TBS.

7. Air dry. Subject to autoradiography.

Far Western Transfer

Like the Western transfer, the "far Western" utilizes the transfer of proteins to a membrane. However, in this system a second, labeled protein (rather than antibody) is used to probe the renatured membrane (30) to determine protein–protein interactions. In this protocol, the protein used as a probe is generated as a glutathione *S*-transferase (GST) fusion protein to facilitate rapid synthesis, purification, and radiolabeling (31–33). The labeled protein can be used to determine protein–protein interactions using a number of approaches, such as screening an expression library or probing a membrane derived from SDS-PAGE resolution of a cell extract or purified proteins. Many of the conditions (e.g., induction and incubation times, lysis and wash buffers, and concentration of labeled protein in the hybridization solution) should be optimized for individual fusion proteins. We provide the conditions used for

T-antigen and p53 (S. Ames and J. Butel, unpublished, 1993) as examples, along with hints for optimization of the conditions for other fusion proteins.

Materials

Generation of [^{32}P]GST Fusion Proteins

Isopropyl-β-D-thiogalactopyranoside (IPTG): 800 mM (filter sterilize)

Dithiothreitol (DTT): 1 M, and 40 mM prepared fresh

Reduced glutathione: Final concentration, 20 mM in elution buffer

[γ-^{32}P]ATP: 6000 Ci/mmol

NETN: 20 mM Tris (pH 8), 100 mM NaCl, 1 mM EDTA, 0.5% (v/v) NP-40, 1% (v/v) aprotinin, 0.2 mM leupeptin; see note 1F

Heart muscle kinase (HMK) buffer (10×): 200 mM Tris (pH 7.5), 120 mM MgCl$_2$, 1 M NaCl; store at $-20°$ C

HMKD buffer (10×): 10× HMK buffer with DTT; add 10 μl of 1 M DTT to 990 μl of 10× HMK buffer

HMK stop buffer: 10 mM Na$_2$HPO$_4$ (pH 8), 10 mM Na$_4$P$_2$O$_7$, 10 mM EDTA, bovine serum albumin (1 mg/ml)

Elution buffer: 100 mM Tris (pH 8), 120 mM NaCl; add reduced glutathione to 20 mM just before use

Disruption buffer (see Protein Labeling, Extraction, and Immunoprecipitation, above)

L broth containing ampicillin

Transformation-competent bacteria (e.g., strains XL-1 Blue, RR1, or DH5α)

Recombinant plasmid expressing GST fusion protein of interest (see note 2F)

Parental plasmid containing only the GST leader sequences

Bovine HMK enzyme: Lyophilized powder, 250 units/vial; see note 3F

Glutathione–Sepharose beads: 1:1 in NETN plus 0.5% (w/v) powdered milk; may be stored at 4° C for 1 month; wash beads with NETN plus 0.5% dry milk prior to use

Preparation of Membranes for Probing

Guanidine hydrochloride (6 M)

Far Western transfer buffer: 25 mM Tris base, 192 mM glycine, 0.01% (w/v) SDS

HBB (10×): 250 mM HEPES–KOH (pH 7.7), 250 mM NaCl, 50 mM MgCl$_2$

Blocking buffer A: 1× HBB, 5% (w/v) dry milk, 1 mM DTT, 0.05% (v/v) NP-40

Blocking buffer B: 1× HBB, 1 mM DTT, 0.1% (v/v) NP-40, 0.08% (v/v) Tween 20, 0.25% (w/v) gelatin

Hyb 75 plus 1% dry milk: 20 mM HEPES (pH 7.7), 75 mM KCl, 2.5 mM MgCl$_2$, 0.1 mM EDTA, 1% (v/v) milk, 1 mM DTT (added just prior to use), containing 0.1% (v/v) NP-40 and 0.08% (v/v) Tween 20

Wash buffer: Hyb 75 + 1% milk containing 0.5% (v/v) NP-40, 0.01% (w/v) SDS, 0.08% (v/v) Tween 20, 0.25% (w/v) gelatin

Polyvinylidine difluoride (PVDF) transfer membrane: 0.45-μm pore size

Notes

1F. Add the protease inhibitors leupeptin and aprotinin to the appropriate solutions just prior to use.

2F. Do not maintain bacterial stocks on plates or use colonies more than 2–3 days old. Keep bacterial stocks transformed with recombinant plasmids frozen at $-70°$C in 15% glycerol. Never thaw frozen bacterial stocks (even partially); handle on dry ice.

3F. HMK is labile. Aliquot dry powder when first opened; store desiccated at $-20°$C. Open each vial only once; resuspend in 12.5 μl of 40 mM DTT, and let sit 5–10 min at room temperature to help dissolve.

4F. Prepare unlabeled GST leader from parental plasmid (nonfusion), following the procedure for the GST fusion proteins; freeze for stock. This extract will be used in the blocking buffer prior to adding the labeled probe.

5F. For fusion proteins that are detrimental to cell growth, a longer incubation may be required before induction.

6F. Fusion proteins that are toxic should be induced for 1–2 hr or less at 37°C, using a late-stage bacterial culture for inoculation.

7F. Proteins that induce poorly, are highly insoluble, or are unstable can often be harvested in a smaller volume (e.g., 0.02 vol of NETN) with an increased yield of fusion protein on reaction with glutathione beads. Such lysates may be stored in aliquots of ~200 μl at $-70°$C.

8F. If cells have not lysed, freeze for 30 min at $-70°$C; thaw quickly, but without warming the extract. Some fusion proteins are unstable and should not be harvested by freeze–thaw.

9F. The incubation time may be extended to 60 min to increase yields of fusion proteins present in low abundance due to poor induction, insolubility, or instability.

10F. The optimal number of counts may be empirically determined by titration of the probe.

11F. The appropriate volume depends on protein recovery and should be determined by titration of the lysate.

12F. If blocking or renaturing multiple membranes, all can be processed in one container. Volumes of buffers will depend on the number and size of the membranes being processed. The volume should be great enough to cover all the membranes and allow free movement in the container.

13F. All agitations of membranes should be gentle, with minimal movement to keep membranes moving in the container.

Methods

Generation of Fusion Proteins

1. Grow an overnight culture of GST–fusion recombinant [L broth plus ampicillin (300 μg/ml); see notes 2F and 4F].

2. Dilute the overnight culture 1:10 in fresh L broth plus ampicillin (100 μg/ml). Grow for 2–6 hr at 37°C, with shaking (see note 5F).

3. Induce expression of the fusion proteins with IPTG at a final concentration of 100 μM; grow the culture overnight at 23°C, in a shaking incubator at 150 rpm (see note 6F).

4. Pellet the bacteria (5000 rpm, 5 min, 4°C; *keep cold from this point on*). Resuspend the bacteria in 0.1 vol of NETN plus lysozyme (3 mg/ml) plus 1% aprotinin plus 200 μM leupeptin (see note 7F). Swirl occasionally, and keep on ice for 30 min (see note 8F).

5. Sonicate (1–10 ml of bacterial suspension per tube) on ice with three 10-sec bursts. Pellet the sonicate at 10,000 g for 5 min at 4°C. Aliquot 1 ml/tube, and freeze at −70°C.

Preparation of Fusion Proteins for SDS-PAGE and for Production of Probe

1. Prepare glutathione–Sepharose solution: resuspend the beads several times in NETN plus 0.5% dry milk, each time letting the beads fall by gravity and aspirating the supernatant (removes "fines"). Then resuspend 1:1 (v/v) in NETN plus 0.5% dry milk.

2. Microcentrifuge the thawed lysate for 3–4 min. Decant the supernatant into a fresh tube with aprotinin and leupeptin (as above); add 30 μl of glutathione–Sepharose solution per 1 ml of lysate (assuming lysis was in 0.1 vol of NETN in step 4, above).

3. Using a mechanical shaker, gently agitate the supernatant with glutathione–Sepharose; incubate for 15–30 min (see note 9F).

4. Wash the beads three times with 1.5 vol of NETN (per volume of lysate), transfer to microcentrifuge tube(s) after two washes. (For analysis by SDS-PAGE, add disruption buffer, and heat for 10 min at 100°C. Proceed with SDS-PAGE; see Preparation of Membranes for Probing, below.)

5. For probe labeling, wash the beads once with 1 × HMK buffer (diluted fresh, without DTT).

6. Resuspend the beads in 30 μl of reaction mix (3 μl of 10× HMKD, 1 μl of HMK, 2 μl of [γ-^{32}P]ATP, 24 μl of distilled H$_2$O). Incubate the reaction for 30 min at 4°C, on a mechanical shaker to keep the beads suspended.

7. Add 1 ml of HMK stop buffer, mix, and pellet the beads. Aspirate the supernatant, using a 23-gauge needle and syringe or a gel-loading pipette tip. Wash five times in NETN; aspirate the supernatant each time.

8. Add 1 ml of freshly prepared glutathione solution; agitate for 5–10 min at 4°C. Pellet the beads; carefully aspirate and save the supernatant.

9. Determine the specific activity of the ^{32}P label. Add 2 × 10^5 counts/ml of hybridization solution to blocked membranes (see note 10F).

Preparation of Membranes for Probing

1. Separate proteins by SDS-PAGE, using 0.001–1.0 ml of lysate reacted with beads per gel lane (see note 11F).

2. Transfer proteins from the SDS-PAGE gel to a PVDF membrane, using far Western transfer buffer at 0.1 A overnight, 0.2 A for 20–30 min, and 0.3 A for 20–30 min. (Amperage and time should be determined for each fusion protein. These conditions work well for several 50- to 90-kDa proteins.)

3. Following transfer, stain the membrane with Ponceau S for 5–10 min. Destain in distilled H$_2$O for 5 min and mark the lanes. Place the membrane protein side up in blocking buffer A (perform at 4°C from this point on; see note 12F); agitate for 1 hr (see note 13F). Decant the blocking solution.

4. Add 250 ml of 1× HBB plus 6 M guanidine hydrochloride plus 1 mM DTT; agitate for 10 min. Discard this solution.

5. Add a fresh 250 ml of 1× HBB plus 6 M guanidine hydrochloride plus 1 mM DTT; agitate for 10 min. Decant this solution into a 500-ml graduated cylinder, add 250 ml of 1× HBB plus 1 mM DTT, and mix well.

6. Add 250 ml of this 1:1 dilution to membranes; agitate for 10 min. Remove this solution from the membranes and discard the solution.

7. Again add 250 ml of 1× HBB plus 1 mM DTT to the graduated cylinder and mix well. Repeat step 6 four times (i.e., the membranes will have gone through 6, 6, 3, 1.5, 0.75, 0.375, and 0.187 M guanidine hydrochloride washes to renature the proteins gradually). Discard the solution.

8. Add 250 ml of 1× HBB plus 1 mM DTT to the membranes; agitate for 10 min; repeat once. Quickly rinse membranes in ~50 ml of blocking buffer B.

9. Add ~100 ml of blocking buffer B; incubate for 1 hr on a mechanical shaker (see note 13F).

10. Separate the membranes and place in Hyb 75 plus 1% dry milk/unlabeled GST lysate (1:1 ratio; see note 4F). Use enough volume to cover each membrane while shaking, and incubate for 30 min.

11. Add 2 × 10^5 counts of probe per milliliter, and incubate overnight at 4°C. Wash the membranes three times, 15 min each, in wash buffer.

12. Air dry the membranes for 10 min. Cover the membranes in plastic wrap and subject to autoradiography at $-70°$ C, using intensifying screens.

Direct RNA-PCR Sequencing of Murine *p53*

To determine *p53* mutations associated with tumorigenesis, the majority of investigators have sequenced only the highly evolutionarily conserved regions II–V, which include exons 5–9 of the murine cDNA (reviewed in Refs. 1 and 5). However, mutations have been reported outside of these regions. Direct polymerase chain reaction (PCR) sequencing alleviates the need to clone the cDNA and subsequently reduces the probability of detecting bases misincorporated by the *Taq* polymerase in the PCR reaction. We have used one biotinylated and one nonbiotinylated primer to amplify *p53* cDNA sequences by PCR. The biotin-linked PCR sequences can be captured with streptavidin-coated beads, allowing the purification of single-stranded DNA for use in a typical dideoxy chain-termination sequencing reaction (7, 12, 13, 34).

Materials

NH$_4$OAc (7.5 *M*), absolute ethanol, 70% ethanol
Phenol–chloroform–isoamyl alcohol (25:24:1, v/v), chloroform–isoamyl alcohol (24:1, v/v)
TBS and TE buffer [10 m*M* Tris-HCl (pH 8), 1 m*M* EDTA]
Reverse transcription reaction buffer (10×): 100 m*M* Tris-HCl (pH 8.8), 10 m*M* dNTPs, 50 mM MgCl$_2$, 500 mM KCl, 1% (v/v) Triton X-100; see note 1G
PCR buffer (10×): 100 m*M* Tris-HCl (pH 9.0), 15 mM MgCl$_2$, 500 mM KCl, 1% (v/v) Triton X-100
Total RNA samples (many RNA extraction kits are commercially available)
PCR primers (see Table I): 100 pmol/μl
Oligo[dT]$_{15}$ primers
Avian myeloblastosis virus (AMV) reverse transcriptase
Taq polymerase
Dynabeads M-280 streptavidin (Dynal, Inc., Great Neck, NY)
Dynal MPC-E magnet (magnetic particle concentrator; Dynal, Inc.)
DNA sequencing kit (commercially available)

Notes

1G. Commercially available RNA-PCR kits contain all of the needed reagents, with the exception of the PCR primers.

2G. The primer pairs 34A plus 27B and 21A plus 30B should be employed to generate the PCR products to be used as sequencing templates. These primers yield products of 961 and 918 bp, respectively.

3G. Annealing temperatures of 55–65°C may be appropriate, depending on the melting temperature of the primer pairs. We find that 62°C works well for specific amplification of *p53*, using the primers described. Nonspecific amplification will impede the synthesis of clean sequencing ladders.

4G. Depending on the efficiency of the PCR reaction, two sequencing reactions may be obtained from a single 100-μl PCR amplification.

Methods

RNA–PCR Reactions

1. Extract total RNA from whole tissue or tissue culture cells (we prefer to use a guanidine thiocyanate method).

2. Denature 1 μg of RNA for 5 min at 65°C; quench on ice.

3. Incubate the RNA samples with 15 units of AMV reverse transcriptase and 0.5 μg of oligo[dT]$_{15}$ primers in 1× reverse transcription reaction buffer for 15 min at 42°C, in a total volume of 20 μl. Inactivate the reaction buffer by heating for 5 min at 99°C. (We use a thermocycler for these incubations as well as for PCR.)

4. Amplify the *p53* cDNA by adding 50 pmol of each PCR primer pair (see Table II [34a] and note 2G), 2 units of *Taq* polymerase, and 1× PCR buffer in a 100-μl reaction volume. Overlay each sample with 75–100 μl of mineral oil.

5. Preincubate the samples for 5 min at 94°C, followed by 30 cycles as follows: 30 sec, 94°C; 1 min, 62°C; 2 min, 72°C (see note 3G). Postincubate the samples for 7–15 min at 72°C.

TABLE II   Primers for PCR and Direct Sequencing of Murine *p53* cDNA

Primer	Sequence[a]	Exon	Sense or antisense	Nucleotides
34A	5' Biotin TCA GTT CAT TGG GAC CAT CC 3'	1	Sense	43–62
21A	5' Biotin CAG TCT GGG ACA GCC AAG TC 3'	4	Sense	491–510
29B	5' TCG GTG ACA GGG TCC TGT 3'	4	Antisense	364–381
25B	5' CTC CCA GCT GGA GGT GTG 3'	5	Antisense	595–612
2B	5' CTG TCT TCC AGA TAC TCG GGA TAC 3'	6	Antisense	751–774
27B	5' CTT CTG TAC GGC GGT CTC TC 3'	8	Antisense	985–1004
21B	5' **CAT CGA ATT** CTC CCG GAA CAT CTC GAA GC 3'	10	Antisense	1157–1177
30B	5' AGG ATT GTG TCT CAG CCC TG 3'	11	Antisense	1390–1409

[a] Bold-face bases were added to create an *Eco*RI restriction site. The nucleotide sequence of *p53* is from D. Pennica, D. V. Goeddel, J. S. Hayflick, N. C. Reich, C. W. Anderson, and A. J. Levine, *Virology* **134,** 477 (1984).

6. Verify the size of the PCR product by analysis of 2–3 µl by agarose gel electrophoresis and ethidium bromide staining.

7. Purify the double-stranded DNA (dsDNA) product by extraction: once with phenol–chloroform–isoamyl alcohol (25:24:1), once with chloroform–isoamyl alcohol (24:1).

8. Precipitate the DNA with 0.5 vol of 7.5 M NH$_4$OAc and 2 vol of absolute ethanol for >4 hr at −20°C. Pellet the DNA for 30 min in a microcentrifuge, wash the pellet with 70% ethanol, and vacuum dry the pellet.

Sequencing Reactions

1. Prewash 8 µl of Dynabeads with 20 µl of 1× TBS.

2. Dissolve the DNA pellet in 100 µl of sterile distilled H$_2$O (see note 4G). Capture the DNA by adding 11 µl of 10× TBS and transfer the DNA to the microcentrifuge tube containing the washed beads. Mix well and incubate for 30 min at room temperature with occasional mixing.

3. Add 225 µl of 0.15 N NaOH to elute the complementary strand of DNA, and incubate for 5 min at room temperature. Collect the biotin–streptavidin complexes for 30 sec, using a Dynal MPC-E magnet.

4. Aspirate and discard the supernatant, and wash the pellet with 200 µl of 1× TBS. Resuspend the complexes in 8 µl of H$_2$O.

5. Process each sample as prescribed by the DNA sequencing kit, using 0.5 pmol of the appropriate primer to prime the sequencing reaction (Table II).

6. After terminating the elongation reactions with the dNTP/ddNTP mixes, stop the reaction by adding 50 µl of TE buffer to each sample.

7. Collect the DNA template and labeled reaction products, using the Dynal MPC-E magnet and aspirate the supernatant. Resuspend the pellet in 5 µl of stop solution/dye and incubate for 30 min at 37°C, to elute the complementary, labeled strand.

8. Immobilize the biotin-bound, unlabeled DNA using a Dynal MPC-E magnet. Add 10 µl of sterile distilled H$_2$O and aspirate the solution containing the labeled cDNA strand to a fresh tube. Store at −20°C.

9. Analyze the reactions (4 µl/lane) on a 7% polyacrylamide, 8 M urea gel (34).

Mutant-Specific PCR and PCR-Restriction Fragment Length Polymorphism Detection of *p53* Mutations

It is possible to screen a large number of samples for a specific *p53* mutation using one or both of the following techniques, which are based on PCR amplification from cDNA or genomic DNA. Mutant-specific PCR relies on the ability to design a PCR primer that is specific only to a given mutant nucleotide sequence and not to the wild-type sequence (13); the second primer of the pair need not be mutant specific. The mutant-specific PCR primer should be synthesized such that the extreme 3' nucleotide(s) is homologous to the specific mutant base(s). Obviously, drastic mutations (i.e., insertions or deletions) are more likely to be specifically amplified. The addition of a destabilizing reagent [e.g., Stratagene (La Jolla, CA) Perfect Match] may or may not give added specificity, depending on the mutation. The sensitivity of this technique has been estimated to detect a mutation present at 10% of a wild-type background (M. Ozbun and J. Butel, unpublished observations, 1993).

A less sensitive approach utilizes the probability that a point mutation may cause a change (loss or gain) of a restriction enzyme site [restriction fragment length polymorphism (RFLP)] in the DNA/cDNA (13, 35). The PCR is performed as usual without the need for mutant-specific primers, and the product is subjected to restriction enzyme digestion and gel electrophoresis. The sensitivity is limited to the amount of DNA that can be visualized by ethidium bromide staining or Southern hybridization (see Fig. 3).

Materials

Mutant-Specific PCR

Mutant-specific and wild-type *p53* PCR primers of similar melting temperatures
Other PCR reagents as described above

PCR–RFLP Screening of p53 Mutations

Ethidium bromide: 0.5 μg/ml in 1× TBE
Polyacrylamide resolving gel (20%): Acrylamide–bisacrylamide (29:1, v/v), 1× TBE, 0.05% (w/v) ammonium persulfate, 3.3 mM TEMED
Polyacrylamide stacking gel (7%): Acrylamide–bisacrylamide (29:1, v/v), 169 mM Tris (pH 6.6), 0.125% (w/v) ammonium persulfate, 10.2 mM TEMED
TBE buffer (1×): 90 mM Tris (pH 8.3), 90 mM borate, 2 mM EDTA
Restriction enzyme and appropriate enzyme buffer

Notes

1H. As a positive control for the reverse transcription reaction, we typically divide each reaction in half and use 50-μl PCR amplification volumes. One-half of the cDNA is PCR amplified using control PCR primers (e.g., specific for wild-type *p53*); the other half of the cDNA is used in the mutant-specific PCR amplification (13).

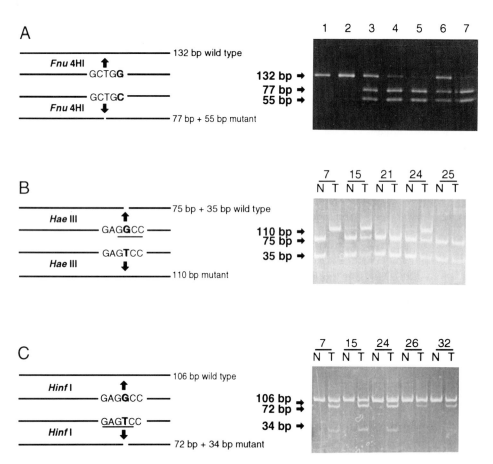

FIG. 3   Analysis of missense mutations at murine *p53* codon 135 (A) and human *p53* codon 249 (B and C) by PCR-RFLP. (A) The mutation (G to C) in murine *p53* codon 135 creates a unique *Fnu*4HI restriction site. When digested with *Fnu*4HI, the 132-bp PCR product is cleaved to 77- and 55-bp fragments. The samples were from inbred (BALB/c) mouse cells with wild-type *p53* (lanes 1 and 2) and mouse cells expressing various ratios of the wild-type and *p53* codon 135 mutation (lanes 3–7). The mutation (G to T) in human *p53* at codon 249 results in the loss of a *Hae*III restriction site (B) and creates a new *Hin*fI restriction site (C). Digestion of wild-type *p53* with *Hae*III results in cleavage of the 110-bp PCR product to 75- and 35-bp fragments. *Hin*fI digestion of the *p53* PCR product containing the codon-249 mutation produces fragments of 72 and 34 bp. A *Hin*fI site at base 106 of the PCR products from both wild-type and mutant *p53* causes a 4-bp fragment that cannot be resolved on the gel. The samples were from matched normal (N) and tumor (T) DNA; sample pairs 7, 15, and 24 illustrate the loss of a *Hae*III site (B) and the gain of a *Hin*fI site (C) in the tumor as compared to the normal. Tumor samples 26 and 32 show the gain of a *Hin*fI site (C), whereas this mutation is not detected in tumor samples 21 and 25 (B). Note that the resolving gel in (A) was under a stacking gel, which minimized the "smiling" effect on the samples compared to those samples resolved on the gels in (B) and (C).

2H. If the PCR reactions are not extremely efficient, the PCR products may be ethanol precipitated prior to restriction enzyme digestion.

3H. A thin (~1 cm) stacking gel poured over the 20% resolving gel will help to minimize the "smiling" effect of the DNA samples (Fig. 3).

Methods

Mutant-Specific PCR

1. Reverse transcription and PCR is performed essentially as described above, with minor modifications. With the exception of the *Taq* polymerase and the mineral oil, combine the PCR reagents as described above.

2. Following an initial denaturation at 94°C for 5 min, briefly centrifuge the samples to remove condensate from the top of the tubes. Do not allow the samples to remain at room temperature longer than necessary, as nonspecific priming will occur.

3. While holding the samples for 5 min at 62°C, add one unit of Perfect Match and 2.5 units of *Taq* polymerase to each 100-μl PCR reaction. Overlay each sample with 100 μl of mineral oil.

4. Thermal cycle for 30 rounds of PCR as follows: 72°C for 1 min; 95°C for 1 min; 62°C for 2 min. Postincubate the samples for 7–15 min at 72°C.

5. Analyze 10–15 μl of the PCR product by electrophoresis on a 20% nondenaturing polyacrylamide gel with an ~1 in. 7% stacking gel in 1× TBE buffer.

6. Visualize the DNA by ethidium bromide staining (0.5 μg/ml).

PCR–RFLP Screening of p53 Mutations

1. Amplify the cDNA product by RNA-PCR essentially as described in the section Direct RNA-PCR Sequencing of Murine *p53*, above.

2. Phenol–chloroform extract and ethanol precipitate the DNA product as described above (alternatively, some restriction enzymes do not require removal of PCR reagents prior to digestion; see note 2H).

3. If ethanol precipitation is used, resuspend the product in the original volume of the PCR reaction (i.e., 100 μl).

4. Digest 10 μl of the PCR product with the restriction enzyme of choice, using the appropriate buffer. Keep the final digestion volume to ≤20 μl.

5. Analyze 10–15 μl of the digest by electrophoresis on a 20% nondenaturing polyacrylamide gel with an ~1-cm high 7% stacking gel in 1× TBE buffer (see note 3H).

6. Visualize the DNA fragments by staining in ethidium bromide (0.5 μg/ml).

Acknowledgments

Research in the authors' laboratory that involved the development and use of the described methods was supported by research Grants CA22555, CA25215, CA33369, and CA54557. We gratefully acknowledge the many contributions of previous trainees in the laboratory.

References

1. A. J. Levine and J. Momand, *Biochim. Biophys. Acta* **1032**, 119 (1990).
2. J. S. Butel and D. L. Jarvis, *Biochim. Biophys. Acta* **865**, 171 (1986).
3. J. M. Pipas, *J. Virol.* **66**, 3979 (1992).
4. E. Fanning and R. Knippers, *Annu. Rev. Biochem.* **61**, 55 (1992).
5. M. Montenarh, *Crit. Rev. Oncogen.* **3**, 233 (1992).
6. R. E. Lanford and J. S. Butel, *Virology* **97**, 295 (1979).
6a. J. S. Butel, *Cancer Surv.* **5**, 343 (1986).
7. M. A. Ozbun, D. J. Jerry, F. S. Kittrell, D. Medina, and J. S. Butel, *Cancer Res.* **53**, 1646 (1993).
8. M. Santos and J. S. Butel, *J. Virol.* **49**, 50 (1984).
9. J. S. Butel, C. Wong, and B. K. Evans, *J. Virol.* **60**, 817 (1986).
10. S. A. Maxwell, M. Santos, C. Wong, G. Rasmussen, and J. S. Butel, *Mol. Carcinogen.* **2**, 322 (1989).
11. E. T. Sawai and J. S. Butel, *J. Virol.* **63**, 3961 (1989).
12. D. J. Jerry, M. A. Ozbun, F. S. Kittrell, D. P. Lane, D. Medina, and J. S. Butel, *Cancer Res.* **53**, 3374 (1993).
13. M. A. Ozbun, D. Medina, and J. S. Butel, *Cell Growth Differen.* **4**, 811 (1993).
14. H. R. Soule, R. E. Lanford, and J. S. Butel, *Int. J. Cancer* **29**, 337 (1982).
15. M. Santos and J. S. Butel, *Virology* **120**, 1 (1982).
16. D. L. Jarvis, R. E. Lanford, and J. S. Butel, *Virology* **134**, 168 (1984).
17. R. E. Lanford and J. S. Butel, *Virology* **105**, 303 (1980).
18. R. E. Lanford and J. S. Butel, *Cell (Cambridge, Mass.)* **37**, 801 (1984).
19. D. L. Jarvis and J. S. Butel, *Virology* **141**, 173 (1985).
20. C. Brandt-Carlson and J. S. Butel, *J. Virol.* **65**, 6051 (1991).
21. A. J. Levine, J. Momand, and C. A. Finlay, *Nature (London)* **351**, 453 (1991).
22. J. W. Yewdell, J. V. Gannon, and D. P. Lane, *J. Virol.* **59**, 444 (1986).
23. J. V. Gannon, R. Greaves, R. Iggo, and D. P. Lane, *EMBO J.* **9**, 1595 (1990).
24. L. Banks, G. Matlashewski, and L. Crawford, *Eur. J. Biochem.* **159**, 529 (1986).
25. S. W. Kessler, *J. Immunol.* **115**, 1617 (1975).
26. H. R. Soule and J. S. Butel, *J. Virol.* **30**, 523 (1979).
27. J. J. Langone, M. D. P. Boyle, and T. Borsos, *J. Immunol. Methods* **18**, 281 (1977).
28. B. L. Slagle, R. E. Lanford, D. Medina, and J. S. Butel, *Cancer Res.* **44**, 2155 (1984).
29. M. Santos and J. S. Butel, *J. Cell. Biochem.* **19**, 127 (1982).
30. C. R. Vinson, K. L. LaMarco, P. F. Johnson, W. Landschulz, and S. L. McKnight, *Genes Dev.* **2**, 801 (1988).

31. M. A. Blanar and W. J. Rutter, *Science* **256,** 1014 (1992).
32. W. G. Kaelin, Jr., W. Krek, W. R. Sellers, J. A. DeCaprio, F. Ajchenbaum, C. S. Fuchs, T. Chittenden, Y. Li, P. J. Farnham, M. A. Blanar, D. M. Livingston, and E. K. Flemington, *Cell (Cambridge, Mass.)* **70,** 351 (1992).
33. D. S. Smith and L. M. Corcoran, *in* "Current Protocols in Molecular Biology" (F. M. Ausubel, R. Brent, R. E. Kingston, D. D. Moore, J. G. Seidman, J. A. Smith, and K. Struhl, eds.), p. 16.7.1. John Wiley & Sons, New York, 1992.
34. D. J. Jerry and J. S. Butel, *U.S. Biochem. Ed. Commun.* **19,** 106 (1993).
34a. D. Pennica, D. V. Goeddel, J. S. Hayflick, N. C. Reich, C. W. Anderson, and A. J. Levine, *Virology* **134,** 477 (1984).
35. K. A. Scorsone, Y.-Z. Zhou, J. S. Butel, and B. L. Slagle, *Cancer Res.* **52,** 1635 (1992).

[20] Genetic Analysis of Infection by DNA Viruses

Michael Reed, Yun Wang, and Peter Tegtmeyer

Introduction

Viruses provide opportunities for comprehensive mutational analyses that are not possible in most organisms. Their genomes are relatively small and easy to manipulate, and their functions are focused on relatively simple objectives. The ultimate purpose of viral genetic studies is to discover the functions of viral sequences, to determine how different functions are related, and to dissect structure–function mechanisms. We present a number of strategies and methods for the genetic analysis of DNA viruses. These include traditional "knockout" mutagenesis and newer approaches to study autonomous domains within regulatory and protein-coding sequences.

DNA virology is the source of many genetic tools now commonly used in virology and cell biology. The small double-stranded DNA virus SV40, in particular, has served as a major model system for the development of techniques for genome mapping, DNA recombination, and site-directed mutagenesis. We will use SV40 here as an example of how powerful genetic strategies can contribute to an understanding of viral infection. Figure 1 briefly reviews the roles of SV40 T antigen, the viral origin of replication, and the cellular p53 protein in viral DNA replication. T antigen assembles double hexamers around the SV40 origin of replication (1, 2), melts origin DNA (3, 4), and recruits cellular proteins to form replication complexes (5, 6). After melting the origin, T antigen also acts as a bidirectional helicase to extend the primary replication bubble in opposite directions (7). To ensure the availability of cellular replication proteins, T antigen binds the cellular p53 and Rb proteins that act as suppressors of cell proliferation (8, 9). These protein–protein interactions inactivate cellular suppressor functions and prime the cell for DNA synthesis.

The genetic strategies presented here are intended to provide reasonably complete information. Because both positive and negative effects of mutations are important for interpretation, the principle of "more is better" applies both to the number of mutants isolated and also to the number of specific functional assays available for characterization of the mutant properties. Furthermore, both regulatory sequences and proteins often have a number of functions crowded into small and even overlapping spaces. Because knockout mutations have multiple effects that may be difficult to interpret, it is important to couple the knockout approach with an approach that identifies isolated DNA or protein sequences with autonomous functions. Perhaps the most important ingredient in any genetic analysis is the choice of appropriate assays for mutant functions. In the case of viruses, the optimum standard is usually

Methods in Molecular Genetics, Volume 4

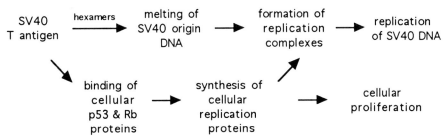

FIG. 1   T antigen functions in SV40 DNA replication. T antigen participates directly in the process of DNA replication and also primes the host cell for DNA synthesis by sequestering cellular suppressors of proliferation.

an assay of broad biological activity in intact cells. Once the general biological effects of mutations have been established, then appropriate assays for molecular activities *in vitro* can be chosen.

Genetic Studies of SV40 Origin of Replication

Simian virus 40 replication is a consequence of the interaction of SV40 T antigen with the viral origin of replication. The success of this interaction in intact cells can be determined by introducing plasmids containing viral origin DNA into cells that constitutively express T antigen. Replication of the input methylated plasmid DNA produces unmethylated progeny DNA that can be identified by virtue of its susceptibility to appropriate restriction enzymes. Deletion analysis with exonuclease showed that a unique 64 base-pair (bp) region of SV40 DNA is sufficient for the initiation of DNA replication (10, 11). Because the techniques used in these preliminary studies are well established, we will not review them here. The core origin of replication is a good example of an autonomous region that is an ideal target for detailed analysis. We wanted to know how it is organized into domains, to identify the functions of those domains, and to determine the relationships among the origin domains. To do this, we first used knockout mutations to map potential domains and then tested isolated domains for autonomous molecular activities. Because of the small size of the core origin, we used *de novo* oligonucleotide synthesis to accomplish both goals (10, 12).

Knockout Mutations Using Oligonucleotide Cassettes

Following the "more is better" principle of mutagenesis, we mutated the majority of the base pairs in the core origin of replication one at a time. We chose to make single base-pair transversions to maximize the effects of the mutations and the resolution of the domain maps. As shown below, this strategy was ideal for analysis of the core origin. It should be noted, however, that the simultaneous mutation of blocks of adjacent base pairs is sometimes required to knock out functions of regulatory regions. For example, Zenke *et al.* (13) mutated the SV40 enhancer sequences in blocks of three consecutive base pairs. Because there is significant redundancy in the enhancer, even multiple mutations failed to abolish enhancer function completely.

Figure 2 shows our strategy for saturating the core origin with knockout mutations. We synthesized the wild-type core origin as small oligonucleotide cassettes (the rectangles at the top of Fig. 2) that would reconstitute the entire origin or parts of the origin when annealed and ligated. The terminal overhangs allowed the insertion of ligated oligonucleotide cassettes into unique restriction sites in appropriate positions in plasmid DNA. The creation of replication origins with base substitution mutations required the synthesis of two mutant oligonucleotides with complementary base substitutions, for example, those with a shaded base pair in Fig. 2, for ligation with appropriate wild-type oligonucleotides. From 50 to 90% of the resulting cloned plasmids had the expected mutations. This strategy is versatile because it allows the creation of insertion, deletion, and inversion mutations in addition to base changes. A protocol for this technique is included in the Appendix to this chapter.

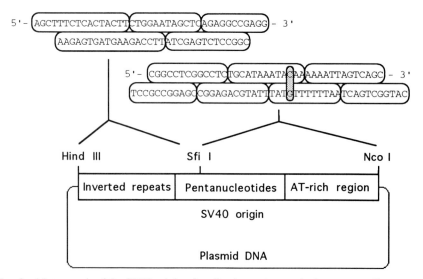

FIG. 2   Mutagenesis of the SV40 origin of replication, using synthetic cassettes. Cassettes of 5, 7, or 12 overlapping oligonucleotides were used to reconstruct the replication origin or portions of the replication origin. Two of the oligonucleotides had complementary mutations (e.g., those shown by shading). These two mutant oligonucleotides were annealed with appropriate wild-type oligonucleotides and inserted between unique sites of a plasmid to create an SV40 origin with mutation of a single base pair. [Adapted from S. Deb, A. L. DeLucia, A. Koff, S. Tsui, and P. Tegtmeyer, *J. Virol.* **6,** 4578 (1986). Adapted with permission of the American Society for Microbiology.]

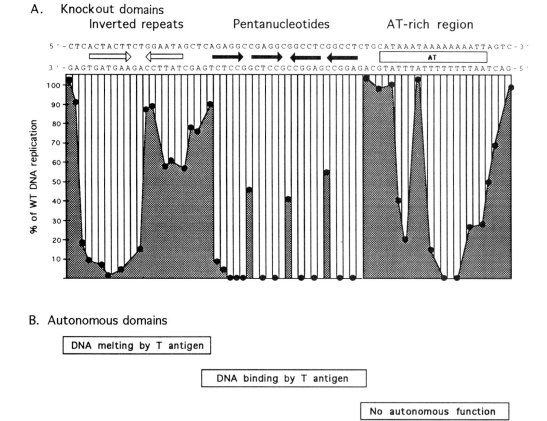

FIG. 3   Domain structure of the SV40 origin of replication. (A) Domains determined by knockout mutagenesis. The sequences at the top show three distinctive motifs: an inverted repeat, four pentanucleotide repeats, and an adenine plus thymine-rich region. The histogram compares the replication efficiency of each mutant (shown by dots) with that of the wild-type origin. (B) The autonomous functions of isolated subdomains are indicated. [Adapted from R. Parsons, M. E. Anderson, and P. Tegtmeyer, *Mol. Cell. Biol.* **6,** 4578 (1986). Reprinted with permission of the American Society for Microbiology.]

The histogram in Fig. 3A shows the results of this approach (10, 14, 15). The replication efficiency of 44 origin mutants with single-base substitutions (shown by dots in the histogram) is compared with the replication efficiency of the wild-type origin in isogenic plasmid DNAs. We show the histogram here to support our contention that "more is better" in genetic analyses. The histogram demonstrates a clear domain pattern. We operationally define a domain in a regulatory sequence as a contiguous or nearly contiguous group of sequences in which mutations cause significant functional changes. A comparison of the domain pattern with the sequences at the top of Fig. 3A led to the following possible interpretations. There is a two-part domain at the left of the origin, each part corresponding to one arm of an inverted repetition (the inverted repeat domain). In the center of the origin, there is a four-part domain, each part corresponding to one of four 5′ GAGGC 3′ pentanucleotide repeats (the pentanucleotide domain). At the right end of the origin, there is a two-part domain, the parts of which correspond to continuous tracts of three and eight adenines in an adenine plus thymine-rich region. If these interpretations were correct, we reasoned that the isolated domains might have autonomous functions.

Autonomous Functional Domains

We synthesized and cloned segments of the SV40 core origin corresponding to each of the three domains described above (16). None of the isolated domains were capable of initiating replication in cells that express T antigen. We then used purified SV40 T antigen to identify possible autonomous molecular interactions with the putative origin domains (Fig. 3B). T antigen melted DNA in the inverted repeat domain but, surprisingly, did not protect it from nuclease. T antigen protected the pentanucleotide domain from DNaseI but did not induce structural changes in the pentanucleotides or adjacent DNA. T antigen failed to interact with the isolated adenine domain. These findings indicate that T antigen can indeed interact independently with two of the three putative origin domains. Nevertheless, the lack of an observable interaction with the adenine domain suggested to us that we had not subdivided the core origin in the most ideal way. Therefore, we further subdivided the origin into combinations of two of the three domains and into half origins (17). The simplest and most functional subdivision of the origin was into two halves divided in the center of the pentanucleotide domain. Using these two half-origins, we were able to reconstitute most of the interactions that T antigen has with an intact origin. T antigen assembled hexamers around both half-origins and protected each from nuclease. Furthermore, T antigen melted the half-origin with the inverted repeat domain and distorted DNA in the half-origin with the adenine domain. The only missing function was the ability to move as a helicase away from origin DNA. Presumably these activities require interactions between T antigen hexamers bound to the two half-origins (18).

Conclusions

The use of synthetic oligonucleotides is an efficient, economical, and versatile way to saturate a small regulatory region with knockout mutations for analysis of domain structure. The same approach is an ideal way to recreate small domains to demonstrate autonomous functions. Meaningful conclusions, however, require the analysis of many mutants using a complete array of functional assays. Even under favorable circumstances, second- and third-generation mutations may be needed to test and refine preliminary conclusions.

Knockout Mutations of SV40 T Antigen

Simian virus 40 T antigen, both as a testing ground for site-directed mutagenesis and as the object of intense biological interest, may be the most mutated protein in history. Mutants include naturally occurring evolutionary variants, randomly mutated temperature-sensitive mutants, deletion mutants engineered by recombinant DNA techniques, protein chimeras, and too many site-directed base substitutions to count. Because these mutants were isolated over the course of years during the development of many techniques for mutagenesis, T antigen may not be the best paradigm for an orderly, integrated approach to the genetic analysis of a protein. Furthermore, we know so little about protein structure that there is no single preferred strategy for protein mutagenesis. The complete scanning of a protein with blocks of mutations, such as alanine mutations (19), might be one consistent way to subject a protein to comprehensive analysis. However, no protein of significant size and complexity has yet been completely analyzed by this or by a similar strategy.

Several small regions of T antigen have been singled out for a relatively rigorous analysis using modern techniques. Among these is a single zinc finger motif in SV40 T antigen (20). Although it is risky to focus on such a small region within a complex structure, a review of mutants with changes made in this region may serve to demonstrate the process of designing and characterizing mutants and interpreting the results of such studies. The SV40 zinc finger motif consists of amino acids 302–320 in the 708-amino acid protein, is highly conserved among related viruses, and has a classic arrangement of cysteines and histidines found in zinc fingers (21). It maps outside the T antigen region needed for origin-specific DNA binding but within a region required for the origin unwinding and helicase activities of T antigen.

Because this T antigen motif is likely to have a specific function, we saturated it with single amino acid substitutions and determined the effects of the mutations on all of the known functions of T antigen (20, 22). Once we had chosen a small target in T antigen for saturation mutagenesis, we had to decide which amino acid substitutions to make. We could have scanned the finger region with alanine substitutions. However, a single substitution for the existing wild-type amino acids would have made changes with varying degrees of conservation. That variation would have made it difficult to identify possible domains within the finger region. If such mutations had affected the finger function to different extents, the severity of the mutation rather than domain structure might have been the major determining factor. Therefore, we attempted to make mutations of similar severity at each position on the finger. To do so, we used a published matrix (23) that scores the conservation of amino acids in families of proteins. We chose amino acid substitutions with a similar score at every position within the finger region. We made the mutations according to the method of Kunkel (24). Details of that technique are not included in the Appendix because it is well described in many other sources.

Figure 4 shows the effect of the zinc finger mutations on the ability of the mutated virus to reproduce and form plaques in cell culture. As in the case of the genetic analysis of the SV40 origin of replication, a clear pattern of mutational effects was evident. Surprisingly, mutations at most positions had no apparent effect on viral replication. In contrast, mutations of cysteines and histidines in the finger severely inhibited replication functions. This finding is consistent with the original assump-

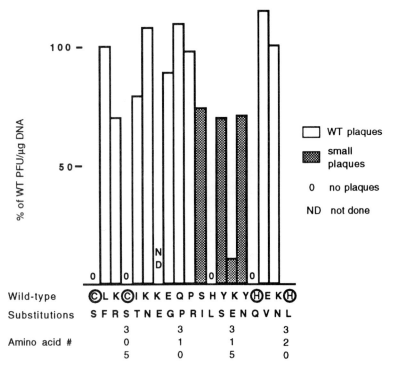

FIG. 4   Plaque formation by wild-type and zinc finger mutant T antigens. Wild-type amino acids, single amino acid substitutions in mutants, and the numerical positions of the amino acids in T antigen are indicated at the bottom. Putative zinc-binding amino acids are circled. The single-letter code for amino acids is used. The histogram compares the plaque-forming efficiency of zinc finger mutant T antigens with that of wild-type T antigen and indicates the sizes of the plaques. [Adapted from G. Loeber, R. Parsons, and P. Tegtmeyer, *J. Virol.* **63,** 94 (1989). Reprinted with permission of the American Society for Microbiology.]

tion that these amino acids are crucial for the finger structure. Interestingly, mutations of a contiguous block of amino acids at positions 312–316 significantly reduced the size and number of plaques formed and reduced DNA replication to various degrees. This clustering suggested that the region is an active site important for a specific function in DNA replication. A careful analysis of purified T antigens with these mutations that affect plaque formation and DNA replication in cells revealed consistent physiological effects *in vitro* (22). These mutant proteins bind to the central pentanucleotide domain of the core origin of replication but fail to melt or distort origin DNA and do not form hexamers. The consistency of these changes among defective mutants led us to propose that the zinc finger motif of SV40 T antigen contributes to protein–protein interactions important for assembly of T antigen hexamers and that hexamers are needed for subsequent alterations in the structure of origin DNA.

Conclusions

As in the case of the origin of replication, a saturation mutagenesis approach coupled with a thorough functional analysis of zinc finger mutants of T antigen identified distinct domain structures associated with well-defined functions. Because of the small size of the zinc finger motif, we did not attempt to assign autonomous functions to the finger region. Domain swapping experiments with related zinc finger proteins, however, might provide a complementary approach for further investigations.

Autonomous Domains of T Antigen-Associated Cellular p53 Protein

T antigen not only directly participates in viral DNA replication but also controls SV40 DNA replication indirectly by binding and sequestering p53, a suppressor of cellular proliferation (Fig. 1). This interaction primes the cell for viral DNA replication by making cellular replication proteins available for incorporation into replication complexes with T antigen. In some cells, such as murine cells, T antigen interacts with p53 and stimulates cellular DNA synthesis but fails to form replication complexes with the murine DNA polymerase α (25). In this case, infection does not lead to viral DNA replication and the host cells are not killed. If viral DNA is integrated in the host DNA, constitutive expression of T antigen causes oncogenic transformation of the cell (26). Therefore, we undertook a genetic analysis of the p53 protein to gain a better understanding of this pathway to the stimulation of cellular proliferation (27). We include examples of our approach here to demonstrate a strategy for the identification of autonomous domains in proteins using a biological assay.

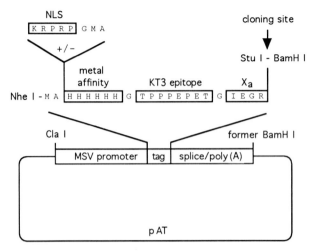

FIG. 5   Vector for the expression of p53 and autonomous domains of p53. The vector contains the Moloney Sarcoma Virus (MSV) promoter, a cassette for insertion of protein segments, an RNA splice signal, and a polyandenylation signal. Insertion of coding sequences in frame in the cassette creates a tagged protein. The tag contains an optional E1A nuclear localization signal (NLS), six histidines for metal affinity purification, the KT3 epitope, and a site for cleavage by proteinase X_a. [Adapted from M. Reed, Y. Wang, G. Mayr, M. E. Anderson, J. F. Schwedes, and P. Tegtmeyer, *Gene Expression* **3**, 95 (1993) with permission.]

We designed an expression vector that would be useful for the characterization of p53 domains *in vivo* and eventually for biochemical studies as well. The vector has an *Nhe*I to *Bam*HI cassette in pBluescript to facilitate the cloning, expression, identification, and purification of segments of p53 or of other polypeptides (Fig. 5). The cassette has unique *Stu*I and *Bam*HI cloning sites that accept appropriately designed polymerase chain reaction (PCR) generated DNA segments of the wild-type *p53* gene in a single orientation. The cassette also encodes a 22-amino acid tag N terminal to cloned *p53* or segments of *p53*. At the left end of the cassette, a 5′ GCC*ATG*G 3′ sequence provides a favorable context for the initiation of translation (28). The initiation codon is followed by a block encoding six histidines to create synthetic metal-binding sites for purification of tagged polypeptides by metal chelate affinity chromatography (29). Next, the cassette encodes an epitope from SV40 large T antigen recognized by the KT3 monoclonal antibody (30) for purification of expressed proteins by immunoaffinity chromatography. Finally, the cassette encodes a factor X_a proteinase site (31) for removal of the N-terminal protein tag from the p53 domains, should that be desirable. We designate this vector pAT. We used vector pAT[n] to add a strong E1A nuclear localization signal to the N terminus of tagged p53 segments lacking natural nuclear localization signals (32).

We wanted to identify autonomous functional domains of p53 that would suppress or enhance cell proliferation. We amplified desired segments of p53 using PCR primers that would produce blunt DNA on one end of the DNA segment and a *Bam*HI overhang at the opposite end for directional insertion into our expression vector. Clones expressing the fragments just downstream from the N-terminal 22-amino acid tag were cotransfected into primary rat embryo fibroblasts with plasmids that express the *ras* and *E1A* oncogenes. The Ras and E1A proteins alone transform rat cells. Wild-type p53 suppresses transformation by Ras and E1A, and some mutant p53s enhance transformation (33). Suppression was measured by dividing the number of foci produced by Ras and E1A by the number of foci produced by Ras, E1A, and p53. Enhanced transformation was determined by dividing the number of foci produced by Ras, E1A, and p53 segments by the number of foci produced by Ras and E1A.

Figure 6 compares the biological functions of wild-type and mutant p53s. As in the other examples of systematic mutagenesis cited above, the functional analysis revealed a consistent pattern of domain activity. The segments of p53 expressed in each of the plasmids are shown in Fig. 6A, and their functional activities, without an added E1A nuclear localization signal, are shown by the shaded bars in the histogram in Fig. 6B. Note that the scales measuring suppression and transformation in Fig. 6B are different. Transfection of plasmids expressing *E1A* and *ras* in the absence of p53 induced transformation of an average of 49 foci/assay. Wild-type p53 suppressed transformation about 25-fold whereas the prototypical mutant p53[Val-135] increased transformation more than 3-fold. None of the segments of p53, including the largest segments consisting of amino acids 1–320 and 80–390, suppressed cellular transformation to a significant extent. In contrast, the substitution of the N-terminal *trans*-activating domain of p53 with the *trans*-activating domain of VP16 (34) resulted in a chimeric protein that suppressed transformation somewhat better than did wild-type p53. Interestingly, all segments of p53 that include amino acids 280–390 enhanced transformation, and the transformation frequency increased as the central regions of p53 were removed. No segments of p53 lacking amino acids 280–390 increased transformation. Some p53 segments did not include all the nuclear localization signals (NLSs) of p53. Shaulsky *et al.* (32) have presented evidence that amino acids 312–321 and 366–381 are important for the nuclear localization of p53. Therefore, we added a sequence encoding a strong NLS from the adenovirus E1A protein to the N terminus of the *p53* gene in our expression vectors. The adenovirus E1A NLS has been shown to transport certain fusion proteins into the nucleus within 15 min after microinjection (35, 36). The addition of the artificial NLS did not influence the suppression or transforming activities of any p53 tested (solid bars in the histogram of Fig. 6B). These independently constructed plasmids also serve to demonstrate the reproducibility of our results. Nuclear localization signals were not added to p53[Val-135] or to segments 280–390, 80–390, 1–110/280–390, and 180–390, because these all have natural p53 NLSs.

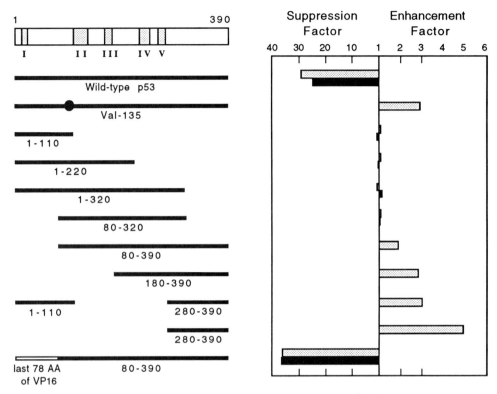

FIG. 6   Suppression and transformation by p53 and segments of p53. (A) Structures of the p53 segments. Wild-type p53 has five conserved regions (I–V). p53 with a Val-135 mutation is a prototype for transforming mutants. The solid lines represent p53 segments expressed by different plasmids. (B) Suppression and transformation activities of the segments shown in (A). Levels of suppression and transformation were determined as described in text. [Adapted from M. Reed, Y. Wang, G. Mayr, M. E. Anderson, J. F. Schwedes, and P. Tegtmeyer, *Gene Expression* **3,** 95 (1993) with permission.]

The pattern of autonomous domains described above makes sense. Although most of p53 is needed for the suppression function, a relatively small segment is capable of enhancing transformation by other oncoproteins. We know from other studies that purified p53 forms tetramers and that the p53 segment 280–390 includes amino acids sufficient for the formation of tetramers (37–39). These findings strongly suggest that the 280–390 domain of p53 oligomerizes with endogenous wild-type p53 in rat cells and, like T antigen binding, blocks its suppression functions. More detailed studies (not shown) have indicated that amino acids 320–360 are sufficient for cellular transformation.

Conclusions

A comprehensive genetic analysis of p53, coupled with relevant biological assays, demonstrates an internally consistent pattern of autonomous domain function. Any segment containing p53 amino acids 280–390 has transforming activity. The results were consistent with the known oligomerization functions of p53 and their proposed role in dominant negative inhibition of wild-type p53 (40). The identification of autonomous domains in p53 sets the stage for further analyses with knockout mutations.

Summary

We have used SV40 to present examples of integrated and comprehensive strategies for the genetic analysis of a variety of events in virus-infected cells. The examples include strategies for the identification of functional domains in an origin of replication, in a viral protein, and in a cellular protein that associates with a viral protein. The strategies couple knockout mutations with the isolation of autonomous segments of either DNA or proteins. We have demonstrated that, when sufficient mutations are characterized by either biological or biochemical assays, a domain pattern is often apparent. The value of domain identification is that it compels the investigator to assign functions to each domain and to relate biochemical and biological functions of various domains. The goal is to divide and conquer. We have emphasized strategies over techniques here because technical approaches are in a constant state of evolution. Nevertheless, we describe selected protocols for the use of oligonucleotide and PCR synthesis in mutagenesis in the Appendix below.

Appendix

Protocol for Mutagenesis Using Synthetic Oligonucleotide Cassettes

1. Preparation of oligonucleotides: We use detritylated and deprotected oligonucleotides (5–10 OD units) made by an Applied Biosystems (Foster City, CA) synthesizer without purification other than extraction with n-butanol as described by Sawadogo and Van Dyke (41). The oligonucleotides are stored in 0.1 ml of 10 mM Tris-HCl (pH 8.0), 1 mM EDTA at $-20°$C. Because the oligonucleotides are diluted and subsequent hybridization steps select for complete and correct oligonucleotide sequences, no further purification is necessary.

2. Phosphorylation of oligonucleotides: Oligonucleotide (5 μl) is mixed with 5 μl of kinase buffer [700 mM Tris-HCl (pH 7.6), 100 mM MgCl$_2$, 50 mM DTT], 5 μl

of 10 mM ATP, 30 μl of distilled water, and 3 μl of T4 polynucleotide kinase (2000 units/ml; New England BioLabs, Beverly, MA). After incubation for 60 min at 37°C, the kinase is inactivated by heating for 10 min at 70°C. The samples are stored at −20°C.

3. Preparation of vector DNA: A plasmid (5 μg) with unique sites designated for accepting cassettes of ligated oligonucleotides is cut at two sites with two single-cut restriction enzymes in the appropriate buffers. After extraction with phenol–chloroform, precipitation with ethanol, and vacuum drying in a desiccator, the DNA pellet is resuspended in 15 μl of 10 mM Tris-HCl, 1 mM EDTA at pH 8.0.

4. Dephosphorylation of vector DNA: Three microliters of vector DNA (about 1 μg) is mixed with 27 μl of 50 mM Tris-HCl at pH 8.0 and 0.2 μl of calf intestinal alkaline phosphatase (22,000 units/ml; Boehringer Mannheim, Indianapolis, IN) for 60 min at 37°C. The reaction is stopped with 6 μl of 200 mM EGTA and heating for 10 min at 65°C. After extraction with phenol–chloroform, precipitation with ethanol, and vacuum drying in a desiccator, the DNA pellet is resuspended in 5–10 μl of distilled water (about 1 pmol/μl to be mixed with 0.5 pmol of each oligonucleotide).

5. Annealing and ligation of oligonucleotides and vector DNA: The phosphorylated oligonucleotides are diluted 20-fold in distilled water, and 1 μl of each oligonucleotide is mixed with 1 μl of 10 mM ATP and 1 μl of kinase buffer (see above) for 1 hr at 37°C and 1 hr at room temperature. One microliter of dephosphorylated vector DNA, 1 μl of T4 DNA ligase (400,000 units/ml; New England BioLabs), and 5 μl of distilled water are added to the reaction mixture for overnight incubation at 12°C.

6. Standard procedures are used for transformation of bacteria, the isolation of bacterial clones, purification of plasmid DNA, and verification of mutations by sequencing.

7. Comments: Using the oligonucleotides shown in Fig. 2 in groups of five or seven, 50–90% of cloned plasmid DNAs have the correct mutated DNA sequences. Oligonucleotides with inverted repeats, however, have a tendency to form hairpin structures that can interfere with hybridization and ligation of oligonucleotide cassettes. It is important to design the oligonucleotides to avoid inverted repeats and to ensure that they will anneal with only the appropriate oligonucleotides.

Protocol for Expression of Autonomous Domains Using Polymerase Chain Reaction Synthesis

1. Plasmid construction: We construct the pAT plasmids to express tagged proteins in animal cells. Plasmid pAT is made by inserting the Moloney sarcoma virus (MSV) promoter (nucleotides 2380–2990), *p53* cDNA (sequences −67 to −11), a cloning and expression cassette, the SV40 small t antigen splicing signals (nucleotides 4710 to 4100), and the SV40 polyadenylation signal (nucleotides 2770–2533)

in that order between the *Cla*I and *BAM*HI sites of pBluescript KS+ (Stratagene, La Jolla, CA). The cloning cassette encodes unique *Stu*I and *Bam*HI sites for the insertion of PCR-synthesized DNA. The *Nhe*I site in the MSV promoter and the *Bam*HI site of the pBluescript polylinker are intentionally destroyed by oligonucleotide-directed mutagenesis during construction of pAT. The plasmid pAT[n] is identical to pAT except for the addition of the adenovirus E1A nuclear localization signal (36) to the N terminus of the cloning and expression cassette.

2. Synthesis of p53 domains: We copy the entire wild-type *p53* sequence (40) or segments of it by using PCR. The PCR primers match the N and C termini of chosen segments of *p53*. N-terminal primers are blunt ended, and C-terminal primers include a termination codon and a *Bam*HI site. The PCR fragments are cut with *Bam*HI and phosphorylated at their 5' ends. These fragments are directionally ligated into the dephosphorylated *Stu*I and *Bam*HI sites in the expression and cloning cassettes of pAT and pAT[n]. The entire sequence of wild-type p53 and the *Stu*I N-terminal junctions of all cloned segments are verified by sequencing. The correct sequence of the entire *p53* gene indicates that our PCR technique does not introduce mutations with significant frequency. To exclude further the possibility of mutations within p53 segments, we isolate and test two independent clones expressing each p53 segment.

3. Primer design: N-Terminal synthetic oligonucleotides are completely complementary to p53 template DNA. Typically, this primer consisted of 18 to 23 nucleotides with a G or C combination on the 3' end to facilitate polymerase recognition and synthesis. The C-terminal PCR primer includes a noncomplementary stop codon and *Bam*HI site. It is important to note that many restriction enzymes require adjacent DNA for efficient cleavage. The enzyme of choice dictates the number of base pairs to be added. To ensure proper annealing, the C-terminal primer has a 2:1 ratio of complementary to noncomplementary DNA. This ratio requires oligonucleotides of between 30 and 35 nucleotides in length.

4. DNA polymerase: We use Vent DNA polymerase (New England BioLabs), which has a $3' \rightarrow 5'$-exonuclease activity for proofreading. The increased fidelity of Vent over nonproofreading polymerases is an important consideration in the replication of large DNA segments. Furthermore, unlike *Taq* polymerase, Vent polymerase does not add nucleotides in a non-template-directed manner to the 3' ends of DNA chains. By using Vent, we are able to avoid "polishing" the PCR products before cloning.

5. Reaction conditions: The reaction conditions used for Vent polymerase are as described by the supplier (New England Biolabs, Beverly, MA), with minor modifications. Each 100-μl reaction mixture contains 75 μl of sterile H_2O, 10 μl of 10× amplification buffer (100 m*M* KCl, 100 m*M* $(NH_4)_2SO_4$, 200 m*M* Tris-HCl at pH 8.8, 20 m*M* $MgSO_4$, 1% Triton X-100), 8 μl of 2.5 m*M* dNTP mix, 1 μl of 100 m*M* $MgSO_4$, 1 μl of acetylated bovine serum albumin (BSA; 10 mg/ml), 1 μl of plasmid template DNA (50 μg/ml), 1 μl of 0.5 m*M* oligonucleotide 1, 1 μl of 0.5 m*M* oligonucleotide 2, and 2 μl of Vent DNA polymerase (2000 U/ml). Reac-

tions are carried out in Sarstedt PCR tubes, which do not require the overlaying of the reaction mixture with mineral oil.

6. Thermocycling conditions: DNA is denatured for 1 min at 96°C, annealed for 90 sec at 50°C, and extended for 1 min at 72°C. Twenty-five cycles are sufficient to generate microgram quantities of product under these conditions.

7. PCR product preparation for cloning: We routinely purify three identical 100-μl PCR reactions by using the Magic PCR purification kit (Promega, Madison, WI). The purified product is digested with *Bam*HI and 5' phosphorylated. Because PCR reactions often give rise to unwanted products through mispriming, we routinely gel purify the product in low-melt agarose and isolate the DNA product of correct size.

Acknowledgment

This work was supported by PHS grants awarded by the National Cancer Institute, DHHS: CA-18808 and CA-28146.

References

1. F. B. Dean, J. A. Borowiec, T. Eki, and J. Hurwitz, *J. Biol. Chem.* **267,** 14 (1992).
2. I. A. Mastrangelo, P. V. C. Hough, J. S. Wall, M. Dodson, F. B. Dean, and J. Hurwitz, *Nature (London)* **338,** 658 (1989).
3. J. A. Borowiec and J. Hurwitz, *EMBO J.* **7,** 3149 (1988).
4. F. B. Dean, P. Bullock, Y. Murakami, R. Wobbe, L. Weissbach, and J. Hurwitz, *Proc. Natl. Acad. Sci. USA* **84,** 16 (1987).
5. I. Dornreiter, A. Hoss, A. K. Arthur, and E. Fanning, *EMBO J.* **9,** 3329 (1990).
6. S. T. Smale and R. Tjian, *Mol. Cell. Biol.* **6,** 4077 (1986).
7. R. Wessel, J. Schweizer, and H. Stahl, *J. Virol.* **66,** 804 (1992).
8. J. A. DeCaprio, J. W. Ludlow, J. Figge, J.-Y. Shew, C.-M. Huang, W.-H. Lee, E. Marsilio, E. Paucha, and D. M. Livingston, *Cell (Cambridge Mass.)* **54,** 275 (1988).
9. D. P. Lane and L. V. Crawford, *Nature (London)* **278,** 261 (1979).
10. S. Deb, A. L. DeLucia, C.-P. Baur, A. Koff, and P. Tegtmeyer, *Mol. Cell. Biol.* **6,** 1663 (1986).
11. A. L. DeLucia, S. Deb, K. Partin, and P. Tegtmeyer, *J. Virol.* **57,** 138 (1986).
12. T. Grundstrom, W. M. Zenke, M. Wintzrith, H. W. D. Matthes, A. Staub, and P. Chambon, *Nucleic Acids Res.* **13,** 3305 (1985).
13. M. Zenke, T. Grundstrom, H. Matthes, M. Wintzerith, C. Schatz, A. Wildeman, and P. Chambon, *EMBO J.* **5,** 387 (1986).
14. S. Deb, A. L. DeLucia, A. Koff, S. Tsui, and P. Tegtmeyer, *Mol. Cell. Biol.* **6,** 4578 (1986).
15. S. Deb, S. Tsui, A. Koff, A. L. DeLucia, R. Parsons, and P. Tegtmeyer, *J. Virol.* **61,** 2143 (1987).

16. R. Parsons, M. E. Anderson, and P. Tegtmeyer, *J. Virol.* **64,** 509 (1990).

17. R. Parsons, J. E. Stenger, S. Ray, R. Welker, M. E. Anderson, and P. Tegtmeyer, *J. Virol.* **65,** 2798 (1991).

18. R. Parsons and P. Tegtmeyer, *J. Virol.* **66,** 1933 (1992).

19. B. C. Cunningham and J. A. Wells, *Science* **244,** 1081 (1989).

20. G. Loeber, R. Parsons, and P. Tegtmeyer, *J. Virol.* **63,** 94 (1989).

21. A. Klug and D. Rhodes, *Trends Biochem. Sci.* **12,** 464 (1987).

22. G. Loeber, J. E. Stenger, S. Ray, R. Parsons, M. E. Anderson, and P. Tegtmeyer, *J. Virol.* **65,** 3167 (1991).

23. R. Staden, *Nucleic Acids Res.* **10,** 2951 (1982).

24. T. A. Kunkel, *Proc. Natl. Acad. Sci. USA* **82,** 488 (1985).

25. Y. Murakami, T. Eki, and J. Hurwitz, *Proc. Natl. Acad. Sci. USA* **89,** 952 (1992).

26. M. Botchan, W. C. Topp, and J. Sambrook, *Cell (Cambridge, Mass.)* **9,** 269 (1976).

27. M. Reed, Y. Wang, G. Mayr, M. E. Anderson, J. F. Schwedes, and P. Tegtmeyer, *Gene Expression* **3,** 95 (1993).

28. M. Kozak, *Cell (Cambridge, Mass.)* **44,** 283 (1986).

29. J. Porath, J. Carlsson, I. Olsson, and G. Belfrage, *Nature (London)* **258,** 598 (1975).

30. H. MacArthur and G. Walter, *J. Virol.* **52,** 483 (1984).

31. C. V. Maina, P. D. Riggs, A. G. Grandea, B. E. Slatko, L. S. Moran, J. A. Tagliamonte, L. A. McReynolds, and C. di Guan, *Gene* **74,** 365 (1988).

32. G. Shaulsky, N. Goldfinger, A. Benzeev, and V. Rotter, *Mol. Cell. Biol.* **10,** 6565 (1990).

33. C. A. Finlay, P. W. Hinds, and A. J. Levine, *Cell (Cambridge, Mass.)* **57,** 1083 (1989).

34. I. Sadowski, J. Ma, S. Triezenberg, and M. Ptashne, *Nature (London)* **335,** 563 (1988).

35. C. Dingwall and R. A. Laskey, *Trends Biochem. Sci.* **16,** 478 (1991).

36. R. E. Lanford, C. M. Feldherr, R. G. White, R. G. Dunham, and P. Kanda, *Exp. Cell Res.* **186,** 32 (1990).

37. J. E. Stenger, G. A. Mayr, K. Mann, and P. Tegtmeyer, *Mol. Carcinogen.* **5,** 102 (1992).

38. H.-W. Sturzbecher, R. Brain, C. Addison, K. Rudge, M. Remm, M. Grimaldi, E. Keenan, and J. R. Jenkins, *Oncogene* **7,** 1513 (1992).

39. E. Shaulian, A. Zauberman, D. Ginsberg, and M. Oren, *Mol. Cell. Biol.* **12,** 5581 (1992).

40. I. Herskowitz, *Nature (London)* **329,** 219 (1987).

41. M. Sawadogo and M. W. Van Dyke, *Nucleic Acids Res.* **19,** 674 (1991).

[21] Mapping Sites of Replication Initiation and Termination in Circular Viral Genomes Using Two-Dimensional Agarose Gel Electrophoresis

Randall D. Little and Carl L. Schildkraut

Introduction

A variety of approaches have been developed to localize sites of DNA replication initiation and termination in eukaryotes (for review see Ref. 1). Two-dimensional (2D) agarose gel electrophoretic techniques, in particular, have permitted the detailed characterization of replicon dynamics in a number of systems. Two complementary methods have been described: the neutral/neutral 2D gel method developed by Brewer and Fangman (2) (reviewed in Ref. 3) provides information about the structures of steady state replicating molecules, and the neutral/alkaline 2D gel method developed by Nawotka and Huberman (4) (see also Refs. 5–7) provides information about the direction of replication fork movement within a DNA segment and the size of nascent strands at particular sites within the segment. Although the majority of studies using 2D gels have focused on the replication of lower eukaryotes and mammalian viruses, the use of various strategies for the enrichment of replicative intermediates prior to 2D gel analyses has made possible the analysis of multicopy and even single-copy chromosomal loci in mammalian cells (8–10).

Several studies on the replication of mammalian viruses have utilized 2D gel replicon mapping techniques. The neutral/alkaline approach, for example, has been used (4) to confirm the location of the origin of replication and the region of replication termination in simian virus 40 (SV40). The neutral/neutral 2D gel technique has also been used to examine the replication of portions of the polyoma virus genome (9). In addition, the location of replication initiation and termination sites in bovine papillomavirus (BPV) has been studied in our laboratory (11) and by Yang and Botchan (12), using neutral/neutral 2D gel techniques. These studies showed that initiation of replication in BPV genomes occurs at a single site-specific origin and that replication forks proceed bidirectionally, terminating in a zone approximately 180° from the origin. We have also studied, using neutral/neutral 2D gels, the replication of extrachromosomal plasmids containing the *oriP* region of the Epstein–Barr (EBV) genome which serve as models for the latent replication of EBV. In contrast to BPV, these plasmids replicate in an asymmetric, bidirectional manner (13). Initiation occurs at a specific site within the origin of replication, *oriP,* and proceeds bidirectionally until one replication fork is stalled at a replication barrier within the genetically defined origin but approximately 1 kb from the site of initiation. The other replication fork traverses the remainder of the plasmid and termination occurs when the stalled

Methods in Molecular Genetics, Volume 4

fork is encountered. Finally, neutral/alkaline 2D gel patterns consistent with a rolling circle mode of replication were observed for the geminivirus, African cassava mosaic virus (14). The 2D gel methods thus provide a powerful approach with which to determine sites of replication initiation and termination as well as the mode of replication for mammalian viruses.

We describe below a series of factors that bear consideration prior to initiating 2D gel analysis of viral replication intermediates. We then describe in detail several relevant protocols for 2D gel replicon mapping.

Viral Life Cycle and Culture

Because the life cycle of certain viruses may involve different modes of replication, an understanding of the infectious process and general features of the viral life cycle is essential. Epstein–Barr virus, for example, can persist in either a latent or a lytic state. In another example, SV40, the timing of harvest during lytic infection must be carefully monitored. Thus, one must be aware of the state of infection prior to any analysis of replication patterns.

The copy number of the viral genome may determine whether or not one or more enrichment steps are necessary in order to provide adequate sensitivity. The following enrichment steps have been successfully used prior to the analysis of viral DNA replication using 2D gels: (a) the isolation of particular S-phase populations of cells by centrifugal elutriation (13; and our unpublished data, 1993), (b) the enrichment for virus genomes using the Hirt extraction method (15; and see Refs. 11 and 16 for examples of its use in 2D gel analysis), (c) enrichment for replicative intermediates by isolating DNA associated with the nuclear matrix (our unpublished data, 1993), and (d) the use of benzoylated naphthoylated DEAE (BND) cellulose chromatography to enrich for molecules containing single-stranded regions, a characteristic of replication forks (4, 11, 13).

The culture method used to propagate the virus is also an important factor. Many viruses, such as EBV, BPV, and SV40, can be propagated in appropriate cell lines, but a culture system has not yet been developed for human papillomavirus (HPV). However, we have demonstrated that intact HPV replication intermediates can be isolated directly from human laryngeal papillomas (17). This represents an important advantage in the study of viral replication *in vivo*.

Two-dimensional gel analysis of the products from *in vitro* replication assays (i.e., for SV40) can also be used to determine features of the resulting replicative intermediates. For example, a role for the EBV-encoded protein, Epstein–Barr nuclear antigen 1 (EBNA1), in the formation of a replication barrier has been determined by cloning portions of the EBV barrier region into a vector containing the SV40 origin and studying the progression of replication forks initiated *in vitro* in the presence of T antigen (18).

Characteristics of Viral Genome

Various features of the viral genome affect the choice of DNA isolation procedure as well as interpretation of 2D gel results. These include (a) the size of the viral genome, (b) the configuration of the viral genome (circular and/or linear), (c) the presence of episomal and/or integrated forms of the viral genome, and (d) the existence of oligomeric forms of the viral genome.

For some viral genomes, specific genetic elements that permit autonomous replication of plasmids in cultured cells have been identified. In addition, transcriptional regulatory elements, such as enhancers, have been located within or near some viral replication origins (for review see Ref. 19). These observations may provide information to suggest the location of putative origins of replication. Also, in some instances, information regarding the mode of replication (i.e., rolling circle, unidirectional, bidirectional, etc.) may be available, thus providing specific predictions for the types of replicative intermediates present. Finally, because variability between strains exists for some viruses (for an example of EBV variation, see Ref. 20), the analysis may be more complete if several strains are directly compared.

Methods

DNA Preparation

The choice of a DNA isolation procedure is influenced by several factors, the most important being the copy number of the particular DNA sequences being analyzed. In some cases, enrichment for the viral DNA directly, or for replicative intermediates, in general, may be necessary. The Hirt method (15) has been successfully used to enrich for supercoiled molecules such as the SV40 (4) and BPV (11) genomes, and for plasmids containing the EBV latent origin of replication, *oriP*, prior to 2D gel analyses (16). Because replicative intermediates are fragile and sensitive to nuclease action and branch migration, several DNA isolation procedures have been developed to minimize these problems. First, we describe a method for isolating total cellular DNA that avoids the harsh treatments commonly encountered in conventional DNA isolation methods that employ proteinase K digestion followed by phenol extraction. Second, we describe our modifications of a method for the isolation of DNA associated with the nuclear matrix originally developed by Dijkwel *et al.* (10). This DNA isolation procedure has the advantage of stabilizing replicative intermediates and, importantly, enriches 10- to 20-fold for replicating molecules. It has been suggested that because replication may occur at or in close association with the nuclear matrix, molecules in the act of replicating could copurify with the matrix (21). It is possible, however, that for certain viral genomes, replicative intermediates may not be enriched in DNA associated with the nuclear matrix. A high-quality DNA preparation is perhaps the most important factor in obtaining interpretable 2D gel patterns.

High Molecular Weight Total Cellular DNA Isolation

The high molecular weight total cellular DNA isolation procedure described below was originally developed for the isolation of intact chromosomal yeast DNA in solution (22) and was adapted for use in the preparation of high molecular weight mammalian DNA for cloning (23) as yeast artificial chromosomes (YACs). We have modified this technique for the isolation of total cellular DNA for 2D gel replicon mapping.

Approximately 5×10^7 cells are used for each gradient. Centrifuge the cells for 20 min at 1600 rpm in the JS·4.2 rotor of the Beckman (Fullerton, CA) J6B centrifuge (all of the centrifugations described below utilize this rotor and centrifuge combination except where noted). Wash the cell pellet once with phosphate-buffered saline (PBS). Collect the cells again by centrifugation.

Resuspend the cell pellet completely in 3.5 ml of SCE [1 M sorbitol, 0.1 M trisodium citrate, 60 mM ethylenediaminetetraacetic acid (EDTA), final pH 7]. Add the cells very slowly into 7 ml of lysis buffer [0.5 M Tris-HCl (pH 9), 3% (w/v) Sarkosyl, 0.2 M EDTA, pH 9] containing proteinase K (100 μg/ml) (Boehringer Mannheim, Indianapolis, IN) in a 200-ml Erlenmeyer flask at room temperature. Make sure the mixture is fairly homogeneous before proceeding by swirling the flask very gently. This portion of the procedure may require up to 20 min.

Heat the lysate at 50°C for 15 min with occasional gentle mixing, then place in a water bath for 10 min at room temperature. During this incubation a sucrose step gradient is prepared. Three sucrose solutions are used [15, 20, and 50% (w/v) sucrose each in 0.8 M NaCl, 20 mM Tris-HCl (pH 8), and 10 mM EDTA]. First, place 11 ml of the 20% sucrose solution in a centrifuge tube compatible with a Beckman SW 28 rotor, then add 11 ml of the 15% sucrose solution without special precaution; an interface is not necessary. Then, carefully underlay with 3 ml of the 50% sucrose solution.

Pour the lysate slowly on top of the gradient; if pipetting is necessary use a wide-bore pipette to minimize shearing of the DNA. Centrifuge the gradient at 26,000 rpm for 3 hr at 20°C in a Beckman SW 28 rotor. Carefully aspirate the top 25–30 ml of the gradient and discard. When nearing the bottom of the tube carefully observe the DNA, and stop when it is about to enter the suction apparatus. Vacuum aspiration is better than removal by pipetting; less protein, RNA, and membranous debris are carried down the gradient. Viscosity throughout the gradient indicates overloading.

Collect the remaining 3–6 ml with a large-bore 10-ml pipette. Begin close to the bottom of the tube and remove the solution slowly. Slowly drip the collected solution into a tube, watching for extremely viscous material. Collect this material in a separate tube; it should remain as one contiguous viscous mass of DNA. The remaining slightly viscous material is smaller in size, contains little DNA, and is discarded.

Transfer the DNA to a dialysis bag and dialyze overnight at 4°C against TE7.5 [10 mM Tris-HCl (pH 7.5), 1 mM EDTA] with one change [include 0.05 mM phenylmethylsulfonyl fluoride (PMSF) in the first dialysis]. The volume will increase significantly if excessive space is left in the bag. Leave just enough air space for the bag to float.

To concentrate the DNA, surround the dialysis bag with solid sucrose at 4°C, replacing it with fresh sucrose as the sucrose becomes saturated with buffer. During concentration, occasionally rinse the bag with water and adjust the dialysis bag clamps closer together until the volume is approximately 0.5–1 ml (this may take 2–6 hr). Redialyze (with the dialysis bag clamps as close as possible) overnight in two changes of TE7.5 at 4°C.

For restriction endonuclease digestion, pipette about 300–500 μl of DNA into a separate tube (because the DNA is extremely viscous this is conveniently done by placing the DNA on a sheet of Parafilm, then pipetting the appropriate amount with a wide-bore tip and pressing the pipette against the Parafilm to release the remaining DNA). Place the DNA in a tube and equilibrate in restriction buffer on ice for about 1 hr in a final volume of 3 ml. Add restriction endonuclease (3000 units for each enzyme used; New England BioLabs, Beverly, MA), RNase T1 (270 units/ml; Worthington Biochemical Corporation), and RNase A (15 μg/ml; Worthington Biochemical Corporation). Typically, digestions are for 4–6 hr with an additional 3000 units of restriction enzyme added after 2–3 hr. It is important to gently mix the tube occasionally to help disperse the clumps of DNA. After the digestion, precipitate the DNA and redigest for 2–3 hr if necessary. Determine the amount of DNA spectrophotometrically (it is difficult to remove a suitable aliquot for dilution prior to restriction digestion). About 150–200 μg of DNA can be applied to a BND cellulose column (see below); using much more DNA results in overloading and distortion of the 2D gel pattern. This procedure has been used to detect bubbles from DNA fragments up to 12 kb.

Nuclear Matrix-Associated DNA Preparation

The isolation of nuclear matrix-associated DNA stabilizes and enriches for replicative intermediates and has been successfully used for 2D gel analyses of chromosomal loci (8, 10) as well as for EBV genomic sequences (R. D. Little and C. L. Schildkraut, unpublished results, 1993). The following protocol contains minor modifications of the procedure described by Dijkwel *et al.* (10). Conditions here are described for DNA isolation from 1.5×10^8 cells; we have scaled up the protocol successfully to isolate matrix-associated DNA from 9×10^8 cells. Also, the method described here is for the isolation of DNA from cells grown in suspension culture; modifications in the initial steps for monolayer cells are described elsewhere (10). All of the manipulations are performed on ice except where noted. The centrifugation steps described below utilize the JS·4.2 rotor of the Beckman J6B centrifuge, except where noted.

Centrifuge the cells at 1600 rpm for 20–30 min at 4°C in the JS·4.2 rotor of the Beckman J6B centrifuge. Drain the medium carefully, removing excess medium. Keep the bottles on ice. Add 15 ml of cold (4°) cell wash buffer [CWB: 5 mM Tris-

HCl (pH 7.4), 50 mM KCl, 0.5 mM EDTA, 0.05 mM spermine, 0.125 mM spermidine, 0.5% (v/v) thiodiglycol, 0.25 mM PMSF] and resuspend the cells. Carefully transfer the cell suspension to a 50-ml tube. Rinse the centrifuge bottles with an additional 10 ml of CWB and add to the cell suspension. Centrifuge the cells at 1600 rpm for 15 min at 4°C and remove the supernatant completely.

Dislodge the cell pellet by tapping the tube and resuspend in 40 ml of CWB containing 0.1% (w/v) digitonin. Gentle heating of the 2.5% (w/v) digitonin stock solution in a microwave aids in solubilization. Pour the suspension into the barrel of a 60-ml syringe fitted with a 21-gauge needle. Force the suspension through the needle three times, collecting in another 50-ml tube. Carefully layer 10-ml aliquots over 4 ml of CWB containing 12.5% (v/v) glycerol and 0.1% (w/v) digitonin in 15-ml tubes. Centrifuge at 1900 rpm for 10 min at 4°C and remove the supernatant as completely as possible. Thoroughly resuspend each nuclear pellet in 5 ml of CWB containing 0.1% (w/v) digitonin. Resuspension may require repeated pipetting; clumps of nuclei are to be avoided. Pool the four 5-ml suspensions and pass through a 30-ml syringe fitted with a 21-gauge needle. Centrifuge at 1500 rpm for 7 min at 4°C. Remove the supernatant completely.

Resuspend the nuclei in 2 ml of CWB containing 0.1% (w/v) digitonin. Repeated pipetting may again be necessary; do not proceed until an even suspension is obtained. Pour the suspension into the barrel of a 5-ml syringe fitted with a 21-gauge needle and eject into 8 ml of cold stabilization buffer [5 mM Tris-HCl (pH 7.4), 50 mM KCl, 0.625 mM CuSO$_4$ 0.05 mM spermine, 0.125 mM spermidine, 0.5% (v/v) thiodiglycol, 0.25 mM PMSF, 0.1% (w/v) digitonin] in a 15-ml tube (be sure that the stabilization buffer is well mixed before using). Incubate for 20 min on ice, mixing occasionally. During this incubation the lithium diiodosalicylate (LIS) buffer [LIS buffer: 10 mM lithium diiodosalicylate, 100 mM lithium acetate, 0.1% (w/v) digitonin, 20 mM HEPES–KOH (pH 7.4), 1 mM EDTA, 0.25 mM PMSF, 0.05 mM spermine, 0.125 mM spermidine, final volume of 110 ml] is prepared at room temperature. Add the individual components in the order shown, mixing well after each addition; otherwise a precipitate may form. Pour the stabilized nuclei into the barrel of a 10-ml syringe fitted with a 21-gauge needle and eject into 110 ml of LIS buffer at room temperature. Incubate for 20 min at room temperature, mixing occasionally. Near the end of the incubation, aliquot 30 ml to each of four 50-ml plastic tubes.

Centrifuge the samples at 3000 rpm for 15 min at 4°C and pour the supernatant carefully into a clean beaker. The pellet appears as an opaque gelatinous mass. Pool two pellets and add 30 ml of the appropriate restriction buffer containing 0.1% (w/v) digitonin to each. Do not break up the pellet, the purpose being to equilibrate it with restriction endonuclease buffer. Leave the mixture on ice for 5 min, and then carefully pour off the supernatant, retaining the gelatinous material. This is sometimes difficult, as the pellet may float. It is best to avoid centrifugation to collect the pellet because some digitonin may be in suspension and will contaminate the pellet (gentle heating of the digitonin stock solution may help in dissolving). Repeat the washes four or five times with 30 ml of restriction buffer lacking digitonin. Centrifuge at 3000 rpm for 15 min at 4°C to collect the pellet between washes. Remove the supernatant from the last wash and add 1 ml of restriction buffer.

Pipette the pellet repeatedly with a wide-bore pipette tip until it is relatively homogeneous. The gelatinous nature of the pellet makes this somewhat difficult, but if clumps exist, restriction digestion is often incomplete. Combine the pellets into a 50-ml tube and adjust the volume to 7.5 ml with restriction buffer. Pipette repeatedly again until the solution is homogeneous. Remove about 200 μl for examination by fluorescence microscopy. To view the sample on the microscope, mix an equal volume of the sample with ethidium bromide (8 μg/ml) diluted in restriction buffer. A "halo" of fluorescence around the nucleus should be visible. The LIS buffer extracts histones and other nonhistone chromatin proteins. Any DNA that is not directly attached to the nuclear matrix is looped out, forming a diffuse halo that surrounds the nucleus.

Add 2500 units of restriction endonuclease to the samples, mix, and incubate for 30 min at 37°C. Note that not all enzymes cleave to completion; we have had success with the restriction endonucleases *Eco*RI, *Hind*III, *Bam*HI, and *Xba*I. Mix the solution occasionally; the viscosity should decrease significantly. Centrifuge at 3000 rpm for 10 min at 4°C. Collect the supernatant into a fresh tube (this is the first "loop" fraction) and incubate at 37°C. Resuspend the pellet (the first "matrix" fraction) in another 7.5 ml of restriction buffer by pipetting repeatedly. Add another 2500 units of restriction endonuclease to the matrix fraction and place at 37°C for 15 min, mixing occasionally. Add RNase A to a final concentration of 15 μg/ml and RNase T1 to a final concentration of 270 units/ml and mix well. After 15 min, EDTA (pH 8) is added to both the loop and matrix fractions to a final concentration of 20 mM. Centrifuge the matrix at 3000 rpm for 10 min at 4°C in the JS·4.2 rotor of the Beckman J6B centrifuge. Combine the first and second supernatants (the loop fraction), add 2 vol of ethanol, and place at -20°C. Resuspend each matrix pellet in 0.5 ml of resuspension buffer [100 mM NaCl, 20 mM EDTA, 10 mM Tris-HCl (pH 8)]. Pipetting repeatedly may be necessary. Remove a sample for fluorescence microscopy and treat as above. After restriction endonuclease digestion, the halo should disappear and the fluorescent intensity of the remaining nuclear remnant should decrease because the majority of the DNA has been removed in the loop fraction.

Add 7.5 ml of proteinase K buffer [60 mM Tris-HCl (pH 7.9), 460 mM NaCl, 40 mM EDTA, 1% (w/v) Sarkosyl]. Mix well; the solution should clear somewhat. Add proteinase K (Boehringer Mannheim) to 250 μg/ml. Incubate for 1.5–2 hr at room temperature with occasional mixing. Transfer the matrix sample to a dialysis bag and dialyze overnight in 1 liter of precooled dialysis buffer [10 mM Tris-HCl (pH 7.9), 300 mM NaCl, 1 mM EDTA, 0.05 mM PMSF] at 4°C. Change the dialysis buffer once. Transfer the matrix material to a 50-ml tube. Centrifuge at 2000 rpm for 10 min at 4°C in the JS·4.2 rotor of the Beckman J6B centrifuge to remove debris. Add 2 vol of ethanol to the supernatant and place at -20°C for at least 2 hr. Centrifuge both the matrix and loop DNAs using the SW28 rotor of a Beckman ultracentrifuge at 12,000 rpm for 1 hr at 4° C. Dissolve the loop pellet in 1 ml of TE8 [10 mM Tris-HCl (pH 8), 1 mM Na$_2$EDTA] and the matrix pellet in 0.5 ml of TE8. Quantitate the amount of DNA in the matrix and loop fractions; 5–10% of the total DNA should fractionate with the matrix.

The DNA can then be digested with additional restriction enzymes prior to 2D gel analyses. Partial digestion during the matrix isolation procedure is occasionally encountered with certain restriction endonucleases and redigestion is thus necessary. It is important, however, to minimize the time of incubation (we routinely digest for 2–4 hr; restriction endonucleases with optimal activity above 37° C are to be avoided). If partial digestion during matrix-associated DNA isolation is encountered, the enrichment for replicative intermediates is reduced and may require the use of more DNA for 2D gel analysis.

Benzoylnaphthyl-DEAE Cellulose Chromatography

BND cellulose chromatography provides a simple and convenient method for the enrichment of DNA molecules containing single-stranded regions, a characteristic of replication forks. The enrichment for replicating molecules varies between 10- and 20-fold, depending on the type of DNA sample being used. It is important to eliminate RNA prior to chromatography because it also binds strongly to BND cellulose and thus could compete for the binding of DNA. The following protocol outlines a column procedure described in Ref. 10; a batch absorption method has also been described (24).

Place 8 g of BND cellulose (Serva, Heidelberg, Germany) in a 50-ml tube. Add 40 ml of 5 M NaCl, mix vigorously for 5 min, and centrifuge in the JS·4.2 rotor of the Beckman J6B centrifuge at 3000 rpm for 10 min. Discard the supernatant. Repeat the salt wash three more times. Wash with water once. Wash with 0.3 M NET8 [0.3 M NaCl, 10 mM Tris-HCl (pH 8), 1 mM Na$_2$EDTA] twice. The BND cellulose in suspension can be stored at 4° C for several months.

Aliquot the DNA (about 50–100 μg of matrix-associated DNA or up to about 200–300 μg of total DNA) to an Eppendorf tube. Adjust the volume to 0.5 ml with TE8 and adjust to 0.3 M NaCl.

Mix the BND cellulose well before packing a column (Poly-Prep columns; Bio-Rad, Richmond, CA) to a 1- to 1.4-ml bed volume. Wash the column three times with 1 ml of 0.3 M NET8. Apply the DNA to the column slowly, discarding the flow-through. Wash the column three times with 0.3 ml of 0.3 M NET8, again discarding the flow-through. Wash the column three times with 1 ml of 0.8 M NET8 [0.8 M NaCl, 10 mM Tris-HCl (pH 8), 1 mM Na$_2$EDTA]. Collect these fractions and pool together as the "salt wash," which contains primarily linear double-stranded DNA. Wash the column three times with 1 M NET8/1.8% (w/v) caffeine [1 M NaCl, 10 mM Tris-HCl (pH 8), 1 mM EDTA, 1.8% (w/v) caffeine]. This solution may precipitate at room temperature; make sure it is in solution before use by warming at 37° C for a few minutes. Collect these fractions together as the "caffeine

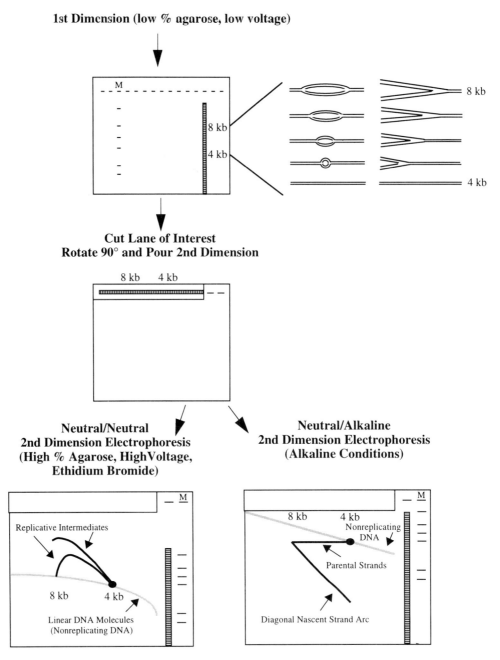

1st Dimension (low % agarose, low voltage)

M

8 kb

4 kb

8 kb

4 kb

Cut Lane of Interest
Rotate 90° and Pour 2nd Dimension

8 kb   4 kb

Neutral/Neutral
2nd Dimension Electrophoresis
(High % Agarose, HighVoltage,
Ethidium Bromide)

Neutral/Alkaline
2nd Dimension Electrophoresis
(Alkaline Conditions)

M

Replicative Intermediates

8 kb   4 kb

Linear DNA Molecules
(Nonreplicating DNA)

M

8 kb   4 kb

Nonreplicating
DNA

Parental Strands

Diagonal Nascent Strand Arc

Southern Blot and Hybridization

FIG. 1   Schematic diagram illustrating the steps involved in 2D gel electrophoresis. The first dimension of the 2D gel uses low-percentage agarose gels and a low voltage gradient. Under these conditions, replicative intermediates are separated primarily according to their mass. The gel diagram at the top shows the first dimension of a DNA sample as a striped rectangle. The

wash," which contains the DNA population enriched in replicative intermediates. Add 2 vol of ethanol to the salt and caffeine washes and mix well. Place at $-20°$C for at least 2 hr.

Centrifuge the DNAs in a Beckman SW 41 rotor at 12000 rpm for 1 hr. Pour off the supernatant and carefully wipe the inside of the tubes with a Kimwipe. It is not necessary to dry the pellet. Dissolve the pellet in 0.5 ml of TE8 and transfer to an Eppendorf tube. Adjust to 0.3 M NaCl, mix, and add 2 vol of ethanol. Place at $-20°$C for at least 1 hr. Centrifuge in a microfuge for 10 min at 4°C. Carefully remove the supernatant as completely as possible and dry the samples under vacuum. Finally, dissolve the salt wash precipitate in 100 μl of TE8 and the caffeine wash precipitate in about 9 μl of TE8. (It is important to make sure that the volume for dissolving the caffeine wash sample is appropriate for the size of the well for the first-dimension gel.) Leave on ice for at least 30 min, mixing occasionally, before loading the gel.

Two-Dimensional Agarose Gel Electrophoresis

Neutral/Neutral Two-Dimensional Gel Electrophoresis

Two complementary methods for 2D gel electrophoresis have been developed. In the first, termed *neutral/neutral,* information about the structures of molecules in the process of replication is obtained. Three classes of replicative intermediates are routinely observed: molecules that contain a single replication fork that originates from an external origin and progresses from one end of a fragment to the other (simple Y forms), molecules that contain a bubble due to an internal initiation site, and molecules in which two opposing replication forks converge (double Y forms).

Figure 1 shows a schematic flow chart for the steps involved in 2D gel electrophoresis. DNA molecules (including replicative intermediates) are separated primarily

positions at which segments of 4 and 8 kb would migrate are indicated. The series of horizontal lines at the top indicate the positions of the wells. On the right, populations of different replicative intermediates of a 4-kb segment are shown, separated according to their mass in the first dimension. The lane is excised, rotated 90°, and placed at the top of the second-dimension gel (middle gel diagram). Second-dimension conditions for the neutral/neutral 2D gels (left bottom) include high-percentage agarose, high voltage gradient, and the presence of ethidium bromide. These conditions maximize the effect of shape on the electrophoretic mobility of a molecule. Branched molecules, such as the replicative intermediates shown at the top right, migrate slower (solid arcs, bottom left) than linear molecules (shaded arc, bottom left). Molecules with different shapes migrate in a characteristic arc (see text and later figures). The second dimension for the neutral/alkaline 2D gels is performed under alkaline conditions. Nascent strands are separated from parental strands and migrate according to their size as a diagonal arc (bottom right); parental strands are the same size for each replicating molecule and migrate as a horizontal line (bottom right). Nonreplicating linear DNA molecules are represented as shaded lines in both 2D gel diagrams. The positions of bacteriophage λ DNA cleaved with *Hin*dIII used as a size marker are indicated in the lanes marked M.

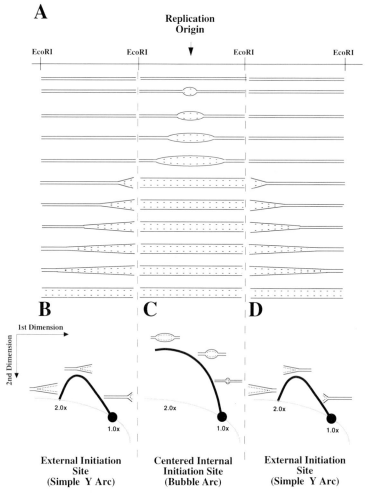

FIG. 2   Replicative intermediates from three segments exhibiting different patterns of replication and their corresponding neutral/neutral 2D gel patterns. (A) A hypothetical region of
DNA is shown as three *Eco*RI restriction fragments (distinguished by vertical dashed lines).
The middle segment contains a centered origin of replication that generates replication forks
that proceed bidirectionally. A series of replicative intermediates are shown below each segment with progressively elongated replication forks. The dashed lines indicate nascent DNA
strands; the solid lines indicate parental DNA strands. Because the middle segment contains a
centered origin of replication, forks proceed bidirectionally from within the fragment outward,
generating a bubble arc as shown in the middle 2D gel diagram (C). To detect bubble arcs by
the neutral/neutral 2D gel method, the initiation site should lie within the central third of the
fragment analyzed (discussed in M. H. K. Linskens and J. A. Huberman, *Nucleic Acids Res.*
18, 647). The fragments on the left and the right are replicated from an external origin; thus,

by size in the first dimension. As a fragment is replicated, its mass doubles, so that the replicative intermediates derived from that segment are distributed over a portion of the first-dimension gel lane. The migration of DNA molecules in the second dimension of neutral/neutral 2D gels is influenced primarily by shape; branched intermediates containing replication forks thus migrate slower than linear molecules. Because each type of replicative intermediate has a different shape, the population from each family (i.e., Y forms, bubbles, and double Y forms) migrates in a unique arc or pattern. Figures 2 and 3 graphically depict populations of replicative intermediates with various shapes and show the expected neutral/neutral 2D gel patterns for those segments.

It has been shown that in order to detect bubble arcs, the site of initiation of DNA replication should be positioned in the central third of the segment analyzed (25). Thus, a series of overlapping restriction fragments must be analyzed in order to survey a genomic region completely. For the initial scanning of a region for initiation sites, it may be beneficial to begin by analyzing relatively large segments (10–15 kb) so that the complete region can be examined more rapidly.

Neutral/neutral 2D gel electrophoresis of DNA segments between 2 and 5 kb is performed as previously described (2). The first-dimension gel is 0.4% (w/v) agarose (SeaKem ME; FMC Bioproducts, Philadelphia, PA) in $1 \times$ TBE [89 mM Tris-HCl (pH 8), 89 mM boric acid, 2 mM EDTA]. The width of the wells for the first dimension should be relatively small. The combs we use have teeth that are 3 mm wide and 1 mm in length, so that 9 μl of caffeine wash plus 2 μl of loading buffer can be conveniently added to the well. The gels are 28 cm long, 15 cm wide, and about 9 mm thick. The electrodes for all electrophoresis apparatuses described below are spaced 44 cm apart. The gels are run at 1 V/cm at room temperature for 15–19 hr, depending on the size of the fragment being analyzed. Include a molecular weight marker on one side of the gel, separated from the edge of the gel by several wells. The sample at the other side of the gel is also separated from the edge of the gel by several wells (we routinely run up to four different samples on one gel). Make sure to leave enough space between the samples so that excision is possible (we routinely leave two empty lanes between each sample). As a precaution, place a plastic or glass plate wrapped with aluminum foil over the electrophoresis apparatus to minimize nicking of DNA by exposure to ultraviolet light during electrophoresis.

After electrophoresis, the gel is stained in $1 \times$ TBE (the TBE from the first dimension can be used) containing ethidium bromide (0.3 μg/ml) for about 15 min (protect from light). The region of the gel containing the marker lane is then excised from the remainder of the gel and photographed (do not expose the sample lanes to ultraviolet

---

forks traverse these segments from right to left or from left to right. Populations of replicating molecules of this type typically migrate on neutral/neutral 2D gels as simple Y arcs as shown in the left (B) and right (D) 2D gel patterns. $1X$ and $2X$, masses of an unreplicated and a fully replicated segment, respectively.

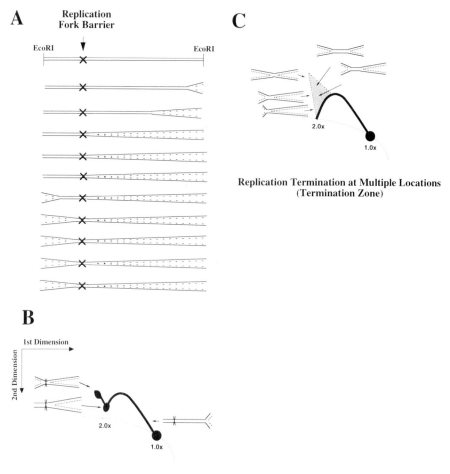

A

Replication
Fork Barrier

EcoRI   EcoRI

C

Replication Termination at Multiple Locations
(Termination Zone)

B

1st Dimension

2nd Dimension

2.0x

1.0x

Replication Fork Barrier
(Site-Specific Replication Termination)

FIG. 3   Neutral/neutral 2D gel electrophoresis of a segment containing a replication fork barrier and a segment in which termination occurs throughout the segment at multiple locations.
The DNA segment depicted on the left (A) contains a barrier to replication fork movement
(shown as an ×). A leftward moving replication fork from an external origin is stalled at the
barrier, resulting in an accumulation of simple Y forms. This accumulation is represented as a
spot of greater intensity along the simple Y arc shown in (B). A rightward moving fork then
enters the other terminus of the fragment and stalls on the opposite side of the barrier, resulting
in the accumulation of molecules containing two convergent forks (double Y forms). This 2D
gel pattern is the predicted result if the leftward moving fork reaches the barrier before the
rightward moving fork enters the fragment and if forks are stalled completely. Additional patterns may be observed if either of these conditions is altered. A neutral/neutral 2D gel pattern
typical of replication termination in a zone at multiple locations throughout a segment is
shown in the 2D gel diagram in (C). A diagram illustrating the population of replicative inter-

light). Each sample lane is then excised with the adjacent empty lanes still attached, yielding a strip about 1.5 cm wide by 13–14 cm long. Cut the gel so that the region of interest is in the middle third of the lane, that is, if the restriction fragment is 4 kb, place the region corresponding to 4–8 kb in the middle of the slice. Rotate the slices 90° and place at the top of another gel tray (15 × 15 cm). The second-dimension gel [0.8–1% (w/v) agarose in 1× TBE plus ethidium bromide (0.3 μg/ml) is then poured; include two lanes on one side of the first-dimension slice for a salt wash control (1–2 μl) and a molecular weight marker. The second dimension is run in a cold room (4°C) in precooled circulating 1× TBE containing ethidium bromide (0.3 μg/ml) at 4.8 V/cm for 4–7 hr.

For restriction fragments between 5 and 19 kb we have modified the conditions described by Krysan and Calos (16). Because inappropriate gel conditions can result in the distortion of 2D gel patterns (26), it is important that suitable control experiments be performed with any new electrophoresis parameters. DNA restriction fragments are first separated in 0.28% (w/v) agarose gels (Bio-Rad high-strength analytical grade) in 1× TBE at room temperature. The voltage and times are varied for different sized fragments as follows: fragments 5–9 kb are separated at 0.66 V/cm for 40–48 hr, fragments 9–12 kb are separated at 0.45 V/cm for 65–72 hr, and fragments 12–19 kb are separated at 0.34 V/cm for 90–96 hr. Gels are stained with ethidium bromide (0.3 μg/ml) in TBE for 15 min, and the lane containing molecular weight markers is photographed. The sample lane is excised without exposure to ultraviolet light, rotated 90°, and placed at the top of a gel tray (15 × 28 mm), and a 0.58% (w/v) agarose gel (Bio-Rad high-strength analytical grade) containing ethidium bromide (0.3 μg/ml) in 1× TBE is poured. Second-dimension conditions for all fragment sizes are 0.89 V/cm for 40–48 hr at room temperature in TBE containing ethidium bromide (0.3 μg/ml). Standard Southern transfer techniques are then used to transfer the DNA to a hybridization membrane.

Figure 4 illustrates the use of the neutral/neutral 2D gel in the analysis of the replication of EBV (data reproduced from Ref. 13). From these data we have found that a plasmid containing the EBV *oriP* region replicates in an asymmetric bidirectional manner. This mechanism of replication is unlike that observed for any other mammalian virus to date. Replication initiates at or near a dyad symmetry (DS) element within the latent origin of replication, *oriP*. Replication forks proceed bidirectionally until one encounters a replication fork barrier at a family of repeated sequences (FR) within *oriP* about 1 kb away from the site of initiation (the DS). The other replication fork then proceeds through the remainder of the plasmid until it reaches the stalled fork and replication termination occurs. The EBV origin of replication thus contains both the initiation and termination sites for viral DNA replication.

---

mediates as in (A) is not shown for this segment. The filled triangular region of hybridization results from a population of replication intermediates containing two converging replication forks that meet at multiple sites within the segment.

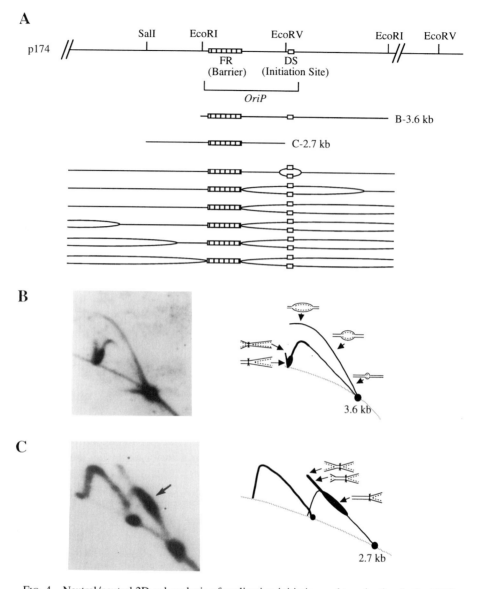

FIG. 4 Neutral/neutral 2D gel analysis of replication initiation and termination in the EBV origin of replication, *oriP*. In (A), a map of a portion of the EBV-containing plasmid p174 is shown with relevant restriction sites. The location of *oriP* is shown with its two essential components, the family of repeated sequences (FR) and the dyad symmetry element (DS). Below the map, the 3.6-kb *Eco*RI segment and the 2.7-kb *Sal*I/*Eco*RV segment analyzed in (B) and (C) are shown. Below these fragments, a schematic of the proposed mechanism of replication in this region is diagrammed. An autoradiograph showing a neutral/neutral 2D gel pattern and a graphic representation are shown in (B) for the 3.6-kb *Eco*RI fragment. This segment contains the DS element in a central location. As a result of initiation from a site at or near the DS element, a bubble arc is observed. Leftward moving forks are stalled at the FR

Neutral/Alkaline Two-Dimensional Agarose Gel Electrophoresis

The second 2D gel method we describe is termed *neutral/alkaline*. This technique is used to determine the direction of replication fork movement within a segment as well as to provide an indication of the sizes of nascent strands within a segment. Because the 2D gel methods provide complementary information, it is often useful to perform both types of analyses on restriction fragments. Identical first-dimension conditions are used for the neutral/alkaline 2D gel technique, as for the neutral/neutral technique. The second dimension, however, is performed under alkaline conditions so that nascent strands are separated from parental strands and migrate according to their size. Probes from multiple locations within the segment analyzed are used to detect the size of nascent strands near the site of each probe. Figure 5 illustrates the predicted neutral/alkaline 2D gel patterns for hypothetical populations of replicative intermediates containing various structures.

To obtain intense nascent strand arcs, more DNA is required than for the neutral/neutral 2D gels, because parental strands are separated from nascent strands and thus do not contribute to the hybridization signal along the nascent strand arc. The technique is also more sensitive to nicks in the DNA because the alkaline second dimension denatures the DNA strands. It is important to hybridize with probes from the ends of the fragment as well as internal to the fragment to assess the mode of replication accurately.

The neutral/alkaline 2D gel conditions described below were modified from Nawotka and Huberman (4). The first-dimension conditions are identical to the neutral/neutral 2D gel method. The gels are treated as above, except that 0.8–1.2% (w/v) agarose in water is used for the second-dimension gel. The solidified gels are placed in an electrophoresis apparatus containing circulating alkaline electrophoresis buffer (40 mM NaOH, 2 mM EDTA) and incubated at room temperature for 1 hr. Molecular weight markers (1 μg of bacteriophage λ DNA digested with *Hin*dIII and

---

and rightward moving forks proceed past the right end of the segment. The shape of maturing replicative intermediates thus changes into large Y forms that accumulate due to the barrier (the intense spot at the left end of the simple Y arc). As forks proceed around the plasmid, they eventually enter the other side of this segment and converge on the stalled fork. These intermediates are molecules with two forks (double-Ys) and are indicated in the diagram. In (C), the 2.7-kb *Sal*I/*Eco*RV fragment places the FR (the barrier region) near the center of the segment. The 2D gel pattern for this segment contains an intense elongated signal (arrow in the autoradiograph and filled oval in the schematic) corresponding to molecules with a leftward moving fork stalled at multiple locations within the FR region. The diagonal spike emanating from this accumulation corresponds to molecules in which forks enter the other side and converge on the stalled fork. The probe used in (C) also detects a Y arc from a 4.7-kb *Eco*RV segment. [These results are reproduced from T. A. Gahn and C. L. Schildkraut, *Cell (Cambridge, Mass.)* **58,** 527 (1989) with permission from *Cell* press. The data are described in detail in the above reference.]

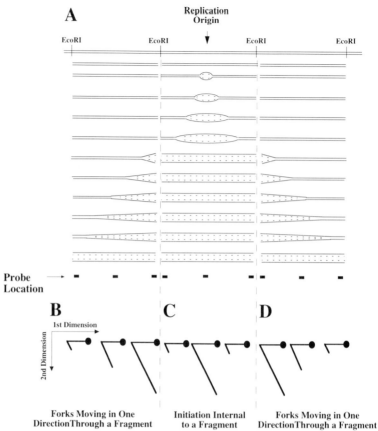

Fig. 5 Replicative intermediates from three segments exhibiting different patterns of replication and their corresponding neutral/alkaline 2D gel schematic patterns. The upper portion (A) is the same as in Fig 2. At the bottom, neutral/alkaline 2D gel patterns of the three segments are shown, using three different hybridization probes for each fragment [the positions of the probes along the DNA map shown in (A) are indicated by the filled rectangles]. Because the fragments on the left and the right are replicated from an external origin, forks traverse these segments either from right to left or left to right. For the left segment, a probe located at the right terminus will detect nascent strands of all sizes, because forks are entering the fragment close to the location of the probe and progressing across to the other terminus. For this probe, the nascent strand diagonal arc extends to small sizes (B). A probe located in the center of the fragment will detect nascent strands from forks that have progressed as far as the location of the probe and from those that continue past the probe location. The nascent strand diagonal arc detected by this probe stops at about half of the replicated mass of the fragment. The probe on the left terminus detects only large nascent strands because forks enter from the opposite side and must proceed through the entire segment to reach the location of the probe. Similar patterns are observed for the right segment (D), except that the orientation is reversed because forks travel from left to right. The middle segment contains a centered origin; thus, replication forks move bidirectionally from within the fragment outward. Probes located close to the position of the origin will detect nascent strands of all sizes, whereas probes in a more terminal position will detect only relatively large nascent strands (C).

2 μg of double-stranded ϕX174 DNA digested with *Hae*III) and salt wash controls (1 μl) are added to about 9 μl of alkaline electrophoresis buffer. Add the DNAs to the wells and wait for 5 min before turning on the power. Electrophoresis is performed at 0.57 V/cm for 24–48 hr, depending on the fragment size. To visualize the molecular weight markers, gels can be stained with ethidium bromide (0.5 μg/ml) in neutralization solution [0.5 M Tris-HCl (pH 7.5), 1.5 M NaCl] for 45 min. Alternatively, radioactively labeled marker DNA can be included in the hybridization solution. After photographing, gels are rinsed with two or three changes of water for 15 min prior to standard Southern transfer.

Summary

The 2D gel methods described above have provided extremely powerful approaches for the study of DNA replication in a wide array of organisms and gene loci. The main limitations seem to be the quality of the DNA that can be isolated from a particular organism and the signal intensity obtainable. As discussed above, various enrichment strategies and specific DNA isolation procedures have been developed to permit a greater degree of sensitivity. The use of these techniques in studies of viral DNA replication will provide a more detailed understanding of mammalian DNA replication.

Acknowledgments

The authors are grateful to Dr. Bernardo Schvartzman for helpful discussions and for comments on the manuscript. This work was supported by the National Institutes of Health Grant RO1 GM45751 and by Fellowships to R.L. from the American Cancer Society. Support was also provided by Cancer Center Core Grant NIH/NCI P30-CA13330.

References

1. L. T. Vassilev and M. L. DePamphilis, *Crit. Rev. Biochem. Mol. Biol.* **27**, 445 (1992).
2. B. J. Brewer and W. L. Fangman, *Cell (Cambridge, Mass.)* **51**, 463 (1987).
3. B. J. Brewer and W. L. Fangman, *BioEssays* **13**, 317 (1991).
4. K. A. Nawotka and J. A. Huberman, *Mol. Cell. Biol.* **8**, 1408 (1988).
5. M. H. K. Linskens and J. A. Huberman, *Mol. Cell. Biol.* **8**, 4927 (1988).
6. J. Zhu, C. S. Newlon, and J. A. Huberman, *Mol. Cell. Biol.* **12**, 4733 (1992).
7. J. Zhu, C. Brun, H. Kurooka, M. Yanagida, and J. A. Huberman, *Chromosoma* **102**, S7 (1992).
8. R. D. Little, T. H. K. Platt, and C. L. Schildkraut, *Mol. Cell. Biol.* **13**, 6600 (1993).
9. J. P. Vaughn, P. A. Dijkwel, and J. L. Hamlin, *Cell (Cambridge, Mass.)* **61**, 1075 (1990).
10. P. A. Dijkwel, J. P. Vaughn, and J. L. Hamlin, *Mol. Cell. Biol.* **11**, 3850 (1991).

11. J. B. Schvartzman, S. Adolph, L. Martin-Parras, and C. L. Schildkraut, *Mol. Cell. Biol.* **10,** 3078 (1990).
12. L. Yang and M. Botchan, *J. Virol.* **64,** 5903 (1990).
13. T. A. Gahn and C. L. Schildkraut, *Cell (Cambridge, Mass.)* **58,** 527 (1989).
14. K. Saunders, A. Lucy, and J. Stanley, *Nucleic Acids Res.* **19,** 2325 (1991).
15. B. Hirt, *J. Mol. Biol.* **113,** 414 (1967).
16. P. J. Krysan and M. P. Calos, *Mol. Cell. Biol.* **11,** 1464 (1991).
17. K. Auborn, R. D. Little, T. H. K. Platt, M. Vaccariello, and C. L. Schildkraut, *Proc. Natl. Acad. Sci. U.S.A.* in press (1994).
18. V. Dhar and C. L. Schildkraut, *Mol. Cell. Biol.* **11,** 6268 (1991).
19. M. L. DePamphilis, *Cell (Cambridge, Mass.)* **52,** 635 (1988).
20. G. H. Hatfull, A. T. Bankier, B. G. Barrell, and P. J. Farrell, *Virology* **164,** 334 (1988).
21. J. P. Vaughn, P. A. Dijkwel, L. H. F. Mullenders, and J. L. Hamlin, *Nucleic Acids Res.* **18,** 1965 (1990).
22. M. V. Olson, K. Loughney, and B. D. Hall, *J. Mol. Biol.* **132,** 387 (1979).
23. F. E. Abidi, M. Wada, R. D. Little, and D. Schlessinger, *Genomics* **7,** 363 (1990).
24. J. A. Huberman, L. D. Spotila, K. A. Nawotka, S. M. El-Assouli, and L. R. Davis, *Cell (Cambridge, Mass.)* **51,** 573 (1987).
25. M. H. K. Linskens and J. A. Huberman, *Nucleic Acids Res.* **18,** 647 (1990).
26. O. Hyrien and M. Mechali, *Nucleic Acids Res.* **20,** 1463 (1992).

[22] Generation of Human T Cell Lines Using Lymphotropic Herpesviruses

Helmut Fickenscher and Bernhard Fleckenstein

Introduction

During the past decades, molecular and biochemical studies of human T lymphocytes were limited for mainly two reasons. First, T lymphocytic tumor lines such as the Jurkat line (1) have a strongly altered phenotype in comparison to primary cells with respect to signal transduction (2) and gene regulation. Second, the appropriate primary cell cultures are limited in their life span. It is laborious to grow primary lymphocytes to large cell numbers, and it requires considerable effort to amplify the lymphocytes in periodic response to a specific antigen with accessory cells expressing the appropriate major histocompatibility complex (MHC) restriction elements. Moreover, impurities due to the addition of irradiated feeder cells may cause difficulties in interpretation of results. Conventional systems for immortalizing lymphocytes have not been successful for studying T cell biology: the technique of human T cell fusion hybridomas suffers from genomic instability of the clones; lymphocytes transformed by human T cell leukemia virus (HTLV-I) often lose their T cell receptor complex and produce HTLV-I virions regularly (reviewed in Ref. 2). Here, we wish to summarize procedures developed for using cell-free *Herpesvirus saimiri* for the targeted transformation of specific human T lymphocytes to antigen-independent growth and to an activated phenotype (3).

Herpesvirus saimiri (4) is the prototype of the genus of γ_2-herpesviruses or rhadinoviruses (5), of which no human member has been found so far. *Herpesvirus saimiri* is not pathogenic in its natural host, the squirrel monkey (*Saimiri sciureus*), and can easily be isolated from the peripheral blood of most individuals (6). In other South American monkeys such as common marmosets (*Callithrix jacchus*), cotton top marmosets (*Saguinus oedipus*), and owl monkeys (*Aotus trivirgatus*), as well as in rabbits (7), *H. saimiri* causes fulminant polyclonal T cell lymphomas and acute lymphatic leukemias (reviewed in Refs. 8–10). Furthermore, marmoset T lymphocytes can be immortalized *in vitro* by *H. saimiri* (11). The nucleotide sequence of strain A-11 has been entirely determined (12). The viral genome consists of 112,930 bp of low density DNA, AT rich (L-DNA) with 76 tightly packed major open reading frames, flanked by some 35 noncoding high density DNA GC rich (H-DNA) repeats of 1444 bp in tandem orientation. The genome organization is colinear to the human γ_1-herpesvirus Epstein–Barr virus (EBV). However, the conserved gene blocks are rearranged and the transformation and persistence-associated genes of EBV (EBNA, LMP) are lacking (12). Several open reading frames of *H. saimiri* display strong sequence homologies to known cellular genes, among them thymidylate synthase, dihydrofolate reductase, complement control proteins, the surface protein CD59, cyclins, and G protein-coupled receptors (reviewed in Ref. 12). Their role during T lymphocyte transformation and during the lytic cycle remains to be determined.

The genomic region that is essential for transformation of monkey T lymphocytes and for oncogenicity was mapped to the left end of the genome (13–17). *Herpesvirus saimiri* strains of different subgroups (A, B, and C) vary in oncogenicity and left terminal sequence (18–21). The left genomic terminus of the virus strains from these distinct subgroups encodes different *H. saimiri* transformation-associated proteins (STPs). The respective proteins of subgroups A and C (STP-A and STP-C) are able to transform rodent fibroblasts. STP-C-transformed fibroblasts cause invasive tumors in nude mice (22). Moreover, transgenic mice expressing STP-C of strain C-488 develop epithelial tumors of the salivary gland, bile ducts, and thymus (23). All STP proteins have a hydrophobic C terminus. STP-C of strain C-488 further contains a stretch of 18 collagen repeats and a specific N-terminal sequence of 17 amino acids (21). STP-C was shown to reside in the perinuclear compartment of transformed rodent fibroblasts (24) and to be phosphorylated in the transformed fibroblasts on a serine residue close to the amino terminus (25). Viral genomes lacking the oncogene region were used as expression vectors (13, 26, 27) in order to analyze the T cell transforming functions of HTLV-I X-region proteins. This approach made it possible to assign this transforming activity to the Tax protein of HTLV-I. Phenotypically, the *tax* recombinant-transformed cultures resembled the known HTLV-immortalized cell lines, but did not shed infectious virus (10, 28–30).

Human primary T lymphocytes are readily transformed by wild-type subgroup C strains of *H. saimiri* to continuously proliferating T cell lines with the phenotype of mature CD4- or CD8-positive cells. The transformed lymphocytes often remain interleukin 2 (IL-2) dependent, but do not need a periodic restimulation with antigen or mitogen. They carry multiple nonintegrated viral episomes and do not produce virion particles (3). In transformed cells, viral gene expression could not be found from many gene regions. However, the viral oncogene *stp-C* is strongly expressed at the transcript and protein level (31). When antigen-specific T cell clones are transformed by *H. saimiri,* the MHC-restricted antigen-specific reaction is retained (2, 32, 33, 34). The transformed T lymphocytes maintain early signal transduction patterns of primary cells (2, 34), express activation surface markers, may show unspecific cytotoxicity (3, 34), and are hyperreactive to CD2/LFA-3 contacts (35). Thus, the observation that *H. saimiri* C-488 is able to transform human T lymphocytes provides novel tools to study T cell biology.

Methods

Definitions

Transformation describes changes in cell morphology and growth behavior at a given moment, whereas *immortalization* reflects proliferation over prolonged time periods.

Both terms overlap and may be used as synonyms after about 1 year of observation. Virologists and molecular biologists use *cell line* for any culture growing for long time periods with stable properties. The immunological terms *cell line* and *cell clone*, however, are used to describe polyclonal or monoclonal cultures, respectively. *H. saimiri* is frequently used as the abbreviation for *Herpesvirus saimiri*, while *HVS* gives rise to confusion with *HSV* (herpes simplex virus). *Herpesvirus saimiri* is termed *saimirine herpesvirus 2* (SHV-2) by the official nomenclature of the International Committee for the Taxonomy of Viruses (5).

Basic Protocols for Working with Herpesvirus saimiri

Owl Monkey Kidney Cells

Epithelial owl monkey kidney (OMK) cells from healthy *A. trivirgatus* (owl monkeys) can be kept in primary culture over long time periods (36) if they are not contaminated with lytically growing viruses, such as *Herpesvirus aotus* type 1 or 3. The American Type Culture Collection (ATCC, Rockville, MD) offers strain OMK (637-69) in the fourteenth passage under catalog number ATCC CRL 1556. This strain was isolated in November 1970 and later submitted to the ATCC by M. D. Daniel (New England Regional Primate Research Center, Harvard Medical School, Southborough, MA). OMK-637 has been used in several laboratories intensively over decades without any hints of viral contaminations or latent agents. The monolayer cells are maintained in Earle's minimal essential medium (MEM) or Dulbecco's modified Eagle's medium (DMEM) supplemented with 10% (v/v) heat-inactivated fetal calf serum (FCS), glutamine, and—if desired—antibiotics (gentamicin or penicillin/streptomycin). OMK cell growth does not require exogenous CO_2 supplementation to the atmosphere. The cells are trypsinized and split only once a week by $1:2$. This maintains the original status of the primary OMK cells and does not enforce selection for fast-growing subtypes. The medium is changed on the fourth day after splitting. The cells should not be used for more than 50 passages. OMK cells are the typical propagation system for herpesviruses of nonhuman New World primates, such as *H. saimiri* and *Herpesvirus ateles* (for review see Ref. 8).

Virus Cultures

The tissue culture medium of confluent OMK monolayer cultures is removed on day 2 to 4 after splitting and infectious virion suspension is added in a minimal volume (e.g., 2 ml for a 25-cm² flask, 5 ml for an 80-cm² flask). Adsorption is allowed to take place at 37°C for 1 to 2 hr. The monolayer should not suffer from drying out during this time. Afterward, medium is added and the incubation is continued. After 1 to 14 days, initial cytopathic changes are found. Typically, focal rounding of cells is observed. Plaques appear later, surrounded by rounded cells. Several days later, the whole cell layer will be lysed by the virus.

After completion of the CPE, infected OMK cultures, including the OMK cell debris, are used as *H. saimiri* stocks, without any further treatment. The virus stocks are stable and can be stored at 4°C for several months without loss of infectivity. Small volumes of supernatant should be frozen in liquid nitrogen or at −80°C; the titer will decrease by at least one order of magnitude. Addition of 10% (v/v) dimethylsulfoxide (DMSO) is recommended by some researchers for freezing virus stocks.

The lytic system of OMK cells is used to estimate concentration of infectious virion particles from the supernatant of lytic cultures (see Virus Titration, below) and also to monitor *H. saimiri*-transformed T lymphocytes for their (non)producer status. Typically, about 10^6 transformed lymphocytes are added to a 25-cm² flask (5 ml of MEM or DMEM with 10% FCS, but without IL-2) of fresh confluent OMK cells. Cocultivation allows the infection of OMK cells with small amounts of cell-associated virions from the lymphocytes, as close cell contact is achieved. Infectivity from the supernatant of lymphocyte cultures is usually not detectable; it was suspected that transformed and virus-producing monkey lymphocytes pass on virions by cell-to-cell contact. Additionally, the transformed human lymphocytes are activated by the OMK cell contact, presumably by the CD2–LFA-3 interaction (35). As a side effect of the activated status, especially the $CD8^+$-transformed lymphocytes may sometimes exhibit a strong unspecific cytotoxic effect on OMK cells. Thus it occurs that the OMK cells of the first cocultivation passage are killed by unspecific cytotoxicity. The cytotoxic activity will decrease during further passage of cocultivation on fresh OMK cells. One cocultivation passage can be kept in the incubator for 14 days. After 1 week, a small volume of complete medium may be added. Cocultivations are routinely followed up during three passages, which is about 4 to 6 weeks. Doubtful observations of cytopathic effects (CPEs) can be clarified by transferring an aliquot of sterile-filtered supernatant to a fresh OMK culture, where a typical CPE should be visible after a few days. If there is virus growth in later passages, low titers of infectivity have been amplified. Several weeks after infection, virion production from human T lymphocyte cultures can no longer be demonstrated. The sensitivity of the cocultivation method was estimated to be 1 virus-producing cell per 10^6 cells (37).

Virion Purification

The supernatant and the cell remnants from lytic virus cultures in OMK both contain high amounts of virion particles (about 10^9 virion particles per milliliter in the supernatant). Cultures with complete CPE (e.g., 1 or 2 liters) are bottled into Sorvall-GSA (DuPont, Wilmington, DE) centrifuge flasks (with canted necks to avoid any spilling) with 200 ml each. A first centrifugation is carried out for 15 min at 4°C and 3000 rpm to collect the cell-associated fraction of the virus. The cell sediment is resuspended in 3 ml of hypotonic virus standard buffer [VSB: pH 7.8, 50 mM Tris, 12 mM KCl, 5 mM ethylenediaminetetraacetic acid (EDTA)]. The supernatant is transferred to clean GSA bottles and spun for a second time at high speed (12,000 rpm) at 4°C for 4 hr. Usually, clearly visible virus pellets are found. Virions are resuspended in 3 ml of VSB. Sucrose gradients are prepared in ultraclear Beckman (Fullerton, CA) ultracentrifuge tubes (SW27), using a gradient mixer with 15 ml of 30% (w/w) sucrose in VSB solution (stirring) close to the outlet and 16 ml of 15% (w/w) solution in the distal container. The rotor and the buckets are cooled to 4°C prior to centrifugation, whereas the sucrose gradients are not. Both virus and cell debris suspensions are homogenized in a Dounce glass homogenizer (Kontes, Vineland, NJ) by 20 strokes. About 3 ml of the homogenized material can be loaded on one gradient. The buckets are weight adjusted with VSB and centrifuged in a SW27 rotor for 30 min at 4°C and 20,000 rpm. In case the virus band is not clearly visible, a strong and focused apical light source should be used. The band material is collected into a Beckman SW27 polyallomer tube, diluted with VSB, and spun for 90 min at 4°C and 17,000 rpm. The virions are then resuspended in a small volume of VSB, for example, 100 to 500 μl. This method is based on the procedures described earlier by Fleckenstein and Wolf (38) and Fleckenstein *et al.* (39). The small final volume should contain nearly all infectious particles from the starting material and is concentrated by several orders of magnitude. Therefore, special care should be taken for the biosafety of all laboratory staff working in the respective rooms. Virion purification of oncogenic *H. saimiri* strains should be performed in a P3 facility.

Purification of Virion DNA

Purified virion DNA of oncogenic *H. saimiri* strains is infectious. When virion DNA had been injected intramuscularly into marmoset monkeys, animals died from lymphoma after a short time (40). Large-scale purification of virion DNA starts with concentrated virion suspensions as described above. Sodium laurylsarcosinate [200 μl of a 20% (v/v) solution] is added to 1.8 ml of purified virions in VSB. The lysis of virions is performed for 1 hr at 60°C. The lysate (2 ml) is diluted with 6 ml of VSB, and 10 g of CsCl is added. The solution should have a refractive index of 1.412 at 25°C. The DNA/CsCl solution is loaded without ethidium bromide in 50 TI Beckman Quickseal tubes (polyallomer), which are filled to the top with paraffin. The gradients are spun at 35,000 rpm and 20°C for 60 hr. Two syringe needles are placed into the gradient tube: the first one on top into the paraffin to allow pressure compensation, the second one at the bottom. The flow rate can be reduced, when a syringe filled with paraffin is plugged to the upper needle. About 20 fractions of 10 drops each are collected in Eppendorf caps. One microliter of each fraction is diluted to 10 or 15 μl with water and colored loading buffer and run slowly on a conventional agarose test gel. A single, clean, high molecular weight band should be visible. The pooled DNA-containing fractions are dialyzed against 20 mM Tris, pH 8.5, and may be extracted with phenol–chloroform–isoamyl alcohol and chloroform–isoamyl alcohol, ethanol precipitated, and resuspended in water. Concentration is determined by spectrophotometry. The yields are about 40 μg from 200 ml of supernatant. Virion DNA isolated from the cleared supernatant will be relatively pure (less than 5% contamination by cellular DNA). The cell-bound virus may lead to varying quality and amounts of DNA yields. However, large amounts of pure virion DNA are often obtained from the cell fraction. These procedures are modified from Fleckenstein and Wolf (38), Fleckenstein *et al.* (39), and Bornkamm *et al.* (41).

A short protocol has proved useful for minipreparations of viral DNA. When only small amounts of virion DNA are needed, 1.5 ml of cell-free supernatant is centrifuged at 16,000 rpm and 4°C in a Sorvall SA600 rotor. Using screw-cap Eppendorf tubes is recommended to avoid aerosols. The virus sediment is dissolved in 50 μl of TE buffer (10 mM Tris-HCl, pH 8.0, 1 mM EDTA) with 2% sodium lauryl sarcosinate, incubated at 56°C for 15 min, diluted with TE, phenol extracted, and precipitated. The yield from 1.5 ml is about 50 ng and sufficient for Southern analyses (26).

Virus Titration

The limiting dilution method is the simplest way to estimate virus titers. OMK cells are trypsinized and split by 1:5 per area, for example, into 24-well dishes, and incubated under CO_2. Cell-free virus suspension is diluted in MEM at 10^{-1} to 10^{-9}. To each plate well, 1 ml of virus dilutions is added 1 day after splitting. It is important to perform the assay at least in triplicate and to run control cultures. The plates are observed for at least 14 days. The progress of CPE should be monitored daily.

The more laborious plaque assay is also based on serial dilution. Methyl cellulose [4.4 g; e.g., Fluka (Ronkonkoma, NY) 64630 Methocel MC 4000] is suspended in 100 ml of H_2O in a 500-ml bottle and autoclaved (45 min, 121°C). One hundred milliliters of 2× MEM, 20 ml of FCS (heat inactivated), antibiotics, and glutamine are added. The mixture is stirred at 4°C overnight. Subconfluent OMK cultures in 6-cm dishes are infected, each with 1 ml of cell-free virus dilutions (10^{-4} to 10^{-8} in MEM) to 3 dishes each, 1 day after splitting the cells. After 1 hr at 37°C, the supernatant medium is aspirated, and 5 ml of 2% methyl cellulose in MEM is added. The cultures are fed with 3 ml of 2% methyl cellulose in MEM after 1 week. For plaque purification, single plaques are picked with a pipette tip and transferred to a fresh OMK culture. For titer estimations, the medium is poured off carefully after plaque formation has been observed. The cells are washed mildly with phosphate-buffered saline (PBS) and stained with 1% (v/v) crystal violet in 20% ethanol/80% water (v/v). Subsequently, the surplus dye is washed out carefully with water without destroying the cell layer. The cells are air dried, and the plaques are counted. Routine OMK cell cultures should reach titers between 10^5 to 10^7 plaque-forming units (pfu)/ml (42).

Demonstration of Episomes and Virion DNA

To demonstrate and distinguish persistent nonintegrated episomal (slow migrating) or lytical linear (higher mobility) viral DNA, *in situ* lysis gel electrophoresis can be applied. The method was first described by Gardella *et al.* (43). A simplified version of that protocol is presented (7). Typical examples for Gardella gels are shown in various reports (3, 7, 43, 44). The gel system is a vertical 1% (w/v) agarose gel in 1× Tris–borate–EDTA (TBE) with gel size 20 × 20 cm, 0.5 cm thick, at 4°C.

The slots have a 0.5 × 0.5 cm area and are 1 cm deep; they are separated by 0.3- to 0.5-cm agarose teeth. The thick gel tends to slip out from the gel plates during assembly of the gel apparatus. A large, 3% (w/v) agarose block is put into the lower buffer chamber, so that the gel and the gel plates can rest on it. The gel system is placed into a refrigerator or cold room; the upper chamber is still left without buffer. TBE (1×) for the chambers is precooled. Frozen aliquots of buffers A and B (composition listed in Table I) without enzymes have been prepared earlier, as Ficoll can be very sticky. Enzymes (RNase, protease K) are added just before use. About 2 × 10^6 lymphocytes (e.g., transformed with *H. saimiri* C488) are washed in PBS, resuspended in 50 μl of probe buffer A (blue), and loaded into the very bottom of the slots. As controls, conventional λ-markers (*Hind*III) and a small amount of the virions are used, the latter obtained by centrifuging 1.5 ml of virus stock at 16,000 rpm in a Sorvall SA600 rotor, as described for the minipreparation protocol for virion DNA. Such a small virion preparation should be enough for several gels. The controls are also loaded in buffer A. Green buffer B (120 μl) is laid over the sample, nearly filling the gel slots. Some 1× TBE buffer is added carefully to fill the slots completely, and more buffer to fill the upper chamber. Electrophoresis is slowly performed at 4°C. Proteins and RNA are allowed to degrade enzymatically during a first period of at least 3 hr at 10 V. Subsequently, electrophoresis is continued for at least 12 hr at 100 V, until xylene cyanol reaches the bottom of the gel. The agarose gel is stained in TBE containing ethidium bromide and photographed. However, ethidium bromide staining is usually not sufficient for demonstrating episomal bands. The DNA is transferred onto a nylon membrane (e.g., Hybond, Amersham), using the alkali transfer protocol (protocol for alkali transfer supplied by Amersham with purchase of membrane). Viral DNA bands can be visualized by stringent hybridization and subsequent autoradiography. There are several possibilities for the choice of hybridization probe. If maximal sensitivity is necessary, a cloned H-DNA fragment should be used (45). H-DNA is highly repetitive in the nonintegrated episomes and amplifies the viral signals. If

TABLE I   Buffer Composition[a]

Buffer	Substance	Final concentration	Stock solution	For 10-ml volume
A	Ficoll 400	200 mg/ml	Powder	2 g
	TBE	1×	10×	1 ml
	BPB	0.25 mg/ml	10 mg/ml	0.25 ml
	RNase A	50 μg/ml	5 mg/ml	0.1 ml
B	Ficoll 400	50 mg/ml	Powder	0.5 g
	SDS	1% (w/v)	10% (w/v)	1 ml
	XC	0.25 mg/ml	10 mg/ml	0.25 ml
	Protease K	1 mg/ml	10 mg/ml	1 ml

[a] BPB, Bromophenol blue; XC, Xylene cyanol.

higher specificity is intended, for example, to prove that the correct virus strain was used, a fragment from the variable genomic left end should be applied (21). If band intensities must be compared from cultures infected with different virus strains, it might be useful to take a probe from the less variable right genomic end. Insert DNA should be used to probe for recombinant viral vectors.

Expression Cloning of Transforming Genes into Herpesvirus saimiri

Foreign genes can be inserted efficiently into the right-terminal genomic junction region of *H. saimiri* L- and H-DNA, where no viral transcription has been found (45, 46). The neomycin resistance marker gene was inserted into that position by Grassmann and Fleckenstein (26). Cloning into *H. saimiri* is achieved via homologous recombination. As herpesviruses are assumed to replicate according to the rolling circle model, a single cross-over suffices for insertion of the foreign gene (26). Therefore, the respective recombination plasmids must provide in series the right-terminal genomic sequences, the gene of interest to be inserted, and a selection marker. Two kilobases of homologous right-terminal sequences is sufficient for homologous recombination (27). Alt *et al.* (27) constructed the recombination vector plasmid pRUPHy containing 2 kb of homologous sequence, a multilinker stretch for the insertion of the gene of interest, and the hygromycin B resistance gene, which was under the transcription control of the human cytomegalovirus major immediate early enhancer-promoter and had the simian virus 40 (SV40) termination signals. The targeting plasmid (4 μg) is linearized upstream of the viral segment (*Sma*I and *Nsi*I sites of plasmid pRUPHy), mixed at about 200-fold molar excess with purified infectious virion DNA (200 ng), and cotransfected into permissive OMK cells by the calcium phosphate method. From the supernatant of cultures with complete CPE (after approximately 3 weeks), virion DNA is prepared on a small scale and tested by Southern hybridization. Clean recombined virus stocks are obtained using selection media (200 μg of hygromycin B per milliliter) and by plaque purification. Under selection pressure, *H. saimiri* recombinants with neomycin resistance were shown to infect a broad spectrum of cell types (47). Cloned virus stocks may be used to infect T cells to introduce foreign genes, for example, the transforming genes of HTLV-I (28, 29).

Immortalization of Human T Lymphocytes

Infection of primary human T lymphocytes with *H. saimiri* subgroup C strains, but not with strains of subgroups A or B, yields continuously proliferating mature T cell lines that exhibit the CD4$^+$/CD8$^-$ or the CD4$^-$/CD8$^+$ phenotype (3). Permanently growing T cell lines have been obtained from primary cells of different purity and various sources. Mononuclear cells from adult peripheral blood or from cord blood, from thymus or bone marrow, as well as characterized T cell clones are suitable for the procedure. Polyclonal preparations, antibody-sorted populations, or clonal cultures can be transformed. The CD4/CD8 and T_h1/T_h2 phenotypes do not influence the susceptibility to growth transformation. All cell lines obtained had an α/β T cell receptor (TCR), but so far never γ/δ-TCR. When γ/δ cell clones or enriched populations had been infected, immortalization never succeeded. Thus, the α/β-TCR may be a prerequisite for the transformation with wild-type *H. saimiri.* Growing cells are needed with good viability and with the morphology of activated T lymphocytes. The primary cells may be purified by applying Histopaque density gradients (1.1 g/ml; Pharmacia, Uppsala, Sweden, or Biochrom, Berlin, Germany) or the dextran sedimentation protocol [selective sedimentation of erythrocytes at 37°C and 1.2% dextran 250,000 (w/v)/30 mM NaCl for 45 min]. Prestimulation of fresh primary cells for 1 day with phytohemagglutinin (PHA) (0.5 to 10 μg/ml; Burroughs Wellcome, Research Triangle Park, NC) or OKT-3 (10 to 200 ng/ml; Ortho, Raritan, NJ) gives no significant advantage over untreated samples. Antigen-dependent T cell clones should be restimulated with antigen and feeder cells 3 to 5 days prior to infection. If the antigen is not known, an unspecific mitogen stimulation should be performed 3 to 5 days before infection. The T cell cultures are kept at a density of about 1 (0.5 to 1.5) $\times$ 10^6 cells/ml. Different types of cell culture plastic ware have been tested. The cells retain best viability in 25-cm^2 flasks (3 to 10 ml with 3 to 10 $\times$ 10^6 cells) or 96-well round-bottom microwell plates (100 to 200 μl with 10^4 to 10^5 cells). The culture flasks should be incubated with a slightly elevated top (at an angle of about 15°) to allow close cell contacts in the lower edge of the culture flask. Various medium formulations have been used. Lines CB-15, PB-W, Lucas (3), and CB-23 (48) were isolated without prestimulation, without IL-2 supplementation, and with a standard medium formulation [80% (v/v) RPMI-1640, 20% (v/v) FCS, 50 μM 2-mercaptoethanol, glutamine, gentamicin]. During further experiments, the addition of supplements enhanced the transformation efficiency considerably. The quality of FCS batches, even from the same supplier, varies remarkably. It seems to be important to take a batch with low endotoxin level and to compare different serum batches. Fetal calf serum is inactivated at 56°C for 30 min to avoid mainly complement-dependent problems with both virus and lymphocytes. The addition of IL-2 (at 40–50 U/ml in the medium stock) activates cell proliferation and enhances the transformation frequency. As usually about one-third of the culture medium is added or replaced twice a week, a final concentration in the culture of less than 20 U of IL-2 per milliliter is presumed. In our hands, comparison of many different IL-2 batches from different suppliers revealed consistently that pretested batches with low endotoxin levels of the human recombinant IL-2, isolated from *Escherichia coli* and offered by Boehringer Mannheim (Indianapolis, IN; Cat. Nos. 1011456 and 1147528), are most active and provide optimal conditions for a high transformation efficiency. For financial reasons it may be used mainly for the initial transformation procedure itself and for maintaining small culture volumes. When large cell culture volumes of transformed lines are required, the less expensive Proleukin (EuroCetus, Frankfurt/ Main and Amsterdam) is recommended. It is a human recombinant IL-2 for clinical use, carrying 1-des-Ala and Ser 125 amino acid substitutions. Efforts to enhance transformation efficiencies led to the observation that adding 45% of CG medium (Vitromex GmbH, D94474 Vilshofen, Germany) was beneficial (Table II). CG (cell

TABLE II   Composition of CG medium [a, b]

Component	Concentration (mg/liter)	Component	Concentration (mg/liter)
Amino acids		Other components	
L-Alanine	19.87	Cholesterol	13.88
L-Arginine hydrochloride	84.00	D-Glucose	4500.00
L-Asparagine	17.22	HEPES	3574.50
L-Aspartic acid	23.48	Sodium pyruvate	87.42
L-Cysteine	28.00	Phenol red	15.00
L-Cysteine hydrochloride	62.57	Bovine serum albumin	1000.00
L-Glutamic acid	59.60	Soybean lipids	66.60
L-Glutamine	584.00	Human transferrin	32.00
Glycine	30.00	2-Mercaptoethanol	$50 \mu M$
L-Histidine monohydrochloride monohydrate	42.00	Inorganic salts	
L-Isoleucine	105.00	$CaCl_2$	200.00
L-Leucine	105.00	$FeCl_3 \cdot 6 H_2O$	0.075
L-Lysine hydrochloride	146.00	$Fe(NO_3)_3 \cdot 9 H_2O$	0.10
L-Methionine	30.00	KCl	400.00
L-Phenylalanine	66.00	$MgSO_4$	97.67
L-Proline	31.79	NaCl	6400.00
L-Serine	42.00	$NaHCO_3$	2500.00
L-Threonine	95.00	$NaH_2PO_4 \cdot H_2O$	125.00
L-Tryptophan	16.00	Na_2O_3Se	0.022
L-Tyrosine disodium	103.79		
L-Valine	94.00		
Vitamins			
Biotin	0.013		
Calcium D-pantothenate	4.00		
Choline chloride	4.00		
Folic acid	4.00		
iso-Inositol	7.20		
Niacinamide	4.00		
Pyridoxal hydrochloride	4.00		
Riboflavin	0.40		
Thiamin hydrochloride	4.00		
Vitamin B_{12}	0.013		

[a] From Ref. 48 with permission.
[b] CG medium is only available from Vitromex GmbH (D94474 Vilshofen, Germany).

growth) medium is a synthetic medium originally designed for the serum-free culture of B cell hybridomas (49). Several types of synthetic media were tested as supplement, but only CG medium improved transformation efficiency significantly. Recent experiments suggest that AIM-V medium (Gibco BRL, Gaithersburg, MD) might be an alternative to CG medium. Unfortunately, Gibco BRL does not provide a list of contents for AIM-V. Using pretested IL-2 (Boehringer Mannheim) and CG medium, about 90% of primary T cell cultures were transformed to antigen- and mitogen-independent growth. The failure rate of about 10% is explained mostly by trivial reasons such as low viability of the cells prior to infection. In our hands, the optimal medium formulation is as follows: RPMI-1640, 45% (v/v); CG, 45% (v/v); FCS, 10% (v/v); glutamine; gentamicin (if desired); and IL-2 (Boehringer Mannheim, 40 U/ml). Most experience is based on using the *H. saimiri* strain C-488. However, strains C-484 (18–20) and C-139 are also able to immortalize human T cells. Strain C-139 is an isolate from our laboratory (B. Biesinger, I. Müller-Fleckenstein, and B. Fleckenstein, unpublished, 1993). Typically, 10% by volume of infectious OMK supernatant is added to a vital lymphocyte culture. That is, for example, 500 μl of 10^6 pfu/ml (5×10^5 pfu) to a 5-ml culture volume with 1×10^6 cells/ml. During the following weeks and sometimes months, the cells need patient care; medium should be partially exchanged at regular intervals, about twice a week. Good negative controls are uninfected cells and cultures infected with a virus strain of the same titer, such as A-11 or B-SMHI, which are not able to immortalize human T cells. During the first weeks of culture, the normal growth of T lymphocytes is observed. This mitotic phase should not be misinterpreted as transformation events. It may last up to 2 months until the viral-infected T cells start to proliferate quickly. About 10^5 viable cells, at least, are necessary for obtaining a transformed culture. The efficiency seems to vary considerably between different donors. When using a few million vital cells for infection, the efficiency rate reaches 100%. Several criteria were found to indicate a transformed or immortalized phenotype: (a) doubling of the cell number constantly one to four times a week over several months, independent of antigen and mitogen, (b) morphology of T lymphoblasts, enlargement of cells, good contrast on microscopy, and irregular shape, (c) death of control cultures that have been treated identically to the transformed ones, (d) persistence of viral nonintegrated episomes without virion production (Gardella gel and cocultivation test on OMK), and (e) hyperreactivity against LFA-3 (35) or foreign cells and unspecific cytotoxicity.

Cultures that seem to be growth transformed can be subjected to a gradual withdrawal of the CG medium component. Addition of 25 mM 4-(2-hydroxyethyl)-1-piperazine-ethanesulfonic acid (HEPES), pH 7.4 (GIBCO/BRL, Gaithersburg, MD) helps to avoid cell degeneration caused by low pH. With CD4-positive cells, it is also often possible to reduce IL-2 supplementation gradually and even to terminate it. However, withdrawing IL-2 from CD8$^+$-transformed cell lines for longer periods was not successful. When the cells are growing stably, the pretested IL-2 may be replaced by a less expensive product.

Basically, the infection of T lymphocyte cultures by *H. saimiri* vectors follows the same rules. Using HTLV-*H. saimiri* recombinants, CD4-positive cell lines were obtained from cord blood cells and from thymocyte cultures. These transformation experiments were carried out using a standard medium formulation [RPMI-1640, 90% (v/v); FCS, 10% (v/v); glutamine; gentamicin; IL-2 (Boehringer Mannheim), 20 U/ml] after 72 hr of stimulation with PHA (10 μg/ml). Cultures decreased in proliferation 2 weeks after infection. After some 4 weeks, cell aggregation and proliferation increased in the *tax*$^+$ recombinant-infected cultures and led to a transformed phenotype (29).

Discussion

Persisting Herpesviral Vector Based on Herpesvirus saimiri

Nononcogenic *H. saimiri* strains have been used as eukaryotic expression vectors (26). Recombinant viruses can infect and persist in human T cells and in a broad range of hematopoietic, mesenchymal, and epithelial cells under selection conditions (47). The X region of the human T cell leukemia virus type I (HTLV-I) was inserted into *H. saimiri* A-11 S4 (28). This deletion variant does not cause malignant disease in animals and does not transform human lymphocytes in culture, because it is lacking the left-terminal oncogene region (13, 16). The HTLV-I X region in the recombinant herpesvirus vector was able to transform primary human T lymphocytes to IL-2-dependent growth, similar to HTLV-I-transformed cells (28). The recombinant transformed cells expressed lymphocyte activation genes (50), the CD4 surface molecule, MHC class II antigens, and the IL-2 receptor α chain in large amounts, and they contained episomes of the recombinant viruses in high copy number (28). Deletion variants of the HTLV-I X region were introduced into the viral vector. The Rex protein of HTLV-I did not show transforming properties. The broad transcriptional activator p40^{tax} was found to be necessary and sufficient for lymphocyte immortalization in the context of the herpesviral vector (29; reviewed in Refs. 10 and 30). The *H. saimiri* vector system (27) may be useful for studying other transforming genes. It found further application in studying c-*fos* function (51). Overexpression of the protooncogene c-*fos* is known to induce transformation in a broad range of primary cells of various mammalian species. c-*fos* recombinant herpesviral vectors expressed large amounts of the oncoprotein on persistent infection of human neonatal fibroblasts. However, these primary mesenchymal cells did not show any sign of transformation (51). Both the recombinant herpesviral vector system and the wild-type *H. saimiri* subgroup C strains are replication competent. Reactivation from transformed human T cells has not been observed, but cannot be excluded. The technique of homologous recombination could be applied for constructing replication-defective, but transforming, deletion variants that preclude reactivation. Furthermore, additional genes could be introduced into C strain viruses by homologous recombination to express and to study those gene products in the transformed lymphocytes.

Persistence of Herpesvirus saimiri

Herpesvirus saimiri, a primate virus, can infect human cells. An early analysis of *H. saimiri* strain SMHI revealed weak productive growth on primary human fetal cells (52). Selectable *H. saimiri* recombinants, derived from strain A-11, were later used to study the spectrum of cells that can be infected by the virus (47). A broad range of epithelial, mesenchymal, and hematopoietic cells became infected and carried nonintegrated episomal DNA of the recombinant viruses. The pancreatic carcinoma line PANC-1 and human foreskin fibroblasts produced even infectious virus under selection conditions. These findings suggest that the receptor used by *H. saimiri* is widely distributed among various tissues. The receptor seems to be well conserved, as rabbit T cells can also be infected and transformed by *H. saimiri* strains (7, 20). Cell lines that had been infected with the recombinants under selection pressure retained the viral nonintegrated episomes after withdrawal of the selecting drug for long time periods. The lack of counterselection against cells with persisting viral episomes may suggest that the virus persists with mostly suppressed viral gene expression (47). This model is also supported by the observations that the persisting nonintegrated viral episomes are heavily methylated (53) and may carry extensive genomic deletions (11). These observations were made using *H. saimiri* strain A-11, mainly. It is likely that they are also valid for subgroup C strains and for transformed human T cells. Although monkey T lymphocytes produce *H. saimiri* particles in many cases, it was not possible to isolate virus from transformed human T cell cultures that carry nonintegrated viral episomes in high copy number (3). Even after treatment with phorbol esters, nucleoside analogs, and other drugs known to cause reactivation of other viruses like EBV, or after specific or unspecific stimulation of the T cells, virion production could not be demonstrated (H. Fickenscher, I. Müller-Fleckenstein, and B. Fleckenstein, unpublished observations, 1993). Nevertheless, it will be difficult to provide formal proof that the virus can never be reactivated from immortalized human T lymphocytes.

Viral Expression in Human T Cells Transformed by Herpesvirus saimiri

Numerous genomic regions of the virus have been used to search for viral transcripts in the transformed cells. Viral transcription could not be demonstrated from any tested genomic region except from the left terminus, where the viral oncogene *stp-C* is transcribed. In contrast, virion-producing monkey lymphocytes show broad viral transcription. For example, gene 14 (reviewed in Ref. 12) is a major immediate early gene of the lytic cycle and shows sequence homology to known murine superantigens. For several reasons, a superantigen function for this gene seems to be unlikely: no transcription of gene 14 could be found during transformation, there is no early mitogenic effect of the infection, and there is no preference for expression of particular TCR Vβ genes early after infection with either transforming or nontransforming virus strains. Several months after infection, a predominance of a reduced number of TCR Vβ gene families was found, which varied from culture to culture. The most likely explanation is that the originally polyclonal T cell cultures are overgrown by those clones that proliferate most rapidly. This is a general phenomenon in long-term cell culture that occurs in the absence of any specific stimulation (54). The mRNA of the viral oncogene *stp-C* is subject to cellular regulation and is translated to a cytoplasmic phosphoprotein with a predicted size of about 10 kDa, which migrates at 20 to 22 kDa on sodium dodecyl sulfate (SDS) protein gels (31). The oncoprotein STP-C was shown to transform rodent fibroblasts, leading to invasive tumors in nude mice (22). STP-C transgenic mice (23) develop epithelial tumors, namely salivary gland adenomas and adenocarcinomas and bile duct and thymus epitheliomas. However, when the transformed human T lymphocytes expressing *stp-C* were tested for tumorigenesis in nude or SCID mice in a conventional implantation experiment, no tumor formation could be observed (H. Fickenscher, unpublished observation, 1993). In heavily conditioned mice, acute xenogeneic graft versus host reaction and vasculitis was induced, when primary peripheral blood leukocytes or *Herpesvirus saimiri*-transformed human T cells were implanted (55).

Phenotype of Human Growth Transformed T Lymphocytes

Human T lymphocytes that are transformed to stable growth by *H. saimiri* C strains are mostly IL-2 dependent and do not need restimulation with antigen or mitogen. Cell number increases by factors of two to four per week. The cells show the morphology of T blasts with irregular shape and express surface molecules that are typical for activated mature T lymphocytes. The cell lines exhibit the CD4$^+$/CD8$^-$ or the CD4$^-$/CD8$^+$ phenotype. Mixed populations may occur when polyclonal populations are infected. The antigens CD2, CD3 (α/β T cell receptor), CD5, CD7, CD25 (IL-2 receptor α chain), CD30, MHCII, and CD56 [a natural killer cell (NK) marker] are expressed, whereas the NK markers CD16 and CD57 are lacking (3). The RO and RB isoforms of the membrane-bound phosphatase CD45, which are typically found on mature memory T cells, were both present on the CD4-positive transformed T lymphocytes. The CD8-positive transformed cell lines, however, expressed the CD45 isoform RA additionally, which is typical for naive T cells and their precursors (B. Simmer and E. Platzer, unpublished observations, 1993).

The *H. saimiri*-immortalized human CD4 T cells provide a new productive lytical system for T lymphotropic viruses such as human herpesvirus type 6 (F. Neipel and

B. Fleckenstein, unpublished observation, 1993) and HIV (48). The prototype viruses HIV-1_{IIIB} and HIV-2_{ROD} replicated rapidly in the cells and caused cell death within 14 days. Also, a poorly replicating HIV-2 strain and primary clinical isolates grew to high titers. *Herpesvirus saimiri*-transformed human CD4 T cells may be used for poorly growing HIV strains with narrowly restricted host cell range (48).

Several cytokines are produced by the transformed human T blasts after activation. Interleukin 2 (35) and IL-3 (33) are secreted by the cells in response to mitogenic or antigenic stimuli. Antibodies against IL-2 receptor α chain (35) and against IL-2 and IL-3 (33) suppressed the growth rate; both cytokines seem to support autocrine growth. Interleukin 4 and IL-5 are secreted only at low rates by transformed T helper type 2 (T_h2) cells (33). Transformed CD4-positive T cells secrete interferon γ (IFN-γ), tumor necrosis factor α (TNF-α), and TNF-β after specific or unspecific stimulation (2, 32, 33). Both T_h1 and T_h2 clones were transformed by *H. saimiri*. The cytokine pattern of the T_h1 clones was enhanced, whereas the profile of T_h2 clones switched to a mixed phenotype after transformation (33). The viral-transformed human T cells show a strong unspecific cytotoxic activity. When tested on K562 cells, CD8-positive lines, and to a lesser extent CD4-positive cells, showed NK-like cytolytic activity (3). The lectin-dependent cytolytic activity of T_h1 clones against P815 target cells was enhanced after transformation, whereas T_h2 clones showed this activity only in the transformed state (33).

The karyotypes of a series of cell lines have been analyzed in detail and found to be normal (56). When early signal transduction properties of the transformed cells were compared with those of the uninfected parental cells, no significant differences were encountered. After stimulation with IL-2, anti-CD3, and/or anti-CD4, similar patterns of tyrosine phosphorylation and calcium mobilization were observed in primary clones or in transformed lines (2, 34). In contrast, Jurkat cells (1) behaved differently (2). The herpesvirus-transformed cell lines were strongly stimulated by LFA-3, which is expressed on cells of various, even xenogenic, origin. This effect was mediated by CD2/LFA-3 interaction and led to IL-2 production and enhanced proliferation (35). The functionality of CD3, CD4, and the IL-2 receptor was shown after stimulation by signal transduction parameters, by proliferation, and by IFN-γ production (2). The IL-2-dependent proliferation of transformed lymphocytes was strongly inhibited by soluble CD4 antibodies. This effect could be overcome by high doses of IL-2. In parallel, the activity and abundance of the CD4-bound fraction of the tyrosine kinase p56^{lck} was diminished by the anti-CD4 treatment (44).

Characterized T helper cell clones reacting specifically to myelin basic protein (32), tetanus toxoid (2), Ni^{2+} ions (34), bovine 70-kDa heat-shock protein (HSP70), *Lolium perenne* group I antigen, *Toxocara canis* excretory antigen, and purified protein derivative from *Mycobacterium tuberculosis* (33) were successfully transformed and retained their MHC-restricted antigen specificity after immortalization. The high basal proliferation activity, which is probably due to contact with LFA-3-bearing cells (35), may interfere with the demonstration of antigen specificity, because the antigen-presenting cells alone cause a strong stimulation of the transformed T cells. This antigen-independent proliferation of the transformed antigen-specific clones could be reduced by using monoclonal antibodies against LFA-3 and CD2 and MHC-transfected mouse L cells (32) or by starving the cells prior to antigen presentation (2). In all cases, clear responses to antigen contact were noticed, measured by proliferation and cytokine production. Berend *et al.* (57) observed maintained HLA-restricted cytolytic function and antigen specificity of transformed EBV specific CD8$^+$ cytotoxic T lymphocytes. These results show that a number of experimental approaches became feasible with the availability of a new immortalization system for human T lymphocytes.

Acknowledgments

Original work underlying this chapter was supported by the Deutsche Forschungsgemeinschaft (Forschergruppe DNA-Viren des Hämatopoetischen Systems), the Bayerische Forschungsstiftung (Forschungsverbund Biologische Sicherheit), and the Johannes-und-Frieda-Marohn-Stiftung. We thank Ingrid Müller-Fleckenstein and Drs. Brigitte Biesinger, Barbara Bröker, Bryan Cullen, Armin Ensser, and Ralph Grassmann for critically reading the manuscript.

References

1. U. Schneider, H.-U. Schwenk, and G. Bornkamm, *Int. J. Cancer* **19,** 621 (1977).
2. B. Bröker, A. Y. Tsygankov, I. Müller-Fleckenstein, A. Guse, N. A. Chitaev, B. Biesinger, B. Fleckenstein, and F. Emmrich, *J. Immunol.* **151,** 1184 (1993).
3. B. Biesinger, I. Müller-Fleckenstein, B. Simmer, G. Lang, S. Wittmann, E. Platzer, R. C. Desrosiers, and B. Fleckenstein, *Proc. Natl. Acad. Sci. USA* **89,** 3116 (1992).
4. L. V. Melendez, M. D. Daniel, R. D. Hunt, and F. G. Garcia, *Lab. Anim. Care* **18,** 374 (1968).
5. B. Roizman, R. C. Desrosiers, B. Fleckenstein, C. Lopez, A. C. Minson, and M. J. Studdert, *Arch. Virol.* **123,** 425 (1993).
6. L. Falk, L. G. Wolfe, and F. Deinhardt, *J. Natl. Cancer Inst.* **48,** 1499 (1972).
7. D. V. Ablashi, S. Schirm, B. Fleckenstein, A. Faggioni, J. Dahlberg, H. Rabin, W. Loeb, G. Armstrong, J. W. Peng, G. Aulahk, and W. Torrisi, *J. Virol.* **55,** 623 (1985).
8. B. Fleckenstein and R. C. Desrosiers, *in* "The Herpesviruses" (B. Roizman, ed.). Vol. 1, p. 253. Plenum Press, New York, 1982.
9. R. C. Desrosiers and B. Fleckenstein, *in* "Advances in Viral Oncology" (G. Klein, ed.), Vol. 3, p. 307. Raven Press, New York, 1983.
10. B. Biesinger and B. Fleckenstein, *in* "Malignant Transformation by DNA Viruses, Molecular Mechanisms" (W. Doerfler and P. Böhm, eds.), p. 207. Verlag Chemie (VCH), Weinheim and New York, 1992.
11. S. Schirm, I. Müller, R. C. Desrosiers, and B. Fleckenstein, *J. Virol.* **49,** 938 (1984).
12. J. C. Albrecht, J. Nicholas, D. Biller, K. R. Cameron, B. Biesinger, C. Newman, S. Witt-

mann, M. A. Craxton, H. Coleman, B. Fleckenstein, and R. W. Honess, *J. Virol.* **66,** 5047 (1992).

13. R. C. Desrosiers, R. L. Burghoff, A. Bakker, and J. Kamine, *J. Virol.* **49,** 343 (1984).

14. J. M. Koomey, C. Mulder, R. L. Burghoff, B. Fleckenstein, and R. C. Desrosiers, *J. Virol.* **50,** 662 (1984).

15. R. C. Desrosiers, A. Bakker, J. Kamine, L. A. Falk, R. D. Hunt, and N. W. King, *Science* **228,** 184 (1985).

16. R. C. Desrosiers, D. P. Silva, L. M. Waldron, and N. L. Letvin, *J. Virol.* **57,** 701 (1986).

17. S. Murthy, J. Trimble, and R. Desrosiers, *J. Virol.* **63,** 3307 (1989).

18. R. C. Desrosiers and L. A. Falk, *J. Virol.* **43,** 352 (1982).

19. P. Medveczky, E. Szomolanyi, R. C. Desrosiers, and C. Mulder, *J. Virol.* **52,** 938 (1984).

20. M. M. Medveczky, E. Szomolanyi, R. Hesselton, D. DeGrand, P. Geck, and P. G. Medveczky, *J. Virol.* **63,** 3601 (1989).

21. B. Biesinger, J. Trimble, R. C. Desrosiers, and B. Fleckenstein, *Virology* **176,** 505 (1990).

22. J. U. Jung, J. J. Trimble, N. W. King, B. Biesinger, B. Fleckenstein, and R. C. Desrosiers, *Proc. Natl. Acad. Sci. USA* **88,** 7051 (1991).

23. C. Murphy, C. Kretschmer, B. Biesinger, J. Beckers, J. Jung, R. C. Desrosiers, H. K. Müller-Hermelink, B. Fleckenstein, and U. Rüther, *Oncogene* **9,** 221 (1994).

24. J. U. Jung and R. C. Desrosiers, *J. Virol.* **65,** 6953 (1991).

25. J. U. Jung and R. C. Desrosiers, *J. Virol.* **66,** 1777 (1992).

26. R. Grassmann and B. Fleckenstein, *J. Virol.* **63,** 1818 (1989).

27. M. Alt, B. Fleckenstein, and R. Grassmann, *Gene* **102,** 265 (1991).

28. R. Grassmann, C. Dengler, I. Müller-Fleckenstein, B. Fleckenstein, K. McGuire, M.-C. Dokhelar, J. G. Sodroski, and W. A. Haseltine, *Proc. Natl. Acad. Sci. USA* **86,** 3351 (1989).

29. R. Grassmann, S. Berchtold, I. Randant, M. Alt, B. Fleckenstein, J. G. Sodroski, W. A. Haseltine, and U. Ramstedt, *J. Virol.* **66,** 4570 (1992).

30. H. Fickenscher, R. Grassmann, B. Biesinger, U. Ramstedt, W. A. Haseltine, R. C. Desrosiers, and B. Fleckenstein, *in* "Virus Strategies" (W. Doerfler and P. Böhm, eds.), p. 401. Verlag Chemie (VCH), Weinheim and New York, 1993.

31. H. Fickenscher, B. Biesinger, A. Knappe, S. Wittmann, and B. Fleckenstein, submitted (1994).

32. F. Weber, E. Meinl, K. Drexler, A. Czlonkowska, S. Huber, H. Fickenscher, I. Müller-Fleckenstein, B. Fleckenstein, H. Wekerle, and R. Hohlfeld, *Proc. Natl. Acad. Sci. U.S.A.* **90,** 11049 (1994).

33. M. De Carli, S. Berthold, H. Fickenscher, I. Müller-Fleckenstein, M. M. D'Elios, Q. Gao, R. Biagiotti, M. G. Giudizi, J. R. Kalden, B. Fleckenstein, S. Romagnani, and G. Del Prete, J. Immunol. **15,** 5022 (1993).

34. H. W. Mittrücker, I. Müller-Fleckenstein, B. Fleckenstein, and B. Fleischer, *Int. Immunol.* **5,** 985 (1993).

35. H. W. Mittrücker, I. Müller-Fleckenstein, B. Fleckenstein, and B. Fleischer, *J. Exp. Med.* **176,** 909 (1992).

36. M. D. Daniel, D. Silva, and N. Ma, *In Vitro* **12,** 290 (1976).

37. J. Wright, L. A. Falk, D. Collins, and F. Deinhardt, *J. Natl. Cancer Inst.* **57,** 959 (1976).

38. B. Fleckenstein and H. Wolf, *Virology* **58,** 55 (1974).

39. B. Fleckenstein, G. W. Bornkamm, and H. Ludwig, *J. Virol.* **15,** 398 (1975).

40. B. Fleckenstein, M. D. Daniel, R. Hunt, J. Werner, L. A. Falk, and C. Mulder, *Nature (London)* **274,** 57 (1978).
41. G. W. Bornkamm, H. Delius, B. Fleckenstein, F. J. Werner, and C. Multer, *J. Virol.* **19,** 154 (1976).
42. M. D. Daniel, L. V. Meléndez, and H. H. Barahona, *J. Natl. Cancer Inst.* **49,** 239 (1972).
43. T. Gardella, P. Medveczky, T. Sairenji, and C. Mulder, *J. Virol.* **50,** 248 (1984).
44. B. M. Bröker, A. Y. Tsygankov, H. Fickenscher, N. A. Chitaev, I. Müller-Fleckenstein, B. Fleckenstein, J. B. Bolen, and F. Emmrich, *Eur. J. Immunol.* in press (1994).
45. A. T. Bankier, W. Dietrich, R. Baer, B. B. Barrell, F. Colbere-Garapin, B. Fleckenstein, and W. Bodemer, *J. Virol.* **55,** 133 (1985).
46. T. Stamminger, R. W. Honess, D. F. Young, W. Bodemer, E. D. Blair, and B. Fleckenstein, *J. Gen. Virol.* **68,** 1049 (1987).
47. B. Simmer, M. Alt, I. Buckreus, S. Berthold, B. Fleckenstein, E. Platzer, and R. Grassmann, *J. Gen. Virol.* **72,** 1953 (1991).
48. S. Nick, H. Fickenscher, B. Biesinger, G. Born, G. Jahn, and B. Fleckenstein, *Virology* **194,** 875 (1993).
49. A. B. Lang, U. Schürch, F. Zimmermann, and U. Bruderer, *J. Immunol. Methods* **154,** 21 (1992).
50. K. Kelly, P. Davis, H. Mitsuya, S. Irving, J. Wright, R. Grassmann, B. Fleckenstein, Y. Wano, W. C. Greene, and U. Siebenlist, *Oncogene* **7,** 1463 (1992).
51. M. Alt and R. Grassmann, *Oncogene* **8,** 1421 (1993).
52. M. Daniel, D. Silva, D. Jackmann, P. Sehgal, R. Baggs, R. Hunt, N. King, and L. Melendez, *Bibliotheca Haematol.* **43,** 392 (1976).
53. R. C. Desrosiers, C. Mulder, and B. Fleckenstein, *Proc. Natl. Acad. Sci. USA* **76,** 3839 (1979).
54. H. Fickenscher, N. Rebai, B. Simmer, F. Emmrich, B. Fleckenstein, R. Sékaly, and B. Bröker, unpublished (1993).
55. W. Huppes, H. Fickenscher, B. 'tHart, B. Fleckenstein, and D. Van Bekkum, submitted (1994).
56. B. Troidl, B. Simmer, H. Fickenscher, I. Müller-Fleckenstein, F. Emmrich, B. Fleckenstein, and E. Gebhart, *Int. J. Cancer* **56** 433 (1994).
57. K. Berend, J. Jung, T. Boyle, J. DiMaio, S. Mungal, R. Desrosiers, and K. Lyerly, *J. Virol.* **67,** 6317 (1993).

[23] Use of Pseudorabies Virus for Definition of Synaptically Linked Populations of Neurons

J. Patrick Card and Lynn W. Enquist

Introduction

The neurotropic properties of alpha herpesviruses that permit these pathogens to invade the nervous system have been exploited to define functionally related populations of neurons. The utility of this approach is based on the assertion that these viruses invade and replicate within neurons in a predictable fashion and then pass transneuronally through synapses to infect neurons that are presynaptic to the infected cell (1, 2). Thus, with advancing time, virus passes through chains of synaptically linked neurons, and the organization of these circuits can be defined by immunohistochemical localization of virus in fixed tissue (3). Although both human and animal strains of virus have been successfully employed for this type of analysis, the most extensive and well-characterized studies have utilized a swine alpha herpesvirus known as pseudorabies virus (PRV). The extensive host range of this virus (4) and the availability of well-defined viral mutants that differ in virulence (5–7) have made it an attractive tool for characterization of both neuronal circuitry and virally induced pathogenesis in rodents.

In early studies employing PRV, Dolivo and collaborators demonstrated that PRV replicated within multisynaptic circuits in both the peripheral and central nervous system (8–12). Subsequently, there have been numerous demonstrations of projection-specific transport of PRV through functionally related populations of neurons in the central nervous system (CNS). In particular, PRV has been used effectively to define circuits that modulate the output of both the sympathetic (13–18) and parasympathetic (19, 20) nervous system. In addition, specific transport of PRV through central visual circuitry following injection of virus into the vitreous body of the eye has been reported (7, 21). Nevertheless, the contention that virus infects these neurons by passing through synapses rather than via nonspecific release into the extracellular space has remained a point of debate. This is due in large part to the clear demonstration that virulent strains of the virus produce lethal infections of CNS neurons (22, 23) and the observation of infected glia in the vicinity of chronically infected neurons (16, 19). To address this issue we have conducted an extensive analysis of the response of glia and blood-borne cellular elements to infection of neurons with PRV (24, 25). These studies demonstrated a temporally organized mobilization of astrocytes, microglia, and peripheral macrophages to the site of neuronal infection, effectively limiting nonspecific dissemination of virus through the extracellular space. However, the ability of virions to cross synapses and thereby gain access to afferent neurons synapsing on these cells defeated the efforts of the nonneuronal response to isolate the infection, and resulted in the transport of virus to other areas of the nervous system (24, 25). Considered with the many reports of unique, projection-specific patterns of transport of PRV in the nervous system, these data argue convincingly that the transport of attenuated strains of PRV through the nervous system reflects specific passage of virus through synaptically linked chains of neurons.

We review here the important issues that must be considered in using PRV for definition of neuronal circuitry, and detail the methodology that can be applied to characterize these circuits at the light microscopic level. Issues related to the use of other strains of herpesvirus (i.e., human herpes simplex virus) or other methods of detection are not considered here, but are available in another review (3).

Application of Method

Safety Precautions

Safety is a major consideration in the conduct of these studies. In general, alpha herpesviruses are Centers for Disease Control/National Institutes of Health (CDC/NIH) Class II infectious agents and all experiments employing these viruses must be conducted in a laboratory sanctioned for Biosafety Level 2 (BSL-2) experiments. It is well documented that humans are not a host for PRV (4), but appropriate precautions should be exercised in handing both the virus, infected animals, and infectious waste. Specific regulations required for BSL-2 experiments are available in Health and Human Services Publication 88-8395, entitled *Biosafety in Microbiological and Biomedical Laboratories*. In addition, the application of these regulations as they pertain to the use of viruses for definition of neuronal circuits has been reviewed previously (3, 19). Nevertheless, it should be emphasized that it is the responsibility of each investigator to apply these standards rigorously to any experimental analysis of this type and to exercise appropriate controls in the disposal of contaminated waste.

Selection of Strain of Pseudorabies Virus

Knowledge of the strain of virus and its history are important considerations in designing a study. There is clear documentation that neuroinvasiveness and neurovirulence vary substantially among even closely related strains of PRV. For example, we have shown that deletion of a single gene from the PRV genome alters the ability of the mutant virus to infect visual circuitry. Injection of the virulent Becker strain of PRV infects all retinorecipient neurons in the CNS, but it does so in two temporally separated waves that discriminate among functional subdivisions of the visual system (21). The first wave of infection targets visual projections to the dorsal geniculate nucleus (visual perception) and the tectum (visual reflexes). In contrast, visual centers involved in circadian timing of behavior and neuroendocrine secretion are not infected until the second wave, which occurs approximately 24 hr later. This pattern of infection is strikingly different from that produced when animals are infected with an attenuated, vaccine strain of PRV known as Bartha (6). Rats inoculated with the Bartha strain exhibit a restricted neurotropism in which visual circuits involved in circadian timing are selectively infected, and retinorecipient neurons involved in visual perception (dorsal geniculate neurons) and reflexes (tectum) never become infected. Subsequent studies have shown that deletion of viral genes encoding the gI or gp63 envelope glycoproteins (Fig. 1), either alone or in combination, also produces the restricted tropism produced by the Bartha strain (7, 26). Furthermore, deletion of either of these genes dramatically reduces the virulence of the infection. Thus, even slight alterations in the viral genome can dramatically alter both the invasiveness and neurovirulence of the infecting virus.

The significance of the above observations is even more apparent when one considers the fact that the majority of track-tracing studies employing PRV have used the Bartha strain of virus. This is primarily because the reduced neuropathogenicity produced by this attenuated strain permits analysis of viral transport over a longer survival interval, reduces the chances that circuit-specific transport will be compromised by nonspecific release of virus into the extracellular space, and labels the somata and dendrites of infected neurons in a "Golgi-like" fashion. Nevertheless, it is important to consider the limitations that mutant strains of virus may place on a circuit analysis. For example, the restricted neurotropism exhibited by the Bartha strain of PRV in the visual system has obvious advantages for those interested in studying neuronal circuitry involved in the regulation of circadian rhythmicity. However, it is also easy to envision circumstances under which reduced neurotropism of a strain of virus could become a serious impediment to experimental analysis or could produce misleading data. All studies to date indicate that the Bartha strain of PRV is an effective tool for delineating circuits influencing the activity of the autonomic nervous system (ANS). Loewy's group in particular has used PRV to characterize multisynaptic circuits extending from the forebrain to spinal cord that modulate the outflow of the sympathetic component of the ANS (14–18). However, their work has also shown that PRV is less efficient in infecting the somatic nervous system that it is in infecting the sympathetic neurons (18). Following injections of the gastrocnemius muscle, approximately 90% of the animals exhibited infected preganglionic sympathetic neurons in the intermediolateral cell column of the spinal cord, whereas only 60% of the animals displayed infected motor neurons in the ventral horn. Furthermore, they were unable to detect infected sensory neurons in the appropriate dorsal root ganglia, even though polymerase chain reaction (PCR) analysis revealed viral DNA in these cells. Consequently, it is apparent that the pattern of neuronal infection resulting from viral in-

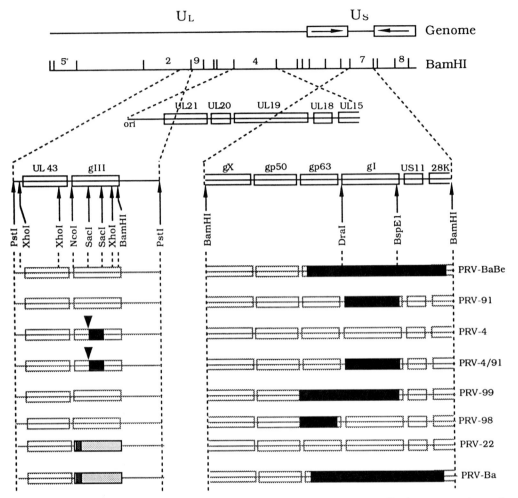

FIG. 1   Pseudorabies virus genome and map of selected mutations affecting neurotropism and virulence. A schematic of the PRV genome is indicated on the first line. The open rectangles with arrows represent the large inverted repeats at the right end of the genome that define the U_L and U_S regions. U_L and U_S correspond to the unique long and unique short regions, respectively. The second line is the *Bam*HI restriction map of PRV strain Becker. Several relevant unique *Bam*HI DNA fragments are numbered. On the basis of work with the Bartha strain of PRV, Ben-Porat and colleagues (see ref. 4) defined three regions that were involved in virulence and these are indicated in the third and fourth lines as expansions of particular *Bam*HI fragments. The region spanning the *Bam*HI-2 and 9 fragment contains the gIII gene, which is defective in strain Bartha. The *Bam*HI-4 fragment contains several genes involved in capsid formation and one or more of these are involved in the reduced virulence of strain Bartha. The *Bam*HI-7 fragment contains several genes, many of which encode viral envelope glycoproteins including gp50, gp63, and gI. A representative set of isogenic mutations constructed in strain Becker is also shown. Deletions are indicated by the extent of the black boxes. The deletion in the PRV Bartha strain (PRV-Ba) and the defective gIII gene are indicated on the last line. Relevant restriction sites used in construction of the mutants are indicated. The names of viruses carrying each mutation are given on the right.

oculation is dependent on properties of the virus (i.e., neuroinvasiveness) as well as the permissiveness of the host neuron. Thus, each of these factors should be carefully considered in evaluating the results of these studies, and the patterns of infection should be compared to data acquired with conventional track tracers to ensure that the pattern of viral transport represents an accurate and complete picture of the circuit.

Virus Concentration

It is extremely important to know the concentration (titer) of the viral inoculum. Viral titer is measured routinely in plaque-forming units (pfu) per milliliter and is determined by examining the number of plaques that an aliquot of virus will form on a monolayer of cells grown in tissue culture. In studies of PRV transport through sympathetic circuitry, Loewy's group have reported infection rates as low as 20% of animals injected with a total of 10^4 pfu of the Bartha strain of PRV (14–16). In contrast, our early studies with both virulent and attenuated strains of PRV produced 100% infectivity following injections of a total of 10^6 pfu or greater (7, 19–21, 24–26). These findings led us to conduct a systematic analysis of the effect of viral titer on the infectivity of virulent and attenuated strains of PRV (27). That analysis demonstrated that the ability of either strain of PRV to infect visual circuits following injection of the vitreous body of the eye was directly dependent on the titer of the injected virus. Injection of 10^5 pfu routinely produced 100% infectivity, whereas the percentage of infected animals dropped substantially when lower concentrations of virus were injected. In addition, the virulent strain consistently infected a higher percentage of animals across a broader range of titers than was achieved with the attenuated strain. These data argue that 10^5 pfu of virus, or greater, is required to achieve high and reproducible rates of infection with the Becker and Bartha strains of PRV.

The way that virus is stored can have a dramatic effect on titer. Pseudorabies virus is an enveloped virus and integrity of the envelope is required for infectivity. The envelope can be disrupted by many agents, including the formation of ice during freezing and thawing of stocks. It is therefore important to freeze virus in protective media (usually tissue culture medium with serum is sufficient). Virus should be stored in frozen aliquots that will be thawed only at the time of inoculation. Unused virus should never be refrozen as this can substantially reduce the infectious titer and introduces unnecessary variability in any experimental paradigm. We routinely store our viral stocks in 50- to 100-μl aliquots at $-80°$C and use a single aliquot for each experiment. In addition, we attempt to use the same lot of virus for each set of experiments to standardize further the infectivity and rate of transport of virus through a neuronal circuit. Detergents and bleach, both commonly used to clean syringes and surgical instruments, inactivate virus and can severely compromise virus titer. Consequently, care should be taken to ensure that all syringes and other instruments used in the surgery and inoculation are free of any of these reagents. Finally, viral stocks should be stored in the portion of the freezer that experiences the least variation in temperature (i.e., the bottom of a chest freezer) and the temperature of the freezer should be carefully monitored to identify variations in temperature. Unpublished observations in our laboratory have shown that a rise in temperature to $-40°C$ can reduce the titer of a viral stock and decrease neuronal infectivity.

Route of Inoculation

Pseudorabies virus has been shown to infect neurons following injection into peripheral organs (i.e., eye, adrenal gland, viscera) or directly into the CNS. There are both cell- and virus-specific factors that influence the direction of viral transport through a cell in each of these paradigms, and these factors should be carefully considered in choosing the site of viral inoculation. The following general principles have been established in animal studies of viral transport and can be used to guide the design of experimental studies. The most consistent finding confirmed in all studies to date is that once attenuated strains of virus infect a neuron in the central nervous system, all subsequent transport is via retrograde passage of virions through sites of afferent synaptic contact. We believe this results from the differential sorting of newly replicated virus to the somata and dendrites of the infected neuron, which effectively prevents newly replicated virus from even entering the anterograde transport pathway into the axon of the cell (25). Although the fundamental mechanisms underlying this phenomenon have not been elucidated, it is consistent with the demonstration by Dotti and Simmons (28) of polarized sorting of viral glycoproteins in cultured hippocampal neurons. Irrespective of the mechanism, this differential sorting of virions is the single most important component of the transneuronal passage of virus that makes it a useful tool for examining neuronal circuitry. In the absence of this specific sorting mechanism, virus would also be transported to the projection fields of infected neurons and it would be impossible to discriminate functional pathways after two or more orders of transneuronal passage.

The transport of virulent strains of PRV injected directly into the CNS provides a prominent exception to the "retrograde-only" rule of viral transport. After injection of the Becker strain of virus into the rat prefrontal cortex (PFC) we observe a pattern of first-order neuronal infection that is consistent with both anterograde and retrograde transport of virus from the injection site (29). In contrast, only neurons projecting to the PFC are infected when the attenuated Bartha strain of PRV is injected in an identical manner. This interesting observation raises the possibility that the genetic defects in strain Bartha prevent anterograde transport of virions through axons of central neurons, but do not compromise spread of virus via retrograde transsynaptic routes. This conclusion assumes that different factors mediate the anterograde transport of the Bartha strain of PRV through retinal ganglion cells and peripheral sensory neurons. However, this is not an unrealistic possibility given the clear demonstration that virus causes lysis in central neurons, but has the ability to go latent in peripheral neurons. This is obviously a complex issue that requires further investigation, but it vividly illustrates the necessity of carefully documenting the behavior of virus in each circuit rather than generalizing from other systems.

Light Microscopic Methods for Detecting Virus

Immunohistochemical localization of virus in infected neurons by light microscopy is easily the most straightforward and sensitive method of detecting viral replication and transport. Both fluorescent and peroxidase-based methods of detection have been successfully applied to these type of analyses, and each possess specific attributes that recommend them for different experimental applications. For example, peroxidase methods are probably the most useful for initial characterization of a circuit because they provide a permanent record of viral localization. This is particularly important because viral transport often produces complex patterns of neuronal labeling, and the fading characteristic of fluorescent labels makes them incompatible with prolonged observation. In contrast, fluorescent methods are useful for characterizing the chemical content of infected circuits because different-colored fluorophors can be used to label both the virus and peptide or neurotransmitter in the same section. In the following sections, we detail immunohistochemical methods that have been successfully applied to viral tract-tracing studies and suggest approaches that permit the most efficient use of experimental material.

Tissue Fixation

We fix all tissue for immunohistochemical localizations with a modification of the paraformaldehyde–lysine–periodate (PLP) fixative developed by McLean and Nakane (30). This fixative was initially designed to preserve both antigenicity and structure of tissue in immunoelectron microscopic analyses through stabilization of carbohydrate moieties. It consists of paraformaldehyde, lysine hydrochloride, and sodium metaperiodate in 0.1 M sodium phosphate buffer. Stabilization of carbohydrates is thought to be the product of oxidation and cross-linking by periodate and lysine, respectively. We have found this fixative to be excellent for preserving the antigenicity of virus as well as a variety of peptides and neurotransmitters, and therefore routinely employ it in all of our immunohistochemical applications. It is prepared by heating 0.1 M sodium phosphate buffer (pH 7.4) to approximately 80°C and adding 4 g of paraformaldehyde for each 100 ml of buffer to create a final concentration of 4%. This solution is stirred vigorously until it clears and is then cooled to room temperature with ice. Immediately prior to use, lysine and sodium metaperiodate are added to a final concentration of 0.075 and 0.01 M, respectively. Addition of these reagents reduces the pH of the solution to approximately 6.5, and it is not readjusted prior to use.

Tissue is fixed by transcardiac infusion of 400–600 ml of the fixative, preceded by 50–100 ml of physiological saline to wash the blood from the vascular tree. All solutions are infused at controlled pressure using a peristaltic pump. Aldehydes inactivate the virus, so we perfuse the entire animal rather than clamping the descending aorta, as is often done when perfusion of the brain is the ultimate goal. On completion of fixative infusion, the brain and other tissues to be included in the analysis are removed, postfixed in the primary fixative, washed in 0.1 M sodium phosphate buffer (pH 7.4), and placed in 20% phosphate-buffered sucrose for cryoprotection. All of these steps are carried out at 4°C. The length of the postfixation varies with the application. If localization of virus is the sole objective of the analysis, tissue can be postfixed overnight to further optimize structural preservation. However, if double-labeling procedures are going to be used to localize virus along with a peptide or neurotransmitter it is best to restrict the postfixation to 1 hr. The buffer washes generally consist of two or more changes of the buffer over a period of 1 or 2 hr, and cryoprotection of the tissue is judged adequate when it sinks to the bottom of the vial (generally about 24 hr).

Preparation of Tissue

As noted previously, experiments of this nature require special facilities and are labor intensive. Furthermore, these pathogens can compromise animal health at long post-inoculation survival intervals. Consequently, we go to great lengths to organize studies so that we acquire the maximum amount of data from each experimental animal. This is primarily accomplished by sectioning tissue into multiple wells of buffer and transferring excess tissue to a cryoprotectant solution that permits storage at −20°C for long periods of time with no evident loss of antigenicity or morphology (31). Sectioning of tissue is accomplished by freezing it onto the chuck of a sliding microtome, using dry ice or a commercially available freezing unit (Lipshaw, Detroit, MI). Sections (35 μm) through the rostrocaudal extent of the neuraxis are then collected in consecutive bins of buffer, and a single bin of tissue is immediately processed for immunohistochemical localization of virus (see the following procedure) to determine the extent of viral transport and to evaluate the preservation of tissue morphology. The remaining bins are stored at 4°C and can be used to analyze viral immunoreactivity in a more frequent series of sections, or to localize other antigens either alone or in combination with the virus. Any tissue that is not processed within the first 2 days following sectioning is transferred to cryoprotectant and stored at −20°C. Studies by Watson and colleagues (31) have demonstrated that this method of storage provides an effective means of preserving tissue antigenicity for a variety of peptides, and our studies have demonstrated that viral immunoreactivity is also preserved for extended periods of time. The cryoprotectant is made by mixing 200 g of sucrose and 10 g of polyvinylpyrrolidone-40 in a large beaker and then slowly stirring in 500 ml of 0.1 M sodium phosphate buffer. Thereafter, another 100 g of sucrose is added to the solution followed by 300 ml of ethylene glycol and enough distilled water to create a final volume of 1 liter. The solution is then stored in the freezer.

Immunoperoxidase Localization of Virus

Localization of virus with immunoperoxidase methods is a straightforward endeavor that can be accomplished with polyclonal antisera generated against entire virus, or with monospecific antisera generated against individual viral envelope glycoproteins. However, polyclonal antisera are preferable because they are more readily available and produce a more robust signal. In the past we have used the avidin–biotin modification (32) of the peroxidase/antiperoxidase procedure (33). This procedure (3) provides a sensitive means of detecting virus localization, using dilutions of rabbit polyclonal antisera in the range of 1:1000 to 1:2000. However, the introduction of more sensitive methods of detection by Vector Laboratories (Burlingame, CA) now permits the use of dilutions of primary antisera extending to 1:300,000 with no apparent loss of immunoreactivity. The following procedure, perfected for detection of virus in the CNS by J. Speh in the laboratory of R. Y. Moore at the University of Pittsburgh (Pittsburgh, PA) uses the Vectastain Elite modification of the avidin-biotin immunoperoxidase procedure.

Protocol 1: Light Microscopic Immunoperoxidase Detection of Pseudorabies Virus

1. Antibody incubation: Free-floating sections of tissue are immersed in a solution containing primary antibody, normal serum, and Triton X-100 and stored at 4°C for 24–48 hr. The amount of each reagent necessary to achieve a variety of dilutions is indicated in Table I. We routinely use a 1:10,000 dilution to achieve robust and reproducible staining.

2. Preparation for incubation in bridge antibody: Tissue is brought to room temperature over a period of 30–60 min with agitation and is then washed in multiple changes of a 10mM solution of sodium phosphate-buffered saline (PBS, pH 7.6) over 30 min.

3. Incubation in bridge antiserum: Tissue sections are placed in biotinylated secondary antibody raised in another species. for example, if the primary antiserum is generated in rabbit, an appropriate secondary would be biotinylated donkey anti-rabbit IgG. A 1:200 dilution is made as follows: 5 μl of secondary antibody, 20 μl of normal serum, 30 μl of 10% Triton X-100, and 950 μl of PBS. Tissue is incubated 60 to 90 min at room temperature.

TABLE I   Dilution Charts for Primary Antibodies[a]

Final dilution	Stock (μl)	Serum (μl)	Triton X-100 (μl)	PBS
Using 1:10 stock				
1:100	100	10	30	860
1:500	20	10	30	940
1:1,000	10	10	30	950
1:2,000	5	10	30	955
1:5,000	2	10	30	958
Using 1:100 stock				
1:1,000	100	10	30	860
1:5,000	20	10	30	940
1:10,000	10	10	30	950
1:15,000	6.67	10	30	954
1:20,000	5	10	30	955

[a] These charts are based on the following assumptions: (a) The final volume of primary antibody will be 1 ml; (b) a 10 mM phosphate-buffered saline (PBS) solution (pH 7.6) is used for all dilutions and washers, (c) the normal serum is a 10% solution that matches the host of the bridge (secondary) antibody; and (d) a 10% solution of Triton X-100 is used for dilution, giving a final concentration of 0.3%.

4. Preparation of ABC reagents: The avidin–biotin complex is prepared 60 to 90 min prior to use. Two aliquots designated A and B are available in the Vectastain Elite kit sold by Vector Laboratories. Five microliters of each of these reagents is combined in the bottom of an Eppendorf tube and incubated at room temperature with agitation.

5. Washing tissue in PBS: Tissue is washed in multiple changes of PBS over 30 min.

6. Incubation tissue in ABC reagent: The components of A and B mixed in step 4 are combined with 960 μl of PBS and 30 μl of 10% Triton X-100. Tissue is incubated in this mixture for 60 to 90 min at room temperature with agitation.

7. Washing tissue in PBS: Repeat step 5.

8. Reacting tissue in diaminobenzidine solution: Visualization of the antibody–peroxidase complex is accomplished by preparing a "saturated" solution of diaminobenzidine (DAB) by placing an excess of DAB in Tris buffer (pH 7.2) followed by 5 min of vigorous stirring. The solution is filtered, the tissue incubated in it for 10 min, and 35 μl of hydrogen peroxide is added. Development of the brown reaction product is monitored visually or with the aid of a light microscope. The reaction is terminated with repeated washes of PBS.

9. Mounting tissue on microscope slides: Sections are mounted on gelatin-coated microscope slides and allowed to air dry. After a period of 1 to 2 days, the sections are dehydrated with a graded ethanol series, cleared in xylene, and coverslipped with Permount.

Double-Label Immunoperoxidase Localizations

In many instances it is useful to localize virus and another antigen in the same section. For example, first-order neurons are sometimes closely associated with short-axon, second-order local circuit neurons and it becomes difficult to distinguish the two cell populations. Under these circumstances we inject a classic neuroanatomical tracer such as cholera toxin-conjugated horseradish peroxidase (CT-HRP) along with the virus. The CT-HRP remains confined to the first-order neurons and serves as a marker that can be used to distinguish these cells from the virus-infected second-order neurons. Discrimination of the two antigens (CT-HRP versus PRV) can be accomplished colorimetrically by combining sequential incubation of this tissue in discriminating antisera generated in different species, and the use of the nickel intensification procedure introduced by Adams (34) to make one of the peroxidase labels blue-black. The following procedure details a method that we have successfully applied to localization of PRV in combination with a number of other antigens.

Protocol 2: Immunoperoxidase Detection of Pseudorabies Virus and Another Antigen

1. First antibody reaction: Follow steps 1 through 7 of protocol 1 to localize the first antigen. In the experiment described above we would first localize the CT-HRP with a goat polyclonal antiserum generated against CT and diluted to a final concentration of 1 : 10,000. However, in some experimental paradigms it will be most useful to localize PRV with the nickel intensification method. Irrespective of the antigen, the nickel intensification procedure should be conducted first.

2. Nickel intensification of first immunoperoxidase signal: Prepare the "saturated" solution of DAB in Tris buffer as described in step 8 of the previous protocol. Immediately prior to tissue incubation add 0.3% nickel chloride, preincubate the tissue in this solution for 10 min, and then add 35 μl of hydrogen peroxide. Monitor the development of the blue-black reaction product visually and terminate the reaction with repeated washes of PBS.

3. Second antibody reactions: Repeat steps 1 through 8 of the immunoperoxidase protocol, using the second antibody. This will produce a brown reaction product that can be easily distinguished from the nickel-intensified signal.

4. Mounting tissue on microscope slides: Sections are mounted on gelatin-coated microscope slides and allowed to air dry. After a period of 1 to 2 days the sections

are dehydrated with a graded ethanol series, cleared in xylene, and coverslipped with Permount.

Immunofluorescence Localization of Virus

Modifications of the fluorescence immunohistochemical method of Coons (35) are the preferred means of localizing PRV and other antigens in the same cell. A variety of different-colored fluorophors are available that can be effectively applied for localization of antigens in the same tissue section, and this methodology has been successfully applied to localize PRV in combination with a variety of antigens (14–16, 24). In addition, this method can be effectively applied to localizing PRV in combination with another fluorescent tracer such as Fluoro-Gold (Fluorochrome, Inc.) (36). However, negative data on the colocalization of PRV with peptides and neurotransmitters should be viewed conservatively in light of the pathological changes that PRV causes in neurons. The best approach is to apply the method early in the temporal course of neuronal infection to minimize any adverse effects that viral infection may have on host cell synthesis. Use of colchicine to increase intracellular levels of peptides or neurotransmitters is not a viable approach because it would interfere with the transneuronal passage of virus and also compromise the health of the animal. We have successfully applied the following method for localization of PRV in combination with cell-specific markers.

Protocol 3: Immunofluorescence Detection of Pseudorabies Virus and Other Antigens

1. Tissue preparation: Perfuse and cryoprotect tissue, using the same procedures detailed for the immunoperoxidase method. Mount cryostat-sectioned tissue (10μm/ section) on gelatin-coated slides and store at $-20°$C until antibody incubation.

2. Primary antibody incubation: The primary antibodies should be generated in different species. Tissue can be incubated sequentially or simultaneously. The optimal dilution for each antibody must be determined separately, but we find that it is generally in the range of 1:500. Use the same antibody diluent described for the immunoperoxidase localizations. The tissue sections should be ringed with rubber cement, covered with primary antibody, and incubated overnight at room temperature in a humidified box.

3. Tissue washing: Wash tissue in several changes of sodium phosphate buffer over 30 min.

4. Fluorescent antibody incubation: Incubate tissue in a 1:200 dilution of isotype-specific antibodies tagged with fluorescein isothiocyanate or tetramethylrhodamine isothiocyanate to label antigens green and red, respectively. Other color fluorophors are also available.

5. Mounting tissue on microscope slides: Rinse slides in several changes of buffer and coverslip with Gelmount (Biomeda, Foster City, CA). Examine tissue with a fluorescence microscope, using the appropriate excitation filters. Fluorophors often fade quickly during examination, so it is best to be prepared to photograph the material on first examination.

Interpretation of Results

In general, the same parameters that must be critically considered in evaluating the transport of conventional tracers through the nervous system are relevant to the analysis of viral transport. They include route of administration, spread from the site of injection, lytic release of virus from infected neurons (this often can be equated with the virulence of the infecting virus), and the temporal aspects of viral transport through a multisynaptic circuit. Issues relevant to the route of viral injection were considered earlier in the chapter. Aspects of the other parameters as they apply to viral transport follow.

Spread of Virus from Site of Injection

Available evidence indicates that PRV does not diffuse significant distances following either peripheral or central injection. This was one of the foci of our initial analysis of viral transport through neurons innervating the viscera (19). In that study we injected a virulent strain of PRV (Becker) close to the gastroesophageal junction. The stomach is innervated by preganglionic parasympathetic neurons in the dorsal motor vagal nucleus (DMV) of the brainstem, whereas the motor innervation of the esophagus is provided by a separate population of neurons residing in a cell group known as the nucleus ambiguus. Thus, if there was even a small amount of diffusion of virus from the site of injection one would expect to infect neurons in both of these cell groups. In contrast, when virus was injected into stomach wall we observed infected neurons only in the DMV, and nucleus ambiguus neurons were infected only when the virus was injected into the esophagus. Similarly, following direct injection of PRV into the rat cortex we have found little evidence of virus spread from the injection site (29). These observations suggest that the high affinity of the virus for neurons effectively limits the spread of virus from the injection site and contributes to its effectiveness as a neuronal track tracer. Nevertheless, it is important to document clearly the extent of viral dissemination by examining the injection site and/or injecting conventional tracers in combination with the virus.

The volume of virus injected will also contribute to the extent of diffusion from the injection site. As a general rule, it is best to inject the smallest volume possible. When injecting the vitreous body of the eye in rats, we inject $1-2\ \mu l$ of virus through

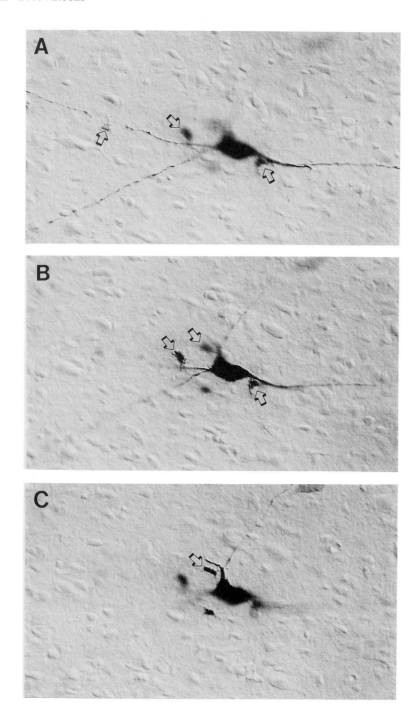

a Hamilton microliter syringe and leave the 26-gauge needle in the globe of the eye for a minimum of 5 min to reduce leakage of virus into the orbit. Successful injections of the viscera also employ small volumes of virus $(1-5 \, \mu l)$ injected into multiple sites in the organ of interest, using a microliter syringe and a surgical microscope. Once again, the needle is left in place for a short period of time to reduce reflux along the needle tract, and the surface of the injection site is washed with saline-soaked cotton swabs to prevent dissemination of virus through the peritoneal cavity. When injecting the virus directly into the CNS, a small cannula is placed in the desired area, using a stereotaxic apparatus, and virus (10–100 nl) is injected at a controlled rate of approximately 10 nl/min. On completion of the injection the cannula is left in place for a minimum of 5 min. These parametrics have produced specific transport of virus through multisynaptic pathways subserving visual, visceral, and cortical function, but are provided only as general reference points for designing a study. Independent parametrics should be rigorously characterized in each experimental application.

Lytic Release of Virus from Infected Neurons

Virus-induced neuropathology and the appearance of infected glia surrounding chronically infected neurons have raised concerns regarding the specificity of transneuronal passage of alpha herpesviruses. It is obvious that lysis of infected neurons would release virions into the extracellular space and that this release could compromise the circuit-specific transport of virus. The extent of this pathological response is obviously dependent on the virulence of the infecting strain of virus. However, even attenuated strains of virus produce pathological changes after prolonged viral replication (Fig. 2), and it is therefore important to characterize fully the temporal aspects of this response and to characterize the consequences of viral release under these circumstances. We have conducted an extensive analysis of the development of neuropathogenesis and the nonneuronal response to infection of neurons with virulent and attenuated strains of PRV (24, 25). These studies have shown that cell-to-cell transmission of virus preferentially occurs at sites of synaptic contact, and have further demonstrated that the response of glia and macrophages to neuronal infection actually contributes to specific transsynaptic passage of virus rather than

---

FIG. 2   A neuron in the lateral hypothalamus infected by injecting the Bartha strain of virus into the vitreous body of the eye is shown in three different focal planes (A–C). The dark immunoperoxidase reaction product resulting from immunohistochemical localization of newly replicated virus is present throughout the cell body and its processes. The chronic replication of virus in this cell has also led to infection of reactive astroglia associated with the infected neuron (arrows). Nevertheless, there is no indication that adjacent neurons have become infected due to liberation of virus into the extracellular space.

reflecting nonspecific dissemination of virus through the extracellular space. Clearly, one must understand the time course of viral replication, transport and nonneuronal response to infection when considering the impact of neuropathogenesis on the specificity of viral transport in a given system.

Temporal Aspects of Viral Transport

An accurate definition of a neuronal circuit is dependent on consideration of the temporal aspects of viral transport. This in turn is dependent on an understanding of the factors that contribute to replication, assembly, and egress of virus in an infected neuron. The most significant feature of the viral life cycle that is relevant to interpreting the transport of virus through a multisynaptic circuit is the fact that, on infecting a neuron, the virus must reproduce before it can pass to the next neuron. Replication of virus in a neuron is a complex process that involves retrograde transport of viral DNA to the cell nucleus, replication of viral DNA and assembly of new virions, and intracellular transport of newly replicated virions to sites of afferent synaptic contact. This collective process requires time, and the length of time necessary for it to occur can vary substantially in different neurons. For example, viral replication and transneuronal passage occur quicker in interneurons than in projection neurons due to the shorter length of the axon. It is therefore important to consider the temporal aspects of the viral life cycle in analyzing both the course of viral transport through a circuit as well as the consequences of neuropathological changes on the specificity of transneuronal transport. This is readily apparent in considering the transport of virus through central circuits following injection of virus into the viscera. After injection of the stomach we noted immunohistochemically detectable viral replication in preganglionic parasympathetic neurons of the DMV as early as 30 hr postinoculation, with robust viral immunoreactivity apparent 45–50 hr after injection (19). In contrast, short-axon, second-order neurons in the overlying nucleus of the solitary tract (NST) and nearby brainstem tegmentum infected by transsynaptic passage of virus from DMV neurons displayed viral immunoreactivity 50–55 hr after injection. These observations, combined with our studies of neuropathogenesis (24, 25), emphasize the need for considering the temporal sequence of viral infection in evaluating the specificity of transneuronal passage of virus. All available evidence indicates that virus begins to cross synapses shortly following replication, within a time frame preceding the appearance of any overt neuropathology in the infected neuron.

Conclusions

In writing this article we have sought to review the critical issues to be considered when using PRV to define specific neuronal circuitry, and we have detailed the methods used to accomplish this type of analysis. The available strains of PRV and methods of detection can be powerful tools for identifying synaptically linked neurons. However, enthusiasm for the application of this method must be tempered with the caveat that PRV is a complicated virus whose specific neurotropism and virulence are understood only in principle. The molecular basis for almost all aspects of CNS infection by PRV eludes us and many observations remain unexplained. For example, we suspect that all neurons and all circuits cannot be infected by PRV, but evidence for this is only anecdotal and negative results are difficult to interpret. Another important limitation of this technology is that PRV infection of rodents is lethal. We have few methods that extend the postinoculation survival interval of infected animals for protracted analysis. Use of antiviral agents, steroids, and passive immunization are all being studied in various laboratories with mixed results. To date, the use of less virulent PRV strains has been the most reproducible method for extending animal survival. However, a common observation is that attenuating mutations in the viral genome often slow transport and/or replication (7, 21).

Many aspects of this technology are still empirical and lack firm understanding at the mechanistic and molecular levels. Consider the specific example of direction and rate of transport. It is a gross simplification to conclude that PRV is a simple retrograde tracer. From our own work we know that the Becker strain of PRV can move in both the retrograde and anterograde directions (7, 19, 21), but the mutants lacking glycoprotein gI or gp63 appear to be defective in anterograde transport in some, but not all, neurons. In addition, retrograde transport of the Becker strain occurs significantly faster than anterograde transport, and the postinoculation survival interval rarely extends beyond 75 hr in animals infected with this strain of PRV. Consequently, this combination of events can mislead the experimenter to believe that the virus moves only in the retrograde direction.

It is encouraging, however, that many experiments indicate that the rate and transport of PRV have a genetic basis. Our strategy in studying this phenomenon has been straightforward. Because PRV can be easily manipulated in the laboratory by recombinant DNA technology, we have been able to prepare well-defined viral mutants and compare their infectivity to that resulting from infection with isogenic, nonmutant parent strains (4, 7, 26, 37). By testing various field isolates as well as attenuated variants of PRV and also by making isogenic mutants in a given strain, we and others are defining the specific viral genes involved in entry, transport, exit, and virulence in neuronal cells. For example, Peeters *et al.* (38) have shown that the PRV glycoprotein gp50 is essential for virus entry but not required for viral spread in mice. These workers and Heffner *et al.* (39) also showed that viruses lacking gp50 had reduced virulence. Peeters *et al.* (40) and Rauh *et al.* (41) also showed that the PRV gII glycoprotein is essential for both virion penetration and membrane fusion.

The techniques of genetic engineering and gene replacement by homologous recombination can also be used to introduce into PRV foreign genes such as *Escherichia coli lacZ* (43), human tissue plasminogen activator (44), HIV envelope glycoproteins (45), herpes simplex virus glycoprotein C (46), enzymatically active *E.*

coli lacZ (47), and luciferase (48). The recombinant viruses have utility in circuit tracing in that different reporter genes can be used to produce antigenically distinct strains of virus so that multiple infections can be done in the same animal. A powerful extension of the technology to introduce foreign genes into PRV involves targeted foreign gene delivery to the brain (42). This technology has been advanced primarily by using human alpha herpesvirus, herpes simplex virus type 1 (HSV-1), but in principle should be directly applicable to PRV. Pseudorabies virus offers a number of distinct advantages for laboratory research. Pseudorabies virus does not cause human disease despite extensive natural exposure of agricultural workers, veterinarians, and meat packers to infectious virus (4). However, despite the restricted natural pathogenesis profile, PRV has a broad host range in tissue culture cells that facilitates many *in vitro* experiments. Laboratory experiments can be less time consuming because standard strains of PRV grow significantly faster than HSV-1 in the laboratory. Certain molecular biology protocols are made easier because the specific infectivity of PRV DNA (plaques per microgram of DNA) can exceed 5–10 times that of HSV-1. As discussed above, PRV can be easily manipulated using standard recombinant DNA protocols and can also be used for site-specific insertion of DNA (49). Nevertheless, the molecular genetics of PRV is not as extensive as HSV and further characterization of the determinants of target cell recognition in the brain, specific and regulated gene expression, neurovirulence, and construction of defective and noncytopathic PRV vectors is vital to expanding the utility of this virus as a general tool for study of the nervous system.

In summary, the natural affinity of herpesviruses for neurons and glial cells makes it possible to use them in many ways to study the mammalian nervous system. As we have detailed here, they are significant tools to map neural circuits. However, they also may be exploited to create specific animal models of disease through targeted elimination of specific neuronal populations, to create local areas of inflammation or damage in the brain, and to alter normal neuronal physiology by introduction of foreign or mutant genes. There is significant need for fundamental research to understand the interaction of viruses and the nervous system at the molecular level. These studies will reveal much about both parasite and host, as well as improve our chances to extend virus technology. The field of neurovirology shows significant promise indeed.

Acknowledgments

It is a pleasure to acknowledge our research associates Mary Whealy and Joan Dubin, whose technical skills, enthusiasm, and intellect were vital for the success of our experiments. We gratefully recognize the constructive criticism, technical skill, and shared experiences of our collaborators Richard Miselis, Linda Rinaman, Pat Levitt, Joan Speh, and Robert Y. Moore. We are indebted to Henry Paulter for his diligence in animal care.

References

1. M. Dolivo, *Trends Neurosci.* **3,** 149 (1980).
2. H. G. J. M. Kuypers and G. Ugolini, *Trends Neurosci.* **13,** 71 (1990).
3. P. L. Strick and J. P. Card, *in* "Experimental Neuroanatomy. A Practical Approach" (J. P. Bolam, ed.), p. 81. IRL Press, Oxford, 1992.
4. G. Wittmann and H.-J. Rziha, *in* "Herpesvirus Diseases of Cattle, Horses and Pigs" (G. Wittmann, ed.), p. 270. Kluwer, Boston, 1989.
5. C. H. Becker, *Experentia* **23,** 209 (1967).
6. A. Bartha, *Magy Allatorv Lapja,* **16,** 42 (1961).
7. J. P. Card, M. E. Whealy, A. K. Robbins, and L. W. Enquist, *J. Virol.* **66,** 3032 (1992).
8. M. Dolivo, E. Beretta, V. Bonifas, and C. Foroglou, *Brain Res.* **140,** 111 (1978).
9. M. Dolivo, P. Honegger, C. George, M. Kiraly, and W. Bommeli, in "Progress in Brain Research" (M. Cuenod, G. W. Kreutzberg, and F. E. Bloom, eds.), Vol. 51, p. 51. Elsevier/North Holland, Amsterdam, 1979.
10. X. Martin and M. Dolivo, *Brain Res.* **273,** 253 (1983).
11. E. M. Rouiller, M. Capt, M. Dolivo, and F. De Ribaupierre, *Neurosci. Lett.* **72,** 247 (1986).
12. E. M. Rouiller, M. Capt, M. Dolivo, and F. De Ribaupierre, *Brain Res.* **47,** 21 (1989).
13. C. F. Marchand and M. Schwab, *Brain Res.* **383,** 262 1983.
14. A. M. Strack, W. B. Sawyer, J. H. Hughes, K. B. Platt, and A. D. Loewy, *Brain Res.* **491,** 156 (1989).
15. A. M. Strack, W. B. Sawyer, K. B. Platt, and A. D. Loewy, *Brain Res.* **491,** 274 (1989).
16. A. M. Strack and A. D. Loewy, *J. Neuroscience* **10,** 139 (1990).
17. A. S. P. Jansen, G. J. Ter Horst, T. C. Mettenleiter, and A. D. Loewy, *Brain Res.* **572,** 253 (1992).
18. D. M. Rotto-Percelay, J. G. Wheeler, F. A. Osorio, K. B. Platt, and A. D. Loewy, *Brain Res.* **574,** 291 (1992).
19. J. P. Card, L. Rinaman, J. S. Schwaber, R. R. Miselis, M. E. Whealy, A. K. Robbins, and L. W. Enquist, *J. Neurosci.* **10,** 1974 (1990).
20. I. Nadelhaft, P. L. Vera, J. P. Card, and R. R. Miselis, *Neurosci. Lett.* **143,** 271 (1992).
21. J. P. Card, M. E. Whealy, A. K. Robbins, R. Y. Moore, and L. W. Enquist, *Neuron* **6,** 957 (1991).
22. R. M. McCracken, J. B. McFerran, and C. Dow, *J. Gen. Virol.* **20,** 17 (1973).
23. R. M. McCracken and C. Dow, *Acta Neuropathol. (Berl.)* **25,** 207 (1973).
24. L. Rinaman, J. P. Card, and L. W. Enquist, *J. Neurosci.* **13,** 685 (1993).
25. J. P. Card, L. Rinaman, R. B. Lynn, B.-H. Lee, R. P. Meade, R. R. Miselis, and L. W. Enquist, *J. Neurosci.* **13,** 2515 (1993).
26. M. E. Whealy, J. P. Card, A. K. Robbins, J. R. Dubin, H.-J. Rziha, and L. W. Enquist, *J. Virol.* **67,** 3786 (1993).
27. J. P. Card, J. R. Dubin, M. E. Whealy, and L. W. Enquist, submitted (1994).
28. C. G. Dotti and K. Simmons, *Cell (Cambridge, Mass.)* **62,** 63 (1992).
29. J. P. Card, L. W. Enquist, and P. Levitt, in preparation (1994).
30. I. W. McLean and P. K. Nakane, *J. Histochem. Cytochem.* **22,** 1977 (1974).
31. R. E. Watson, S. T. Wiegand, R. W. Clough, and G. E. Hoffman, *Peptides* **7,** 155 (1986).
32. S. M. Hsu, L. Raine, and H. Fanger, *J. Histochem. Cytochem.* **29,** 577 (1981).

33. L. A. Sternberger, in "Immunocytochemistry" (L. A. Sternberger, ed), p. 104. Wiley, New York, 1979.
34. J. C. Adams, *Neuroscience* **2,** 141 (1977).
35. A. H. Coons, in "General Cytochemical Methods" (J. F. Danielli, ed.), p. 399. Academic Press, New York, 1958.
36. L. C. Schmued and J. H. Fallon, *Brain Res.* **377,** 147 (1986).
37. J. P. Ryan, M. E. Whealy, A. K. Robbins, and L. W. Enquist, *J. Virol.* **61,** 2962 (1987).
38. B. Peeters, J. Pol, A. Gielkens, and R. Moorman. *J. Virol.* **67,** 170 (1993).
39. S. Heffner, F. Kovacks, B. Klupp, and T. Mettenleiter. *J. Virol.* **67,** 1529 (1993).
40. B. Peeters, N. deWind, M. Hooisma, F. Wagenaar, A. Gielkens, and R. Moormann, *J. Virol.* **66,** 894 (1992).
41. I. Rauh, F. Weiland, F. Fehler, G. Keil, and T. Mettenleiter, *J. Virol.* **65,** 621 (1991).
42. X. Breakfield and N. DeLuca, *New Biologist* **3,** 203 (1991).
43. C. Keeler, M. E. Whealy, and L. W. Enquist, *Gene* **50,** 215 (1986).
44. D. Thomsen, K. Marotti, D. Palermo, and L. Post, *Gene* **57,** 261 (1987).
45. M. E. Whealy, K. Baumeister, A. K. Robbins, and L. W. Enquist, *J. Virol.* **62,** 4185 (1988).
46. M. E. Whealy, A. K. Robbins, and L. W. Enquist, *J. Virol.* **63,** 4055 (1989).
47. T. Mettenleiter and I. Rauh, *J. Virol. Methods* **30,** 55 (1990).
48. F. Kovacks Sz. and T. Mettenleiter, *J. Gen. Virol.* **72,** 2999 (1991).
49. B. Sauer, M. Whealy, A. Robbins, and L. Enquist, *Proc. Natl. Acad. Sci. USA* **84,** 9108 (1987).

[24] Detection of Viral Homologs of Cellular Interferon γ Receptors

C. Upton and G. McFadden

Introduction

Viruses that propagate successfully within vertebrates have, by necessity, evolved a multiplicity of strategies to confront the various arms of the host immune repertoire (1, 2). The larger DNA viruses, especially the herpesviruses, poxviruses, and adenoviruses, have the genomic coding capacity for more functions than are minimally required to permit viral replication and morphogenesis, and have evolved open reading frames (ORFs) devoted specifically to modulate host immunity (3, 4). Given the importance of cytokines in regulating the antiviral responses (5), it is perhaps not surprising that some of these virus genes encode proteins that are targeted directly at key soluble antiviral cytokines such as interferon (IFN), tumor necrosis factor α and β, and interleukin 1(IL-1). The first such anticytokine viral evasion strategy to be described evolved from the observation that the sequence of the T2 ORF of Shope fibroma virus (SFV), a poxvirus of rabbits, exhibited significant similarity to the ligand-binding domain of one of the cellular receptors for tumor necrosis factor (6). Later, the homologous T2 gene from myxoma virus, a related leporipoxvirus, was shown to be expressed as a secreted glycoprotein from virus-infected cells, and a T2-negative recombinant myxoma virus was shown to exhibit attenuated pathogenesis in rabbits (7). The term *viroceptor* was coined to designate viral gene products that mimic cellular receptors for the purpose of binding to, and sequestering, host cytokines critical to the antiviral response (7).

Poxviruses are complex eukaryotic viruses that encode in excess of 200 gene products and replicate in the cytoplasm of infected cells (8, 9). The viral DNA genome consists of a single linear double-stranded DNA molecule (>160 kb) with covalently closed hairpin termini. Essential genes involved in viral RNA transcription, DNA replication, and morphogenesis of progeny virions tend to be highly conserved amongst different poxvirus genera and map within the central portion of the genome (10). In contrast, many of the genes that are located toward the termini of the genome are not required for virus growth in tissue culture and a number of these have been shown to encode proteins that contribute directly to the virulence of the virus (11, 12). Here we describe methods for the identification and detection of the myxoma virus-encoded IFN-γ viroceptor, which specifically binds rabbit IFN-γ and negates its antiviral effects (13). The same strategy can be equally applied to other large DNA viruses, and preliminary data indicate that similar virus-encoded mimics of the cellular IFN-γ receptor are relatively common, at least in the poxvirus family.

Methods in Molecular Genetics, Volume 4

Materials and Methods

Isolation of Proteins Secreted from Virus-Infected Cells

1. African green monkey kidney (BGMK) cells are grown to 90% confluence in roller bottles (3×10^8 cells/bottle) in Dulbecco's modified Eagle's medium (DMEM) supplemented with 10% (v/v) newborn calf serum.

2. Medium is removed and myxoma virus inoculum [3×10^9 plaque-forming units (pfu)/bottle] allowed to absorb for 2 hr at 37°C.

3. Inoculum is removed and infected cells are washed three times with phosphate-buffered saline (PBS) (37°C), taking care to drain well between washes. Ten milliliters of DMEM (minus serum) is added to each bottle and the conditioned medium is collected after a further 24-hr incubation (37°C).

4. The culture supernatant is centrifuged in a Sorvall SS-34 rotor at 10,000 g for 1 hr (4°C) to remove cellular debris and detached cells.

5. The supernatant is concentrated 20-fold, using Centriprep-10 ultrafiltration units (Amicon, Danvers, MA) and stored at 4°C.

Labeling of Interferon γ with [γ-^{32}P]ATP

Although ligands are frequently radiolabeled with ^{125}I, all of the known IFN-γ sequences contain at least one consensus site for phosphorylation by cAMP-dependent protein kinase (14). Figure 1 indicates such consensus sites at the C termini of a variety of IFN-γ species. The recombinant rabbit IFN-γ used for the following labeling protocols was generously provided by Genentech Corp. (San Francisco, CA).

```
                                                         *                      *
Rabbit      E L S N V L N F L S P K S N L K K R K R S Q T L F R G R R A S K Y
Human       E L I Q V M A E L S P A A K T G K R K R S Q M L F R G R R A S Q
Marmoset    E L I Q V M A E L S P A P K I G K R R R S Q T L F R G R R A S Q
Bovine      E L I K V M N D L S P K S N L R K R K R S Q N L F R G R R A S M
Sheep       E L I K V M N D L S P K S N L R K R K R S Q N L F R G R R A S M
Red deer    E L I K V M N D L S P K S N L I K R K R S Q N L F R G R R A S M
Pig         E L I K V M N D L S P R S N L R K R K R S Q T M F Q G Q R A S K
Mouse       E L I R V V H Q L L P E S S L R K R K R S R C
Rat         E L I R V I H Q L S P E S S L R K R K R S R C
```

FIG. 1   Alignment of the C-terminal regions of IFN-γ from different species. The PIR accession numbers for the various sequences are as follows: human (A26968), marmoset (S24824), bovine (A24390), sheep (S12723), red deer (S18172), pig (S10513), mouse (A01844), and rat (A01845). The rabbit sequence is from Ref. 14a. The consensus sequences for cAMP-dependent protein kinase are boxed, and the serine residues that become phosphorylated by the kinase *in vitro* are indicated by an asterisk.

1. One microgram of IFN-γ is labeled in 50 μl [20 mM Tris-HCl (pH 7.4), 1 mM dithiothreital (DTT), 100 mM NaCl, 12 mM MgCl$_2$], using 7.5 units of the catalytic subunit of cAMP-dependent protein kinase (bovine heart; Sigma, St. Louis, MO) for 15 min at 37° C with 200 μCi of [γ-^{32}P]ATP.

2. The labeling reaction is terminated by the addition of 500 μl of bovine serum albumin (BSA) [1 mg/ml in 10 mM sodium phosphate buffer (pH 7.0), 10 mM sodium pyrophosphate, 10 mM ethylenediaminetetraacetic acid (EDTA)].

3. Labeled IFN-γ is separated from unincorporated label with a desalting column (Econo-Pac 10DG; Bio-Rad, Richmond, CA). Typically, the labeled ligand is eluted in 2 ml of PBS.

Consensus Amino Acid Sequence Motifs for Interferon γ

The human and mouse IFN-γ receptors have been cloned and sequenced (15, 16), and are considered members of a larger receptor superfamily (17, 18). The most abundantly secreted myxoma virus-encoded protein, now known to be the product of the T7 gene, was first identified as a member of the IFN-γ receptor family in the following way. N-terminal sequencing of this 35-kDa myxoma protein suggested that it was homologous to the SFV T7 ORF, and subsequent cloning and sequencing of the myxoma counterpart confirmed this (13). A systematic search of the published databanks with the myxoma/SFV T7 ORF and the related vaccinia B8R was performed using the computer program NW_Align. This searching algorithm is an option of the SEQSEE protein analysis package and it performs a complete alignment of the query and all sequences in the databank, together with statistical analysis of alignments with the randomized query sequence. It is especially efficient in the identification of significant homologies that are relatively low and spread out over a large region of protein sequence. In contrast, many alignment programs (e.g., BLAST) are optimized for the detection of short regions of higher similarity. The entire SEQSEE package and its accompanying databases may be obtained through anonymous file transfer protocol (ftp) by logging onto [nunki.biochem.ualberta.ca], using an E-mail address as the password. Further information can be obtained at the following E-mail address: seqsee@procyon.biochem.ualberta.ca. Such alignment searches identified a significant relationship between these viral ORFs and the mouse and human IFN-γ receptors, which themselves are only 50% identical at the amino acid level within their extracellular ligand-binding domains. Using an alignment of the five proteins, a consensus sequence for this family of IFN-γ-binding proteins was deduced (Fig. 2A). However, a minimal motif may be more useful for identifying novel proteins of related function and one such motif is shown in Fig. 2B. This motif correctly identifies all five members of the family and produces no false-positive hits in the PIR databank. The significance or function of the proline–proline sequence that is used in the

A

SY•{7,12}W•{10,15}K•Y•{3,6}W•••[CY]•{6,7}[CY]•{13,14}W•{5•7}G•{4,6}Y•{4,8}C•••

•••PP•{15,17}H•{5,6}G•{10,12}C•{5,7}Y•{19,20}C••••C•{9,13}C•{18,20}VC

B

[CY]•{6,7}[CY]•{30,38}C••••••PP•{30,37}C•{25,30}C••••C•{9,13}C•{18,21}C

FIG. 2   The consensus sequences of the IFN-γ receptor/T7 family of IFN-γ-binding proteins. (●) Any amino acid; [CY], cysteine or tyrosine; {5,7}, range of five to seven repeats of preceding amino acid; amino acids shown are identical in all five proteins. (A) The conserved amino acid sequences derived from the alignment of mouse and human receptors, T7 proteins of SFV and myxoma, and the B8R ORF of vaccinia virus. (B) The minimal motif required to identify the IFN-γ receptor/T7 family from the PIR database.

motif to differentiate this family from many other cysteine-rich proteins is unknown, but has proved to be useful for the assignment of members to this subfamily.

Cross-Linking of ^{32}P-Labeled Rabbit Interferon γ with Myxoma T7 Protein

1. The following reagents are mixed (and made up to a total volume of 20 μl with 10 mM sodium phosphate, pH 7.0) and incubated at room temperature for 2 hr.

   Secreted proteins (concentrated from mock- or myxoma virus-infected cells)
   Rabbit [^{32}P]IFN-γ (~5 ng)
   Rabbit IFN-γ (~2.5 μg: 500-fold excess cold competitor) or PBS (no cold competitor)

2. Cross-linking reagent 1-ethyl-3-(3-dimethylaminopropyl)carbodiimide (EDAC) (Sigma) is prepared fresh (1.0 M in 0.1 M potassium phosphate, pH 7.5), added to the mixture to a final concentration of 20 mM, and incubated at room temperature for 15 min.
3. A second aliquot of EDAC is added to bring the concentration to 40 mM and the incubation continued for a further 15 min.
4. The cross-linking reaction is quenched by the addition of a 1/10 vol of 1.0 M Tris-HCl, pH 7.5.
5. Samples are centrifuged in a microfuge at 4°C for 15 min and the supernatants analyzed by sodium dodecyl sulfate-polyacrylamide gel electrophoresis (SDS-

PAGE). Gels are stained with Coomassie blue, to display protein size standards, before drying and exposure to X-ray film.

Discussion

The interferons are the most extensively studied of the antiviral cytokines (19–25) and IFN-γ is a key component in the regulation of the host antiviral immune response (26, 27). Like many other cytokines, IFN-γ produces a multiplicity of effects, each brought about by the same initial event, namely that of IFN-γ binding to its ubiquitous specific receptor (15, 28). Interferon γ is primarily produced by T cells and natural killer cells in response to antigen activation, and its antiviral functions include activation of macrophages, upregulation of MHC class I and II molecules, upregulation of the antigen-processing machinery (proteosomes), and induction of an antiviral state similar to that induced by IFN-α and IFN-β (22–26). In many instances IFN-γ does not act alone but in conjunction with other cytokines, so that the observed result is the product of a complex interaction of additive, subtractive, and synergistic effects.

The importance of IFN-γ in the host immune system for defense against poxvirus infection has been affirmed by a number of approaches. First, recombinant vaccinia viruses manipulated so as to express the gene for IFN-γ were found to be attenuated in infected nude mice (29). Second, transgenic mice lacking the IFN-γ receptor, and therefore refractive to effects of IFN-γ, have been shown to be severely defective in mounting an early defense against vaccinia virus (30). These results, together with our observation that several poxviruses encode a soluble counterpart of the host IFN-γ receptor capable of binding to and blocking IFN-γ (13), provide convincing evidence that IFN-γ plays a crucial role in protection against poxvirus infection.

An example of the use of the EDAC cross-linking protocol described here to identify viral IFN-γ-binding proteins is shown for malignant rabbit fibroma virus (MRV) in Fig. 3. The MRV was derived from a recombinant event between two leporipoxviruses, myxoma virus and SFV, and contains two intact copies of the SFV-T7 gene but no myxoma-T7 counterpart (31). As shown for the control virus (v-*myx-lac*, which is essentially wild-type myxoma virus with a β-galactosidase gene incorporated into an intergenic region), a radiolabeled cross-linked complex of 51 kDa (13) was produced between the viral T7 protein and a monomer of the IFN-γ, which was competed away by the addition of 500-fold molar excess of cold IFN-γ. Note that the cross-linking also resulted in a spectrum of monomeric (M), dimeric (D), trimeric (T), and higher oligomeric forms of the 17-kDa IFN-γ polypeptide, which became more apparent after addition of the excess cold ligand in both the mock and *myx-lac* supernatants. The lanes indicating MRV-infected samples also demonstrate a similar heterodimer complex between the MRV-T7 protein and IFN-γ, indicating that the SFV-derived T7 gene of MRV also encodes a binding protein specific for rabbit

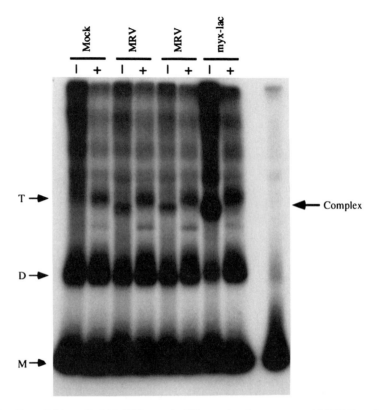

Fig. 3   Cross-linking of rabbit IFN-γ to the T7 protein of myxoma and MRV. Samples are concentrated supernatants from BGMK cells mock infected or following infection with MRV (a recombinant myxoma virus in which the myxoma T7 gene has been replaced with the T7 gene from SFV) or *myx-lac* (a derivative of myxoma virus, strain Lausanne, containing a wild-type T7 gene). After binding and cross-linking, as described in Materials and Methods, samples were analyzed by SDS-PAGE in a 15% polyacrylamide gel. The resulting auto-radiogram is shown. Positions of the monomers (M), dimers (D), and trimers (T) of the labeled IFN-γ are shown, and the 51- kDa heterodimer complexes (13) between T7 and IFN-γ [which is competed away by the presence (+) of excess cold ligand] are indicated.

IFN-γ. Similar experiments with proteins secreted from vaccinia virus-infected cells indicate the presence of a viral protein that binds to ^{32}P-labeled human IFN-γ (not shown).

Similar protocols can also be used to assess for viral proteins that bind to other host cytokines as well. For example, vaccinia virus and cowpox encode secreted proteins that recognize and bind IL-1 (32, 33), and it is likely that other such examples of viroceptors remain to be discovered. Many host cytokine receptors can exist in either membrane-bound or soluble form as a consequence of alternative splicing or the proteolytic cleavage of membrane-spanning precursor molecules (34). Although the function of these secreted host cytokine receptor-derived proteins is unclear, many do in fact bind their cognate ligand and may represent mechanisms for reducing the effective availability of a particular cytokine. Thus it appears that poxviruses have borrowed from a normal host strategy for cytokine regulation and use viroceptors to bind and sequester host cytokines and to abrogate a normal antiviral immune response. Several such examples are now known (3, 4), and this strategy may prove to be a common theme for virus evasion of immune functions that depend on or are regulated by cytokines.

Acknowledgments

We thank Shari Kasinec for help in the preparation of the manuscript. The SEQSEE computer package was developed by D. Wishart and R. Boyko in the laboratory of B. D. Sykes at the University of Alberta, Edmonton, Canada. G. M. is a Medical Scientist of the Alberta Heritage Foundation for Medical Research. This work was funded by the National Cancer Institute of Canada.

References

1. J. L. Whitton and M. B. A. Oldstone, *in* "Virology" (B. N. Fields, D. M. Knipe, *et al.*, eds.), p. 369. Raven Press, New York, 1990.
2. S. Specter, M. Bendinelli, and H. Friedman, eds., "Virus-Induced Immunosuppression." Plenum Press, New York, 1989.
3. L. R. Gooding, *Cell (Cambridge, Mass.)* **71**, 5 (1992).
4. P. M. Murphy, *Cell (Cambridge, Mass.)* **72**, 823 (1993).
5. I. L. Campbell, *Curr. Opin. Immunol.* **3**, 486 (1991).
6. C. A. Smith, T. Davis, D. Anderson, L. Solam, M. P. Beckmann, R. Jerzy, S. K. Dower, D. Cosman, and R. G. Goodwin, *Science* **248**, 1019 (1990).
7. C. Upton, J. L. Macen, M. Schreiber, and G. McFadden, *Virology* **184**, 370 (1991).
8. B. Moss, *in* "Virology" (B. N. Fields and D. M. Knipe, eds.), p. 2079. Raven Press, New York, 1990.
9. F. Fenner, R. Wittek, and K. R. Dumbell, eds., "The Orthopoxviruses." Academic Press, San Diego, 1989.
10. P. Traktman, *Cell (Cambridge, Mass.)* **62**, 621 (1990).
11. P. C. Turner and R. W. Moyer, *Curr. Topics Microbiol. Immunol.* **163**, 125 (1990).
12. R. M. Buller and G. J. Palumbo, *Microbiol. Rev.* **55**, 80 (1991).
13. C. Upton, K. Mossman, and G. McFadden, *Science* **258**, 1369 (1992).
14. A. Rashidbaigi, H. Kung, and S. Pestka, *J. Biol. Chem.* **260**, 8514 (1985).
14a. C. T. Samudzi, L. E. Burton, and J. R. Rubin, *J. Biol. Chem.* **266**, 21791 (1991).
15. M. Aguet, *J. Interferon Res.* **10**, 551 (1990).

16. R. D. Schreiber, M. A. Farrar, J. Fernandez-Luna, G. K. Hurshey, and P. W. Gray, *Prog. Leukocyte Res.* **2,** 337 (1991).
17. J. F. Bazan, *Proc. Natl. Acad. Sci. USA* **87,** 6934 (1990).
18. J. F. Bazan, *Cell (Cambridge, Mass.)* **61,** 753 (1990).
19. D. J. Maudsley, A. G. Morris, and P. T. Tomkins, *in* "Immune Responses, Virus Infections and Disease" (N. J. Dimmock and P. D. Minor, eds.), p. 15. IRL Press, Oxford, 1989.
20. P. Staeheli, *Adv. Virus Res.* **38,** 147 (1990).
21. J. L. Taylor and S. E. Grossberg, *Virus Res.* **15,** 1 (1990).
22. W. K. Joklik, *in* "Virology" (B. N. Fields and D. M. Knipe, eds.), p. 383. Raven Press, New York, 1990.
23. J. Vilcek, *in* "Peptide Growth Factors and Their Receptors" (M. B. Sporn and A. B. Roberts, eds.), p. 3. Springer-Verlag, New York, 1991.
24. C. E. Samuel, *Virology* **183,** 1 (1991).
25. G. C. Sen and P. Lengyel, *J. Biol. Chem.* **267,** 5017 (1992).
26. S. Landolfo and G. Garotta, *J. Immunol. Res.* **3,** 81 (1991).
27. M. Barinaga, *Science* **259,** 1693 (1993).
28. J. A. Langer and S. Pestka, *Immunol. Today* **9,** 393 (1988).
29. I. Ramshaw, J. Ruby, A. Ramsay, G. Ada, and G. Karupiah, *Immunol. Rev.* **127,** 157 (1992).
30. S. Huang, W. Hendriks, A. Althage, S. Hemmi, H. Bluethmann, R. Kamijo, J. Vilcek, R. M. Zinkernagel, and M. Aguet, *Science* **259,** 1742 (1993).
31. C. Upton, J. L. Macen, R. A. Maranchuk, A. M. DeLange, and G. McFadden, *Virology* **166,** 229 (1988).
32. M. K. Spriggs, D. E. Hruby, C. R. Maliszewski, D. J. Pickup, J. E. Sims, R. M. L. Buller, and J. VanSlyke, *Cell (Cambridge, Mass.)* **71,** 145 (1992).
33. A. Alcami and G. L. Smith, *Cell (Cambridge, Mass.)* **71,** 153 (1992).
34. R. Fernandez-Botran, *FASEB J.* **5,** 2567 (1991).

[25] *In Vitro* Phosphorylation of Hepatitis B Virus P Gene Product: A General Method for Radiolabeling of Proteins

Ralf Bartenschlager, Martin Weber, and Heinz Schaller

Introduction

The hepatitis B viruses (HBV) are a group of small enveloped DNA viruses that replicate via protein-primed reverse transcription of an RNA pregenome (1–3). Although this process takes place within a nucleocapsid formed by the viral core protein, all major enzymatic steps are performed by the viral P (*pol*) gene product. Genetic analysis predicts a 90-kDa P protein organized in functional domains as outlined in Fig. 2A: aminoterminally the DNA terminal protein, the primer for the initiation of hepadnaviral reverse transcription; this is followed sequentially by a non-essential, largely deletable spacer region, and then by the two catalytic domains for reverse transcriptase/DNA polymerase and RNase H (3, 4). In addition to carrying multiple functions for DNA synthesis, the HBV P protein also plays a central role in the viral assembly process in that it initiates, as a structural protein, the formation of replication-competent nucleocapsids [through binding to the encapsidation signal in the RNA genome (5)].

Use of standard methods to detect directly and to characterize P protein(s) at the various steps of the HBV life cycle has so far been hindered by its low abundance in the infected cell, or within the virion. Therefore, most analyses of P protein structure and function have had to rely on indirect determinations, such as the more sensitive detection of the covalently linked DNA product (3, 6, 7). To overcome this problem we have developed, initially by employing overexpression of HBV P proteins by recombinant vaccinia viruses, a sensitive detection method involving the *in vitro* phosphorylation of appropriately modified P proteins with [γ-^{32}P]ATP and protein kinase A (PKA). Compared to standard immunological methods, this technique increases the sensitivity of P protein detection by at least two orders of magnitude, thus allowing for the first time the unambiguous detection of the virus-associated P protein as such (8). Here we describe the method in detail and discuss its potential applications for other fields of research.

Materials and Methods

Protein A–Sepharose is purchased from Pharmacia (Uppsala, Sweden); [γ-^{32}P]ATP with a specific activity of 3000 Ci/mmol is obtained from Amersham (Amersham,

A

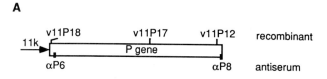

B

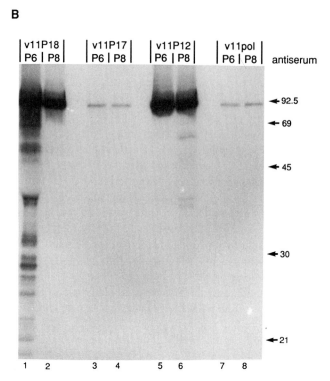

FIG. 1   Detection of vaccinia virus-expressed HBV P protein after *in vitro* phosphorylation. (A) Schematic representation of the P gene, showing the positions of the PKA target sequences introduced into the various P gene recombinants as well as the target sequences for the anti-peptide antisera αP6 and αP8. (B) Detection of the radiolabeled P proteins after gel electrophoresis and autoradiography. Immunoprecipitated proteins were ^{32}P labeled by PKA, reprecipitated with anti-peptide antisera, and separated by SDS-PAGE. Numbers next to the gel refer to the size of marker proteins in kilodaltons.

Arlington Heights, IL). The catalytic subunit of PKA, isolated from bovine heart, is purchased from Sigma (Munich, Germany). The enzyme is provided as lyophilized powder and has a phosphorylating activity of 30–65 units/μg protein (1 unit is the activity required to transfer 1.0 pmol of phosphate from [γ-^{32}P]ATP to hydrolyzed, partially dephosphorylated casein per minute at pH 6.5 at 30° C). For reconstitution, autoclaved distilled water, containing dithiothreitol (DTT) (6 mg/ml), is added to a concentration of approximately 0.05 mg of protein/ml, and the solution is incubated for 10 min at room temperature prior to use. In this form, the enzyme is stable for 3 days if stored at 4°C. For long-term storage (up to 8 days without significant loss of phosphorylating activity) we add glycerol and ovalbumin as carrier protein to a final concentration of 50% (v/v) and 1 mg/ml, respectively.

P gene mutants containing the consensus target sequence for the protein kinase A (RRXSX; Ref. 9) are constructed by site-directed mutagenesis (10) and introduced into an HBV P gene cloned into the vaccinia virus recombination vector pATA-18 (11). Production and selection of recombinant vaccinia viruses are done as described previously (8, 12). Recombinants are plaque purified, amplified, and titrated using HU TK$^-$ 143 cells (13).

For synthesis of P proteins, HepG2 cells (14), grown in Dulbecco's modified minimal essential medium containing 10% (v/v) fetal calf serum (FCS) at 5% CO_2, are incubated with recombinant vaccinia viruses at a multiplicity of infection of 10 in FCS-free medium for 1 hr at 37°C. After removal of the inoculum, fresh medium containing FCS is added and 16 hr postinfection cells are lysed in TNE [10 mM Tris hydrochloride (Tris-HCl) (pH 8.0), 100 mM NaCl, 1 mM ethylenediaminetetraacetic acid (EDTA)] with 1% (v/v) Nonidet P-40 (NP-40). The lysate is clarified by centrifugation and proteins contained in the supernatant are precipitated by the addition of 2% (w/v) sodium dodecyl sulfate (SDS) and 5% (v/v) trichloroacetic acid (the addition of SDS is required to keep the NP-40 in solution). After 30 min at 0°C, precipitated proteins are collected by centrifugation (5 min at 6000 g) and resolved in protein sample buffer [200 mM Tris-HCl (pH 8.0), 5 mM EDTA, 2% (w/v) SDS, 1% (v/v) 2-mercaptoethanol, 10% (w/v) sucrose, and 0.1% (v/v) bromophenol blue]. After 5 min of boiling, samples are diluted in radioimmunoprecipitation assay (RIPA) buffer [phosphate-buffered saline (PBS), 1% (v/v) NP-40, 0.5% (v/v) sodium deoxycholate, 0.1% (w/v) SDS] by adding 20 vol of RIPA buffer without SDS and containing 2 mM Phenylmethylsulfonyl fluoride and 2 kallikrein inhibitor units of Trasylol (Bayer Leverkusen, Germany). Finally, 15 μl of packed protein A–Sepharose, containing preadsorbed immunoglobulin (corresponding to 3–6 μl of antiserum), is added and the samples are incubated overnight at 4°C with vigorous agitation. It should be noted that denaturation of proteins prior to immunoprecipitation (IP) often enhances the efficiency of phosphorylation (15), most likely by increasing the accessibility of the PKA target sequence, and reduces the background of vaccinia virus proteins with high affinity for protein A.

For *in vitro* phosphorylation, immunocomplexes are washed three times with RIPA buffer and twice with protein kinase buffer [TMN: 20 mM Tris-HCl (pH 7.0), 10 mM MgCl$_2$, 100 mM NaCl]. After adding 30 μl of TMN, supplemented with 1 mM DTT, 20 μCi of [γ-^{32}P]ATP, and 4 units of PKA, samples are incubated for 15 min at 30°C, followed by the addition of 4 units of the enzyme and further incubation for 15 min. Immunocomplexes are washed four times with PBS to remove the bulk of unincorporated radioactivity and 40 μl of protein sample buffer is added.

Samples are boiled for 5 min, diluted with RIPA buffer, and subjected to IP as described above. Finally, precipitated proteins are analyzed by sodium dodecyl sulfate–polyacrylamide gel electrophoresis (SDS-PAGE).

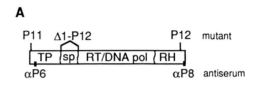

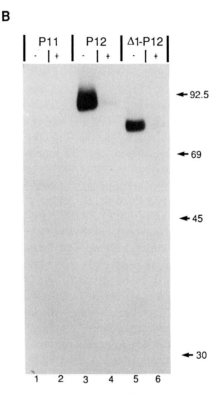

FIG. 2   Identification of the nucleocapsid-associated P protein. (A) Schematic presentation of the P gene, indicating the various functional domains (TP, terminal protein; sp, spacer; RT/DNA pol, reverse transcriptase/DNA polymerase; RH, RNase H). The positions of the introduced mutations and the peptides used to elicit P-specific antisera are given. (B) Detection of the encapsidated P protein after *in vitro* phosphorylation. Constructs shown in (A) were transfected and P proteins were isolated from the purified, denatured core particle by immunoprecipitation. Following the phosphorylation step, radiolabeled P proteins were reprecipitated in the presence (+) or absence (−) of the homologous antigenic peptides.

To detect the particle-associated HBV P protein, the efficiently phosphorylatable PKA target sequence P-12 is transferred into plasmid pCH3097, which contains an HBV genome transcribed under the control of the strong human cytomegalovirus immediate early (CMV IE) promoter (5). Hepatitis B virus nucleocapsids produced from this construct on transfection into the human hepatoma cell line Huh-7 are isolated from cell lysates by immunoprecipitation with an anti-core antiserum and denatured by boiling in an SDS-containing buffer. P proteins are isolated by immunoprecipitation, using a mixture of two anti-peptide antisera (αP6 and αP8; for positions in the P protein, see Fig. 2A) and, following *in vitro* phosphorylation by PKA and [γ-^{32}P]ATP, reprecipitated with the same antiserum mixture and analyzed by SDS-PAGE.

To detect particle-associated P proteins of the duck hepatitis B virus (DHBV), constructs containing a DHBV full-length genome under control of the CMV IE promoter are transfected into a chicken hepatoma cell line (LMH; Ref. 16). After 4 days, DHBV nucleocapsids are isolated from 0.8 ml of cell lysate (from 5×10^7 cells) by sedimentation through a 20% (w/v) sucrose cushion (0.2 ml) in a Beckman (Fullerton, CA) TL-100 ultracentrifuge (rotor TLA100.2, 80000 rpm, 1 hs, 20° C). Denaturation, immunoprecipitation, phosphorylation, and protein analysis are performed as described above for the HBV P protein. The specificities of the antibodies used are shown in Fig. 3.

Results

Because of the lack of other systems allowing the production of substantial amounts of authentic P protein, we chose recombinant vaccinia viruses to express the HBV P gene in eukaryotic cells, and used the proteins produced for determination of the efficiency of radiolabeling by PKA phosphorylation *in vitro*. To convert the protein into a good substrate for radiolabeling at distinct positions, we introduced by site-directed mutagenesis the PKA consensus target sequence Arg-Arg-X-Ser(P)-X into

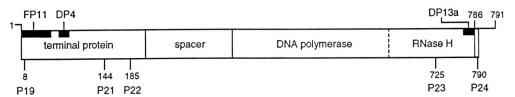

FIG. 3   Schematic representation of the DHBV P protein, showing the positions of the PKA target sequences introduced into the various P gene mutants (position of the phosphorylatable serine within the target sequence is indicated). In addition, the target sequences for the different antibodies used for immunoprecipitation (αFP11, αDP4, αDP13a; amino acids FP11 55 to 50, DP4 70 to 86, DP13a 771 to 786, respectively) and functional domains of the P protein are indicated.

TABLE I   P Gene Mutations and Effects on
Phosphorylation Efficiencies

Recombinant[a]	Amino acid change (wild-type → mutant)	Specific activity[b] (Ci/mmol)	Detection limit[c] (pg)
v11pol		6	750
v11P18	RRLLL → RRLSL	300	15
v11P17	ARLSS → RRLSS	10	450
v11P12	GRTSL → RRTSL	600	8

[a] See Fig. 1A.

[b] Values obtained by liquid scintillation counting of the 90-kDa proteins shown in Fig. 1 and normalization for the input protein as estimated by Western blot analysis and comparison with a dilution series of a purified P protein fragment.

[c] Assuming that overnight exposure of an X-ray film will readily detect 50 cpm of ^{32}P-labeled protein.

various positions of the P gene (Fig. 1A), thereby taking care that the primary sequence of the P gene product was not drastically changed (Table I). The variant P genes were introduced into a vaccinia virus vector directing protein synthesis under control of the strong 11k late promoter (11), and the recombinant viruses generated were used to infect a human hepatoma cell line. P proteins produced in these cells were collected by immunoprecipitation, radiolabeled by *in vitro* phosphorylation, and then analyzed by SDS-PAGE followed by autoradiography (8).

As shown in Fig. 1B, labeled P proteins were obtained from all recombinant viruses as well as from a control virus producing the wild-type protein. However, the strength of the signals obtained depended drastically on both the presence of a PKA consensus target sequence, and its position within the P polypeptide chain. This is demonstrated best by a comparison of the specific radioactivities determined (Table I): The wild-type protein, which lacks a PKA consensus target sequence, was only poorly radiolabeled. This background value was increased only marginally (less than twofold) on introducing such a target sequence into the central part of the P protein (P17), indicating that the sequence was probably not very accessible to the enzyme in this position. With equivalent target sequences introduced near the P protein amino terminus (P18) or carboxy terminus (P12), however, radiolabeling was increased 50- or 100-fold, respectively. Using these terminally modified P proteins, as little as 5 pg of P protein, corresponding to about 50 counts per minute (cpm) of ^{32}P, could be clearly detected as an autoradiographic signal on an X-ray film after an overnight exposure. This increases the sensitivity for direct protein detection over currently available standard methods by at least two orders of magnitude.

As anticipated, *in vitro* phosphorylation of modified P proteins also allowed detection and initial characterization of the low amounts of P gene product(s) as present within the nucleocapsid. For this purpose, the efficiently phosphorylatable carboxy-terminal PKA target sequence P12 was transferred into an HBV genome transcribed under the control of the strong human CMV IE promoter (Fig. 2A). Hepatitis B virus nucleocapsids produced from this construct on transfection into a human hepatoma cell line were isolated by immunoprecipitation with an anti-core antiserum. These were then assayed by PKA phosphorylation for the presence of P protein, and additionally also for DNA polymerase activity and HBV RNA encapsidation to determine whether the protein had maintained these activities (see Discussion and Conclusions, below). To detect the particle-associated P protein, nucleocapsids were denatured by boiling in SDS-containing buffer and P proteins were isolated by immunoprecipitation, using a mixture of two anti-peptide antisera (Fig. 2A). Following *in vitro* phosphorylation by PKA and [γ-^{32}P]ATP, P proteins were reprecipitated with the same antiserum mixture and analyzed by SDS-PAGE. As shown in Fig. 2B, this analysis revealed that the HBV nucleocapsid contains the P protein in the form of a single dominant species with a size corresponding to the unprocessed 90-kDa full-length gene product as expressed from the recombinant vaccinia virus. The authenticity of this protein was further demonstrated (a) by its absence from nucleocapsids produced by mutant P11, which carries a stop codon at the very beginning of the P gene (Fig. 2, lanes 1 and 2), (b) by the interference in a competitive immunoprecipitation of the homologous peptides P6 and P8 (Fig. 2, lanes 4 and 6), and (c) by a size shift of the labeled protein from 90 to about 80 kDa on introducing into the P gene a deletion removing 92 amino acids from the nonessential spacer region (mutant $\Delta 1$, Ref. 4; Fig. 2, lanes 5 and 6). Finally, quantitation of these and further complementing data lead to the conclusion that there is probably only a single P protein molecule contained in the HBV core particle (5).

Essentially similar results were obtained in a study identifying the P protein molecules contained in the nucleocapsid of the duck hepatitis B virus (DHBV), an avian hepadnavirus used as a preferred animal model for studying the molecular biology of hepadnaviral replication (M. Weber, unpublished results, 1993). In this case, PKA consensus target sequences were created at five different positions in the P gene (Fig. 3) and introduced into a full-length DHBV DNA genome capable of producing functional nucleocapsids under CMV promoter control, analogous to the genomic HBV constructs used above. As demonstrated by the results listed in Table II, a protein containing a terminally located PKA site (P19, Fig. 3) was again phosphorylated severalfold better than the respective wild-type protein (DHBV-16), whereas internally located sites did not contribute to radiolabeling. Interestingly, the wild-type protein itself was already well phosphorylated despite the absence of a PKA consensus target sequence. These observations again indicate that the presence or absence of PKA consensus sequences in a protein does not necessarily correlate with its phosphorylation efficiency.

TABLE II   Duck Hepatitis B Virus P Gene Mutations and
Effects on Phosphorylation Efficiencies

Recombinant[a]	Amino acid changes and insertions (wild-type → mutant)	Relative[b] phosphorylation efficiency	Relative[c] DNA polymerase activity
P19	LKQS > RRQS	4.00	0.40
P21	KSIS > RRIS	0.50	<0.02
P22	KRIS > RRIS	0.80	0.90
P23	KRYQ > RRYS	0.60	0.75
P24	T > TRRASA[d]	0.35	0.35
DHBV-16	Wild-type	1.00	1.00

[a] Positions given in Fig. 3.
[b] Relative values (DHBV-16 wild-type = 1.00) obtained by quantitation of ^{32}P-labeled 90-kDa protein band with a phosphoimager and normalization for the input core particles in a core ELISA.
[c] Relative values (DHBV-16 wild-type = 1.00) obtained by Cerenkov counting of incorporated radioactivity in the endogenous polymerase assay and normalization for the input core particles as estimated by Western blot analysis.
[d] Carboxy-terminal elongation of DHBV P protein.

Besides its high sensitivity of detection, PKA phosphorylation also provides the additional advantage that radiolabeling occurs usually at known positions in the polypeptide sequence, thus allowing establishment of structural relationships between phosphorylatable gene products. This is exemplified in Fig. 1B (lane 1) for a series of HBV P gene-related radiolabeled polypeptides from recombinant v11P18, which were immunoprecipitated with an antiserum directed against an amino-terminal P protein sequence (αP6). As these products were radiolabeled efficiently (and therefore at the N-terminally located site P18), this result characterizes these fragments as carboxy-terminally truncated proteolytic cleavage products, and suggests that the HBV P protein is proteolytically degraded from the carboxy terminus. Consistently, only the full-length P gene product was immunoprecipitated with an antiserum directed against the last 18 amino acids of the P protein (αP8; Fig. 1, lane 2).

Discussion and Conclusions

Functional simplicity and the possibility of obtaining the enzyme in large quantities on heterologous expression in *Escherichia coli* (17) make protein kinase A a valuable tool for site-specific radiolabeling of proteins (8, 18, 19). Compared to other protein kinases, PKA offers the major advantages that its catalytic subunit is inexpensive and, furthermore, that the enzyme has a relatively low target specificity, which facilitates the introduction of new target sites into proteins of interest. Compared to radiolabeling with iodine-125, the phosphorus isotopes ^{32}P and ^{33}P combine high specific activity and convenient half-lives, emitting β rays of moderate energy. This results in a high efficiency of detection and resolution and reduced radiological load already appreciated in many molecular biological techniques.

Protein kinase A target sequences in natural substrates are defined by the two main sequence patterns Lys-Arg-X-X-Ser(P)-X and Arg-Arg-X-Ser(P)-X, their common motif being a pair of basic amino acids located two or three amino acids in front of the phosphorylated serine residue (20). Despite this rather low substrate specificity, most proteins are relatively poor substrates for radiolabeling by PKA. As exemplified here for the hepadnaviral P proteins, even proteins of 90 kDa may lack PKA target sequences (as in HBV wild-type or DHBV-16). On the other hand, proteins may be substantially phosphorylated at sites not predicted from their primary amino acid sequence (e.g., in DHBV-16). From our experience, it appears therefore to be reasonable to analyze first the protein of interest experimentally for the extent of PKA phosphorylation and then, if necessary, to construct several variants carrying the target sequence at different positions, and finally to select those that combine high radiolabeling with low influence on function. Thus, the most efficiently phosphorylatable P gene variants used by us displayed a moderate reduction of the nucleocapsid-associated DNA polymerase activity on the endogenous DNA template (25 or 40% activity of the wild-type for HBV P-12 or DHBV P-19, respectively), but no loss of the encapsidation function as assayed by RNA protection experiments (Table II, and Ref. 5).

As described here for the P protein variants and also as observed with the HBV core protein and a variant of the HBV X protein (21, 22), we found that the terminal regions of a target protein are usually well phosphorylated. Thus, these parts of a polypeptide chain appear to be particularly well suited for introducing PKA target sequences, either by changing a given amino acid sequence or by adding short terminal extensions, also described for interferon α (18). Obviously, a combination of several phosphorylation sites in a single protein molecule should lead to even higher specific activities.

For setting up an assay system, it should be pointed out that sensitive detection by PKA phosphorylation of minor proteins requires removal of the relatively large excess of radiolabeled ATP and also of contaminating proteins that are either highly phosphorylatable, or present in large quantities, or both. Our present protocol uses for this purpose two immunoprecipitation steps, the one following *in vitro* phosphorylation being essential to remove radiolabeled immunoglobulins and protein A, which themselves carry phosphorylation sites. In contrast, immunoprecipitations may not be required if phosphorylation was performed with purified protein preparations lacking phosphorylatable proteins of similar size. In this case, unincorporated radioactivity could be removed either by trichloroacetic acid (TCA) precipitation (if necessary with nonradioactive carrier protein) or, if denaturation of the phosphorylated protein should be avoided, by gel filtration. In cases in which DTT (present in the enzyme solution and the reaction buffer) may cause unwanted opening of disulfide bonds in the target protein, it could be used at much reduced concentrations or even omitted, provided that the enzyme is used immediately after reconstitution.

In vitro phosphorylation with PKA provides a gentle method to radiolabel proteins at predetermined positions, which should have a variety of applications beyond the detection and quantitation of proteins of low abundance. Radiolabeled proteins may be used as tracers, for instance, for rapid detection of a protein during purification from a crude homogenate. More importantly, phospholabeled proteins could turn out to be useful tools for a highly sensitive ligand screening of expression libraries, or in the direct determination of protein binding (18, 23, 24), or in following cellular uptake of proteins in tissue culture or *in vivo*. Last, but not least, one could imagine that phosphorylated antibodies, for instance directed against tumor-specific antigens, could be used to identify their target cells and receptors. These and other examples may illustrate the potentially broad range of applications for phosphorylatable proteins in clinical and basic research.

References

1. J. Summers and W. Mason, *Cell (Cambridge, Mass.)* **29,** 403 (1982).
2. D. Ganem and H. Varmus, *Annu. Rev. Biochem.* **56,** 651 (1987).
3. R. Bartenschlager and H. Schaller, *EMBO J.* **7,** 4185 (1988).
4. G. Radziwill, W. Tucker, and H. Schaller, *J. Virol.* **64,** 613 (1990).
5. R. Bartenschlager and H. Schaller, *EMBO J.* **11,** 3413 (1992).
6. G. Radziwill, H. W. Zentgraf, H. Schaller, and Bosch, *Virology* **64,** 613 (1990).
7. V. Bosch, R. Bartenschlager, G. Radziwill, and H. Schaller, *Virology* **163,** 123 (1988).
8. R. Bartenschlager, C. Kuhn, and H. Schaller, *Nucleic Acids Res.* **20,** 195 (1992).
9. A. M. Edelman, D. K. Blumenthal, and E. G. Krebs, *Annu. Rev. Biochem.* **56,** 567 (1987).
10. M. J. Zoller and H. Smith, *DNA* **3,** 479 (1984).
11. H. Stunnenberg, H. Lange, L. Philipson, R. van Miltenberg, and P. van der Vliet, *Nucleic Acids Res.* **16,** 2431 (1988).
12. N. Schek, R. Bartenschlager, C. Kuhn, and H. Schaller, *Oncogene* **6,** 1735 (1991).
13. S. Chakrabarti, K. Brechling, and B. Moss, *Mol. Cell Biol.* **4,** 3403 (1985).
14. B. Aden, A. Fogel, S. Plotkin, I. Damjanov, and B. Knowles, *Nature (London)* **282,** 615 (1979).
15. D. B. Bylund and E. G. Krebs, *J. Biol. Chem.* **250,** 6355 (1975).
16. L. D. Condreay, C. E. Aldrich, L. Coates, W. S. Mason, and T.-T. Wu, *J. Virol.* **64,** 3249 (1992).
17. L. W. Slice and S. S. Taylor, *J. Biol. Chem.* **264,** 20940 (1989).
18. B.-L. Li, J. A. Langer, B. Schwartz, and S. Pestka, *Proc. Natl. Acad. Sci. USA* **86,** 558 (1989).

19. H.-F. Kung and E. Bekesi, *in* "Methods in Enzymology," Vol. 119, p. 296. Academic Press, San Diego, California, 1986.
20. D. B. Glass and E. G. Krebs, *Annu. Rev. Pharmacol. Toxicol.* **20,** 363 (1980).
21. F. Birnbaum, Ph.D. Thesis, University of Heidelberg, Germany (1992).
22. M. Fischer, Ph.D. Thesis, University of Heidelberg, Germany (1991).
23. J. A. Langer, A. Rashidbaigi, and S. Pestka, *J. Biol. Chem.* **261,** 9801 (1986).
24. A. Rashidbaigi, H.-F. Kung, and S. Pestka, *J. Biol. Chem.* **260,** 8514 (1985).

Index